## ANALOG DEVICES

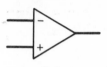

Operational amplifier
or comparator

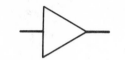

General amplifier

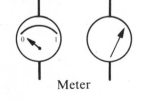

Meter

## LOGIC DEVICES

AND gate

OR gate

Gate with Schmitt-
trigger inputs

Exclusive-OR gate

Inverter

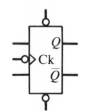

HI→LO triggered
flip-flop

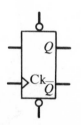

LO→HI triggered
flip-flop

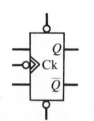

HI→LO triggered
flip-flop with
data lockout

P9-EKC-987

## QUICK GUIDE TO REFERENCE DATA

### Components and Devices

| | |
|---|---|
| Capacitor marking codes | 480–481 |
| Conductor and transmission lines | 305–306, 474–475 |
| IC identification and manufacturers | 486–491 |
| IC lead assignments | 523–526 |
| Resistor color codes | 469 |
| Standard resistor values | 470–471 |
| Switch and relay nomenclature | 137, 139 |
| Time constant vs. percent change | 144 |

### Signals

| | |
|---|---|
| Measures of sine-wave amplitude | rear end paper |
| Noise sources | 215, 410 |
| Peak, average, and rms values | 31 |

### Logic

| | |
|---|---|
| Boolean Theorems | 257 |
| CMOS characteristics and rules | 293 |
| Conventions | 254–255 |
| Gate equivalents | 258 |
| Open-collector gates | 438 |
| TTL characteristics and rules | 287, 289, 308 |

### Digital Data

| | |
|---|---|
| ASCII code | 311 |
| Binary and BCD codes | 82, 374–375 |
| Magnetic storage device characteristics | 351 |
| Serial communication standards | 309 |

# Electronics and Instrumentation for Scientists

# Electronics and Instrumentation for Scientists

**Howard V. Malmstadt**
University of Illinois

**Christie G. Enke**
Michigan State University

**Stanley R. Crouch**
Michigan State University

**The Benjamin/Cummings Publishing Company, Inc.**
Reading, Massachusetts • Menlo Park, California
London • Amsterdam • Don Mills, Ontario • Sydney

*Sponsoring Editor:* Mary Forkner
*Production Editor:* Madeleine Dreyfack
*Cover Designer:* Michael Rogondino
*Book Designer:* Elaine Merrick-Wilson
*Copy Editor:* Carolyn Eastman

Copyright © 1981 by The Benjamin/Cummings Publishing Company, Inc.
Phillipines copyright 1981 by The Benjamin/Cummings Publishing
Company, Inc.

All rights reserved. No part of this publication may be reproduced, stored
in a retrieval system, or transmitted, in any form or by any means,
electronic, mechanical, photocopying, recording, or otherwise, without the
prior written permission of the publisher. Printed in the United States of
America. Published simultaneously in Canada.

**Library of Congress Cataloging in Publication Data**

Malmstadt, Howard V    1922-
  Electronics and instrumentation for scientists.
  Bibliography: p.
  Includes index.
  1. Electronic instruments. I. Enke,
Christie G., 1933-    joint author. II. Crouch,
Stanley R., joint author. III. Title.
TK7878.4.M29    621.3815'4    80-26973
ISBN 0-8053-6917-1

abcdefghij-MA-8987654321

**The Benjamin/Cummings Publishing Company, Inc.**
2727 Sand Hill Road
Menlo Park; California 94025

# Preface

The goal of this book is to provide scientists with an efficient means of understanding the principles of modern electronic instrumentation so that they can confidently choose instruments and apply them to their full potential. The subject is therefore modern electronics with an emphasis on its applications in scientific instruments, measurement and control systems, and computers. This book has been written particularly for students and practitioners in all areas of science and engineering. The presentation is both fundamental and practical. The only prerequisites are basic physics, algebra, and an interest in scientific instrumentation.

## New Emphases for the Microcircuit Age

We have now entered a new era in scientific instrumentation; the era of microelectronics. The new microcircuits provide both the basis and the need for a new approach to the study of electronics. From the user's point of view, the availability of complete high-performance electronic functions such as operational amplifiers and decade counters in encapsulated, unalterable integrated circuit (IC) form, makes the detailed study of transistor amplifier and flip-flop design less critical than a knowledge of the functions performed by the ICs and their applications. At the same time, the widespread application of sophisticated electronic functions such as phase-locked loops, analog multipliers, and microprocessors creates the need to understand the principles of the advanced measurement and control techniques that are being used in the most common laboratory instruments.

## Approach

This book is thus organized around system principles and electronic functions rather than the detailed analysis of particular components. The latter are not neglected, since components of all types, from resistors through transistors and integrated circuits, are studied as the devices by which the various functions can be implemented. Because of this approach, new devices will not obsolete the knowledge gained from this study; rather they will be recognized as new ways to achieve the appropriate functions. Analog and digital techniques are integrated through the concept of electrical data domains, which are the ways in which data can be encoded in electrical signals.

The need to study all the basic devices and circuits before any of the important and interesting functions and systems can be presented has been eliminated. Theory and application are developed side-by-side and reinforce each other without undue postponement. The traditional topics of dc and ac circuits, power supplies, amplifiers, and logic devices are developed in the context of measurement and control applications. Instrumentation concepts including feedback control, signal-to-noise enhancement, microcomputer interfacing, data acquisition, and signal processing are presented along with the techniques and devices for their implementation. In our opinion, these latter need not be couched in mystique or advanced math and their omission leaves a student without the knowledge needed to properly operate a computerized data system or even a digital oscilloscope.

## Format

Full advantage has been taken of the wide page format for the placement of figures and margin notes. Extended captions make each figure self-contained. The figures and notes provide a helpful "second level" of explanation. When first defined, new terms are presented in bold-face type. The page where a term is defined is marked "(def)" in the index. Explanations are generally both functional and fundamental. Algebraic derivations supplement explanations (often in a note), but are not used in lieu of an explanation. The problems and experiments at the end of each chapter are useful as a study guide. Applications and examples are distributed throughout the book. A specific applications section concludes most chapters. The appendices provide details of practical components (resistors, capacitors, inductors, and ICs) and techniques (shielding and grounding).

## Application in a Formal Course

This book is useful in a structured course or for self study. The fourteen chapters are of nearly uniform length and complexity so that each comprises a similar unit of study. The problems at the end of each chapter are useful for review. As a text for a college upperclass or graduate course, the fourteen chapters, taken one each week, fit neatly into one semester. On the quarter system, Chapters 1-6, 8-10, and 12 could be presented in an introductory course, while the remaining chapters and some computer programming could be given in a second term. Chapter 7 discusses transistor types, characteristics, and circuits. This chapter is not prerequisite to any that follow so this level of implementation can be skipped if desired. On the other hand, valuable insights into the operation of analog and digital ICs will be gained through its study. An instructor's guide, available from the publisher, outlines several options for the classroom use of this book as well as solutions to all of the problems.

This book is organized to support a parallel laboratory. The early introduction of data encoding and measurement techniques allows the immediate use of modern test instruments in the laboratory. A list of suggested experiments follows each chapter to indicate the areas of practice the chapter has introduced. Experiments along these same lines are available in expanded form as an accompanying workbook entitled, *Experiments in Electronics and Instrumentation*. The intent of each chapter is to be immediately useful in the teaching or practicing laboratory.

## Acknowledgements

We are glad to express our appreciation for the assistance of many others in the preparation of this book. Our editor, Mary Forkner, encouraged us strongly in this work and provided a smooth and effective interface with the publisher throughout the project. The production team, directed by Madeleine Dreyfack and including Caroline Eastman, copy editor, and Edna Conway, proofreader, Michael Rogondino, cover designer, and Elaine Merrick-Wilson, book designer, proved very effective and accommodating. The typesetting and artwork were done under the direction of Chuck Alessio and Bob Fuller. The manuscript was prepared with the expert and unfailing aid of Debra Jahangardi of East Lansing, who typed it in all its versions, sketched many of the figures, and who, with Angele van Rijn, laid out pages of the preliminary drafts used for class testing. Helpful suggestions for manuscript revision were made by James Avery of the University of Illinois, J. F. Holler of the University of Kentucky, Larry Shoer of Lake Forrest College, Patrick Gibbons, W. J. Ross, Jim DeFreese (reviewers), and Bruce Newcome of Michigan State University. Assistance with the problems and solutions was provided by several Michigan State University graduate students, Rytis Balciunas, Frank Curran, Rob Engerer, Hugh Gregg, Nelson Herron, Carl Kircher, and Carl Myerholtz. Working and talking with T. V. Atkinson of Michigan State University has produced many valuable insights. We wish also to acknowledge the use of some concepts and figures conceived by Gary Horlick, coauthor of an earlier work, "Optimization of Electronic Measurements," the fourth module in "Electronic Measurements for Scientists."

Our special appreciation goes to our families and also to our colleagues and students who contributed so much to the environments in which this work took place. This has been a very rewarding book in the writing. We hope you find it so in the reading.

Howard Malmstadt
Chris Enke
Stanley Crouch

*February, 1981*

# Contents

ix

x

# Electrically Encoded Information

# Chapter 1

We live in the midst of an explosion of electronics. Few aspects of our professional, social, and leisure lives have not been significantly influenced by the proliferation of electronic devices. This has come about as a result of the great facility with which modern electronic devices can convert information from one form to another—as in measurement, control, and transmission systems—and can process information—as in computation, correlation, and diagnostic systems. In an electronic system the information being converted or processed is **encoded** as particular characteristics of **electrical signals** (see notes 1-1 and 1-2). Since the number of ways information or data can be encoded electrically is limited, the functional basis of a complex electronic system can be readily understood in terms of the modes of electrical data encoding and transformation used in it. The study of electronic components, in spite of their immense variety and rapid evolution, is also greatly simplified by classifying them in terms of the role they play in the data encoding and transformation processes.

This chapter begins with a description of the main characteristics of each of the modes of encoding data. These modes, or **data domains** as we shall call them (see note 1-3), are classified for electrical signals as analog, time, and digital domains. A review of the electrical quantities of charge, current, voltage, and power is included to clarify the signal characteristics that are used to represent information. The process of electrical measurement is then introduced in terms of the data transformations, or **data domain conversions** (see note 1-4), it involves. The chapter concludes with a discussion of basic meters and multimeters (moving-pointer and digital). Reviews of dc circuits and principles of current and voltage measurements are included in these discussions.

## 1-1 Electrical Data Domains

Electrical data domains are the modes of encoding information as electrical quantities or combinations and variations of electrical quantities. For example, the voltage from a photocell is related to and can be interpreted in terms

**Note 1-1. Encode.**
To transfer information from one system of communication into another.

**Note 1-2. Electrical Signal.**
A detectable electrical quantity, such as voltage, charge, or current, or a variation in an electrical quantity by which information can be conveyed.

**Note 1-3. Data Domain.**
The name of a quantity (or the unit of measurement of that quantity) by which information is represented or conveyed.

**Note 1-4. Data Domain Conversion.**
A change in the domain in which information is represented or conveyed.

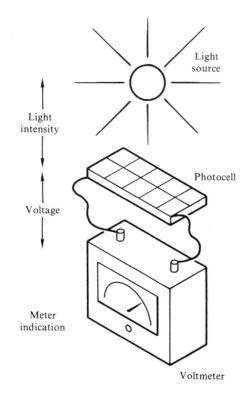

Light source

Light intensity

Photocell

Voltage

Meter indication

Voltmeter

**Fig. 1-1.** Devices and data domains in a simple light-measurement system. Light intensity, voltage, and meter indication are data domains; the photocell and voltmeter are interdomain converters.

**Note 1-5. Interdomain Converter.**
A device for which the data domain at the output is different from the data domain at the input.

**Note 1-6. Transducer.**
A device that converts information from a non-electrical domain to an electrical domain (input transducer) or from an electrical domain to a non-electrical domain (output transducer).

of, the intensity of light falling on the cell as illustrated in figure 1-1. Thus, *voltage* is the data-encoding electrical quantity or domain. Similarly, the frequency of voltage pulses from a Geiger tube can be related in a known way to the level of radiation reaching the tube. In this case, *frequency* is the signal characteristic or electrical data domain that conveys the information.

Of course, not all data domains are electrical. Information can be represented by light, physical position, electrical resistance, temperature, magnetic field strength, chemical composition, air pressure, printed symbols, and many other properties or quantities. Information in these nonelectrical data domains is transformed into one of the electrical data domains by an **interdomain converter** called an **input transducer** (see notes 1-5 and 1-6). Photocells and Geiger tubes are both examples of input transducers. An interdomain converter that transforms electrically encoded information to a nonelectrical domain is called an **output transducer.** Examples of output transducers are numerical displays, electric motors, loud speakers, and TV picture tubes. Among the electrical data domains, there are three classes: analog, time, and digital. Each of these classes is discussed below in terms of the domains in each class and the characteristics they have in common. Understanding and applying the concept of data domains greatly facilitates the study and proper use of modern electronic components, instruments, and systems.

## Analog Domains

Data in the **analog domains** are represented by the magnitude of one of four electrical quantities: charge, current, voltage, or power. These quantities are discussed in the next section. Most input transducers used today convert the measurement data from a nonelectrical domain to one of the analog domains. Many examples of analog input transducers are described in chapter 4. At each instant in time, the data in an analog domain are represented by the amplitude of an electrical signal. Because the smallest unit of charge, the charge on an electron, is so small compared to the total charge stored or transferred by most signal sources, the amplitude of an analog signal is variable in essentially infinitesimal increments. Thus the amplitude of an analog signal is continuously variable.

An output transducer that converts analog data to a number or a position on a scale can be used to measure the electrical quantity. Examples are the current meter and digital voltmeter described in this chapter, the oscilloscope described in chapter 2, and the recorder described in chapter 5. The amplitude of an analog signal can be measured continuously over time or at any instant. When continuous measurements are made, the variations in the signal amplitude can be plotted against time, wavelength, magnetic field strength, temperature, or other experimental parameters as shown in figure

1-2. Note that analog data plots are continuous in both magnitude and time dimensions. Additional information can often be obtained from such plots by correlating amplitudes measured at different times. Such information includes simple observations such as peak height, peak position, and number of peaks, or more complex correlations such as peak area, peak separation, and comparison with data from other plots. The techniques of correlating data taken at different times, i.e., data processing, can be distinguished from the techniques of converting data from one domain to another.

Signals in the analog domains are susceptible to **electrical noise** sources (see note 1-7) contained within or induced upon the circuits and connections of the instrument. The resulting signal amplitude at any instant is the sum of the data and noise components. Data encoded as signals in either a time or digital domain are less affected by electrical noise.

## Time Domain

Measurement data in the **time domains** are contained in the *time relationship* of the signal variations, not in the amplitude of the variations. Typical time domain signals are shown in figure 1-3. These are **logic-level signals;** i.e., their signal amplitude is either in the HI logic-level region or the LO logic-level region. The data are contained in the time relationship between the logic-level transitions, such as the time between successive LO→HI transitions **(period),** the time between LO→HI and HI→LO transitions **(pulse width),** and the number of LO→HI transitions per unit time **(frequency** or **rate).** Examples of input transducers that produce a time-domain

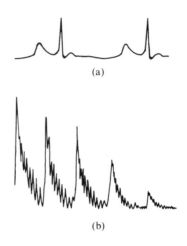

**Fig. 1-2.** Signal amplitude vs. another parameter for analog signal sources. (a) Heart-muscle current vs. time in electrocardiography. (b) Photomultiplier-tube current vs. wavelength of light in spectroscopy.

**Note 1-7. Electrical Noise.**
Any part of an electrical signal that is unwanted, i.e., not related in a known way to the desired information.

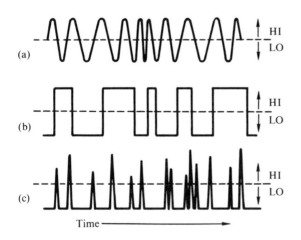

**Fig. 1-3.** Time-domain signals. The data are contained in the relative time of the transition between HI and LO levels. Though the encoding quantity cannot generally be determined by inspection of the waveform, signal (a) could be encoded as the frequency of the sinusoidal waveform, and signal (b) could be a pulse-width encoded signal and signal (c) could be the output of a detector of randomly timed events in which the average number of events per unit of time is sought.

signal are the Geiger tube mentioned earlier and a crystal oscillator that produces a temperature-dependent frequency because of the temperature characteristics of the quartz crystal. An example of a domain converter between an analog domain and a time domain is a voltage-controlled oscillator or voltage-to-frequency converter that provides an output frequency related to an input voltage.

The greater the slope (rate of change) of the signal through the **logic-level threshold** region (see note 1-8), the more precisely the logic-level transition time can be defined. Because the encoding quantity in time domain signals is less amplitude dependent than it is in analog-domain signals, information in a time domain is less affected by electrical noise. A common example of this is the lower noise sensitivity of an FM radio signal (frequency domain) compared to that of the more noise-susceptible AM radio signal (an analog domain). The greater the difference between the average HI or LO signal level amplitude and the logic-level threshold, the less susceptible the signal to noise-induced error. The uncertainty of the time of threshold crossing for a noisy signal is illustrated in figure 1-4.

Signals in the time domains, like those in the analog domains, are continuously variable since the frequency or pulse width can be varied infinitesimally. However, the encoded variable of signals in the time domains cannot be measured continuously with time, nor can it be measured at any instant in time. The minimum time required for conversion of data in a time domain to any other domain is necessarily at least one period or one pulse width.

## Digital Domain

The measurement data in the **digital domain** are contained in a two-level signal (HI/LO, 1/0, on/off, etc.) that is encoded to unambiguously represent a specific integer (or character). The signal in the digital domain may be encoded as a series of logic levels in one channel (**serial digital signal**) or as a set of logic levels on simultaneous multiple channels (**parallel digital signal**). Since a digitally coded signal represents a specific number exactly, the data are actually in numerical form. The unit for the signal characteristic encoding the data in the digital domain is thus *number*. Since this is true for all the many varieties of digital encoding, all digital signals are in a single domain called digital. *Data in the digital domain need no further data domain transformations to be converted to a number.* It is only necessary to decode (interpret) the signal and display the number. Representative digital signal waveforms are shown in figure 1-5.

To minimize uncertainty in digital decoding, the signal amplitudes for logic level HI and LO are carefully defined for each family of logic circuits. For instance, for the popular transistor-transistor logic (TTL) family, LO is

**Note 1-8.  Logic-Level Threshold.**
The signal level above which a signal is interpreted as HI and below which a signal is interpreted as LO. Signal levels near the threshold are avoided in order to minimize the chance of error in logic level interpretation.

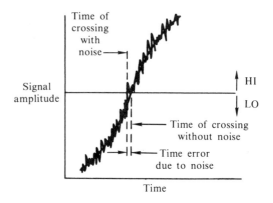

**Fig. 1-4.**  Noise-induced error in time of threshold crossing. The presence of noise can cause a premature or delayed crossing of the threshold. The error is reduced by decreasing noise amplitude and increasing rate of signal amplitude change through the threshold region.

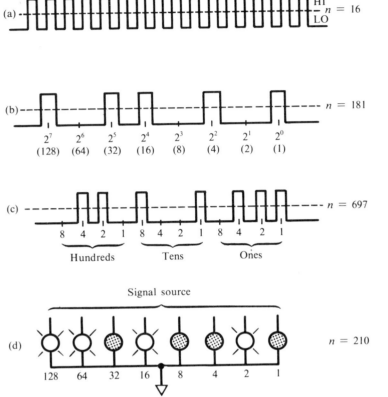

**Fig. 1-5.** Digital-domain signals. In the count serial signal (a) the number is encoded as the number of pulses or HI-LO transitions in the pulse train. The binary-coded serial signal (b) has pulse positions that are weighted according to the base 2 number scheme. The presence or absence of a pulse indicates a 1 or 0 at each position or bit. The binary number represented by the waveform shown is $10\,110\,101_2$, which is $181_{10}$ in the decimal system. The binary-coded decimal (BCD) signal is similar except that the weighting is according to the BCD code. The parallel signal (d) can use any coding but has a separate wire for each bit. An eight-bit parallel data source is shown connected to indicator lights to show the simultaneous appearance of the data logic levels on all eight data lines.

any level from 0.0 to +0.4 V and HI is any level from +2.4 to +4.5 V. Because the response of TTL logic circuits to signal amplitudes between 0.8 and 2.4 V is unspecified, these levels are avoided and transitions through the threshold region should be fast, typically less than 10 ns. The specific characteristics of TTL and other logic circuit families are described in chapter 11.

The **count** waveform (fig. 1-5a) is a series of pulses with a clearly defined beginning and end. The count waveform of figure 1-5 might represent, for instance, the number of photons of a particular energy detected during a single spark excitation. The count form is simple but not very efficient. To provide a **resolution** (see note 1-9) of one part per thousand, the time required for at least 1000 pulses to occur must be allowed for each series of pulses.

In the **binary-coded** serial signal, each pulse position in the series represents a particular binary digit, or **bit** in a **binary number** (see note 1-10). The

**Note 1-9. Resolution.**
The magnitude of the smallest detectable variation in a quantity.

**Note 1-10. Binary Numbers.**
Binary numbers are numbers in the base 2 system. The only numerals are 0 and 1. A binary digit is called a bit. The least significant bit (the one farthest to the right) in a binary number has unit value (0 or 1). The next bit to the left has the value of $2^1$ (0 or 2); the next, a value of $2^2$ (0 or 4) and so on. Counting in binary numbers follows this sequence, beginning with zero: 0, 1, 10, 11, 100, 101, 110, 111, 1000, 1001, 1010, etc.

data are not represented by the exact time of the pulse as in the time domains, but by the logic level present within a given slice of time. A series of 10 pulse times has a resolution of one part in $2^{10} = 1024$, and a 20-bit series has a resolution of better than one part per million. The binary code is the most efficient code possible for signals with only two levels.

The **binary-coded decimal** serial form is somewhat less efficient but very convenient when a decimal numerical output is desired. Each group of four bits represents one decimal digit in a number. Twelve bits thus represent three decimal digits and provide a resolution of one part per thousand.

The principal advantage of parallel digital data connections is speed. An entire **word** (group of bits) can be conveyed from one circuit to another in the time required for the transmission of one bit in a serial connection. Binary coding is shown in figure 1-5d, but binary-coded decimal coding and other coding schemes are used for parallel digital data. Parallel data connections are used in all modern, fast computers. Serial data connections are often used for long-distance communication and for connections where a relatively slow data rate is sufficient.

## 1-2  Electrical Quantities and Properties

The atoms that make up the molecules of matter are composed of negatively charged electrons, equally but positively charged protons, and electrically neutral neutrons. The charge on an electron (and proton) $Q_e$ has been defined as $1.603 \times 10^{-19}$ coulombs (C) (see note 1-11). In a neutral atom or molecule the number of electrons is equal to the number of protons; so the atom or molecule has no net charge. Atoms or molecules sometimes have extra electrons and carry a negative charge, or they may have a deficiency of electrons and carry a positive charge. All electrical phenomena result from the action of charged species on each other and on their environment. All electronic circuits are systems designed to manipulate charge in such a way as to perform specific tasks.

In this section, charge is seen to be the basis of the other **electrical quantities,** current (the motion of charged species), voltage (the potential energy produced by the separation of species of opposite charge), and power (the rate of work performed by moving charged species). The response of matter to the separation of charged species is characterized in the **electrical properties** of matter such as conductivity (the ease of moving charged species through the matter) and capacitance (the ability of a device to store charge).

### Current

Because unlike charges attract each other, energy is required to separate species with opposite charges. When charged species are separated so as to

**Note 1-11.  Coulomb.**
The coulomb (C) is the amount of charge required to produce 0.001 118 00 g of silver metal from silver ions (Ag$^+$).

create regions in matter that have different charges, potential energy differences exist between these regions. Electrical potential energy ($V$ or $v$) is expressed in units of **volts** (V) (see note 1-12). One volt of electrical potential results when one joule (J) of energy has been required to separate one coulomb ($6.23 \times 10^{18}$ electron charges) of charge; that is, volts equal joules per coulomb. An electrical potential difference is often called the **voltage**.

Species with positive or negative charge that are between two regions of different charge experience an attractive force—the positively charged species toward the more negative region and the negatively charged species toward the more positive region. If these charged species are free to move through the matter (for example, electrons in a metal or ions in a solution), they will do so in the direction according to their charge. Movable charged species are called **mobile charge carriers**. The rate of charge transfer that results from the motion of the mobile charge carriers is the electrical **current**, $I$ or $i$. The magnitude of the current is expressed in **amperes** (A). One ampere of current equals one coulomb of charge per second.

## Conductance

Any material that has mobile charge carriers is an electrical **conductor.** Such materials include metals, semiconductors, and electrolytic (ionic) solutions. The **conductance**, $G$ of a conductor depends on the concentration of mobile charge carriers, the mobility of the charge carriers, and the geometry of the conductor and its contacts. In general, the conductance increases with increased density and mobility of the mobile charge carriers. For many conductors, the current through the conductor is proportional to the conductance of the conductor and the voltage across the conductor. That is,

$$I = GV \tag{1-1}$$

This relationship is called Ohm's law for its discoverer. The current-voltage relationship in a conductor can also be expressed in terms of the conductor's **resistance**, $R$ to the flow of charge. Since resistance is the inverse of conductance, $R = 1/G$, Ohm's law can also be written

$$I = V/R \tag{1-2}$$

Other permutations of Ohm's law are: $V = IR$, and $R = V/I$. The unit of resistance is the **ohm** ($\Omega$). A resistance of one ohm allows a current of one ampere with a potential of one volt, i.e., ohms equals volts per ampere. The unit of conductance is the **mho** ($\Omega^{-1}$), which can be expressed as amperes per volt.

A conducting device that is used specifically for its resistance and that obeys Ohm's law is called a **resistor** (see note 1-13). According to Ohm's law, when a resistor with resistance of $R$ ohms is connected between two points

**Note 1-12. Symbols for Variables.**
It is general practice to use a lower case letter, for example, $i, v, q$, as the symbol for the value of a quantity when the variation in that quantity is of interest. Capital letters, for example, $I, V, Q$, are used as symbols for quantities that are considered to be constant in value. For a table of quantities, symbols, and units see the inside back cover of this text.

**Note 1-13. Symbol for a Resistor.**
Resistors are symbolized in diagrams by

———⋀⋀⋀———

Practical resistors and the resistor color code are discussed in appendix B.

**Note 1-14.   Battery Symbol.**
The symbol for a battery or a voltage source is

The longer line conventionally represents the more positive terminal. Practical batteries are described in chapter 3.

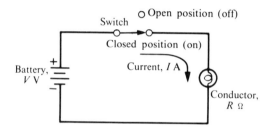

**Fig. 1-6.**   Simple complete circuit. When the switch is closed, the conducting path between the battery terminals is completed, and there is a steady current through the conductor. This is indicated by the current arrow. In accord with the standard convention, the arrow indicates the direction of the flow of *positive* charges.

**Note 1-15.   Circuit Element.**
An element is any component or device such as a resistor, light bulb, or battery that is used to make up an electrical circuit.

that have a potential difference of $V$ volts, charge flows through the resistor at a rate of $V/R$ amperes. Charge continues to move through the resistor at the rate $V/R$ amperes as long as there is a finite potential difference between the points i.e., as long as $V \neq 0$. Thus, there can never be a potential difference between regions of a conductor without an immediate flow of charge acting to reduce the potential difference to zero. However, if a source of energy is present that can maintain a potential difference between the two points (such as at the terminals of a battery), the electrical current in the conductor will be continuous (see note 1-14). Materials with no mobile charge carriers are called **insulators**. Insulators are used to isolate conductors that have different voltages by providing no current path between them.

## A Complete Circuit

A diagram of a conducting path connected between the terminals of a battery is shown in figure 1-6. The conducting path includes switch contacts, a light bulb, and the connecting wires. The connecting wires offer very little resistance to the flow of charge. The filament wire in the light bulb is a poorer conductor. If a continuous conducting path exists between the points of potential difference (e.g., the + and − terminals of the battery), there is a **complete circuit**.

In a simple complete circuit, *the current is the same as that through every other segment at any instant.* This is because unequal currents would result in a charge build-up in some segments of the conducting path and a loss of charge in other segments. As shown above, the potential difference resulting from an uneven charge distribution in a conductor would cause a charge-equalizing current. If the battery voltage and the conductor resistance are known, the current in the circuit can be calculated from Ohm's law. For instance, for the circuit of figure 1-6, assume that the battery voltage is 1.5 V ($V = 1.5$ V), the resistance of the light bulb is 2.5 $\Omega$ ($R = 2.5\ \Omega$), and the resistance of the wires and closed switch are 0 $\Omega$. In this case, the current $I$ is

$$I = V/R = 1.5\ \text{V}/2.5\ \Omega = 0.60\ \text{A}$$

When the switch is open, charge cannot flow through the switch. The circuit is now incomplete and is often referred to as an **open circuit**. Since the current is the same through every **element** (see note 1-15) of the circuit, the current through every element is zero. The connecting wire from the positive terminal of the battery to the switch will be at the positive terminal potential. Similarly, the circuit elements connecting the negative terminal of the battery to the other switch contact are all at the battery's negative terminal potential. The full battery voltage therefore appears between the contacts of the open switch.

## Voltage

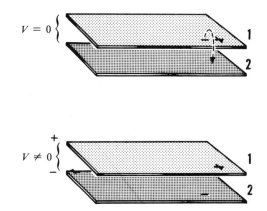

The term voltage refers to a difference in electrical potential. Thus, when two terminals of a voltage source are connected across a resistor, there is a current through the resistor from the more positive to the more negative connection. Some source of energy is required to cause the separation of charge carriers that produces the voltage between the two terminals. Some common methods of producing a voltage are the mechanical separation of charges in an electrostatic generator, the chemical separation of charged species in a cell or battery, and the motion of a conductor through a magnetic field, as in an electromagnetic generator.

**Separation of charge.**    Consider two metallic conductors, 1 and 2, separated by an insulator as in figure 1-7. Assume that to begin with there is no potential difference between them ($V = 0$). This can be assured by connecting them momentarily with a wire. Now, if an electron is removed from conductor 1 and placed in conductor 2, the charge on 2 will be $-Q_e$ and on 1, it will be $+Q_e$. The energy required to transfer *another* electron from 1 to 2 is greater than for the first because the electron must be removed from a region of positive charge and put in a region of negative charge. This charge difference increases the voltage between the conductors. Each unit of charge transferred increases the voltage by the same amount as the first. Thus, the voltage between the conductors is proportional to the charge $Q$ transferred, so that

$$V = kQ \qquad (1\text{-}3)$$

The constant $k$ depends on the geometry of the separated conductors and the nature of the insulator between them.

**Fig. 1-7.** Separation of charge. The transfer of an electron from conductor 1 to conductor 2 increases the net positive charge in conductor 1 and the net negative charge in conductor 2. The charge difference caused by the charge separation produces a difference in electrical potential or a voltage $V$ between the conductors. As charge transfer continues, $V$ increases in direct proportion to the increase in charge difference.

**Capacitance.**    Two conductors of any shape or arrangement that are separated from each other by an insulator can have a potential difference between them. From equation 1-3, the amount of charge $Q$ (coulombs) that must be transferred from one conductor to the other to cause a 1-V difference in potential between the two conductors is $1/k$. This constant is called the **capacitance,** $C$. Thus

$$C = Q/V \qquad (1\text{-}4)$$

From equation 1-4 we see that $C$ has the units coulombs/volt (C/V); 1 C/V has been designated the **farad** (F). If the conductors in figure 1-7 have a capacitance of 1 F and a potential difference of 1 V between them, then from equation 1-4 they must contain (by storing) 1 C of separated charge. Devices that are made and used for their property of capacitance are called **capacitors** (see note 1-16). The capacitance of a capacitor increases with increasing area of the conductors, decreasing separation between the conductors, and

**Note 1-16. Capacitor Symbol.**
The symbols for a capacitor are

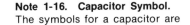    and

increasing dielectric constant of the material between the conductors. However, any pair of conductors has the property of capacitance. The capacitance exhibited by a pair of conductors, like that of a capacitor, increases with an increase in their areas, decreases with an increase in distance between them, and is affected by the nature of the insulator between them. Practical capacitors are described in appendix B.

**Voltage sources.**    Many sources of steady voltage can be described as charged capacitors with a means of maintaining or replenishing their charge as needed. When a **load** (see note 1-17) of resistance $R$ is connected to a steady voltage source of $V$ volts, the current supplied by the voltage source, or the rate of its discharge, is $I = V/R$ coulombs per second. Clearly, the voltage source will not maintain its initial or no-load output voltage if the rate of discharge exceeds the rate at which charge separation can be produced within the source. *There is, therefore, a limit to the current that a voltage source can supply before its output voltage is significantly decreased.* If the load is connected for $t$ seconds, the total charge discharged $Q$ is $It$ coulombs. The **work**, $W$, performed on the load during discharge is

$$W = QV = ItV \qquad (1\text{-}5)$$

where the unit of $W$ is joules. **Power** is defined as the rate of doing work, i.e., work per time, or $W/t$. A work rate of one joule per second, 1 J/s, is equal to 1 **watt** (W) of power ($P$). Therefore,

$$P = W/t = QV/t = ItV/t$$

from which we obtain

$$P = IV \qquad (1\text{-}6)$$

From equation 1-6 the power in watts supplied by the voltage source and dissipated in the load at any instant is equal to the voltage applied to the load times the current in the load at that instant. If the load is a resistance, substitution of Ohm's law (eq. 1-2) in equation 1-6 gives two other equations for the power dissipated in the load. (In a resistor, the dissipated power appears as heat.) These equations are:

$$P = I^2 R \quad \text{and} \quad P = V^2/R \qquad (1\text{-}7)$$

Some voltage sources, such as electromagnetic generators, periodically reverse the direction of the charge separation. This results in a voltage of alternating polarity called an **alternating voltage**, as shown in figure 1-8(a). When an alternating voltage is connected to a load, the current direction reverses with every reversal of the voltage source. Such a current is called an **alternating current**. The abbreviation **ac** is commonly used for alternating current, and the corresponding abbreviation **dc** denotes unidirectional or

**Note 1-17.  Load.**
A load is any conducting device connected to a voltage, current, or power source.

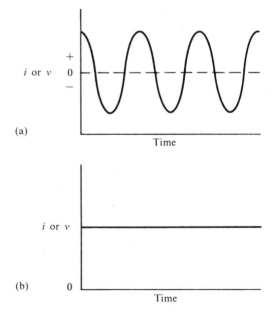

**Fig. 1-8.**  Plots of ac and dc quantities. In an ac, or alternating current or voltage (a), the sign of the charge separation or the direction of the current reverses periodically. By contrast dc, or direct current or voltage (b), maintains the same direction and sign.

**direct current**, as shown in figure 1-8(b). The terms ac and dc are also used to indicate power sources, circuits, and components, that are designed for alternating voltage and/or current operation, or for unipolar operation, respectively. For example, a battery may be referred to as a "dc voltage source," and an "ac radio" is designed to operate from an ac power source such as standard house wiring. The characteristics of electrical quantities and properties in ac circuits are introduced in chapter 2.

## Conductors in Series

A current exists whenever there is a conducting path between two points of different potential, regardless of how circuitous or complex the conducting path may be. The conducting path of figure 1-9 is composed of two resistors and a battery. The two resistors are said to be in **series** because as the conducting path is traced from one battery terminal to the other the resistors are traversed in sequence, or in series. The rate of charge movement in the circuit depends on both resistors. The total resistance to the flow of charge in the circuit is the sum of the resistances of the two resistors, that is, $R_s = R_1 + R_2$, and in general for a series of $N$ resistors,

$$R_s = R_1 + R_2 + R_3 + \cdots + R_N \tag{1-8}$$

In other words, a series combination of resistors, $R_1, R_2, R_3, \ldots, R_N,$ *is equivalent to* a single resistor of value $R_s$. The current through the series of resistors is $I = V/R_s$. In the case of figure 1-7, $I = 6$ V$/(9\ \Omega + 21\ \Omega) = 0.2$ A. As we have already shown, the current is the same at every point in the series circuit at any instant. Therefore, the current everywhere in the series circuit is affected by the values of all the series resistors.

***IR* drop.**   Ohm's law can also be used to calculate the potential difference across each series resistor. A rearrangement of Ohm's law, $V = IR$, indicates that whenever a current $I$ exists in a conductor of resistance $R$, a potential difference $V$ appears across that conductor. This potential difference is sometimes called the ***IR* potential drop** or just ***IR* drop** (see note 1-18). For the resistors of figure 1-9,

$$IR_1 = 0.2 \text{ A} \times 9\ \Omega = 1.8 \text{ V} \quad \text{and} \quad IR_2 = 0.2 \text{ A} \times 21\ \Omega = 4.2 \text{ V}$$

*The sum of the IR drops in the series circuit must equal the potential difference at the battery terminals,* because there are no other elements in the circuit across which a potential difference can exist (see note 1-19). In other words, the source voltage is divided among the series resistors. If one of the series resistors is of negligible resistance compared to the others, its $IR$ drop is negligible compared to the total voltage. Thus, almost all the applied voltage appears across the much larger resistor.

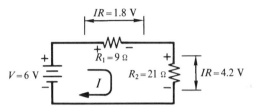

**Fig. 1-9.**  A series circuit. In a series circuit, there is only one path that completes the circuit, and this path includes every element in sequence. Each terminal of each element is in contact with only one other element.

**Note 1-18.  *IR* drop.**
The term *IR* drop is used because the current causes a decrease, or drop, in the voltage across the resistor in the direction of the current, as shown in figure 1-9.

**Note 1-19.  Kirchhoff's Voltage Law.**
Kirchhoff's voltage law states that the algebraic sum of the voltages from all voltage sources and *IR* drops encountered at any instant as a path is traced out around a complete circuit (closed-loop path) is zero.

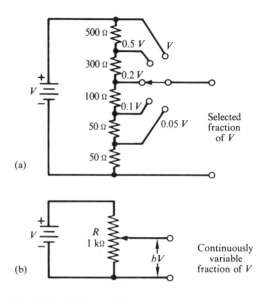

(a)

(b)

**Fig. 1-10.** Voltage dividers. The voltage divider is a series circuit of resistors in which a fraction of the applied voltage $V$ appears across each resistor. The fraction of $V$ obtained at the output is equal to $b$, the fraction of the total resistance across which the output is taken. The fraction $b$ can be varied by fixed resistors and a selector switch as in (a) or by a continuously variable potentiometer as in (b).

**Note 1-20.    Potentiometers.**
The word potentiometer has two meanings. When used to describe a continuously variable voltage divider, it is frequently shortened to "pot." A potentiometer can also mean a complete comparison system for voltage measurements as discussed in chapter 5.

**Voltage divider.**  A **voltage divider** circuit is shown in figure 1-10a. The total source voltage $V$ is distributed through the total circuit resistance. The sum of the divider resistors $R_s = 500\ \Omega + 300\ \Omega + 100\ \Omega + 50\ \Omega + 50\ \Omega = 1000\ \Omega$. The current through each resistor is $V/R_s$. The $IR$ drop across each resistor of value $R$ is then $V(R/R_s)$. In other words, the voltage across any resistor is the same fraction of the total voltage $V$ as the resistance $R$ is of the total resistance $R_s$. The voltage across the 50-$\Omega$ resistor is $V \times (50/1000) = 0.05V$, and the fraction 0.05 is independent of the voltage $V$. For two resistors ($R_1$ and $R_2$) in series, the voltage across $R$ is a constant fraction $b$ of the voltage across the series combination. This fraction is given by $b = R_1/(R_1 + R_2)$. For a fine adjustment of the voltage-divider fraction $b$, a resistor with a continuously variable slider contact called a **potentiometer** (see note 1-20) is used, as shown in figure 1-10b.

It is important to note that the output voltage fraction equals the resistance fraction $b$ only if the current through all the resistors is the same. The connection of a load across the divider output terminals diverts current from the selected resistors and causes an error in the output voltage fraction. Loading errors are discussed later in this chapter, but for now remember that in order for the error to be negligible, the resistance of the load must be very large compared to the resistance of the divider.

## Conductors in Parallel

Often several conducting paths exist between two points in a circuit, providing independent alternate routes for the transport of charge between the two points. The separate conducting paths between two points are said to be connected in **parallel**. Resistors connected in parallel to a voltage source are shown in figure 1-11. Note that a conducting path that is in parallel with other conducting paths has one end of the path connected to one end of all the other parallel paths and the opposite end connected to the opposite ends of all the other parallel paths. Therefore, *the same voltage is applied to each conducting path in parallel*. Since each resistor in figure 1-11 provides a separate conducting path between the points of potential difference $V$, the total parallel conductance $G_p$ is the sum of the conductances of the separate paths:

$$G_p = G_1 + G_2 + G_3 = \frac{1}{R_1} + \frac{1}{R_2} + \frac{1}{R_3} \qquad (1\text{-}9)$$

The resistance equivalent to a number of parallel conductors is $R_p = 1/G_p$. Thus,

$$R_p = \cfrac{1}{\cfrac{1}{R_1} + \cfrac{1}{R_2} + \cfrac{1}{R_3} + \cdots + \cfrac{1}{R_N}} \qquad (1\text{-}10)$$

where $N$ is the number of parallel conductors. For two resistors in parallel,

$$R_p = \frac{1}{1/R_1 + 1/R_2} = \frac{R_1 R_2}{R_1 + R_2} \qquad (1\text{-}11)$$

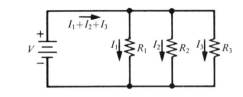

**Fig. 1-11.** A parallel circuit. In a parallel circuit, multiple paths exist to complete the circuit. Charge flows through these separate paths simultaneously. Contacts at both ends of the parallel arms or elements are connected together.

The current through each resistor in figure 1-11 can be calculated from Ohm's law: $I_1 = V/R_1$, $I_2 = V/R_2$, $I_3 = V/R_3$. *The sum of the currents in the parallel conductors must be equal to the current supplied by the battery* (see note 1-21) in order to avoid a net accumulation or depletion of charge at some point in the conducting circuit. The total current, $I_t = V/R_p$ is divided, in proportion to the conductances, among the parallel paths. The fraction of the current in each path is constant.

When two conductors $R_1$ and $R_2$ are connected in parallel, the circuit current $I_t$ is split between them. The fraction of $I_t$ that passes through $R_1$ is

$$\frac{I_1}{I_t} = \frac{V/R_1}{V/R_p}$$

Substituting for $R_p$ from equation 1-11, we have

$$\frac{I_1}{I_t} = \frac{R_2}{R_1 + R_2} \qquad (1\text{-}12)$$

**Note 1-21. Kirchhoff's Current Law.**
Kirchhoff's current law states that the algebraic sum of all currents encountered at any instant at a junction (point where three or more conductors are joined) must be zero.

Thus the fraction of the total current through either resistor is the resistance of the other resistor divided by the sum of the resistance of both resistors. This "current splitting" relationship is frequently useful in calculating the desired values for parallel resistors. Current splitting in parallel circuits is analogous to voltage dividing in series circuits. If one of the parallel paths has a resistance that is much lower than the others, it carries nearly the total current.

## 1-3  Domain Conversions in Electrical Measurement

*Measurement is the determination of a particular characteristic of a sample in terms of the number of standard units for that characteristic.* For example, electrical potential is measured by determining the number of volts of potential difference between the two sampled points. In a data domain sense, this measurement involves converting information in the voltage domain into information in the number (or digital) domain. With many types of voltmeters, the voltage-to-number interdomain conversion is done not in a single step but rather in several conversions using intermediate data domains such as charge, pulse width, or frequency.

In order to keep track of the electrical data domains and interdomain conversions involved in electronic systems and instruments, it is helpful to arrange the data domains in the form of a map, similar to the one shown in figure 1-12. Note that the digital domain includes electrically encoded numbers as well as displayed or printed numbers. Common types of output transducers control the position of a pointer or pen. This data domain is called **scale position** on the map. In this section, we shall study two common electrical measurement devices, the current meter and the digital voltmeter, in terms of the data domains involved in their operation. These devices are the basis of the analog and digital multimeters described in the final sections of this chapter.

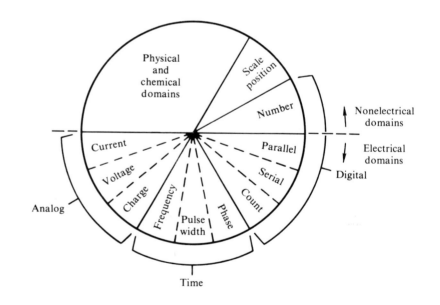

**Fig. 1-12.** Data domains map. Many of the domains by which data may be encoded are arranged in pie-shaped areas in this circular map. Electrical domains are in the lower semicircle and are arranged by class. Although the set of electrical domains is essentially complete, there is no specific identification or classification of the many nonelectrical domains other than number and scale position in this presentation.

## Current Meter

The **moving-coil current meter** is one of the most commonly used magnitude-indicating output devices. The position of a pointer is affected by the magnitude of an electric current. A scale behind the pointer can be calibrated in any desired units. The moving-coil meter remains popular because of its simplicity, low cost, and ability to indicate changes in value by the motion of the pointer.

**The meter movement.**  The D'Arsonval style of moving-coil meter is shown in figure 1-13. A coil of fine wire, wound on a rectangular aluminum frame, is mounted in the air space between the poles of a horseshoe magnet. Hardened steel pivots attached to the coil frame fit into jewelled bearings so that the coil rotates with minimal friction. An indicating pointer is attached to the coil assembly, and springs attached to the frame return the needle (and coil) to a fixed reference point. When a current is applied to the coil, the direction and magnitude of rotation depend on the direction and magnitude of the current.

The moving-coil meter is thus a current-to-scale position interdomain converter. The pointer position can be translated to a number by viewing the pointer against a printed scale and interpolating between scale calibration points. The output of the meter is not in the digital domain because it is not quantized in integers. Rather it is infinitesimally variable in its indication, and it requires the interpretive action of the human reader to produce the number. The data domain transitions involved in a measurement with the moving coil meter are shown by the arrows in figure 1-14.

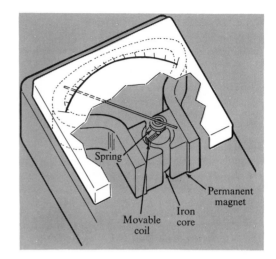

**Fig. 1-13.**  Moving-coil meter. The current to be measured passes through the movable coil. A turning force on the coil assembly results from the interaction of the magnetic fields from the coil and permanent magnet. This force acts against a restoring spring to produce a rotation of the pointer that is proportional to the current.

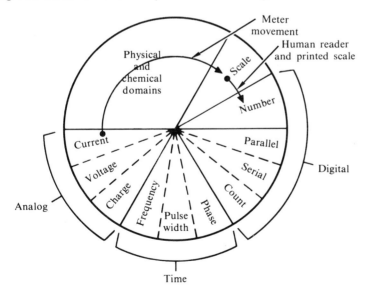

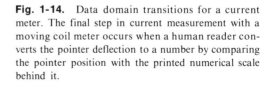

**Fig. 1-14.**  Data domain transitions for a current meter. The final step in current measurement with a moving coil meter occurs when a human reader converts the pointer deflection to a number by comparing the pointer position with the printed numerical scale behind it.

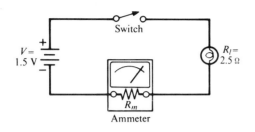

**Fig. 1-15.** Current measurement. To measure the current through a circuit element such as $R_l$, it is necessary to connect the current meter *in series* with the element. In this way, the current through the element and the meter will be identical.

**Note 1-22.   Relative Error of Current Measurement.**
For a maximum error of 1%, the meter resistance $R_m$ should be $0.01R_l$. For 0.1% error, $R_m$ should be $0.001\ R_l$, etc.

**Current measurements.**   Suppose the current in the battery and lightbulb circuit of figure 1-6 is to be measured with a current meter. The resulting circuit is shown in figure 1-15. The figure also shows that the resistance of the current meter $R_m$ is added to the total resistance of the circuit. If the resistance of the meter were only 0.1 $\Omega$, the resulting current would be 1.5 V/ $(2.5\ \Omega + 0.1\ \Omega) = 0.577$ A rather than the 0.600 A calculated when the meter was not present. If it is important that the current meter read the value the current would have in the meter's absence, the meter resistance should be negligible compared to the resistance ($R_l$) in series with it (see note 1-22).

## Digital Voltmeter

In recent years the direct digital readout of voltage and other electrical quantities has become an inexpensive alternative to the moving-coil current meter. As its name implies, the **digital voltmeter (DVM)** is a voltage-actuated device with numerical display of the measured quantity. Basic to all DVMs is the necessity for converting the analog voltage to be measured into an accurate digital representation. This is accomplished by an **analog-to-digital converter (ADC)**, sometimes called a **digitizer** because it determines the number of units or increments that comprise the measured voltage. Once the data are present in the digital domain, one of the several popular types of displays, such as light-emitting diode (LED) displays or liquid crystal displays (LCD), is used to provide a visible, numerical indication to the user.

Accurate digital representation of an analog voltage can be accomplished by several methods. In DVMs digitization is most often performed by converting the input voltage into a number of discrete increments of charge and counting the increments. These digital charge measurement techniques are introduced here and described in detail in chapter 9.

**Dual-slope conversion.**   In one popular digitization technique, known as the **dual-slope converter**, the input voltage $v_{in}$ is converted to a proportional current that is used to charge a capacitor for a precise period of time (fig. 1-16). A charge $q_{in}$, proportional to $v_{in}$ and $t_{in}$ ($q = i_{in}t_{in} = Kv_{in}t_{in}$), accumulates on the capacitor during $t_{in}$. Then the clock disconnects switch $S$ from the input source, connects a constant reference current $I_r$ to discharge the capacitor, and generates the Start signal. When the threshold detector monitoring the capacitor voltage indicates that the capacitor has been discharged, a pulse occurs at the output labelled Stop, and the cycle of charging the capacitor from the unknown source and discharging it from the reference source repeats.

Since the amount of charge required to discharge the capacitor is equal to the charge stored, the time $t_r$ required for discharge with the reference current is also proportional to $q_{in}$ and, thus, to $v_{in}$. Therefore

$$I_r t_r = q_{in} = Kv_{in}t_{in} \quad \text{or} \quad t_r = v_{in}(Kt_{in}/I_r)$$

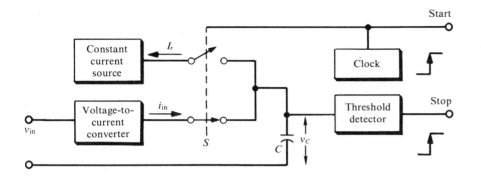

**Fig. 1-16.** Dual-slope conversion technique. The input voltage $v_{in}$ is converted to charge by applying a proportional current $i_{in}$ to capacitor $C$ for an exact time $t_{in}$. During this time, switch $S$ is in the position shown. At the end of $t_{in}$, the start signal is generated, and the constant current source $I_r$ is connected to the capacitor. When the capacitor is exactly discharged, $v_C$ is zero, and the threshold detector generates the Stop signal. The time interval between Start and Stop signals is proportional to $v_{in}$ and can be measured with the time interval meter of figure 1-17.

where $K$ is the proportionality constant between $i_{in}$ and $v_{in}$. The time $t_r$ is the interval between the appearance of pulses at the outputs Start and Stop. The dual-slope converter thus produces a voltage-to-time interval interdomain conversion (see note 1-23).

The time interval $t_r$ can be measured by counting pulses from a standard clock oscillator as shown in figure 1-17. The pulse at Start opens the counting gate and allows clock pulses through to the counter. The pulse at Stop disconnects the clock from the counter. When the counting has ceased, the digital output from the counter is presented to the display. In practice, the various proportionality factors between the number displayed and the input voltage are arranged so that the display reads directly in volts. The various data domain conversions that take place in the dual-slope DVM are illustrated in figure 1-18.

**Note 1-23. Equivalence of Time-Interval and Pulse-Width Domains.**
Time interval and pulse width are equivalent domains because the width of a pulse is the time interval between the LO →HI and HI→LO transitions of the pulse.

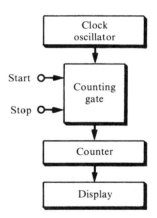

**Fig. 1-17.** Time interval measurement. A digital timer counts clock oscillator cycles (time increments) that occur when the counting gate is open. For time interval measurement, the gate is only open for the interval between the Start and Stop signals.

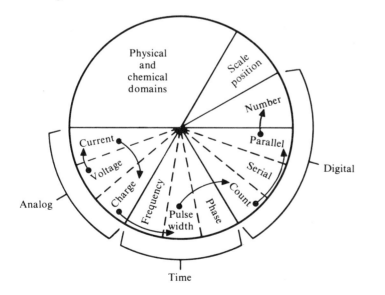

**Fig. 1-18.** Data domain conversions in a dual-slope DVM. There are two analog interdomain conversions prior to the charge-to-pulse width conversion: The pulse width is converted to a count digital signal by the counting gate, and the counter performs a serial-to-parallel intradomain conversion in the digital domain.

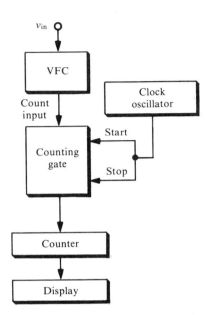

**Fig. 1-19.**   DVM with VFC and frequency meter. Frequency, or cycles per unit time, is measured by counting the cycles that occur during a known unit of time. The VFC supplies a frequency signal proportional to $v_{in}$ and the clock opens the counting gate for an exact time interval. The resulting count is proportional to $v_{in}$.

**Fig. 1-20.**   DVM measurement of voltage. To measure the voltage across a circuit element such as $R_l$, the voltmeter is connected *in parallel* with the element to assure that the voltages across the meter and across the element are identical.

**Note 1-24.   Relative Error of Voltage Measurement.**
In order to measure $V_e$ with a maximum error of 1%, $R_m$ should be at least 100 times $R_e$. For 0.1% relative error, $R_m$ should be greater than 1000 times $R_e$.

**Fig. 1-21.**   Voltmeter connection to source with internal resistance. In the absence of the voltmeter, the terminal voltage is $V_e$. With the voltmeter connected, it is $V_e R_m / (R_e + R_m)$. This loading error is negligible if $R_m$ is very much larger than $R_e$. If the voltage is measured for two known values of $R_m$, both $R_e$ and $V_e$ can be calculated.

**Voltage-to-frequency converter.**   A second popular type of converter for DVMs is the voltage-to-frequency converter (VFC). In this converter the unknown voltage is again converted to a proportional current that charges a capacitor continuously. Increments of charge in the form of pulses with sign opposite to the input charge are generated to maintain a constant voltage across the capacitor.

Note that as with the dual-slope converter, the output of the voltage-to-frequency converter is in one of the time domains; in this case, it is a frequency-domain signal. The frequency is readily measured by counting the number of pulses from the charge generator per unit time. This arrangement of the counter, known as a **frequency meter**, is illustrated in figure 1-19 and discussed in greater detail in chapters 4 and 9.

**Voltage measurement.**   Suppose that the voltage drop across the light bulb in the circuit of figure 1-6 is to be measured with the DVM. The appropriate circuit is shown in figure 1-20. Because the voltmeter is in parallel with the device, it is important for the meter to have an internal resistance much greater than that of the device whose voltage is being measured. This ensures that the $IR$ drop across that device is not changed by the presence of the meter. If the voltmeter is to read the value that the $IR$ drop would have if the meter were absent, the resistance of the device should be negligible compared to that of the meter.

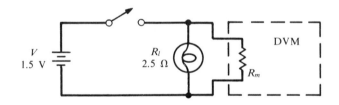

**Loading errors.**   Often it is desired to measure the voltage of a source that has significant internal resistance $R_e$, as shown in figure 1-21. Since the meter resistance and $R_e$ form a voltage divider, the voltmeter must clearly have a resistance very large with respect to the internal resistance $R_e$ if it is to read the correct no-load voltage (see note 1-24).

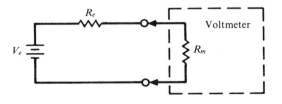

An important theorem in electricity, Thevenin's theorem, states that *any voltage source composed of batteries and resistors that has two output terminals is equivalent to a single battery and a single series resistor* as shown in figure 1-22. The equivalent voltage $V_e$ of the complex source is the voltage between the two terminals when no load is present (no device connected between the terminals). However, in order to measure the voltage of a source, the voltmeter is placed in the circuit as a load as shown in figure 1-21.

The equivalent voltage and resistance of a voltage source can often be determined from its circuit. The network given in figure 1-23 is a voltage divider. The equivalent voltage $V_e$ is the calculated no-load voltage. From the divider equation it is given by $V_e = VR_2/(R_1 + R_2)$. The equivalent resistance $R_e$ is the calculated resistance between the terminals when the voltage sources are replaced by short circuits. In the divider case, the resistance between the output terminals is the parallel combination of $R_1$ and $R_2$, $R_e = R_1R_2/(R_1 + R_2)$.

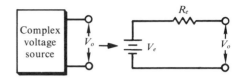

**Fig. 1-22.** Voltage source equivalent circuit. According to Thevenin's theorem, the circuit connecting any two terminals in any network of resistors and voltage sources acts as though it were a single resistor $R_e$ and voltage source $V_e$ in series. The values of $R_e$ and $V_e$ can be determined by analysis of the circuit or by measurement of the voltage across the terminals under no-load and load conditions.

## 1-4 Analog Multimeter

The moving-coil meter discussed in the preceding section is often used to measure resistance, voltage, and other electrical quantities as well as current. When several measurement functions are provided in a single instrument, the device is called a **multimeter**. Another name for the moving coil or analog multimeter is **volt-ohm-milliammeter (VOM)**. The addition of simple resistive networks converts the moving-coil meter into a versatile laboratory multimeter.

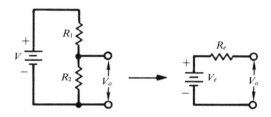

**Fig. 1-23.** Thevenin's theorem applied to the voltage divider. $V_e$ is the no-load divider output voltage, and $R_e$ is the resistance between the terminals with the source $V$ shorted out. Therefore, $V_e = VR_2/(R_1 + R_2)$ and $R_e = R_1R_2/(R_1 + R_2)$.

### Current Ranges

The moving-coil meter itself has a fixed full-scale current range determined by the design of the coil, springs, and magnet.

To allow current measurements over a wide range without use of a separate meter for each range, a technique called shunting is employed. A **shunt** is an alternate, parallel, conducting path around the current meter that causes only a fraction of the circuit current to pass through the meter itself. A simple shunt is shown in figure 1-24. According to the current-splitting relationship (eq. 1-12) the fraction of the total current $i_t$ that exists in the meter $i_m$ is

$$\frac{i_m}{i_t} = \frac{R_{sh}}{R_m + R_{sh}}$$

where $R_{sh}$ is the resistance of the shunt.

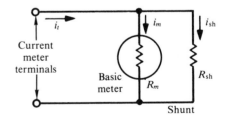

**Fig. 1-24.** Current meter shunt for extended range. If $R_{sh}$ is $1/9$ of $R_m$, $9/10$ of $i_t$ will go through $R_{sh}$ and $1/10$ through $R_m$. If a 1-mA meter movement has a resistance of 46 $\Omega$, a shunt resistance of 5.11 $\Omega$ would allow the meter to read 10 mA full-scale, a tenfold increase in the basic range.

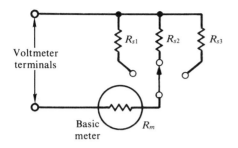

**Fig. 1-25.** Voltage measurement with a current meter. Various values of $R_s$ can be used to set the voltage required to produce the full-scale current in the meter. If a 50-$\mu$A meter movement has a resistance of 5000 $\Omega$ and it is desirable to measure 10 V full scale, the total meter resistance ($R_s + R_m$) would be $R_s + R_m = 10 \text{ V}/(50 \times 10^{-6} \text{ A}) = 200 \text{ k}\Omega$. The series resistance $R_s$ would then be 195 k$\Omega$.

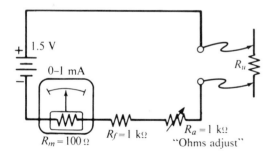

**Fig. 1-26.** Ohmmeter of the series type. When $R_u = 0$ (probes shorted), $R_a$ is adjusted for full-scale deflection. When $R_u$ is connected, the fraction of full scale deflection is $R/(R_u + R)$ where $R = R_m + R_f + R_a$. At midscale, $R_u = R$.

Usually a versatile multimeter has several different full-scale current ranges accessible by switching. Because the value of $R_{sh}$ becomes impractically low for higher current ranges, it is common practice to connect a resistor in series with $R_m$ to increase the effective meter resistance.

## Voltage Measurements

The moving-coil meter can also be readily adapted for voltage measurements. Since the meter has constant resistance, the current through the meter is proportional to the voltage drop across it. The full-scale voltage sensitivity $V_{fs}$ is the full-scale deflection current $I_{fs}$ times the meter resistance $R_m$. To obtain multiple voltage ranges it is necessary only to add series resistance to the meter circuit as in figure 1-25. With a series resistor $R_s$ the full-scale voltage sensitivity is $V_{fs} = I_{fs}(R_s + R_m)$.

Note that the ratio $(R_s + R_m)/V_{fs}$ is a constant for a given meter movement. This constant is the **ohms-per-volt rating** of the voltmeter. The resistance of the voltmeter on a given scale is the ohms-per-volt rating times the full-scale deflection voltage of that scale. Thus a 20 000 $\Omega$/V meter has a resistance of 20 k$\Omega$ on the 1-V full-scale range. The current sensitivity of a meter is the reciprocal of its ohms-per-volt rating. As mentioned above, it is important for accurate voltage measurements that the voltmeter resistance be high compared to the resistance of the circuit being measured.

## Resistance Measurements

The resistance of a device can be determined with a moving-coil meter by the addition of a battery and other resistors. A series type of ohmmeter is shown in figure 1-26. Here the unknown resistance $R_u$ is connected in series with the meter and the range determining resistors. When the test probes are shorted ($R_u = 0$), the "ohms adjust" control is turned so that the current $I_{fs}$ through the series resistors $R = R_m + R_f + R_a$ deflects the meter exactly full scale. When the probes are connected across the unknown resistance $R_u$, the current in the meter is decreased to a value $i$ that depends on the value of $R_u$. Since the 1.5-V battery is across the total resistance in both cases,

$$I_{fs}R = 1.5 \text{ V} \quad \text{and} \quad iR + iR_u = 1.5 \text{ V}$$

Combining these two equations gives the unknown resistance:

$$R_u = \left(\frac{I_{fs}}{i} - 1\right) R$$

Ohmmeters of this and other types have nonlinear resistance scales; that is, they read from $\infty$ to 0 as $i$ goes from 0 to $I_{fs}$. Values of $R_u$ much higher than *R* become crowded on the "infinite ohms" end of the scale, and values very much lower become indistinguishable from zero. For this reason various switch-selectable shunts are used to change the full-scale current sensitivity of the meter and thus to allow several resistance ranges.

A typical analog multimeter is shown in figure 1-27. The function selector switch and the various full-scale ranges are shown. Some functions and ranges require that special input jacks be used.

## 1-5  Digital Multimeter

The **digital multimeter (DMM)** is a versatile laboratory instrument for the measurement of voltage, current, and resistance. Unlike the analog multimeter which is usually based on the current-actuated moving-coil meter, the DMM is based on the voltage-actuated digital voltmeter. Thus, the measurement of current and resistance with the DMM involves conversion of these quantities to a related voltage followed by a DVM measurement.

### Voltage Ranges

The basis of the DMM is the dual-slope or voltage-to-frequency digital voltmeter. Whichever type is employed, most converters in DMMs have one or two fixed full-scale range settings. For example, a dual-slope converter might have fixed range settings of 199.9 mV and 1.999 V full scale. To provide the necessary versatility for measuring voltages of wider dynamic ranges with sufficient resolution, an input voltage divider is normally used to scale the input voltage. With a five-position divider ($\div 1$, $\div 10$, $\div 100$, $\div 1000$, $\div 10\,000$) and a "200 mV" full-scale converter, full-scale voltage settings from 200 mV to 2 kV would be available. Figure 1-28 shows a typical input divider

**Fig. 1-27.**  Typical VOM front panel. Note the nonlinear resistance scales and the multiple scale markings which require care to avoid a reading error.

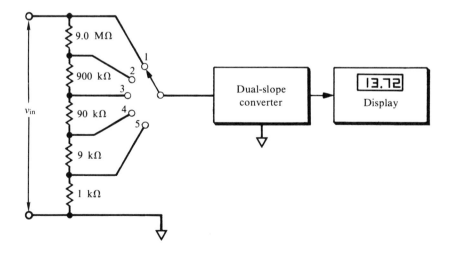

**Fig. 1-28.**  Input voltage divider for DVM. In position 2 the input voltage is divided by 10, and in position 3, by 100, etc. Thus, five full-scale voltage ranges are available with a fixed 10-M$\Omega$ input resistance. The downward—pointing triangular symbols indicate a connection to common—a point of reference for all voltages. The interconnection of common points is indicated in this way.

network, which provides a high input resistance (10 MΩ) to minimize the possible loading of voltage sources. When position 1 of the divider is chosen, the full-scale range of the DVM is 200 mV (actually, 199.9 mV). The divider then gives additional full-scale ranges of 2, 20, 200 and 2000 V.

Some more expensive DMMs have **autoranging** capabilities. That is, the full-scale range setting is changed automatically to provide the highest resolution without overranging. Most DMMs also give a polarity indication and automatically switch on the appropriate decimal point as different full-scale settings are selected.

## Current Measurements

The measurement of current with the DMM is made possible by inserting a current-to-voltage converter ahead of the DVM. One inexpensive type of current-to-voltage converter is a precision resistor network. Here the current through a precision resistor produces an *IR* drop that is measured by the voltmeter. Figure 1-29 shows a typical shunt network for making current measurements.

Modern DMMs have switch-selectable full-scale current ranges from microamperes to hundreds of milliamperes or even amperes. Higher current ranges often require a different connection to the meter than low current ranges to prevent damage to sensitive components.

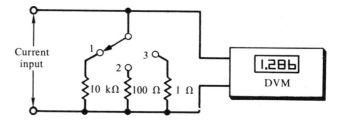

**Fig. 1-29.** Shunt network for DMM current ranges. If the DVM part of the DMM is 200 mV full scale, position 1 has a full-scale range of 20 μA, position 2, 2 mA, and position 3, 0.2 A.

## Resistance Measurements

The DMM can also measure resistance by adding a resistance-to-voltage converter prior to the DVM as shown in figure 1-30. Here a constant current source produces a current *I* in the unknown resistance $R_u$. The unknown resistance develops a voltage drop $IR_u$. The resulting voltage is measured by the digital voltmeter.

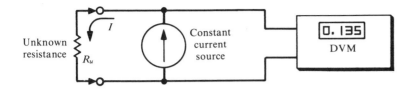

**Fig. 1-30.** Resistance measurement circuit for DMM. The basic DVM measures the *IR* drop across the unknown resistor when a known constant current is applied. The readout is proportional to $R_u$ and can be calibrated to read directly in ohms, kilohms, or megohms.

Modern DMMs use operational amplifier constant current sources that are highly stable over wide variations in resistance. The operational amplifier resistance-to-voltage converter is described in chapter 5. The resistance ranges available with most DMMs are from hundreds of ohms to tens of megohms full scale, providing resolution of a few ohms for a 3½ digit meter.

The front panel of a typical DMM is shown in figure 1-31. The DMM is a very versatile piece of test equipment for a variety of static measurements. In addition, the fact that DMMs can be used to measure the amplitude of ac signals (see chap. 2), makes them an indispensable tool in the laboratory.

**Fig. 1-31.** Typical DMM front panel. Note the range and function push buttons. Also, note the lack of ambiguity of the readout as compared to that of the analog multimeter of figure 1-27. (Reproduced with permission of Keithley Instruments, Inc., Cleveland, OH.)

## Suggested Experiments

**1. Voltage measurement with the DMM.**
Measure the output voltage of various batteries. Observe the effects of voltage scale change and test lead reversal. Describe the meter response for open and shorted input leads.

**2. Resistance measurement with the DMM.**
Measure the values of various types of resistors with the ohm scales of the DMM. Identify a number of resistors by color code. Describe the meter response for open and shorted input leads.

**3. Introduction to the breadboard system.**
Become familiar with the breadboard socket connections. Investigate the power-supply controls and the output voltages provided. Measure several voltages over the full range of the variable voltage source. Practice the technique for connections between devices on the breadboard and those not on it. Measure and calculate the resistance of several combinations of resistors connected in series and parallel.

**4. Current measurement with the DMM.**
Connect a complete circuit, and measure its current. Confirm Ohm's law. Observe the effects of current-scale change and test-lead reversal.

**5. Voltage divider and current splitter.**
Connect a voltage divider and confirm Kirchoff's law for voltages in a loop. Connect a parallel circuit and confirm Kirchoff's law for currents at a junction.

**6. Loading a voltage source.**
Wire a voltage source that has a 20-kΩ output resistance, and measure its voltage with and without a 10-MΩ load. Calculate the error produced by the loading from the DMM. Charge a 5-$\mu$F capacitor and use the DMM to measure the capacitor voltage change with time. Estimate the DMM input resistance from the rate of capacitor discharge.

**7. Logic-level sources and indicators.**
Measure the output voltage levels of binary-switch logic-level sources. Observe the response of LED logic indicators. Observe the BCD code by connecting the BCD switch outputs to LED indicators.

## Questions and Problems

**1.** A resistor of 20.0 kΩ carries a current of 5.00 mA. (a) What is the potential difference across the resistor? (b) If the direction of current is left to right through the resistor, is the right lead of the resistor positive or negative with respect to the left lead?

**2.** What is the resistance of a resistor color-coded with each of the following series of colors: (a) yellow, violet, orange; (b) gray, red, brown; (c) brown, green, yellow? What is the conductance of each resistor? (See appendix B for the color code.)

**3.** A light bulb with a resistance of 9.53 Ω is lit from a 25-V source. How much power does the bulb consume?

**4.** What is the resistance of a 60-W 120-V light bulb?

**5.** Resistors of 1.5 kΩ, 2.7 kΩ and 820 Ω are connected in series. (a) What is the total resistance of the combination? (b) If a 10-V source is connected to the series combination, what is the current?

**6.** A series circuit is composed of a 5-V source, a 1-kΩ resistor and a 4-kΩ resistor. (a) What is the current in the circuit, and what is the $IR$ drop across each resistor? (b) What fraction of the applied voltage appears across each resistor? (c) How much power is dissipated in each resistor?

**7.** If a 5-$\mu$F capacitor has a charge of 12 $\mu$C, what is the voltage across the capacitor?

**8.** A 1-$\mu$F capacitor is being charged by a 5.0-mA constant current source. How long will it take for the voltage across the capacitor to change by 1 V?

**9.** Design a voltage divider of total resistance 10 kΩ that will provide voltages of 0.1, 0.25, 0.5, 1.0 and 2.0 V from a 5-V source.

**10.** (a) What size resistor must be placed in parallel with a 5.0-kΩ resistor to obtain a total resistance of 3.5 kΩ? (b) If the parallel combination of resistors is connected to a 5.0-V source, what is the total circuit current? (c) What fraction of the total current appears in each resistor?

**11.** An analog multimeter has an ohms-per-volt rating of 10 $k\Omega/V$. (a) What is the basic current sensitivity of the meter movement? (b) If the basic meter has a resistance of 100 $\Omega$, what series resistance is used on the 250-mV full-scale voltage range? (c) What is the input resistance of the voltmeter on the 2.5-V full-scale range?

**12.** A current meter of 1-mA full-scale sensitivity has a resistance of 50 $\Omega$. A 5-mA full-scale meter is desired. What shunt resistance should be used?

**13.** (a) What is the maximum resistance of a current meter that is to measure the current in a circuit consisting of a 5-V source and a 10-$k\Omega$ resistor with an error of less than 1%? (b) What is the minimum full-scale current sensitivity required of the meter?

**14.** An ion-selective electrode has an equivalent resistance of $10^{10}$ $\Omega$. With a reference electrode the ion-selective electrode produces a cell voltage of 250 mV in a certain solution. What is the necessary input resistance for a voltmeter that is to measure the cell voltage with an error not to exceed 0.5 mV?

**15.** A 1.5-V alkaline flashlight cell has a maximum output current of 0.9 A. (a) What is the internal resistance of the cell? (b) Compare the internal resistance of the alkaline cell with that of a 12-V lead-acid storage battery that can deliver 500 A (briefly).

**16.** A voltage divider is constructed from a 10-$k\Omega$ resistor, a 20-$k\Omega$ resistor and a 5-V source. The voltage across the 20-$k\Omega$ resistor is to be measured. (a) Draw a Thevenin equivalent circuit, and calculate the equivalent resistance and voltage. The voltmeter used has an input resistance of 100 $k\Omega$. What voltage appears across the 20-$k\Omega$ resistor (b) when the voltmeter is absent and (c) when the meter is present? (d) What is the relative error caused by the connection of the meter?

**17.** The ohmmeter shown in figure 1-26 is used to measure resistance. If full scale is set to 0 $\Omega$ with the "ohms adjust" and a reading of two-thirds of full scale is obtained with the unknown resistance $R_u$, what is the value of $R_u$?

**18.** (a) What value of shunt resistance would be required to give the DMM of figure 1-29 a full-scale current range of 2 A? (b) What percent error (relative to full scale) would result from a contact resistance of 3 m$\Omega$ in the range selector switch? (c) What additional effect would 50 m$\Omega$ of resistance in the leads and lead contacts have?

**19.** The current through the tested component can be a matter of concern when making a resistance measurement. Assuming the ohmmeter section of a DMM uses the scheme shown in figure 1-30, and that the DVM has a full-scale sensitivity of 200 mV, what is the current through the tested component when the ohmmeter is on the 200-$\Omega$ scale?

**20.** (a) Characterize the three electrical-domain classes (analog, time, and digital) according to whether or not they are continuously variable in the encoded quantity and whether or not the data are accessible for decoding at any instant. (b) Comment on the relationship between relative freedom from noise interference and the degree of variability and speed of access.

**21.** (a) In the moving-coil voltmeter of figure 1-25, the series resistor $R_s$ is an interdomain converter between which two domains? (b) What interdomain conversion function is performed by the shunt resistor in the DMM current circuit of figure 1-29? (c) Trace the domain conversions involved in the DMM resistance measurement of figure 1-30 on a domains map.

# Chapter 2

# Periodic Waveforms and the Oscilloscope

All the examples of electrically encoded information described in chapter 1 were either static electrical signals or logic-level signals. Time-varying electrical signals are also of great importance in instrumentation because electrical signals in the analog domains often change with time. With such signals the amplitude at a particular time, or the entire amplitude-time relationship, may be the desired information. Similarly, information in one of the time domains may be encoded as the frequency of a periodic waveform, the time duration of a pulse, or the phase angle between two signals that vary with time. Thus, it is important to understand the basic properties of periodic signals and the techniques used to display them and to convert them into numerical representations.

In this chapter the frequency composition of periodic signals is discussed first in order to show how all periodic waveforms may be considered as the sum of elementary sinusoidal waves. Then the various measures of sine-wave amplitude, such as peak-to-peak, average, and root-mean-square (rms) are described. A brief introduction to reactive components and circuits is presented to provide some experience with the fundamental concepts of impedance and the frequency response of circuits. The digital multimeter is shown to be a useful tool in making quantitative measurements of the amplitude of sinusoidal signals. The final section introduces the oscilloscope, an indispensable tool for displaying periodic signals and for making measurements on the entire amplitude-time waveform. Various functions of the scope are described to demonstrate how it can provide a display of one or two signals versus time or a display of one signal against another.

## 2-1 Frequency Composition of Electrical Signals

For simple time-varying signals such as sinusoidal waveforms, the amplitude-time function can be completely described by a simple equation that includes the frequency of the waveform, the time, and the peak amplitude. In general, signals that result from complex phenomena are not simple sine waves composed of single frequencies, and their description is necessarily more complex. However, all signals, no matter how complex, can be represented as a sum of simple sine waves.

## Sine-Wave Signals

A **sine-wave signal** is a voltage or current that varies sinusoidally with time as illustrated in figure 2-1. Objects whose displacements are sinusoidal functions are said to be undergoing *simple harmonic motion* and are very common in nature. Sinusoidal current is produced by rotating a wire loop in a uniform magnetic field as in a power generator. The sine-wave signal is the only true single-frequency waveform. The repetition interval is the **period** as shown in figure 2-1. Each repetition is called a **cycle.** The period is thus expressed in units of seconds per cycle. The reciprocal of the period is the number of cycles per second, or the **frequency**. The unit cycles per second has been given the name **hertz** (Hz).

The generation of a sine wave from a rotating vector of magnitude $A_p$ is illustrated in figure 2-2. The sine wave results from the projection of the vector on the vertical axis as the vector rotates counterclockwise with a uniform **angular velocity**, $\omega$. If the vector is rotating at the rate of one revolution per second, the sine wave repeats itself periodically once every second. Each revolution of the vector produces one cycle. The period is the time interval $t_{per}$ required to produce one cycle. Since a cycle is generated each time the vector sweeps through 360°, or $2\pi$ radians, the time axis is conveniently expressed as the angle of the vector. Since the vector sweeps one cycle every $t_{per}$ seconds, the frequency $f$ is

$$f = 1/t_{per} \qquad (2\text{-}1)$$

The vector is rotating at a rate $\omega$ or $2\pi$ radians every $t_{per}$ seconds; thus

$$\omega = 2\pi/t_{per} = 2\pi f \qquad (2\text{-}2)$$

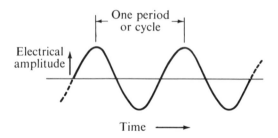

Fig. 2-1. A sinusoidal signal. This representation of signal amplitude versus time is called a **waveform** because of its resemblance to the shape of waves on water. However, it represents the amplitude variation with time at one point rather than the variation of amplitude with distance at a single instant in time.

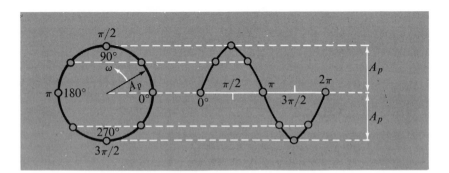

Fig. 2-2. Development of a sine wave from the projection of a rotating vector. The vertical displacement of the point of the rotating vector is plotted against the angular rotation in radians.

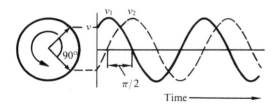

**Fig. 2-3.** Two sine waves with a 90° phase difference. Both vectors must rotate at the same rate (i.e., have the same frequency) to maintain a fixed angle between them. The signal $v_2$ lags (follows after) signal $v_1$ by 90°.

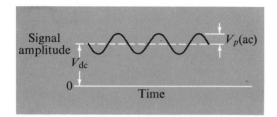

**Fig. 2-4.** Ac and dc signal components. A varying signal has both ac and dc components if the average value of the signal over a long time is not zero. The average value is the dc component, and the variation about that average is the ac.

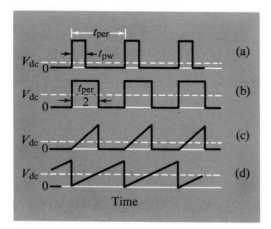

**Fig. 2-5.** Nonsinusoidal periodic signals. (a) Rectangular wave. (b) Square wave. (c) Ramp. (d) Sawtooth. All have the same period, repetition rate, and peak-to-peak amplitude.

If the vector represents a current or voltage, the instantaneous current $i$ or voltage $v$ can be given as

$$i = I_p \sin \omega t = I_p \sin 2\pi ft \qquad (2\text{-}3)$$

$$v = V_p \sin \omega t = V_p \sin 2\pi ft \qquad (2\text{-}4)$$

where $I_p$ and $V_p$ are the peak current and voltage, respectively.

If two voltages have the same frequency but cross zero at different times, as shown in figure 2-3, they are said to be out of **phase**. The difference in time can be conveniently expressed in terms of an angle (fraction of a cycle) called the **phase angle**, $\theta$.

If the two sine waves are shown as rotating vectors with the same origin (fig. 2-3), the phase angle can be seen to arise from the fact that one vector leads the other. A generalized voltage sine-wave signal can be expressed as

$$v = V_p \sin (\omega t + \theta) \qquad (2\text{-}5)$$

where $\theta$ is the phase angle between the sine wave under consideration and a reference sine wave.

Since the sine wave is such a simple waveform, *there are only two pieces of information obtainable from a sine-wave signal; the frequency $\omega$ and the amplitude $V_p$.* If another signal of the same frequency is available as a time reference, the phase angle $\theta$ between them may also carry information.

## AC Signals

An ac voltage or current has been defined as one that changes sign periodically. The voltage and current are almost always of this kind in ac power circuits. Electrical signals, however, are often combinations of a **dc signal level**, or average value, and an ac component that alternates about the dc level. In such cases, it is useful to define the **ac signal** as that part of the signal that varies with time. Figure 2-4, shows a signal with a sinusoidal variation of $V_p$ volts about an average dc signal level of $V_{dc}$. The total signal may be considered the sum of the steady, or dc, voltage and the varying ac voltage. An ac signal is readily isolated from a dc component in an electronic circuit.

**Nonsinusoidal periodic signals.**   If the pattern of current or voltage variation is repeated at regular intervals, the signal is said to be **periodic**. The sine wave is just one example of a periodic waveform. The **rectangular wave** shown in figure 2-5a is a series of regularly occurring pulses. If the width of the positive and negative half cycles are identical, as in figure 2-5b, the waveform is called a **square wave**. **Ramp** and **sawtooth** waves are also shown in figure 2-5c and d. The total variation from the most positive to the most negative voltage is the **peak-to-peak** voltage. The dc level of each of the waveforms in figure 2-5 is different. The average, or dc, level is that value above and below which the voltage-time or current-time *areas* are equal for one complete cycle. If the waveform is asymmetric, i.e., it has dissimilar shapes for the + and − alternations, the dc level is often *not* the midpoint between the peak values.

**Fourier series waveform analysis.**    Any periodic waveform, such as those shown in figure 2-5, has a fundamental frequency, or repetition rate, and contains information at other frequencies. These waveforms can be represented by a **Fourier series expansion** (see note 2-1). A Fourier series of sine waves is a special combination of sine waves that is a summation of multiples of a single fundamental frequency. The multiple frequencies, which may have various amplitudes and phase angles, are known as **harmonics**. Any single-valued, periodic voltage waveform $v(t)$ can be represented by a Fourier series expansion.

As an example of how a Fourier series can be used to represent the waveform of a periodic signal, a square wave can be constructed as shown in figure 2-6. Curves A and B in figure 2-6a show the fundamental frequency and the third harmonic; curve C is the graphical addition of A and B. Other odd harmonics are shown in figure 2-6b and c. Curve G, which is the graphical sum of the fundamental plus the third, fifth, and seventh harmonics, is a fairly good approximation to the square wave. The complete Fourier series representation of the square wave of figure 2-6 would include higher odd harmonics, each in decreasing amplitude. In fact, the harmonic at frequency $3\omega$ has an amplitude one-third that of the fundamental, while that at $5\omega$ is one-fifth the amplitude of the fundamental, etc. A plot of the relative amplitude of the signal at each frequency versus the frequency is shown in figure 2-7. This is a plot of the **frequency spectrum.** As can be seen in figure 2-6, the higher frequency components are required to produce the abrupt changes in the signal at the leading and trailing edges of the square wave. Lower frequency components are needed to produce the flat top and bottom portions of the square wave. Because the square wave contains low and high frequency components, it is a very useful waveform in testing the frequency response of measurement systems.

**Note 2-1.    Fourier Series Expansion.**
A general Fourier series can be written $V(t) = V_{dc} + V_1 \sin(\omega t + \theta_1) + V_2 \sin(2\omega t + \theta_2) + \cdots + V_n \sin(n\omega t + \theta_n)$. The term $V_{dc}$ is the dc level of the signal and is the average value about which the signal varies. The terms $V_1, V_2, \ldots, V_n$ are the peak amplitudes of the fundamental, second harmonic, and so on through the nth harmonic, and the terms $\theta_1, \theta_2, \ldots, \theta_n$ are the corresponding phase angles.

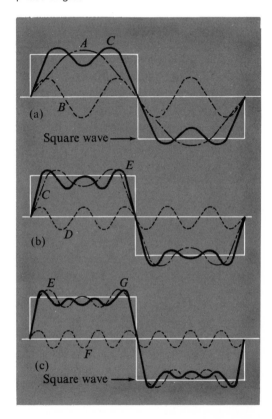

**Fig. 2-6.** Harmonic composition of a square wave. (a) Fundamental *A* and third harmonic *B* are added to give curve *C*. (b) Adding the fifth harmonic *D* gives *E*, the fundamental plus third and fifth harmonics. (c) The seventh harmonic *F* and the fundamental plus third, fifth and seventh harmonics give *G*.

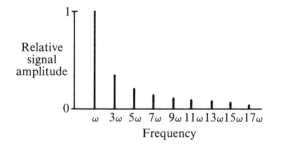

**Fig. 2-7.** Frequency spectrum of square wave. Note that only the odd harmonics occur, but these have substantial amplitudes, even at very high harmonics.

**Fig. 2-8.** Peak measures of common waveforms. The peak-to-peak voltage $V_{p-p}$ is the sum of the maximum positive and maximum negative peak values $V_{p+}$ and $V_{p-}$.

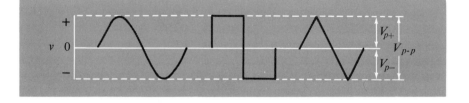

## 2-2   Measures of Periodic Signal Amplitude

The amplitudes of periodic signals are generally described in four ways: the peak value, the peak-to-peak value, the average value, and the root-mean-square (rms) value. The particular description used depends on the application. Measurement instruments such as the digital multimeter and the oscilloscope respond to the peak-to-peak value. Moving-coil meters respond to the average value, and the rms value is very useful for the measurement of power in an ac circuit. In this section these amplitude measures are defined and applied to common waveforms.

### Peak Values

The **peak value** is the voltage or current at the maximum point in the waveform. If the signal has both positive and negative polarity during a cycle, there is a positive peak value $V_{p+}$ and a negative peak value $V_{p-}$ as shown in figure 2-8. If the waveform is centered about zero volts, $V_{p+}$ and $V_{p-}$ are equal in magnitude. The peak value is useful where the maximum positive or negative excursions are of principal interest. Where the total magnitude of the change independent of the zero position is of interest, the **peak-to-peak** voltage $V_{p-p}$ is used. For a waveform that is symmetrical about the zero point, $V_{p-p} = 2V_p$.

### Average Values

The **average** value of a waveform is the average of the *absolute magnitude* of the signal during an entire period. It thus depends on whether a signal is entirely ac or has a dc level associated with it. The following discussion assumes that the waveforms are symmetrical about zero volts (ac). The average ac value is illustrated in figure 2-9 for several waveforms. For the sine wave in figure 2-9a, the average value is obtained by determining the area under the waveform during a half-cycle and dividing by the time of one half-cycle. The result is

$$V_{av} = 0.637V_p \qquad (2\text{-}6)$$

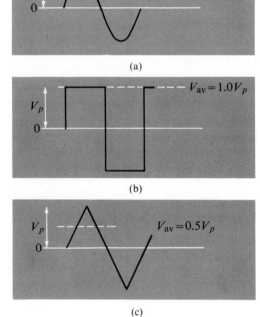

(a)

(b)

(c)

**Fig. 2-9.** Average measures of common ac waveforms. (a) Sine wave. (b) Square wave. (c) Triangular wave. For each waveform, the average absolute amplitude is a different fraction of the peak value.

The average value of the symmetrical square wave shown in figure 2-9b, is equal to $V_p$. On the other hand, the average value of the triangular waveform in figure 2-9c is 0.5 $V_p$ as shown.

## Root-Mean-Square Values

The rms value of a waveform is the effective value used in power calculations. The rms value of a waveform produces the same heating (power dissipation) in a resistor as an identical dc current. In other words, one ampere rms produces the same amount of heat in a resistor in a given time as one ampere of dc current. The power dissipated in a resistor at any time is proportional to the square of the current through the resistor ($P = I^2R$) or of the voltage across the resistor ($P = V^2/R$). Therefore, the average power dissipated by a varying signal is proportional to the average of the square of the current or voltage amplitude over one complete cycle. The steady voltage or current equivalent in power dissipation to the varying voltage or current is the square root of the average of the squared amplitude values. This is called the **root-mean-square (rms)** value. The operation of taking the rms value for a sine wave is shown graphically in figure 2-10. The square of the sine wave ($V_p^2 \sin^2 \omega t$), plotted directly below the sine wave, indicates the heating effect of the current which is directly proportional to the square of the current. Taking the square root of the mean of the square gives

$$V_{rms} = \sqrt{\frac{V_p^2}{2}} = \frac{V_p}{\sqrt{2}} = 0.707 V_p \qquad (2\text{-}7)$$

The relation between the effective (that is, rms) and peak current for a sinusoidal signal is the same as for effective and peak voltage. Thus, $I_{rms} = 0.707 I_p$. Calculations of voltages and currents in ac circuits provide self-consistent values. For example, if rms voltages are used, the currents will have rms values. Note that peak values for sine waves are obtained by multiplying the rms values by the reciprocal of 0.707 ($1/0.707 = 1.414$), so that

$$I_p = 1.414 I_{rms} \quad \text{and} \quad V_p = 1.414 V_{rms} \qquad (2\text{-}8)$$

The rms values for the other waveforms shown in figure 2-9 are given in table 2-1. The bipolar square wave has a mean square voltage of $V_p^2$, which means that for the waveform $V_p = V_{av} = V_{rms}$. The rms value of a triangular wave is obtained by taking the root mean square of the linear voltage relation $v = kt$ over the time from 0 to $V_p/k$. Thus we see that the relationships among the several useful measures of periodic signal amplitude are

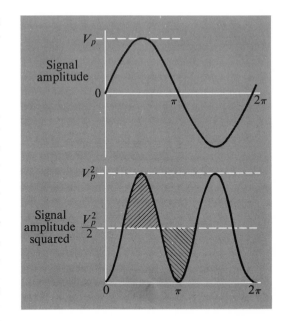

**Fig. 2-10.** Graphical determination of the rms value of a sinusoidal signal. The squared signal is a sine wave that is positive over the entire waveform. The average value is $V_p^2/2$ as demonstrated by the equal areas of the two hatched sections of the waveform.

**Table 2-1.** Peak, average and rms values for ac signals.

|  | $V_p$ | $V_{av}$ | $V_{rms}$ |
|---|---|---|---|
| sine wave | 1 | 0.637 | 0.707 |
| square wave | 1 | 1 | 1 |
| triangle wave | 1 | 0.500 | 0.576 |

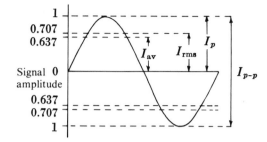

**Fig. 2-11.**   Common measures of sine-wave amplitude.

highly dependent on the wave shape. The importance of this fact will become clear when we discuss how measurement devices are applied later in this chapter.

The various measures of sine-wave amplitude are summarized graphically in figure 2-11 for a sinusoidal current waveform.

## 2-3   Reactive Circuits

When a varying voltage is applied across a conductor, one must consider not only the current that results from the changing voltage at any instant, but also the reaction of the conductor to the variation in voltage. For ac signals the term **impedance**, $Z$ (in ohms), is defined to take both effects into account. Impedance in ac circuits is analogous to dc resistance, and its magnitude is $Z = V_p/I_p$, where $V_p$ and $I_p$ are the peak values of the voltage and current respectively. As the rate of signal variation (its frequency) approaches zero, the impedance of a device approaches its resistance.

For simple resistors the current is always proportional to the instantaneous applied voltage as given by Ohm's law. Since action and reaction are simultaneous in a pure resistance, there is no special reaction to signal variations and $Z = R$ for all frequencies. Thus the application of a sine-wave voltage $v = V_p \sin(\omega t + \theta)$ to a pure resistor gives rise to a sine-wave current $i = I_p \sin(\omega t + \theta)$. Note that the phase $\theta$ does not change; thus for a pure resistance the current and voltage sine waves are in phase for all frequencies. Pure resistance is an ideal that is closely approached by many practical resistors. Deviations from ideality for practical resistors are discussed in appendix B.

Reaction to the rate of change of a signal can be classified as inductive reactance or capacitive reactance. Intentional reactive components are capacitors or inductors. In this section the principles of $RC$ filters, inductance, inductive reactance, and inductive filters are introduced.

### Capacitive Reactance

As described in section 1-2 the quantity of charge $Q$ stored in a capacitor depends on the voltage across the capacitor $v_C$:

$$Q = Cv_C \tag{2-9}$$

If the voltage $v_C$ varies with time as in figure 2-12, we see from equation 2-9 that the changing voltage across the capacitor must correspond to a change in charge, and therefore a current is required. Since current is the rate of charge flow, the current required is directly proportional to the rate of change of the capacitor voltage. The voltage across a capacitor cannot change instantly because an instantaneous voltage change would require an

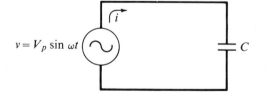

$v = V_p \sin \omega t$

**Fig. 2-12.**   Capacitor with ac voltage source.

infinite current. The capacitor thus reacts against a change in voltage across it. **Capacitance reactance** $X_C$ has the units of ohms and is dependent on frequency as shown by equation 2-10 and note 2-2.

$$X_C = 1/(2\pi fC) = 1/(\omega C) \qquad (2\text{-}10)$$

The higher the frequency, the less reactance a capacitor offers to the flow of charge. Because of the way in which its reactance depends on frequency, a capacitor is an open circuit to a constant voltage: ($X_C \rightarrow \infty$ as $f \rightarrow 0$).

When a sine-wave voltage source, $v = V_p \sin \omega t$ is connected to a capacitor, the resulting current is a cosine wave. There is thus a 90° phase difference between the current and voltage waveforms in a capacitor as shown in figure 2-13.

## Series *RC* Circuits

A series resistor and capacitor (**series *RC* circuit**) is extremely useful as a **filter** circuit. This simple combination of components can be used to reduce the magnitude of low frequency signal components, in which case it is called a **high pass filter**, or to reduce the magnitude of high frequency components, where it is called a **low pass filter**.

A series *RC* circuit with a sinusoidal voltage source is shown in figure 2-14. To understand the frequency dependence of the voltages across the resistor $v_R$ and the capacitor $v_C$, remember that the resistance of a pure resistor is frequency independent and that the reactance of a capacitor varies inversely with frequency. Consider the source voltage $v_s$ to be a slowly varying signal ($\omega$ small). Since the capacitive reactance is very high at low frequencies, a large fraction of the source voltage will appear across the capacitor ($X_C \gg R$). In fact for $\omega = 0$ (a dc voltage source), the capacitive reactance is infinite, and all the source voltage appears across the capacitor $C$. As the frequency increases, $X_C$ decreases and a larger fraction of $v_s$ begins to appear across the resistor. A circuit in which the output is taken across the capacitor is called a low pass filter (see fig. 2-15).

**Note 2-2. Derivation of Capacitive Reactance.**
The current required when the voltage across a capacitor changes is given by

$i = dq/dt = C(dv_C/dt)$

If the capacitor is connected to a sine wave source of voltage $v = V_p \sin \omega t$, then

$i = C(dv_C/dt) = \omega C V_p \cos \omega t$

Since the peak current $I_p = \omega C V_p$, $i = I_p \cos \omega t$. The impedance of a device is $Z = V_p/I_p$, and for a pure capacitor the impedance, called the capacitive reactance $X_C$, is

$X_C = V_p/(\omega C V_p) = 1/(\omega C) = 1/(2\pi fC)$

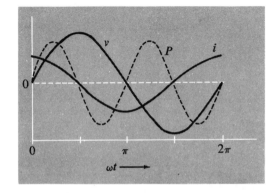

**Fig. 2-13.** Current, voltage and power waveforms in the circuit of figure 2-12. The current wave precedes the voltage by $\pi/2$ radians or 90°. The current thus leads the voltage. The instantaneous power $P = vi$.

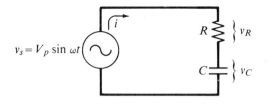

**Fig. 2-14.** Series *RC* circuit. This series arrangement of a resistor and a capacitor is an ac voltage divider whose divider fraction depends on frequency.

**Fig. 2-15.** *RC* low pass filter. The fraction of $v_s$ that appears across *C* increases as the frequency decreases.

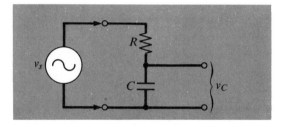

### Note 2-3.   Low Pass Filter.

The ratio output voltage/input voltage of the filter is $v_C/v_s$. The amplitude of this ratio contains no phase information and is given by $V_{pC}/V_p$. This can be written

$$\frac{V_{pC}}{V_p} = \frac{X_C}{Z} \quad \frac{X_C}{\sqrt{R^2 + X_C^2}}$$

The frequency at which $X_C = R$ is the upper cutoff frequency. At this frequency

$$\frac{V_{pC}}{V_p} = \frac{X_C}{\sqrt{2X_C^2}} = \frac{1}{\sqrt{2}} = 0.707$$

### Note 2-4.   Decibel.

Decibel, dB, is a term used primarily to express the ratio of two power levels, such as the signal power output and signal power input where the circuit power gain in decibels is 10 log $(P_{out}/P_{in})$. The term has been extended through the relationship $P = V^2/R$ to express voltage level ratios where the circuit voltage gain in decibels is dB = 20 log $(V_{out}/V_{in})$. Technically, the voltage ratio relationship should be used only when the impedances across which the voltages appear are equal. However, it is common practice to express voltage ratios in decibels regardless of the impedances involved.

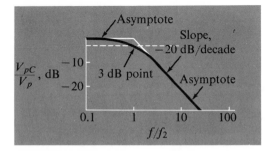

**Fig. 2-16.**   Bode plot of frequency response of low pass filter. The asymptotes are extensions of the linear portions of the curve. At $f = 10\,f_2$ the low pass filter response is $-20$ dB, one-tenth its response at $f_2$.

The series RC network has a total impedance magnitude $Z$ and phase angle $\theta$ given by

$$Z = \sqrt{R^2 + X_C^2} \quad \text{and} \quad \tan\theta = X_C/R \qquad (2\text{-}11)$$

For a low pass filter, such as that shown in figure 2-15 the frequency at which the capacitive reactance equals the resistance, called the **upper cutoff frequency** $f_2$, is readily derived by solving equation 2-10 for the conditions $X_C = R$ and $f = f_2$. The result is

$$f_2 = 1/(2\pi RC) \qquad (2\text{-}12)$$

At the upper cutoff frequency the peak voltage across the capacitor $V_{pC}$ equals $0.707\,V_p$ (see note 2-3), where $V_p$ is the peak voltage of the source. The ratio $V_{pC}/V_p$, which is called the **network transfer function**, is essentially unity at frequencies much lower than $f_2$ since all the source voltage appears across $C$ at very low frequencies. It is convenient to plot the ratio $V_{pC}/V_p$ against frequency in a log-log plot with the transfer function expressed in **decibels** (see note 2-4). Such a plot for the low pass filter of figure 2-15 is shown in figure 2-16. The upper cutoff frequency $f_2$ is also known as the upper **3 dB point**, since this is the frequency at which the filter output has dropped by 3 dB from its very low frequency value. At frequencies much higher than $f_2$ the network transfer function falls by 20 dB for each decade (tenfold) increase in frequency.

A plot such as that shown in figure 2-16 is known as a **Bode diagram**. It can readily be constructed from two limiting regions: at very low frequencies $V_{pC}/V_p$ has a magnitude of 0 dB and a slope versus $f/f_2$ of $\simeq 0$. At very high frequencies the plot has a slope of $-20$ dB/decade. The two limiting slopes meet at the 3 dB point. Note that these two asymptotes represent the true response quite accurately except for the region very near $f_2$.

The high pass filter arrangement of the simple $RC$ circuit is shown in figure 2-17a. Here the output is taken across $R$ instead of $C$. Note in the Bode plot of figure 2-17b, that at high frequencies the network transfer function $V_{pR}/V_p = 1$ (0 dB), since $X_C$ approaches zero as $f$ approaches infinity. At low frequencies the transfer function decreases by 20 dB for each decade decrease in frequency. The point at which $X_C = R$ is known as the **lower cutoff frequency**, or lower 3 dB point, $f_1$ and is given by $f_1 = 1/(2\pi RC)$. Again the Bode diagram can be closely approximated by two limiting slopes, which are approached asymptotically by the actual response.

High and low pass filters also shift the phase of the output signal with respect to the input signal. Recall that with a pure resistor, current and voltage are always in phase and that with a capacitor the current leads the voltage by 90°. Hence $V_{pR}$ always leads $V_{pC}$ by 90°. Plots of $\theta$ versus $f/f_1$ for

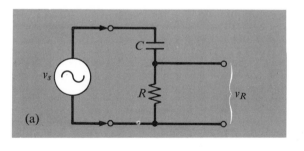

(a)

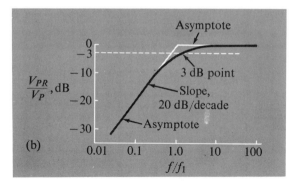

(b)

**Fig. 2-17.** Frequency response of high pass circuit. (a) High pass network. (b) Bode diagram. The reversal of the positions of $R$ and $C$ in the filter reverses the curve about the cutoff frequency.

$V_{pR}$ and of $(90° - \theta)$ versus $f/f_2$ for $V_{pC}$ are shown in figure 2-18 for the high and low pass filter arrangements, respectively. Note that $\theta = 45°$ at the 3 dB frequency where $X_C = R$ and $f = f_1 = f_2$.

## Inductance

**Inductance** is that property of a device that reacts against a change in current through the device. Although all devices have associated electrical inductance, **inductors** (devices designed to have inductance) are often purposely used in circuits to resist changes in current and thus to serve important control functions.

An inductor is based on the principle that a varying magnetic field induces a voltage in any conductor in that field. A practical inductor may simply be a coil of wire as shown in figure 2-19a. A schematic diagram of an inductor connected to an ac voltage source is shown in figure 2-19b. The current in each loop of wire generates a magnetic field that passes through neighboring loops. If the current through the coil is constant, the magnetic field is constant, and no action takes place. If the current increases, however, the magnetic field lines expand, and the changing magnetic flux generates a voltage in each loop. This induced voltage (**electromotive force** or **EMF**) is counter (of opposite polarity) to the change in the applied voltage. This counter emf thus tends to impede, or react against, any change in current

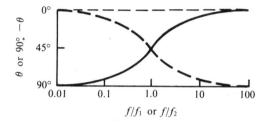

**Fig. 2-18.** Phase shift of low and high pass filters plotted against frequency. (–) $\theta$ vs. $f/f_1$; (---)$90° - \theta$ vs. $f/f_2$. Note that the phase shift is substantial even at frequencies well away from the cutoff frequency.

**Fig. 2-19.** Inductance. (a) Current in a coil produces a magnetic field. The energy stored in the field is proportional to the current. The energy absorbed or released when the current is changed reacts against the change in current. The inductor is then an impedance in an ac circuit (b).

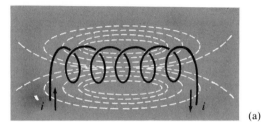

(a)

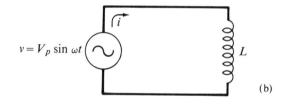

(b)

**Note 2-5.   Derivation of Inductive Reactance.**
The voltage across an inductor $V_L$ is related to the inductance $L$ and the rate of change of current $di/dt$ by

$$V_L = L\, di/dt$$

When the inductor is connected to a sine-wave source of voltage $v = V_p \sin \omega t$,

$$V_p \sin \omega t = L\, di/dt$$

If this equation is solved for $i$ by integration,

$$i = \frac{-V_p}{\omega L} \cos \omega t = -I_p \cos \omega t$$

The impedance is $Z = V_p/I_p$. For a pure inductor the impedance, called the inductive reactance $X_L$, is given by

$$X_L = \frac{V_p}{V_p/(\omega L)} = \omega L = 2\pi f L$$

through the coil. The counter emf is directly proportional to the rate of change of current through the coil. The proportionality constant is the inductance $L$, which has the unit of **henrys** (H). An inductance of 1 H induces a counter emf of 1 V when the current is changing at the rate of 1 A/s.

## Inductive Reactance

In an ac circuit, like that of figure 2-19b, the inductor offers reactance to alternating current. The **inductive reactance** $X_L$ has the units of ohms and is given by equation 2-13 and note 2-5.

$$X_L = \omega L = 2\pi f L \tag{2-13}$$

Note that inductive reactance, like capacitive reactance, is frequency dependent. For inductors, however, the reactance *increases* with increasing frequency. An inductor is said to be a short circuit to direct current since $X_L \to 0$ as $f \to 0$.

When a sine-wave voltage source is connected to an inductor, the current through the inductor lags the voltage across it by 90°. Note that this is opposite to the phase relationships of the current and voltage waveforms in a capacitive circuit.

**Note 2-6.   Impedance and Phase Angle of Series *RL* Circuit.**
The magnitude of the impedance is $Z = \sqrt{R^2 + X_L^2}$ and the phase angle $\theta$ is given by $\tan \theta = X_L/R$.

## Series *RL* Circuit

Like series *RC* circuits, series *RL* circuits are used as filters. (See note 2-6 for impedance and phase angle.) In the low pass filter of figure 2-20a, the output is taken across the resistor. At very low frequencies where $X_L \ll R$ the full source output appears across the resistor. The upper cutoff frequency $f_2$ is reached when $R = X_L$ and is given by

$$f_2 = R/(2\pi L) \tag{2-14}$$

The Bode diagram is the same as for an *RC* low pass filter (fig. 2-16) except that $V_{pR}/V_p$ is plotted against $f/f_2$, and $f_2$ is given by equation 2-14.

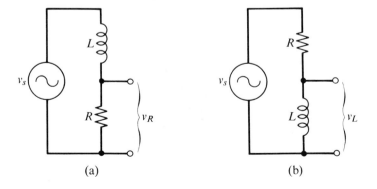

**Fig. 2-20.** *RL* voltage dividers. The increase in the impedance of $L$ with increasing frequency makes circuit (a) a low pass filter and circuit (b) a high pass filter.

(a)                    (b)

The $RL$ high pass filter of figure 2-20b produces the full source voltage at the output only at high frequencies where $X_L \gg R$. At frequencies lower than the lower cutoff frequency $f_1$ the transfer function $V_{pL}/V_p$ decreases by 20 dB for each decade decrease in frequency. Again the Bode diagram is the same as that for the $RC$ high pass filter in figure 2-17b except that $V_{pL}/V_p$ is plotted against $f/f_1$ and $f_1 = R/(2\pi L)$.

## 2-4 AC Measurements with the Digital Multimeter

The digital multimeter is capable of making ac voltage and, in some cases, ac current measurements. Since the DMM functions by measuring dc voltages, an ac-to-dc converter is switched into the circuit prior to the analog-to-digital converter for ac measurements. The output of the ac-to-dc converter is then a dc voltage suitable for measurement by the normal voltmeter part of the DMM. On the ac volts scale an input divider, which must be nearly frequency independent, is used to select the appropriate range for the ac-to-dc converter. On the ac current scale, the input current is sent through switch-selectable resistors to produce an ac $IR$ drop proportional to the current. This ac voltage is then converted to dc and measured in the usual manner. The ac measurement principles upon which the DMM operates are described in this section.

## AC-to-DC Conversion

Even though all ac meters are calibrated in rms voltage or current, very few actually respond to the rms value of the input signal. The type of ac-to-dc converter and the meter itself determine to which ac characteristic the meter responds directly. Moving-coil meters generally convert the ac signal into a pulsating dc signal by a technique known as rectification (rectifier circuits are described in chapter 3). Since the moving-coil meter responds slowly to current variations, this inertia tends to average the fluctuating signal from the ac-to-dc converter. Thus, moving-coil meters respond to the average value of ac current.

In a digital multimeter a filter circuit is normally added after the rectification step. The filter capacitor charges to the peak, or sometimes the peak-to-peak, value of the input voltage. Thus, the ac digital multimeter responds directly to the peak or peak-to-peak voltage.

If the desired amplitude measure is different from the one the meter responds to directly, an ac meter scale can be calibrated with a conversion factor included. For example, in order for moving-coil meters to readout in rms values, a conversion factor of $(0.707/0.637)$ for sine waves is needed since the rms value equals $(0.707/0.637)I_{av}$. By adjusting the current meter shunt resistors and/or the meter scale markings, the appropriate conversion factor can be applied. Since the conversion factor is only good for sinusoidal signals, the average-responding, rms-calibrated moving-coil meter does not read the correct rms amplitude for nonsinusoidal signals. With a digital multimeter that responds to the peak-to-peak voltage, a conversion factor of 0.707/2 is needed to convert to rms values for sine waves. Since 0.707/2 is not the peak/rms factor for most nonsinusoidal waveforms, the rms scale of the digital multimeter is, in general, only accurate for sine-wave voltages.

When an rms measurement is needed on a waveform for which the relationship between the average or peak value and the rms value are not known, a true rms-responding meter must be used. The most common way to obtain the true rms value of a signal is to calculate it by the procedure illustrated graphically in figure 2-10. In one popular type of meter, analog circuits are used to square the signal, to average it over at least one period and finally to take the square root. This produces an output voltage equal to the root mean square of the input voltage. Digital computation techniques are also used in some rms-responding DMMs.

## Complete DMM

A block diagram of the complete digital multimeter is shown in figure 2-21. Switches or push buttons on the front panel select ohms, volts, or amperes as the measured quantity. This has the effect of switching the R-V converter, the voltage divider, or the I-V converter into the circuit. Another front-panel switch selects ac or dc for voltage or current measurements. This switch places the ac-to-dc converter in the circuit (for ac) or out of it (for dc). In addition, range selection controls select the R-V converter sensitivity, the voltage-divider attenuation, or the I-V converter sensitivity. Once the quantity of interest has been converted to a dc voltage, the dual-slope or voltage-to-frequency type analog-to-digital converter presents a parallel digital signal to the front-panel display circuits.

**Fig. 2-21.** Block diagram of complete DMM. Function switches select the appropriate input signal-conversion circuits so that a dc voltage appears at the input to the analog-to-digital converter.

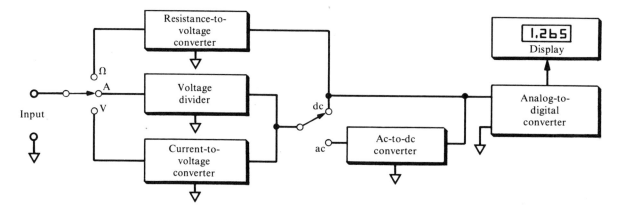

Of course, additional circuitry is present in the DMM for providing current to drive the display, for storing previous results until an updated reading is made and for timing the various operations.

## 2-5  The Oscilloscope

The **oscilloscope** has become an almost indispensable laboratory instrument because of its ability to display voltage information as a function of time $t$, or to plot one voltage $x$ as a function of another voltage $y$. The $x$-$y$ or $x$-$t$ plots are displayed on the face of a **cathode-ray tube** (**CRT**). The CRT is the indicating device for an oscilloscope. The tube consists of the electron gun shown in figure 2-22a and the deflection plates (fig. 2-22c) combined in a vacuum tube with a fluorescent screen on the enlarged end, as shown in figure 2-22b. The purpose of the electron gun is to provide a beam of electrons that is focused to a sharp point at the fluorescent screen. The end of the tube is coated with various phosphors that emit visible phosphorescent radiation at the point of bombardment with electrons. The intensity of the emitted visible light depends on the number of electrons per unit time striking a given area of the screen; the larger the number, the more intense the light. The rate of electron flow through the grid is regulated by partially surrounding the cathode with a metal grid, and applying a voltage such that the grid is negative with respect to the cathode. As the grid potential is made more negative, the repulsion of electrons is greater, the fraction which reaches the fluorescent screen is less, and therefore the visible light spot is less intense. In other words, a variable grid-cathode voltage supply acts as the **intensity control**.

To obtain the point source of electrons on the screen, it is necessary to focus and accelerate the electrons by two anodes that are at high positive potentials with respect to the cathode. For a typical five-inch tube, the second anode is often 2000 to 10 000 V more positive than the cathode, and the first anode is about 350 to 750 V more positive. The electrostatic field between the two hollow cylindrical anodes provides the necessary focusing of the electron beam. The diverging electrons entering the first hollow anode are forced to converge because of the field that exists between the first and second anode. The desired focus point is, of course, the fluorescent screen, and the correct focus can be obtained by varying the voltage on one anode with respect to the other. The **focus control** is another front-panel adjustment on the oscilloscope.

### Electron-Beam Deflection

The electron beam then passes through the first set of deflection plates, which are mounted in the horizontal plane as shown in figure 2-22. These

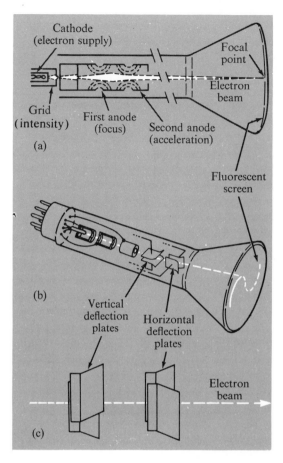

**Fig. 2-22.** Cathode-ray tube. (a) Electron gun. (b) Cut-away sketch. (c) Deflection plates. A beam of electrons is generated by the electron gun and directed to the desired positions on the fluorescent screen by the deflection plates.

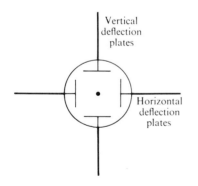

**Fig. 2-23.** Schematic of CRT deflection plates. The spot in the center represents the image on the scope face caused by the undeflected electron beam.

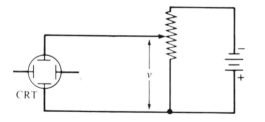

**Fig. 2-24.** Application of a deflection voltage to the vertical CRT plates. As $v$ is increased, the electron beam position is lowered on the CRT face.

plates are referred to as the **vertical deflection plates**. Next the beam passes between two plates mounted perpendicular to the first set, the **horizontal deflection plates**. The CRT deflection plates are shown schematically in figure 2-23.

If one vertical deflection plate is made positive with respect to the other, the electron beam is deflected toward the positive plate. This may be accomplished by applying a voltage as shown in figure 2-24. The greater the magnitude of the applied voltage, the greater the deflection of the electron beam. Horizontal deflection of the electron beam can be accomplished in an analogous manner using the horizontal deflection plates. If voltages are applied to both sets of plates simultaneously, the position of the electron beam in both dimensions of the plane of the CRT screen depends on the sign and the magnitude of the two deflection voltages. In this manner the electron beam can produce a $x$-$y$ plot of one deflection voltage as a function of the other.

## Basic Components of the Oscilloscope

In addition to the CRT and deflection plates, several other circuits are required to make up an oscilloscope, as shown in the block diagram of figure 2-25. The voltage required to deflect the electron beam is usually much higher than the voltage levels that are measured. For this reason, **deflection amplifiers** are used to increase the signal voltage to the level required by the deflection plates. The quality and type of an oscilloscope depend greatly on its amplifiers. The deflection amplifiers of modern oscilloscopes cover the range from dc to 10 MHz for general purpose scopes, or to beyond 500 MHz for special wideband scopes. The internal sawtooth sweep generator described below is used to provide a horizontal deflection that varies linearly with time.

The input resistance of most oscilloscopes is typically 1 M$\Omega$. A probe with an internal resistance, as shown in figure 2-26, is frequently used to

**Fig. 2-25.** Block diagram of basic oscilloscope. The horizontal deflection amplifier can be connected to an external signal or to a sweep generator that causes the beam to move from left to right at a constant rate.

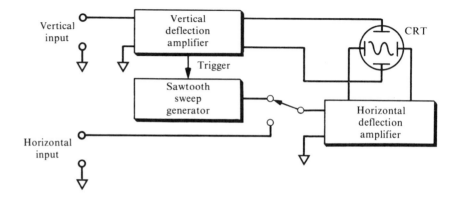

increase the input resistance of an oscilloscope tenfold. The combination of $R_p$ and $R_m$ forms a voltage divider, or **attenuator**, which decreases the maximum sensitivity of the oscilloscope tenfold.

## Linear Sweep for Horizontal Deflection

For most applications of the oscilloscope, the voltage applied between the horizontal deflection plates produces a linear trace; that is, the light spot is moved horizontally across the face of the tube at a uniform rate. The linear trace is important because it provides a uniform time scale against which another voltage can be plotted by applying it across the vertical deflection plates. This is illustrated in figure 2-27. Here a sawtooth voltage provides the linear horizontal movement of the electron beam, the voltage applied to the vertical deflection plates is plotted against time, and the curve is displayed on the screen.

Since the electron beam is "swept" across the screen at a uniform rate, it is customary to refer to the sawtooth waveform as the **sweep**. The oscillator circuit that produces the sweep is called the **sweep generator**. The rate of change of voltage must be very uniform in order to obtain a reliable time base. Good linearity of the sweep is a major requirement for a reliable oscilloscope. In most modern scopes, the time base is calibrated in seconds per centimeter and is accurate to 3%. At the end of the sweep the electron beam is moved back across the screen to the starting point by returning the sweep voltage to zero. The finite time required to change the sweep voltage back to the starting value can cause a visible trace of low intensity, known as the **retrace** or **flyback**. By applying a negative grid voltage during the retrace time interval, it is possible to blank out the retrace.

## Triggered-Sweep Oscilloscope

To provide a continuous, stable display of a repetitive waveform signal on the oscilloscope screen, the starting time of the sweep must coincide with a single point on the waveform of the signal to be observed. For a **triggered-sweep oscilloscope**, this involves using the input signal to start the sweep at the same point on the waveform for each sweep. The triggering feature allows considerable flexibility in choosing the display mode. For instance, it is possible to choose whether the time base is triggered on a positive or a negative slope of the input waveform. It is also possible to select the **triggering level**, that is, the sign and magnitude of signal required to trigger the sweep.

Other advantages of a triggered sweep are that the signal input need not be repetitive to be displayed (i.e., single events can be shown since the event itself starts the sweep), and a calibrated time base (usually seconds per centimeter) is provided, making it possible to perform time interval and

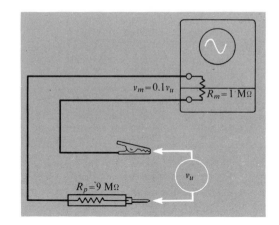

**Fig. 2-26.** A ×10 attenuator probe. Sensitivity is sacrificed for an increase in input resistance.

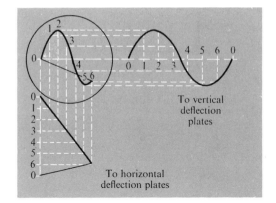

**Fig. 2-27.** Linear horizontal sweep for displaying the plot of vertical deflection voltage against time. To observe waveforms as they are usually drawn, a signal that varies linearly with time must be applied to the horizontal deflection plates.

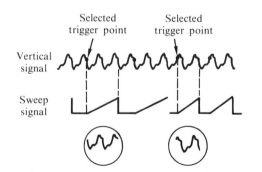

**Fig. 2-28.** Triggered sweep. Both the trigger point and the sweep rate (time window observed) can be independently selected.

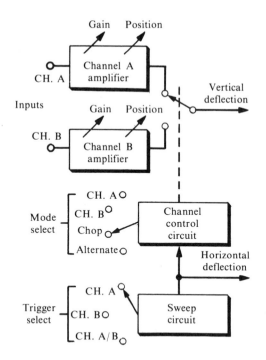

**Fig. 2-29.** Input multiplexer for dual-trace oscilloscope. The rapid alternation of control of the vertical deflection between two signals (A and B) allows both signals to be observed on the CRT. Advantage is taken of the fast writing rate of the CRT.

frequency measurements. The scope displays observed for various triggering conditions are illustrated in figure 2-28.

## Dual-Beam and Dual-Trace Oscilloscopes

In many situations it is desirable to display more than one signal versus time simultaneously. Dual-beam and dual-trace oscilloscopes allow two signals to be displayed simultaneously for comparison purposes.

The **dual-beam oscilloscope** has two separate electron guns and separate pairs of vertical deflection plates for each beam. Because of the need for separate electron guns and deflection plates, dual-beam oscilloscopes tend to be expensive. A more economical approach is time-sharing one electron gun and one pair of vertical deflection plates as is done in the **dual-trace oscilloscope**.

The sharing of a single analog-input measurement and recording system by more than one input signal is accomplished by a signal-switching technique called **multiplexing** (fig. 2-29). Each input channel has its own pre-amplifier with separate range (gain) and offset (position) controls in order to allow the dual-trace display of widely different signal levels. The multiplexer is a two-position solid-state switch with four modes of control. In the CH.A or CH.B positions, the multiplexer output is continuously connected to channel A or B respectively. In the **chop mode**, the multiplexer switch is alternated back and forth between channels A and B at a rate of 100 kHz to 1 MHz. With a horizontal sweep rate of 100 $\mu$s/cm or slower, the traces for both signals appear to be continuous. As the sweep rate is increased (less than 100 $\mu$s/cm), the chopping of each signal becomes apparent. At still faster sweep speeds, the chop mode becomes unusable. In the **alternate mode**, the channel selector switch is alternated after each sweep. Channel A is displayed for one entire sweep, and then channel B is displayed during the next sweep, and so on. This mode is practical when the observed signals are repetitive and when alternate sweeps are frequent enough to appear continuous to the eye. Sweep rates faster than 1 ms/cm generally repeat each trace fifty or more times each second. The retrace portion of the sweep signal is used to alternate the channel selector switch. With many modern oscilloscopes, the position of the time-base selector switch automatically governs whether the scope is in the chop or alternate mode. For fast sweep rates the alternate mode is used; for slow sweep rates the chop mode is selected.

The sweep trigger selector switch on a dual-trace oscilloscope includes provision for triggering from the channel A signal, the channel B signal, or the multiplexed signal. The latter trigger source, which is used only in the alternate mode, allows each signal to determine its own trigger point, but the time correlation between the two traces is lost. Some oscilloscope models have a four-channel input that is an expansion of the two-channel multiplexer shown in figure 2-29.

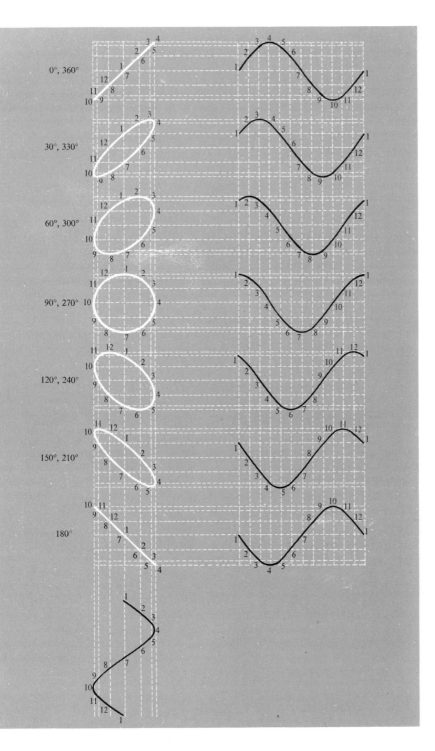

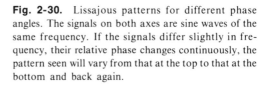

**Fig. 2-30.** Lissajous patterns for different phase angles. The signals on both axes are sine waves of the same frequency. If the signals differ slightly in frequency, their relative phase changes continuously, the pattern seen will vary from that at the top to that at the bottom and back again.

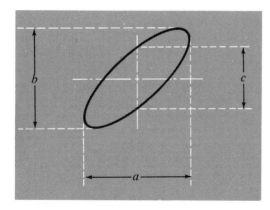

**Fig. 2-31.** Determination of phase angle from Lissajous pattern. The values of $b$ and $c$ are determined from the graduations on the CRT, and the phase angle is calculated from $\sin \theta = c/b$.

**Fig. 2-32.** Lissajous figures for various horizontal-to-vertical ratios: (a) 1:1, (b) 2:1, (c) 1:5, (d) 10:1, (e) 5:3. The frequency ratio is the ratio of the number of nodes along the horizontal extremity to the number of nodes on the vertical extremity.

## x-y Display

When two signals are connected to the horizontal and vertical inputs of an oscilloscope, the oscilloscope becomes an x-y plotter that displays the functional relationship between the two signals. Instead of plotting two signals against time, the oscilloscope plots the value of one signal against the value of the other as both signals vary with time. If two periodic signals are used and the time relationship of the two signals shifts, the pattern changes. Since the time-amplitude relationship of one signal is being plotted against that of the other, the resulting pattern must contain information about the time relationship between the two signals. These patterns, called **Lissajous figures**, are useful for phase-angle and frequency-ratio measurements.

As discussed in section 2-1, the phase angle between two sine-wave signals may encode information. The form of the Lissajous display as a function of phase difference between two signals is shown in figure 2-30. Note that the peak-to-peak amplitudes of the input signals can be obtained from the maximum horizontal and vertical excursions of the trace. For the equal-amplitude signals shown, the phase angle $\theta$, the major measure of phase difference, can be determined by measuring the quantities $b$ and $c$ shown in figure 2-31.

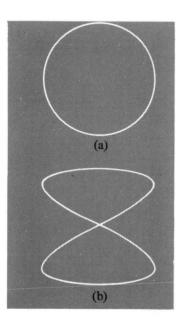

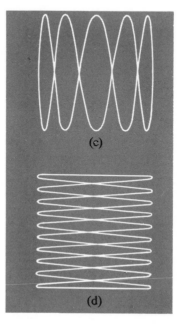

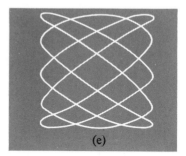

If two signals that have a frequency ratio that can be expressed in a small whole number or a simple fraction are applied to the horizontal and vertical inputs, patterns similar to those in figure 2-32 are obtained. Patterns of this type can be used to determine the frequency ratio of the two signals. For example, two signals of equal frequency give a circle, ellipse or straight line as in figure 2-30. If the horizontal-to-vertical frequency ratio is 2:1 the figure-eight pattern shown in figure 2-32b is obtained. Although the measurement of frequency ratio can also readily be accomplished with digital techniques, the oscilloscope method is still used because of its visual nature and the fact that phase and amplitude information can be obtained simultaneously.

## Suggested Experiments

**1.  Introduction to the oscilloscope and function generator.**
Connect the function generator (FG) output to the oscilloscope vertical input and obtain a stable pattern. Observe the effects of the waveform, gain, frequency, and offset controls of the FG. Observe the effects of the position, sweep rate, and sweep magnifier controls of the scope. Characterize the signal at the TTL output of the FG.

**2.  Oscilloscope triggering.**
Observe the effects of the scope triggering controls (including automatic/adjustable level and $+/-$ slope) on the displays of large and small amplitude triangular and square waveforms.

**3.  Low pass and high pass filters.**
Using a dual-trace oscilloscope, observe the input/output amplitude ratio of a low pass filter for frequencies above and below cutoff. Observe the phase shift for the same frequencies with a Lissajous pattern. Calculate the capacitive reactance for several frequencies. From a measurement of the input/output ratio, determine the cutoff frequency of several high pass filters that have different time constants.

**4.  Frequency effects at the oscilloscope input.**
With the FG connected to the scope, observe the difference in the effect of the FG-offset control when the scope vertical input is in the ac and dc position. Compare the waveform amplitude of a low-frequency signal for ac and dc inputs to obtain the cutoff frequency of the ac input filter. Observe the effect of the ac/dc input switch on a 100-Hz square-wave signal.

**5.  Shielding and induced signals.**
Measure the line frequency noise for an unshielded lead when not connected to a signal source and when connected to a high impedance signal source. Make the same measurements for a shielded cable. Observe the shapes at the rising and falling edges of the TTL signal from the FG when a scope probe is used. Observe the effect of the probe trimmer adjustment.

**6.  AC scales of the digital multimeter.**
Using first a DMM and then the oscilloscope, compare the amplitude of a sine wave with no offset from the FG. Add dc offset to the sine wave and again compare scope and DMM (both polarities of connection) measurements. Repeat with a square-wave source that alternates between zero and some positive voltage.

## Questions and Problems

**1.**  Many ac motors have a speed of rotation proportional to the frequency of the power applied to them. Suppose that a phonograph with such a motor, designed to operate with 60-Hz power, is plugged in to 50-Hz power. (a) What will be the frequency of a note recorded as a concert A (440 Hz)? (b) How long will it take to play the "minute waltz"?

**2.**  (a) What is the angular velocity of a signal of 14 kHz? (b) What is the duration in seconds of a segment of the 14-kHz signal that is $\pi$ radians long?

**3.**  A spectrum analyzer produces a frequency spectrum like that of figure 2-7 for a signal of any periodic waveform. Comment on the use of a spectrum analyzer to determine the signal quality from a sine-wave generator.

**4.**  Sketch a signal that is a combination of a 300-Hz, 5-V rms sine-wave ac signal and a +5-V dc signal. Label both the amplitude and the time axes.

**5.**  Derive a formula for the average dc level of a zero-based pulse signal such as that of figure 2-5a in terms of the maximum voltage $V_{max}$ and the fraction of the period the pulse is at $V_{max}$. This fraction, $t_{pw}/t_{per}$, is called the duty cycle.

**6.**  The power line voltage is nominally 110 V rms, 60 Hz. Calculate (a) the average voltage, (b) the peak voltage, and (c) the peak-to-peak voltage for the power line.

**7.**  If an average-reading voltmeter, calibrated in rms for sine-wave signals, is used to measure a triangular wave of 10 V peak-to-peak, what will the voltmeter reading be?

**8.**  In the Fourier series expansion for a sawtooth waveform, the peak amplitude coefficients $V_1$, $V_2$, $V_3$, . . . are in the proportion $1$, $-1/2$, $+1/3$, $-1/4$, $+1/5$, . . . respectively. Plot the frequency spectrum of a sawtooth signal out to the tenth harmonic.

**9.**  At what frequency will a 0.01-$\mu$F capacitor have a reactance of 2.5 k$\Omega$?

**10.** What is the capacitive reactance of a 0.05-$\mu$F capacitor at frequencies of (a) 100 Hz, (b) 1 kHz, and (c) 10 kHz?

**11.** At what frequency will a 200-mH inductor have an inductive reactance of 1 k$\Omega$?

**12.** What is the inductive reactance of a 20-mH inductor at frequencies of (a) 100 Hz, (b) 1 kHz, and (c) 10 kHz?

**13.** If the total impedance of a series $RC$ circuit is 98 $\Omega$, the phase angle is $-54.5°$, and the capacitive reactance is 80 $\Omega$, what is the value of resistance in the circuit?

**14.** A 100-Hz source is connected to a series $RC$ circuit in which $R = 3$ k$\Omega$ and $C = 0.5$ $\mu$F. Find (a) the magnitude of the impedance and (b) the phase angle between the current in the circuit and the voltage across the load.

**15.** A 10-kHz source of 150-V peak is connected to a series $RC$ circuit in which $X_C = 400$ $\Omega$ and $R = 500$ $\Omega$. Calculate (a) the impedance, (b) the phase angle, (c) the peak voltage across $R$, (d) the peak voltage across $C$, and (e) the power dissipated.

**16.** A series $RC$ circuit is to be used as a high pass filter. If $R = 5$ k$\Omega$, what value of $C$ is necessary to give a lower cutoff frequency of 200 Hz?

**17.** An $RC$ circuit is to be used as a low pass filter. If $R = 2.5$ k$\Omega$ and $C = 0.05$ $\mu$F, what is the upper cutoff frequency?

**18.** A 500-Hz, 25-V peak source is connected to a series circuit in which $X_L = 50$ $\Omega$ and $R = 100$ $\Omega$. Calculate (a) the impedance, (b) the phase angle, and (c) the peak voltages across $R$ and $L$.

**19.** What is the ratio $v_{out}/v_{in}$ for a circuit with a gain of (a) 86 dB? (b) $-18$ dB?

**20.** Draw the Bode diagram for the high pass filter of problem 16. Label the frequency axis in Hz.

**21.** Compare the expected readings of a peak-responding digital voltmeter with that of an average-responding moving-coil voltmeter when both are calibrated in volts-rms for a sine wave and are used to measure a 6.0-V peak-to-peak square wave.

**22.** An oscilloscope screen is divided into ten major divisions on the horizontal axis. What sweep speed (in seconds per division) should be used so that two full periods of a 63-kHz signal can be observed? The sweep speed selector switch has settings of 1, 2, and 5 in each decade.

**23.** (a) If the chopping frequency in a dual-trace oscilloscope is 100 kHz, at what sweep speed will the chopping action produce five samplings of each signal in each horizontal division of the display? (b) If the scope is switched to the alternate mode at this same sweep speed, how often will each input signal be scanned?

**24.** A Lissajous pattern is used to measure the gain and phase shift of an amplifier. The amplifier input signal is connected to the oscilloscope vertical input with a sensitivity of 10 mV/division. The amplifier output is connected to the scope horizontal input with a sensitivity of 2 V/division. The pattern observed looked like that of figure 2-31 with dimensions $a = 4.2$, $b = 7.3$, and $c = 0.6$ divisions. Calculate the gain and the phase shift of the amplifier.

**25.** (a) Sketch the patterns expected if the square and triangular outputs of a function generator are connected to the $y$ and $x$ inputs of an oscilloscope. Assume that the rising and falling edges of the square wave occur at the same time as the midpoint in the triangular waveform. (b) Repeat for the sine and triangular outputs connected to $y$ and $x$ scope inputs.

# Chapter 3

# Power Supplies

Operating electronic equipment requires electrical power supplies that provide the necessary voltages and currents. Portable instruments such as hand calculators are frequently powered by small dc batteries, but most electronic equipment is designed to operate from the conventional ac electrical power distributed through wall sockets. Combinations of various electronic components as described in this chapter can convert ac line power to the desired dc voltages for various applications. The specific characteristics of the dc power supplies, such as precise regulation of the voltage or current and suppression of ripple or transient electrical noise, often determine whether an electronic system operates as required. Equipment troubles are frequently traced to failure of a power supply. The electronic principles of rectification and electronic feedback control are readily introduced through a study of power supplies. The operation and characteristics of transformers, rectifier circuits, filters, and regulator circuits are therefore all discussed in this chapter, as are the types and characteristics of batteries.

## 3-1 Conversion of AC to DC

Power lines into buildings generally provide either 115-volt or 230-volt, 60-Hz alternating current (ac) in the United States and 220-volt, 50-Hz ac in many other countries (see note 3-1). The nominal 115-V ac or 230-V ac depends on user demand and may vary as much as 20%, as from 105 to 125 V ac. On electronic schematic diagrams the input **line voltage** is usually given as the nominal average rms value, as shown in the block diagram of figure 3-1. The functions of the various components and circuits in a typical dc power supply are illustrated in this diagram. Note that the line plug is connected through a suitable on-off switch and some type of protective device, such as a fuse or circuit breaker, to protect electronic circuits from serious overheating and destruction in case of a malfunction. Many modern supplies provide additional protection by automatically shutting the supply down if the output voltage exceeds a preset level (overvoltage protection) or by automatically preventing the load current from exceeding a preset level

**Note 3-1. AC Power Line.**
Line is an abbreviation for "the ac power distribution line." Line voltage is the rms voltage available at the power socket; line frequency is its frequency, and a line cord is the wire and plug used to connect devices to the line power socket.

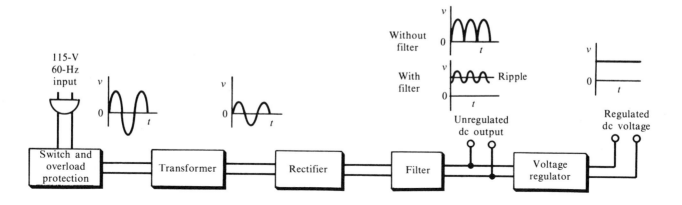

(current limiting). The critical characteristics of a dc power supply are its rated output voltage, its internal resistance, its output voltage variation (noise), and its current limits.

Any power source, since it is made of imperfect conductors, has some internal resistance. Because the current in the power source is the same as the current in the external circuit (the load), an *IR* drop will be developed in the power source. This voltage drop is in opposition to the no-load voltage from the power source. Therefore, the output voltage decreases as the load current increases.

An ideal dc voltage power supply would have: (1) a constant output voltage regardless of variations in the current required by the load (good regulation or low output resistance); (2) a constant output voltage regardless of variations in temperature, ac line voltage, and age of power supply (good stability); and (3) no noise voltage of the line or other frequency superimposed on the dc output (low ripple). In addition to these characteristics, the dc output voltage and the current capability of the power supply must meet the operational requirements of the electronic devices (adequate power).

## 3-2  Overload Protection

The line cord that is to be plugged into a suitable outlet is usually soldered to rugged input terminals within the power-supply chassis, and the cord is made mechanically secure by some type of clamp or knot so that a jerk on the cord will not break the connection within the chassis. A poor cord connection inside the chassis could be dangerous because it could expose the user to line power on bare wires. Inside the chassis the power line is typically connected to a fuse, an on-off switch and an indicator light, and to the transformer primary winding, as illustrated in figure 3-2. Modern line plugs

**Fig. 3-1.** Block diagram showing conversion of ac input to dc output voltage. The transformer converts the line voltage to the approximate desired voltage. A rectifier circuit converts the ac to a pulsating dc, which is smoothed by a filter network. A regulator improves the stability and further reduces the ripple of the output voltage.

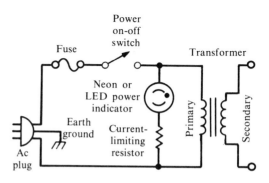

**Fig. 3-2.** Input line connection to fuse, switch, indicator light, and transformer primary.

**Note 3-2.  Ground.**

The term ground, precisely used, refers to earth ground, the universal reference point for potential measurements, power distribution, and radio transmissions. Connection to earth ground is made at a ground rod, metal water pipe, or the power-line common. When a circuit is not connected to ground, it is free to assume any potential with respect to ground. For safety, the metal cases of all devices operated from the power line are, or should be, connected to ground through the power-line common. Ground is also sometimes used to mean the common potential reference point of a circuit even when the circuit is not connected to earth ground. In this book the word **common** will be used unless earth ground is intended. See appendix A for practical grounding techniques.

and cords have three connections—two for the power and one for a connection to **ground** as shown (see note 3-2).

Every power supply should have some provision for overload protection to prevent destruction of electronic components and even severe overheating that could cause a fire. The most common devices for overload protection are fuses and circuit breakers.

## Fuses

A popular type of fuse is illustrated in figure 3-3. Since the fuse is in series with the primary winding of the transformer, the fuse wire overheats and melts if excessive current is drawn by the load. Thus the circuit is broken (forced to an open state), and the electronic equipment is disconnected from the power line until the problem that caused excessive current can be located. The transparent glass tube makes it easy to see if the fuse is blown.

The ratings on this type of fuse range from a fraction of an ampere to several amperes, and they are made to be **slow-blow** or **fast-blow**. The wire in a fast-blow fuse melts if the current exceeds the rated current only briefly, in less than 0.1 s in some fuses. This fast operation is important in some applications, but in other applications brief overloads are expected, and a blown fuse would be a useless inconvenience. In these cases a slow-blow fuse with a wire that heats more slowly is used so that only continued overloads shut down the electronic system. A slow-blow fuse often contains a spring that pulls the fuse wire apart once it does begin to melt.

## Circuit Breakers

One type of thermal **circuit breaker** is illustrated in figure 3-4. Current passes between the points $P$ and $P'$ through a strip of metal that is a thermal element. Excess current bends the metal strip that then operates as a release spring. After the problem that caused shutdown of the electrical system is located and corrected and the metal strip has cooled, the reset button is pushed to reclose the circuit for operation. Magnetic circuit breakers are designed so that the overload current develops sufficient magnetic field in a coil of wire to activate a spring trigger mechanism and pull the contact points apart. An advantage of circuit breakers over fuses is that they can be reset rather than replaced after activation.

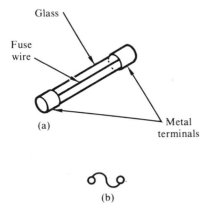

**Fig. 3-3.**  Pictorial (a) and schematic symbol (b) of typical power-supply fuse. A small length of fuse wire is connected between two metal terminals. A small glass or plastic tube supports the metal terminals and protects the fuse wire.

## Switch and Indicator Light

In figure 3-2 a power on-off switch and indicator light are shown. The switch must carry the current demanded by the load and be reliable over many

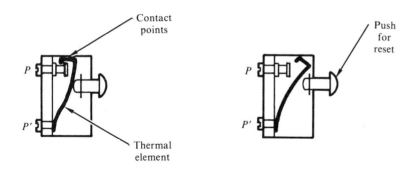

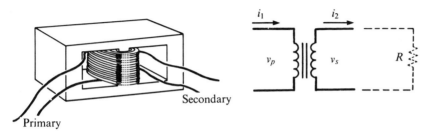

**Fig. 3-4.** Thermal circuit breaker. An overload current through the thermal element causes it to heat and bend, releasing the latch and breaking the contact.

on-off cycles. The indicator light requires only a small current, and a current-limiting resistor allows just enough through the light to indicate clearly when the power supply is on.

## 3-3 Power-Supply Transformers

A transformer is used to provide an efficient multiplication or division of the ac power-line voltage. A schematic symbol for and pictorial representation of a transformer are given in figure 3-5. The operation of the transformer is based on the principles of inductance introduced in section 2-3. A changing current in the **primary** winding produces a changing magnetic flux in the **secondary** coil, which induces a changing voltage across that coil (see note 3-3). The ac voltage $v_s$ produced in the secondary winding is proportional to the ac voltage applied to the primary and the turn ratio $N_s/N_p$ of the secondary and primary coils (see note 3-4), so that $v_s = v_p(N_s/N_p)$.

Only ac voltages are induced in the transformer secondary because the transformer operation depends on a constantly changing magnetic flux. The turns ratio of the power transformer determines whether the ac line voltage is increased or decreased. A single transformer can have several secondary

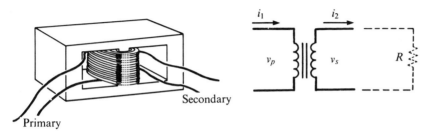

**Fig. 3-5.** Pictorial representation and schematic symbol of transformer.

**Note 3-3. Primary and Secondary.**
The words primary and secondary as applied to a transformer refer to the coil that is connected to the power source (the primary) and the coil that is connected to the load (the secondary).

**Note 3-4. Transformer Turns Ratio.**
The voltage produced in the secondary winding is equal to the number of turns $N_s$ times the rate of change of magnetic flux $\phi$.

$$v_s = -N_s \frac{d\phi}{dt}$$

Since the primary and secondary are intimately wound on a core of high magnetic permeability, the magnetic flux in the primary winding and the magnetic flux in the secondary winding are equal. The rate of change of flux is

$$\frac{d\phi}{dt} = \frac{-v_p}{N_p}$$

where $v_p$ and $N_p$ are the voltage on the primary and the number of turns in the primary, respectively.

Combining the above equations to eliminate $d\phi/dt$, we get

$$\frac{v_s}{v_p} = \frac{N_s}{N_p}$$

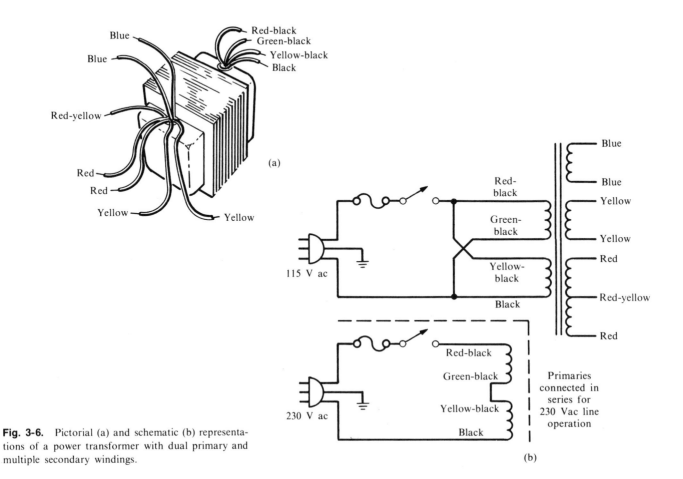

**Fig. 3-6.** Pictorial (a) and schematic (b) representations of a power transformer with dual primary and multiple secondary windings.

**Note 3-5.  Isolation.**
The transformer provides isolation (no direct connection) of the circuit connected to the secondary from the power source connected to the primary. For example, even though the grounded wire of the ac power line is connected to a power transformer primary, the secondary circuit is "floating" or isolated from ground unless some part of it is intentionally connected to ground.

windings to satisfy several different voltage requirements. As illustrated in figure 3-6, some transformers are also built so that dual input primary leads can be arranged in such a way as to accommodate either 115-V or 230-V line voltage. In this case care must be taken to connect the primary leads correctly to avoid serious damage to the transformer, the power supply, and to the equipment connected to the power supply output. If the electronic system is to operate from 115 V ac, the two primary windings are connected in parallel. If the line voltage is 230 V ac, the two primaries are connected in series.

The transformer also provides a degree of **isolation** (see note 3-5) of the secondary load from the primary supply since there is no direct connection between the two. Special isolation transformers are used where there is a problem of conflicting common voltage levels between the primary and secondary circuits.

## 3-4 Rectifier Circuits

The conversion of alternating current to pulsating direct current is accomplished with **rectifier** circuits (see note 3-6), which produce a smooth dc voltage or current. The basic element in all of these circuits is the rectifier diode. Several types of rectifier circuits are found in power supplies. They are classified according to the different configurations of their diodes as half-wave, full-wave, bridge, and voltage-multiplier rectifier circuits. The operation of these circuits is discussed after the concepts and principles of diodes are introduced.

**Note 3-6. Rectifier.**
A rectifier is a device or circuit by which alternating current is converted to direct current.

### Rectifier Diodes

A **diode** conducts current effectively in only one direction. The diode is said to be forward biased when the voltage applied to the **anode** is positive relative to the **cathode** (see note 3-7). When a diode is forward biased, the effective resistance across the diode is very low, somewhat like that of a closed switch. If the diode is reverse biased (that is, the voltage applied to the anode is negative with respect to the cathode) the effective resistance is very high, somewhat like that of an open switch. This is illustrated in figure 3-7 where $R_f$, the **forward resistance**, is always less than $R_b$, the **backward resistance**. It is seen that $i_f = V/R_f$ and $i_b = V/R_b$. Therefore, if $R_f < R_b$ then $i_f > i_b$ or, quantitatively, $R_b/R_f = i_f/i_b$. The ideal diode would be a perfect conductor for forward current and a perfect insulator for reverse current. The ratio $R_b/R_f$ approaches infinity as the diode approaches the ideal. This ratio is used as a figure of merit for rectification devices.

**Note 3-7. Anode and Cathode.**
In a number of conducting devices, the response depends on the direction of the current. To distinguish between the connections to such devices, one connection is called the anode and the other the cathode. The direction of forward current through a diode is from anode to cathode.

### Semiconductor Diodes

Diodes made of semiconductors are the most important rectifier elements used in modern power supplies. Before considering the characteristics of these diodes, however, we shall find it helpful to review some of the properties of semiconductor materials.

As the name implies, a pure **semiconductor** material is neither a good conductor nor a good insulator but is somewhere between. The charge carriers in a semiconductor conduct a current when voltage is applied across it, but since their concentration is low compared with the concentration of mobile electrons in a metal, the resistance of a semiconductor is relatively high. Although the properties of intrinsic (pure) semiconductors are important, the desirable characteristics for use in diode rectifiers and transistors are obtained by purposely adding small amounts of selected impurities.

Assume that enough of a group V element such as antimony is added to a semiconductor such as silicon to make the ratio of antimony to silicon

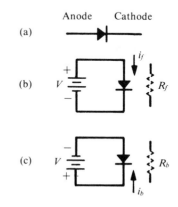

**Fig. 3-7.** Symbol (a) of a rectifier diode. The diode is shown forward biased in (b) and reverse biased in (c). The forward resistance $R_f$ is always much less than the backward resistance $R_b$.

about 1 part per million (ppm). Each antimony atom is therefore completely surrounded by silicon atoms, and because the antimony atom is part of the silicon crystal lattice, four of its five valence electrons form covalent bonds with the four nearest-neighbor silicon atoms. The extra electron is now only loosely bound to the antimony atom; an energy of only about 0.05 electron volts (eV) is required to free it. The antimony atoms are called **donors** because they contribute an excess electron concentration which greatly increases the conductivity of the silicon semiconductor. A semiconductor doped with donor atoms is referred to as an **n-type** semiconductor. It is likewise possible to dope silicon with a group III element, such as indium, to produce an excess **hole** concentration (see note 3-8). The resulting impure crystal is known as a **p-type** semiconductor, and the group III impurity is called an **acceptor**.

**The pn junction.** A pn semiconductor junction is generally made by changing the dominant dopant from acceptor to donor type within a single crystal of semiconductor. The region where the **majority charge carrier** changes from being electrons to being holes is called the **pn junction**. The holes from the p region and the electrons from the n region are free to cross the junction. Since the free energy for electrons is higher in the n region than in the p region, some electrons cross from the n to the p region. Similarly, some holes cross to the n region as a result of the higher free energy for holes in the p region. The transport of positive charge into the n region and negative charge into the p region develops a potential difference between the n and p regions. This potential presents an electrical energy barrier sufficient to offset the free energy difference of the majority carriers on either side of the junction. As a result an equilibrium condition is established in which there is no net flow of charge across the boundary. A **contact potential** thus results with the n region positive with respect to the p region.

The drift of holes across the junction reduces the concentration of electrons in the junction region since the product of the electron and hole concentrations must remain constant. Similarly, the concentration of holes is lower at the junction than in the rest of the p region. This region is thus called the **depletion region**. The depletion region is essentially nonconducting compared to the remainder of the n- and p-doped regions. The contact potential appears across the insulating junction region. The magnitude of the contact potential is about 0.3 V for pn junctions in germanium and about 0.6 V for those in silicon.

The contact potential cannot be measured because in order to attach a voltmeter to the semiconductor, two metal-to-semiconductor contacts must be made as shown in figure 3-8a. Contact potentials are established at the metal-semiconductor junctions that exactly counteract the pn-junction potential. The potential profile through the contacts and junction region is

**Note 3-8. Hole.**
The point where a bonding electron is missing from the crystal lattice is a localized region of positive charge called a hole. Bonding electrons can only move in response to an electric potential if they are adjacent to a hole. The net result of the valence electron motion is a drift of the positive hole in the direction of increasingly negative potential.

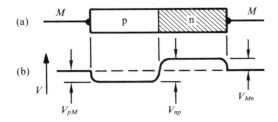

**Fig. 3-8.** A pn junction with metal contacts (a) and potential profile (b). In (b) the potential across the depletion region $V_{np}$ is seen to be exactly compensated by contact potentials between the metal and the p-type material $V_{pM}$ and between the metal and the n-type material $V_{Mn}$.

shown in figure 3-8b. It is assumed here that the metal-semiconductor contacts are ohmic (that is, nonrectifying).

**The biased pn junction.**    A **bias** is an external voltage applied to the
material with the pn junction. Since the connecting wires, the contacts, and
the doped semiconductor are all much better conductors than the depletion
region, essentially all of the applied bias voltage appears as a change in the
voltage across the pn junction. If the external voltage $v$ is set at 0 V as shown
in figure 3-9a, the current $i$ is zero. Since the sum of the voltages around the
loop must be zero, the junction contact potential must again be exactly
compensated by the metal-semiconductor contact potentials.

When the voltage source is connected with the polarity shown in figure
3-9b, the pn junction is said to be **reverse biased**. Holes in the p-type material
and electrons in the n-type material move away from the junction. This
increases the thickness of the depletion layer, and the result is a current that
is nominally zero. Actually a small reverse current $i_b$ does exist because a
small number of electron-hole pairs are generated by thermal energy throughout the semiconductor material. This reverse current is called the **reverse-
bias saturation current**. It is approximately $-40$ $\mu$A for the germanium pn
junction at room temperature and approximately $-40$ nA for silicon. The
reverse current increases rapidly with an increase in temperature.

When the voltage source is connected with the polarity shown in figure
3-9c, the pn junction is said to be **forward biased**. Holes can cross the
junction from the p-type region to the n-type region, and electrons can cross
the junction from the n region to the p region. The depletion layer is reduced
until the fringes of the n and p regions begin to overlap. Since holes traveling

**Fig. 3-9.**    A pn junction under (a) zero bias, (b)
reverse bias, and (c) forward bias. Note that the diode
is essentially nonconducting under zero bias and reverse
bias. The thickness of the depletion region (greatly exag-
gerated) increases in going from (a) to (b). In (c) the
fringe of the p region has begun to overlap the fringe of
the n region, and the diode is a good conductor.

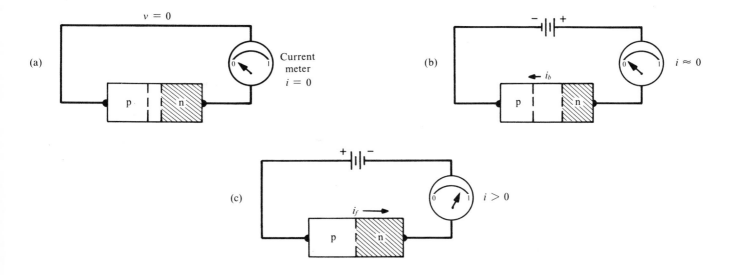

from left to right make a current in the same direction as electrons traveling from right to left, the resulting current $i_f$ at the junction is the sum of the electron current and the hole current.

Plots of the current $i$ through the junction against the applied bias voltage $v$ are shown in figure 3-10 for both germanium and silicon pn junctions. For germanium the 0.3 to 0.4 V required for substantial forward current corresponds to the bias necessary to offset the junction contact potential. For silicon pn junctions a forward bias of 0.6 V is required for substantial forward current. The pn junction is thus a good conductor with a small forward bias applied but a very poor conductor for even large values of reverse bias voltage.

**Reverse-bias breakdown.** The pn junction with reverse or zero bias can be thought of as two conductors separated by a thin layer of insulator (the depletion region). Thus the junction has the property of capacitance. As the reverse bias voltage is increased, the electric potential across the depletion region increases. This has the effect of increasing the thickness of the depletion layer somewhat. The resulting change in capacitance is the basis of operation of voltage-controlled capacitors called **varactors**. The electric potential across the depletion region can be increased to the point that the insulating capacity of the junction region breaks down. Breakdown can

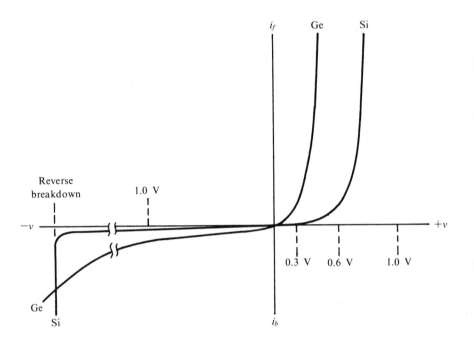

**Fig. 3-10.** Current-voltage curves for germanium and silicon pn junctions.

instability, avalanche multiplica-
...e junction region under reverse
...t which heat is produced more
...unction, the temperature at the
...everse current, which, in turn,
...ted. Beyond some maximum
...ocess proceeds to the thermal
...tion. This is breakdown by

...ge carriers in the depletion
...hey are majority carriers by
...imum velocity attained by
...he voltage and the distance
...other particles. Thus the
...f collision increases with
...harge carriers equals the
...ectron-hole pairs will be
...re accelerated until they
...ns, and so on. The bias
...**alanche-multiplication**
...ckness of the depletion
...lower the bias poten-
...t. In the construction
...olled by the dopant
...ation of majority charge carriers,
the thin ...region, because a smaller fraction of the charge
carriers is required to create the depletion region. The avalanche-breakdown
voltage varies from several hundred volts down to ten volts as the dopant
concentration increases from $10^{15}$ to $10^{17}$ atoms/cm$^3$. Over this same range
of concentration, the depletion layer thickness at breakdown varies from
about 100 to 0.1 $\mu$m. The avalanche breakdown voltage increases somewhat
with increasing temperature. The higher the breakdown voltage, the wider
the range of reverse voltages for which the device is useful. Some devices are
made to be operated at their breakdown voltage (at a safe current). Breakdown diodes and their applications are discussed in section 3-6.

## Half-Wave Rectifier

The simplest rectifier circuit is the **half-wave rectifier** shown in figure 3-11.
This rectifier circuit is so named because only half of the ac current wave is
present in the load circuit (see note 3-9).

In studying rectifier circuits, it is important to consider the ratings of the
rectifiers. The ratings include: (1) the maximum average forward current

**Note 3-9.  Half-Wave Rectifier.**
The half-wave rectifier is simple and requires only
one diode, but it has certain disadvantages. Since
only half of the input wave is used, it is not very
efficient. Also, the current through the transformer
secondary is unidirectional, sometimes causing
the core of the transformer to become magnetized.
This situation, known as dc core saturation, tends
to reduce the inductance and thus the efficiency
of the transformer.

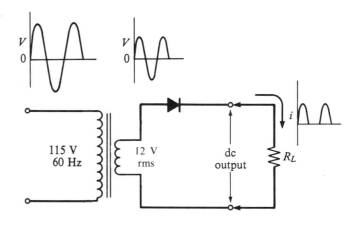

**Fig. 3-11.** Half-wave rectifier circuit. On the positive half-cycle of the ac voltage, the diode can conduct, allowing current to pass through $R_L$. The resistor $R_L$ is the "load" or the circuit that is to be supplied with direct current. On the negative half-cycle, the diode is reverse biased and, therefore, nonconducting.

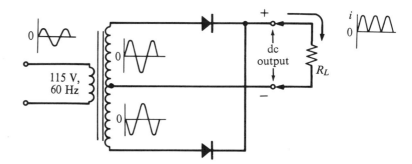

**Fig. 3-12.** Full-wave rectifier circuit. The upper diode conducts during one half-cycle, and the lower diode conducts during the next half-cycle. A center-tapped transformer is required.

rating, which is approximately $\frac{1}{2}\ V_{av}/R_L$, because the diode conducts only half the time; and (2) the **peak inverse voltage (PIV)**, which is the maximum voltage that should be applied to the rectifier when it is reverse biased. The diode of figure 3-11 must withstand a peak inverse voltage of $12 \times 1.4 = 17$ V. If a capacitor is connected across the output of the rectifier circuit in figure 3-11 the diode PIV must be at least $2 \times 17$ V $= 34$ V, because the capacitor holds the output at $+17$ V while the secondary goes to $-17$ V.

The effective resistance of a conducting diode is not constant but depends on the current. However, an estimate of the forward resistance allows the calculation of power loss in the rectifier as approximately $I^2 R_f$.

## Full-Wave Rectifier

Many applications require a rectifier circuit that supplies current during both half-cycles of the ac power and thus provides a more continuous current to the load. A **full-wave rectifier** circuit is shown in figure 3-12. This circuit is essentially two half-wave rectifiers connected in parallel with their inputs at a phase difference of 180°. The voltage output of the full-wave rectifier is equal to the voltage developed by each half of the transformer secondary (see note 3-10). For a 10-V peak output from a full-wave rectifier, one would use a $(20/1.4) + 0.6 = 15$ V center-tapped transformer. The extra 0.6 V represented by the second term compensates for the voltage drop in the rectifiers. Note that the rectifiers must withstand an inverse voltage of twice the peak value of the source. For the case above, the peak inverse voltage is $1.4 \times 15 = 21$ V.

A way to obtain full-wave rectification without use of a center-tapped transformer is shown in figure 3-13. This circuit is called the **bridge rectifier** (see note 3-11). Since two rectifiers are in series with the load, the peak inverse voltage that each rectifier must withstand is equal to the peak value of the supply voltage.

## Voltage-Doubler Rectifier

Two rectifiers can be connected to a single ac source and wired so that their outputs are in series as in figure 3-14. The output voltage available from such

**Note 3-10. Full-Wave Rectifier.**
The full-wave rectifier is more efficient than the half-wave rectifier because it operates on both half-cycles of the secondary voltage. Because the currents in the secondary are in opposite directions during the alternate half-cycles, there is no problem with transformer dc core saturation. The full-wave rectifier does require a center-tapped transformer, which is somewhat more expensive than the transformer in a half-wave rectifier. For a given transformer, the peak voltage is lower than that in the full-wave bridge rectifier.

**Note 3-11. Bridge Rectifier.**
The bridge rectifier is a full-wave rectifier, but it does not require a center-tapped transformer. For a given transformer its output voltage is higher than that of a regular two-diode full-wave rectifier. The PIV across each diode is only the peak voltage rather than twice the peak voltage.

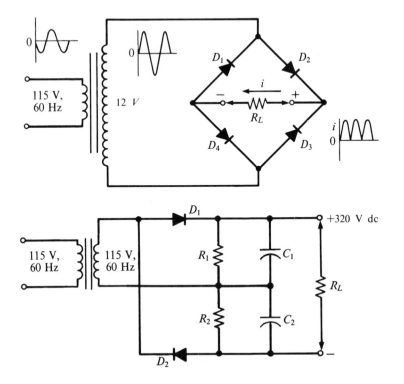

**Fig. 3-13.** Bridge rectifier circuit. On the positive half-cycle, $D_2$ and $D_4$ conduct. On the negative half-cycle, $D_1$ and $D_3$ conduct. In each case the direction of electron flow through the load $R_L$ is the same.

**Fig. 3-14.** Voltage-doubler rectifier. On the positive half-cycle, capacitor $C_1$ is charged to the peak value of the supply voltage (in this case $115 \times 1.4 = 160$ V). On the negative half-cycle, $C_2$ is charged to the same voltage. Since $C_1$ and $C_2$ are in series across the load, the output voltage is twice the peak voltage of the ac source.

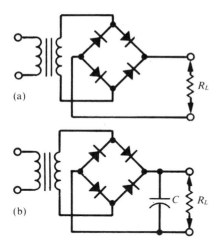

(a)

(b)

**Fig. 3-15.** Bridge rectifier (a) with load resistor $R_L$ and (b) with filter capacitor across load.

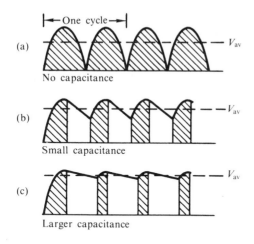

(a) No capacitance

(b) Small capacitance

(c) Larger capacitance

**Fig. 3-16.** Smoothing of output voltage by a capacitor filter. In (a) the output with no capacitor is shown. Parts (b) and (c) show the effect of increasing capacitance. The shaded areas indicate the times during which charging current is supplied by the rectifier. The dashed line indicates the average output voltage. Note that as the filter capacitance increases from zero to a high value, the average voltage increases, and the output fluctuation and charging time decrease.

a circuit is twice that available from the ac source with a half-wave or bridge rectifier. This kind of circuit is therefore called a **voltage-doubler rectifier.** The capacitors are essential to the operation of the circuit because they maintain the voltage developed during one half-cycle so that the voltage developed during the next half-cycle can be added to it. The capacitors $C_1$ and $C_2$ have a filtering action that is described in the next section. Since current is drawn from the transformer during both half-cycles, this voltage-doubler circuit is considered to be full-wave. The peak inverse voltage applied to each rectifier is twice the peak value of the supply voltage, in this case 320 V. If the peak inverse voltage rating of the diode is insufficient, two diodes can be connected in series so that their peak inverse voltage ratings are additive.

## 3-5  Power-Supply Filters

The pulsating dc voltages from the rectifier circuits studied in the previous section are not useful for most electronic applications. Nearly all electronic circuits require a very smooth constant voltage. Therefore rectifier circuits are usually followed by smoothing devices called **filters**, which convert the pulsating dc voltages into the required constant dc voltages.

### Capacitor as a Filter

A rather effective filter is simply a capacitor connected in parallel with the load $R_L$. A bridge rectifier circuit, with and without a capacitive filter, is shown in figure 3-15. Without a filter in the circuit, the voltage across the load is always equal to the pulsating voltage from the rectifier. When the filter capacitor is present, the voltage across the load equals that across the capacitor, which is alternately charged by the pulsating source and discharged by the load. If the discharge between pulses is small compared to the average charge stored in the capacitor, the voltage fluctuations across the load are also relatively small. Another way to think of the capacitor filter is as a low pass filter with an upper cutoff frequency of $1/(2\pi CR_L)$. To be an effective filter, the capacitor must have an upper cutoff frequency lower than the line frequency.

A more detailed picture of the action of a capacitor filter is presented in figure 3-16. The capacitor charges toward the peak value of the input voltage. If $R_L$ were infinite (no load), the voltage across the capacitor would quickly reach a constant value nearly equal to the peak value of the transformer output. In the practical case $R_L$ is not infinite, and the capacitor begins to discharge through $R_L$ as soon as the input voltage decreases below that voltage to which the capacitor has been charged. The capacitor continues to discharge until the next pulse when the rectifier output voltage again exceeds the voltage on the capacitor. Capacitor charging current occurs then, only when the rectifier voltage exceeds the capacitor voltage.

Because of the high capacitance that can be achieved in a small volume with electrolytic capacitors, power-supply filter capacitors are almost always electrolytics. See appendix B for capacitor characteristics. Choosing the appropriate filter capacitor obviously involves compromises among size, expense, and the maximum tolerable output fluctuations.

Another filter device is simply an inductance in series with the load. By opposing changes in current, the inductor tends to maintain a constant load current and thus a constant output voltage. Combinations of inductors and capacitors (*LC* filters) make the best passive filters, but the present trend is toward active filtering and regulating circuits. Active filters are considered in chapter 8.

## Ripple Factor and Frequency

The effectiveness of the filter is called the **ripple factor**, $r$, and is defined as the rms value of the ac voltage component, or ripple, divided by the average dc voltage; that is, $r = I_{ac}/I_{dc} = V_{ac}/V_{dc}$. The ripple factor and ripple frequency are both affected by the choice of rectifier. This is illustrated by the comparison of full- and half-wave rectifiers shown in figure 3-17. For a given discharge rate the ripple amplitude for the half-wave rectifier is almost twice that for the full-wave rectifier.

It can be seen from figures 3-16 and 3-17 that the ripple voltage decreases if $R_L$, $C$, or the frequency is increased. The expression for the ripple factor, $r = 1/(2\sqrt{3}fCR_L)$ bears this out. Here $f$ is the frequency of the main ac component (equal to the line frequency for half-wave rectifiers and twice the line frequency for full-wave rectifiers). It can also be seen that as the ripple increases in magnitude, the average dc output decreases. The dc output is approximately

$$V_{dc} = 1.4V_{rms} - I_{dc}/2fC \qquad (3-1)$$

where $V_{rms}$ is the rms rectifier supply voltage and $I_{dc}$ is the average dc current through $R_L$. Since $I_{dc} \simeq 1.4V_{rms}/R_L$, equation 3-1 can be written

$$V_{dc} = 1.4V_{rms}[1 - 1/(2fCR_L)] \qquad (3-2)$$

Equations 3-1 and 3-2 for $V_{dc}$ and the equation for $r$ are based on assumptions with respect to waveshape and load. They are not very suitable if the load is sufficient to cause the output voltage to decrease by more than about 20% from the peak voltage (1.4$V_{rms}$).

If a system requires a 15-V, 100-mA full-wave rectified power supply with no more than 10% deviation in $V_{dc}$ under full load, the values of $C$ and $V_{rms}$ required can be calculated from equation 3-1. First under no load

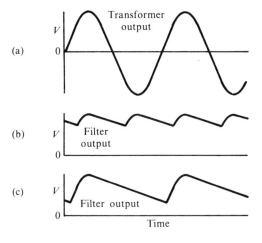

**Fig. 3-17.** A comparison of ripple frequency and amplitude for full-wave rectifier (b) and half-wave rectifier, (c) for a given input frequency (a). Note that the charging pulse frequency of the half-wave rectifier is equal to the input frequency, but that of the full-wave rectifier is twice the input frequency.

$(I_{dc} = 0)$, the dc output voltage should be 15 V. Hence $V_{rms} = 15\,V/1.4 = 10.7$ V. To provide 10.7 V rms after the 0.6 V forward voltage drop across the rectifier diode, the transformer secondary should be 11.3 V for each side of the center tap. If the transformer secondary has a resistance of 4 $\Omega$, $V_{rms}$ is reduced by 0.4 V. The decrease allowable in the filter is thus $1.5 - 0.4 = 1.1$ V. Therefore, 1.1 V $= 0.1\,A/2(120\,Hz \times C)$ from which the minimum value of 379 $\mu$F for $C$ is obtained. The ripple factor under full load can also be obtained: Since $R_L = V_{dc}/I_{dc}$, the equation for $r$ can be written $r = 1/(2\sqrt{3}\,fCV\,/I_{dc})$. Using $f = 120$ Hz, $C = 400$ $\mu$F, $V_{dc} = 15$ V and $I_{dc} = 100$ mA, $r$ is found to be 0.04 or 4.0%. The rms ripple voltage is $V_{rms} = rV_{dc} = 0.04 \times 15$ V $= 0.6$ V.

## 3-6    Voltage Regulation

The dc output voltage from a rectifier-filter power supply is unregulated and varies too much to be useful for many applications. There are two major factors that cause the output voltage to vary—changes in the ac line voltage and changes in the load resistance. The 115-V ac line voltage can range typically from 105 to 125 V from one time of day to another, about a 20% variation. This, of course, would be intolerable if 0.1% or even 1% stability of the dc voltage is required for a particular application. The load resistance in electronic equipment may also fluctuate greatly as one circuit or another is turned on or off, and the change of load resistance can cause significant changes in the output dc voltage.

In this section we shall show how the output dc voltage can be regulated so that it is not significantly influenced by changes of line voltage or load resistance. The Zener shunt regulator is described, and the series regulator and the switching regulator are introduced.

### Shunt Regulator

Any power supply can be represented by the Thevenin equivalent circuit shown in figure 3-18 with the load $R_L$ connected at its output. If the load changes, the current changes, and the effect on voltage $v_o$ depends on the relative values of $R_L$ and $R_s$. Changes in the line voltage also change the voltage across the load. This kind of source is called an unregulated power supply. In other words, to regulate fully (to control) the voltage $v_o$ it is necessary to compensate for changes of $V$ and for changes of load current $i$. One type of voltage regulator is the shunt regulator, illustrated in figure 3-19. One of its most important applications is for low-power voltage reference sources where Zener diodes are used as the automatically variable shunts. A device that could effectively vary $R_v$ in such a way as to hold $v_o$ constant at all times would clearly be a valuable regulator. This is the role of a **Zener diode**. The Zener diode is a type of breakdown diode, as mentioned in section

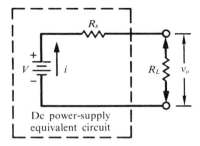

**Fig. 3-18.** Power-supply equivalent circuit. The voltage $v_o$ across the load is $v_o = V - iR_s$. Any change in the current $i$ or voltage $V$ changes $v_o$.

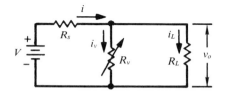

**Fig. 3-19.** Shunt regulation of dc output voltage. In the shunt regulator, a part $i_L$ of the total supply current $i$ goes through the load $R_L$, and a part $i_v$ through a variable resistance $R_v$. Since $v_o = V - iR_s$ and since $i = i_L(R_v + R_L)/R_v$, it follows that

$$v_o = V - i_L \left( \frac{R_v + R_L}{R_v} \right) R_s$$

3-4, that is designed to operate in the reverse-biased breakdown mode and to have a stable breakdown potential of the desired control voltage. The voltage across the diode is regulated at the Zener voltage $V_Z$ as long as the Zener current is greater than $I_{ZK}$ in the region of the knee on the diode $I$-$V$ curve of figure 3-20. The product of the Zener current and the Zener voltage should not exceed the power dissipation rating for the Zener diode ($P_{max} = I_{Z(max)}V_Z$).

The series resistor $R_s$ should allow enough current through the reverse-biased Zener diode so that the device operates in the Zener breakdown region. When the applied dc voltage $V$ at the input is higher than the Zener breakdown voltage $V_Z$, the voltage across the series resistor $V_{Rs}$ equals the difference between the input voltage and the Zener voltage. That is, $V_{Rs} = V - V_Z$. A load connected across the output requires a current $i_L$ that is determined by its resistance and the output voltage. The current through the Zener diode $i_Z$ and $i_L$ both occur in $R_s$: $i_{Rs} = i_Z + i_L$ and $V_{Rs} = R_s(i_Z + i_L)$.

Thus, the value of $R_s$ is chosen so that $i_Z$ is large enough both to be in the Zener breakdown region and to allow the required value of $i_L$ to exist in the load. Note that as $i_L$ increases, $i_Z$ decreases, and vice versa. The result is to hold $i_Z$ plus $i_L$ almost constant and maintain a constant output voltage. Since $V$ is unregulated, its value fluctuates and its minimum value $V_{min}$ must be higher than $V_Z$. When $V_Z$ is subtracted from $V_{min}$, the difference is the minimum voltage that is dropped across $R_s$. If this difference voltage is divided by the maximum load current, the required value of $R_s$ can be determined:

$$R_s = \frac{V_{min} - V_Z}{i_{L(max)}}$$

The Zener diode shunt regulator is commonly used as a reference voltage source for comparator circuits and feedback control regulator circuits. In such applications, the load is light and essentially constant. Output voltage precision of one part in 10 000 or better is obtainable from carefully designed Zener reference sources.

## Series Regulator

Another method of providing voltage regulation is to introduce a variable resistor $R_v$ in *series* with the load, as shown in figure 3-21. A comparator/servo system (discussed in chap. 5) monitors the output voltage, compares it to a reference voltage, and automatically changes the effective resistance of a series element such as a transistor. This type of series regulator can provide excellent voltage regulation. With the recent developments in integrated circuits, the cost of providing series regulation is low. Because of the low cost and excellent voltage regulation, the series-type regulator is now widely used.

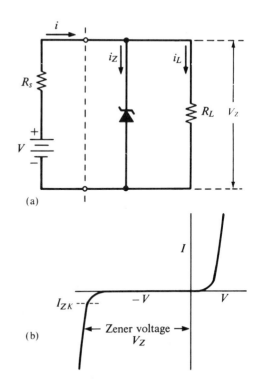

**Fig. 3-20.** Zener-regulated supply (a) and characteristic curve (b). Since the reverse breakdown voltage of the Zener diode remains essentially constant for a wide range of reverse currents, the effective resistance of the diode $R_Z = V_Z/i$ changes to compensate for changes of supply voltage $V$ or load current.

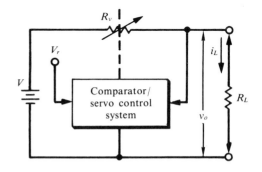

**Fig. 3-21.** Series regulator circuit. The output voltage is $v_o = V - i_L R_v$. The resistance $R_v$ is varied automatically by the servo system to compensate for changes of $V$ or $i_L$.

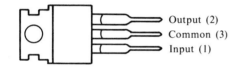

**Fig. 3-22.** Top view of IC regulator, model 7805. This particular model provides regulation at 5 volts with less than 1% variation in output for input voltages from 7 to 25 V and output currents up to 1.5 A. It also provides internal thermal-overload protection and internal short-circuit current limiting. Only an external capacitor is required. The numbers next to the pins are often used in schematic diagrams to avoid having to name each lead.

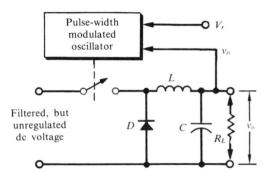

**Fig. 3-23.** Simple switching regulator. The series switch is controlled by an oscillator of constant frequency but variable pulse width. The fraction of the time that the switch is closed is controlled to maintain $v_o$ equal to $V_r$.

A typical three-terminal integrated-circuit (IC) voltage regulator is shown in figure 3-22. It is a small, simple-looking package with only three leads, but it contains dozens of components including nearly twenty transistors.

The current limitation in a series regulator is the result primarily of the power that the series element must dissipate. This power is the product of the excess input voltage (which must be at least 2–3 V for the regulator to work) and the load current. Thus a 5-V regulator with a 1-A load and an input voltage of 10 V dissipates 5 W. A heat sink is usually required to prevent overheating.

The series regulated power supply is further discussed in chapter 7 in conjunction with programmable power supplies. An example of a series regulated power supply is the dual-output supply described in the last section of this chapter.

## Switching Regulator

In a switching regulator the series control element is a switch that is fully on part of the time rather than a resistive device that is partially on all of the time. One type of switching regulator is shown in figure 3-23. The series switch of this type of regulator is turned on and off by an oscillator with a pulsed output. The oscillator has a constant repetition rate (typically 20 kHz), but the on-time of the oscillator pulse is variable. During the pulse the switch is closed, and the current from the unregulated supply is delivered to the $LC$ filter and the load. When the switch is open, diode $D$ conducts, and the energy stored in the inductor is delivered to the capacitor and load. The width of the pulse is controlled by the pulse-width modulated oscillator to maintain a negligible difference between the reference voltage $V_r$ and the regulator output. In other words, the charge delivered to the capacitor and load by each pulse of the switch is exactly equal to the charge consumed by the load during one period of the oscillator and is exactly enough to maintain $v_o$ at the desired voltage.

Another type of switching-regulated supply makes use of direct rectification of the 60-Hz line voltage as shown in figure 3-24. The major advantages of switching regulators over series regulators are their smaller size and

**Fig. 3-24.** Switching regulator with high-frequency transformer. Switches apply pulses of current of alternating direction to the primary of a high-frequency transformer. A full-wave rectifier and $LC$ filter produce a dc voltage from the current in the transformer secondary. The pulse-width modulated oscillator controls the length of time the switch is on to maintain the desired output voltage.

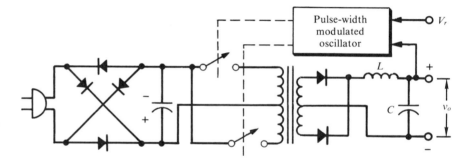

greater efficiency. A series regulator delivers about 50% of the power it consumes to the load; switching regulators have efficiencies of 75–85%. The smaller size of the switching regulator is due to the fact that high-frequency filters and transformers are only about one-tenth the size of their line-frequency counterparts and have lower heat-dissipation requirements. Inexpensive integrated-circuit pulse-width modulated oscillators have made switching regulators economically competitive for many applications.

## 3-7 Current Regulation

There are many transducers, light sources, detectors, and other devices that require a constant current through them, rather than a constant voltage across them, for suitable operation. It is the purpose of this section to introduce current control.

The concept of current regulation can be understood from consideration of the schematic diagram in figure 3-25. The purpose is to control the current $i_c$ through the load $R_L$ at a desired constant value. The load resistance can vary during operation (as indicated by the arrow through $R_L$), and the dc voltage $V$ from the unregulated supply can vary with time. Since the load is in a simple series circuit the current $i_c = V/(R_L + R_f + R_c)$. Therefore, if $R_L$ or $V$ changes in value, the current also changes unless the value of a variable series resistance $R_f$ is altered to compensate for them. In principle, any time the current tends to deviate from the desired value because of changes of $R_L$ or $V$, the deviation is detected by observing the voltage $v_c$ across a control resistor $R_c$ of fixed and known resistance ($i_c = v_c/R_c$). The control voltage $v_c$ is continuously compared by the voltage comparator to a known reference voltage $V_r$. If there is any difference between $V_r$ and $v_c$, the electronic servo control system immediately adjusts the resistance $R_f$ of the feedback control element so that the desired value of $i_c$ is maintained. The control system can be designed to respond in the micro- to millisecond range giving excellent current stability and effective suppression of ripple from the

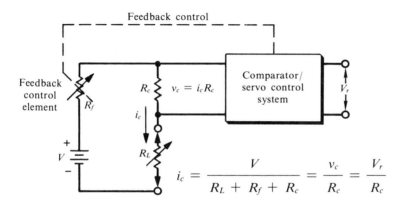

**Fig. 3-25.** Regulated constant-current power supply. The current $i_c$ through load $R_L$ is controlled by varying the feedback element $R_f$. The voltage drop $v_c = i_c R_c$ across the control resistor $R_c$ is kept equal to the reference voltage $V_r$ by the servo control system.

unregulated supply. Correctly choosing values of $V_r$ and $R_c$ allows the controlled value for $i_c$ to be selected. This assumes, of course, that the voltage source $V$ has sufficient voltage and current capability to supply the desired current and the $IR$ drop across $R_L$, $R_c$, and $R_f$.

## 3-8    Batteries

Despite the availability of line-operated power supplies for the conversion of ac to dc in essentially any desired voltage and current range, batteries have many applications in modern electronic equipment. The use of batteries as power sources for instruments is essential in remote places, such as outer space, underseas, and in mines, where a central line voltage is unavailable. Batteries are called for when mobility, portability, and extremely high reliability are required. For some applications, a battery is less expensive than a line-operated power supply of equivalent stability. Batteries also have the advantage of freedom from line frequency noise.

A **battery** consists of several electrochemical cells connected to provide the necessary voltage and current capability. However, it is not unusual to hear the term battery used in reference to a single cell. For many years only two kinds of batteries were readily available that were suitable for light- or moderate-duty applications. These were the familiar carbon-zinc dry cell and the lead-acid storage battery. Today the designer or user of battery-operated devices has a choice of several different kinds of batteries in a wide variety of sizes and voltages. Each of these battery types has characteristics that particularly suit it to certain uses.

### The Carbon-Zinc Dry Cell

The most widely used of all the so-called dry cells is the carbon-zinc dry cell. Its structure is shown in figure 3-26. Compact dry batteries of voltages greater than 1.5 V are generally made by stacking a carbon plate, a layer of electrolyte paste, and a zinc plate, alternately, as many times as necessary to give the desired voltage. The most common dry batteries have voltages of 1.5, 3, 6, 9, 22.5, 45, 67.5, and 90 V.

The carbon-zinc dry cell is one of several kinds of **primary cells**, that is, nonrechargeable cells. The **service life** of a battery is the number of hours that a fresh battery will satisfactorily operate the actual circuit under normal operating conditions. The service life of a battery can vary widely depending on the following factors: the quality of the battery; the length of time it has been stored before use; its temperature during the storage period; the rate at which it is discharged; the number and duration of the off periods; its temperature during discharge; and the lowest voltage for satisfactory operation of the circuit. After actuation all batteries discharge internally at some

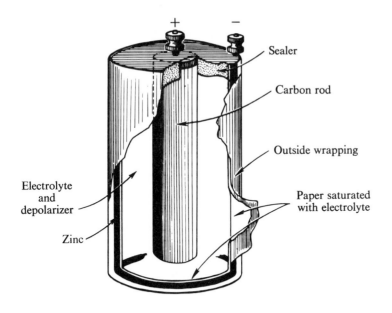

+  −

Sealer

Carbon rod

Outside wrapping

Electrolyte
and
depolarizer

Paper saturated
with electrolyte

Zinc

**Fig. 3-26.** Carbon-zinc dry cell. During discharge the zinc metal of the can is converted to a zinc salt in the electrolyte, leaving two electrons in the zinc can for each atom of zinc dissolved. Manganese dioxide ($MnO_2$) is reduced at the carbon electrode to $Mn_2O_3 \cdot H_2O$. This chemical action establishes a voltage of 1.5–1.6 V between the two electrodes with the carbon electrode positive.

rate. This limits the time a dry cell can be stored before use. At room temperature a dry cell could be stored for about a year, but increasing the temperature shortens the shelf life considerably. The total energy available from a given battery decreases with increasing discharge rate, or "drain," through the load. For a carbon-zinc cell a service life of less than ten hours is considered a heavy drain. Under conditions of heavy drain, periodic rest periods extend the capacity, and thus the service life, of the battery.

The dissolution of the zinc tends to weaken the structure of the cell. Furthermore, during discharge or storage a pressure of evolved hydrogen gas builds up. This can lead to a rupture of the zinc and leakage of the corrosive electrolyte into the instrument. *Instruments using these dry cells should therefore not be stored with the batteries installed.*

## The Lead-Acid Storage Battery

The familiar car battery is a series combination of several cells that consist of lead and lead/lead dioxide electrodes immersed in sulfuric acid. Under discharge the lead is converted to insoluble lead sulfate, releasing two electrons to the lead electrode for each molecule of lead sulfate formed. The lead dioxide in the other electrode is reduced to lead sulfate by accepting two electrons per molecule of lead sulfate formed. During this discharge process the sulfuric acid is converted to water. When fully charged, each cell has a potential of 2.06–2.14 V. The lead-acid battery is a **secondary cell**; that is, it can be recharged. When a reverse, or charging, current is forced through the cell, the lead sulfate is converted back to lead and lead dioxide returning the cell to very nearly its original state.

The dilution of the sulfuric acid causes the cell voltage to decrease during discharge. The decrease in voltage is slow at first but becomes quite rapid during the last one-third of the battery's service life. The internal resistance of the lead-acid battery is so low that it can be ignored at normal discharge rates. The high current, which is free from ac ripple, makes the lead-acid battery the most economical power source for certain applications. However, it is bulky and heavy and requires considerable care if it is to give proper service.

## The Nickel-Cadmium Battery

In recent years the nickel-cadmium battery has come into widespread use. One reason for its growth in popularity has been the development of the sealed nickel-cadmium cell. This is a completely sealed, rechargeable unit that requires no attention other than charging. One form of this cell is shown in figure 3-27. The electrolyte is not involved in the electrode reaction, and thus the voltage is fairly constant over the service life. The capacity of the nickel-cadmium battery is reduced very little even at very high discharge rates. The batteries can be stored in a charged or discharged condition without harm, and the sealed units do not leak. They have the lowest self-discharge rate of any secondary cell. Sealed nickel-cadmium batteries, available in many sizes and voltages, are interchangeable with common primary batteries. Nickel-cadmium batteries offer the advantages of high current capability, long service life, reasonably constant voltage, and the possibility of recharge. These characteristics are particularly suited to instruments in which heavy drain and frequent use would require frequent battery replacement. Their price is many times that of a similar-sized carbon-zinc dry cell, and there is some concern about cadmium pollution in the environment. During the charging process the nickel oxide is reoxidized to its higher oxidation state, and the cadmium oxide is reduced. The sealed cell is designed so that it can be overcharged without the buildup of a large gas pressure.

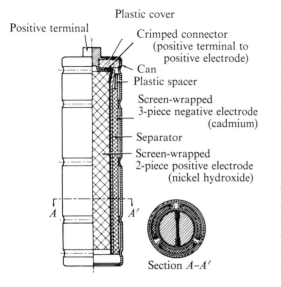

Plastic cover
Positive terminal
Crimped connector (positive terminal to positive electrode)
Can
Plastic spacer
Screen-wrapped 3-piece negative electrode (cadmium)
Separator
Screen-wrapped 2-piece positive electrode (nickel hydroxide)

A        A'

Section A–A'

**Fig. 3-27.**   Rechargeable nickel-cadmium cell. Under discharge the cadmium is oxidized and supplies electrons; at the positive electrode nickel oxide is reduced to a lower oxidation state by accepting electrons. The open-circuit voltage of this cell is 1.3 V.

## Battery Charger

Rechargeable batteries such as the nickel-cadmium battery can be readily charged from the ac power line by use of a half-wave rectifier circuit as shown in figure 3-28. The battery is charged by reversing the electrode reactions that occur when the battery is a voltage source. A dc charging current is obtained by introducing a diode in series with the battery and applying a sufficiently high voltage at the secondary of the transformer to

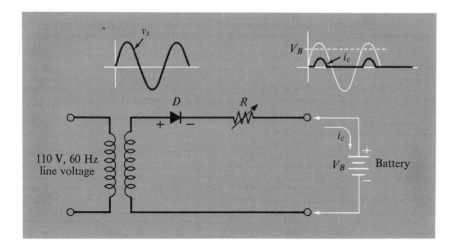

**Fig. 3-28.** Battery charger for rechargeable cells. When the magnitude of the ac voltage $v_s$ exceeds the battery voltage $V_B$, the diode conducts and produces a charging current $i_c$ in the correct direction for recharge. When the ac voltage is less than $V_B$, the diode is reverse biased to prevent discharge.

cause the diode to conduct during part of the ac cycle. By varying the resistance $R$ or the sine-wave amplitude, the rate of battery charging can be adjusted.

## Solar and Fuel Cells

Although not widely applicable yet, both **solar cells** and **fuel cells** hold great promise for the direct conversion of sunlight and chemical energy, respectively, into electrical power. These cells are especially useful for instrumentation in remote locations.

The radiation from the sun that strikes the earth each day represents an amount of energy that is many times our yearly consumption of electrical energy. It is estimated that on a summer day the solar radiant power is 1 $kW/m^2$ at the earth's surface. Through hydroelectric generators and fossil-fuel combustion, solar radiation has long been used indirectly to obtain electrical power, but until recently there has been no efficient way of directly converting this radiant power into electrical power.

In recent years semiconductor solar cells have been designed for direct conversion of solar radiation into electrical power. These cells provide a high power capacity per unit weight and have been used successfully on space vehicles and communication satellites. A representation of a silicon photovoltaic cell is shown in figure 3-29a. Such silicon wafers, 1 cm by 2 cm in area, have been used to build up single-panel arrays of many thousands of wafers to provide a capacity of several hundred watts. A typical array is shown in figure 3-29b.

In the silicon cells of figure 3-29a, a very thin layer of p-type semiconductor is formed on an n-type silicon strip (or vice versa) to form a pn

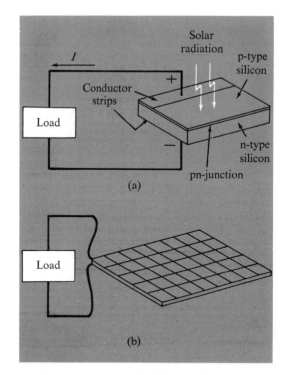

**Fig. 3-29.** Solar cells. A silicon photovoltaic cell is shown in (a). A series-parallel array of forty 1 × 2-cm cells (b) could provide about 1 W of electrical power for 10 W of incident sunlight.

junction and to provide the light-sensitive face. A narrow conducting strip serves as the collector terminal. The bottom of the cell is nickel plated and tinned. The diffusion of majority carriers from both the n- and p-type materials causes a combination of holes and electrons that sets up a barrier potential across the junction. The p- and n-type materials contain many valence electrons that are not affected by the impurity doping. However, a photon can interact directly with a valence electron and provide enough energy to promote it to a conduction electron. The free electron and the hole are now attracted by the barrier potential at the pn junction and travel in the opposite direction to the majority semiconductor carrier. That is, the photoelectrons travel toward the positively charged n-type silicon, and the holes move toward the negatively charged p-type silicon. The result is to provide a current $I$ through a load as shown in figure 3-29a. The arrays used in satellites generally have an output of about 10 W per pound. In view of the huge supplies of silicon and sunlight, increased application of solar cells can be expected as the technology improves.

The direct conversion of chemical energy in a fuel into electrical power is now practical on a small scale in hydrogen-oxygen fuel cells. The Gemini spacecraft utilized $H_2$-$O_2$ fuel cells of the type shown in figure 3-30. Their theoretical efficiency is greater than 90%, and efficiencies of over 80% have been attained with laboratory cells. Low cost, quiet operation, and the absence of noxious byproducts are the major reasons for the great interest in this type of direct chemical-to-electrical converter.

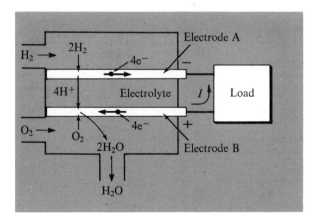

**Fig. 3-30.**  Hydrogen-oxygen fuel cell. The hydrogen gas ($H_2$) in the top chamber diffuses through electrode A and reacts to produce hydrogen ions ($H^+$) in the electrolyte and electrons ($e^-$) in electrode A. The hydrogen ions move through the electrolyte to electrode B. There they combine with oxygen ($O_2$) and electrons to form water ($H_2O$). The reactions at electrodes A and B cause a potential difference across, and thus a current through, the load.

## 3-9 Application: A Dual-Output Power Supply

Many electronic systems contain circuits that require more than one power-supply voltage. Analog circuits, for example, typically require +15 V and −15 V; digital circuits usually require +5 V. Commercial power supplies are available with dual (±15 V) outputs and triple outputs (±15 V, +5 V) to meet these requirements.

Figure 3-31 illustrates a typical dual-output power supply. One center-tapped transformer is used in conjunction with the four diodes to produce the positive and negative dc outputs. The dc outputs are filtered by 1000-$\mu$F capacitors and separately regulated. Additional circuitry could be added to provide overvoltage protection and additional short-circuit protection. Triple-output supplies often use a separate transformer, rectifier, filter and regulator to obtain +5 V because many +5 V supplies provide a higher output current than ±15 V supplies.

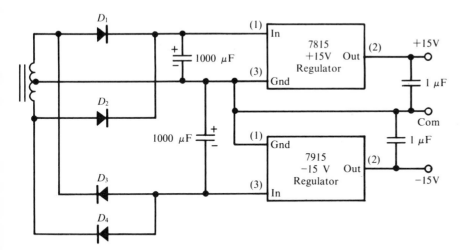

**Fig. 3-31.** Dual-output (±15 V) power supply. Diodes $D_1$ and $D_2$ form a full-wave rectifier for the positive supply; $D_3$ and $D_4$ are the negative supply rectifiers. The filtered outputs are regulated by separate three-terminal regulators. The output 1-$\mu$F capacitors help to stabilize the regulators. The coil on the left of the schematic is the secondary of the power transformer.

## Suggested Experiments

**1.  Diode characteristics.**
Determine the current-voltage curves for forward- and reverse-biased signal and Zener diodes. Use an ohmmeter to determine which end of a diode is the cathode, and note the effect of resistance scale change on the effective diode resistance. Use an ohmmeter to determine the internal arrangement of diodes in a prepackaged diode bridge.

**2.  Power transformer voltages.**
Measure the ac voltages at the secondary of a power transformer from the center tap to each side and from one side to the other. Use the dual-trace scope to observe the phase difference between the signals at either side of the secondary winding when the center tap is connected to common.

**3.  Rectifier circuits.**
For half-wave, full-wave, and bridge rectifier circuits, observe the output waveform with resistive load, the amplitude relative to the transformer secondary, the peak reverse voltage across the diodes, and the forward voltage drop due to the rectifier circuit.

**4.  Capacitive filters.**
Connect an $RC$ filter to the output of one of the rectifiers from experiment 3, and measure the ripple for various loads and values of $C$. Observe that the ac input of the scope allows the sensitive measurement of small ac signals on a relatively large dc level.

**5.  Voltage regulation.**
Connect a three-contact integrated circuit regulator to the above rectifier and filter circuit to produce a regulated output. Measure the ripple voltage and dc output for several values of the load. Note also the temperature of the regulator by carefully touching it. Connect a Zener diode to produce a Zener-regulated reference voltage source. Determine an appropriate value of the series resistor from the current-voltage curve obtained in experiment 1.

## Questions and Problems

**1.**  In the United States the ac power-line voltage is nominally 115 V rms. What are the peak voltage, the peak-to-peak voltage, and the average voltage?

**2.**  In Europe the line voltage is nominally 230 V rms. This higher voltage is also used in the United States for operating appliances and equipment with high power requirements. What are its peak voltage, peak-to-peak voltage, and average voltage?

**3.**  A transformer has a turns ratio of $1/50$. The 115-V ac line voltage is connected to the primary. What is the peak-to-peak secondary voltage? What are the peak, the average, and the rms secondary voltages?

**4.**  A transformer on a power pole transforms the voltage from an efficient transmission value of 4400 V to a safe working value of 115 V. What turns ratio must the transformer have? What is the peak-to-peak voltage transmitted by power lines having a 4400-V rms value?

**5.**  (a) For the half-wave rectifier of figure 3-11 with $R_L = 10\,\text{k}\Omega$, what is the average dc output current? (Assume the diode to be a silicon diode with a forward voltage drop of 0.6 V.) (b) If $R_L$ in figure 3-11 is 100 Ω, what is the average dc output current?

**6.**  For the full-wave rectifier of figure 3-12, the secondary voltage is 6 V rms from one end of the transformer to the center tap. If $R_L = 100$ Ω, what is the average dc output current?

**7.**  A silicon diode is used in a half-wave rectifier circuit with a large filter capacitor placed across the load. If a transformer with a turns ratio of $1/4$ is used to reduce the 115-V rms line voltage, what is the minimum peak inverse voltage (PIV) ratio that the diode should have?

**8.**  The same transformer as in problem 7 is used in a half-wave rectifier circuit with a silicon diode and a load of 150 Ω. What average forward current rating should the diode have?

**9.**  (a) A full-wave rectifier is employed with a center tap transformer to reduce the line voltage. The transformer has a turns ratio of $1/25$ from the center tap to one end. What are the minimum PIV ratings for the diodes if a large filter capacitor is used? (b) What are the minimum PIV ratings for the diodes if the same transformer is used in a bridge rectifier circuit with a large filter capacitor? (c) What are the minimum PIV ratings for the diodes and the necessary turns ratio for a transformer to be used in a bridge rectifier circuit that supplies the same voltage to the same filtered load as in part (a)?

**10.** A transformer that has 15 V rms on its secondary is considered for use with a capacitor-filtered half-wave rectifier, a full-wave bridge rectifier, and a voltage doubler. (a) What are the maximum output voltages available from the three rectifier circuits if the transformer and diodes are ideal and the load is negligible? (b) Consider the voltage drops across the diodes, and recalculate the maximum output voltages.

**11.** (a) The half-wave rectifier of figure 3-11 is used with a capacitor filter of 10 $\mu$F. Calculate the dc output voltage and the current through the 10-k$\Omega$ load. (b) Find the ripple factor $r$ for the half-wave rectifier of figure 3-11 with a 10-$\mu$F filter capacitor.

**12.** (a) The filter capacitor in the power supply described in problem 11 is changed to 100 $\mu$F. Recalculate the dc output voltage, the current through the load and the ripple factor. (b) Repeat part (a) for a 1000-$\mu$F filter capacitor.

**13.** (a) In the full-wave bridge rectifier circuit of figure 3-15 (b), the input is the 60-Hz line, the secondary voltage is 10 V rms, $C = $ 10 $\mu$F and $R_L = $ 10 k$\Omega$. Find the dc output voltage, the dc current through $R_L$, and the ripple factor. (Remember that the frequency of the full-wave rectified output is double the line frequency.) (b) Repeat part (a) for $C = 100$ $\mu$F and $C = 1000$ $\mu$F.

**14.** For the full-wave bridge rectifier of problem 13 with $C = $ 100 $\mu$F, find the dc output voltage for the following load currents: 1 mA, 5 mA, 10 mA, 50 mA, 100 mA, 200 mA, 250 mA. (a) Plot dc output voltage against load current. (b) What load resistances would be necessary to give the load currents found above? (c) Find the ripple factors for each of the load currents in part (a).

**15.** In the full-wave rectifier of figure 3-15(b) with a 60-Hz line voltage, it is desired to choose the filter capacitor. The secondary voltage is 12 V rms and the ripple is to be no more than 1% for load resistances as low as 100 $\Omega$. Find the minimum filter capacitance that can be used. What is the output voltage with this value of $C$ for no load? What is it for a 100-$\Omega$ load?

**16.** The ripple voltage and the dc output voltage of an unregulated supply are, of course, related. (a) Show that the relationship is $V_{dc} = 1.4$ $V_{rms} [1 - r/\sqrt{3}]$. (b) What ripple factor corresponds to a 5% reduction in the dc output voltage?

**17.** In the voltage-doubler circuit of figure 3-14, the load resistance is 5 k$\Omega$. If $C_1 = C_2$, what is the minimum capacitance necessary to insure that voltage drop between charging pulses is less than 5% of the dc output voltage?

**18.** In the voltage doubler circuit of figure 3-14, the secondary voltage is 50 V rms. What is the dc output voltage?

**19.** In the Zener diode regulator of figure 3-20, the series resistance $R_s = 1000$ $\Omega$, the supply voltage $V = 25$ V, and $V_Z = 15$ V. The maximum rated current through the Zener is 150 mA. (a) Over what range of $R_L$ values is the regulator useful? (b) For a constant load of $R_L = 5.0$ k$\Omega$, over what range of input voltages can regulation be achieved?

**20.** In the Zener diode regulator shown in figure 3-20, $V_Z = 5.1$ V, and the Zener diode has a maximum power rating $P_{max} = 500$ mW. Find the required value of $R_s$ if $V = 10$ V, $R_L = 100$ $\Omega$ and the power dissipation is to be no more than 150 mW.

**21.** A current regulated power supply similar to that shown in figure 3-25 is to be used to control the current through a load. The current is to be controlled at 1.00 mA through loads that can vary from 200 $\Omega$ to 9.5 k$\Omega$. If $V = 10.00$ V and the reference voltage $V_r = 100$ mV, find the appropriate value of $R_c$. (a) What range of resistances must the feedback control element $R_f$ be capable of producing? (b) If the lowest value of $R_f = 100$ $\Omega$, over what range of load resistances does the system regulate?

**22.** A lead-acid storage battery has a voltage of 6 V and can deliver 500 A. A carbon-zinc dry cell has a voltage of 6 V and can deliver 500 mA. Compare the internal resistance of the lead-acid storage battery with that of the carbon-zinc dry cell.

**23.** The battery charger of figure 3-28 is used to recharge a Ni-Cd cell with an open circuit voltage of 1.3 V. If the transformer secondary puts out 15 V rms, what value of series resistance $R$ should be used to ensure that the peak charging current is no more than 500 mA?

# Chapter 4

# Input Transducers and Measurement Systems

The specific quantities about which a scientist or engineer seeks quantitative information are usually nonelectrical ones such as temperature, pressure, light intensity, strain, or pH. In order to utilize the elegant electronic measurement devices that now exist, the information of interest must be converted (transduced) into an electrical signal. It is the function of the input transducer to perform this important task in a known and reproducible manner, so that the final number obtained from the measurement can be related directly to the quantity of interest. Because of the importance of the input transducer in the overall measurement scheme, an understanding of the operating principles and the input/output characteristics (such as linearity and transfer function) of transducers is basic to the effective use of modern instrumentation. Among the input transducers characterized and described in this chapter are thermocouples, photovoltaic cells, photomultiplier tubes, photodiodes, thermistors, and strain gauges.

Integrated circuits, which perform an increasing variety of sophisticated electronic functions inexpensively, have opened many new avenues for the conversion of electrically encoded information to a related number. Many of these routes involve counting electrical pulses from transducers that detect discrete events or objects and from converters that divide an analog signal into discrete quanta. Other arrangements of the basic counting function provide precise time measurements or measurements of the rate at which discrete events occur. The convenient digital display provides an unambiguous readout of the final measurement result.

Since all measurement systems, whether simple or sophisticated, involve similar principles, this chapter begins with an exploration of the measurement process itself.

## 4-1 Measurement Principles

The measurement process was defined in chapter 1 as the *determination of a particular characteristic of a sample in terms of a number of standard units of that characteristic.* A measurement seeks an answer to the question "How many standard units are equal to the unknown quantity?" For example, how many standard length units (meters, inches, etc.) are equal to the length of this page, or how many standard voltage units are equal to the voltage of a battery? Two quite general principles result from this definition of measurement. First, the comparison of the sample characteristic with standard units of that characteristic is implicit in the definition. A second, more subtle point is that what we obtain from a measurement is a discrete, quantized value, that is, a number of units such as 0.22 m or 8.52 V.

## Measurement System Classes

The general block diagram of figure 4-1 summarizes the measurement process. The quantity to be measured $Q_u$ is compared with a reference standard quantity $Q_r$. The difference $(Q_u - Q_r)$ is converted by the difference detector to another form (domain) such as scale position. The total value then indicated for the measured quantity $Q_o$ is the sum of the standard units from the reference standard $Q_r$ and the difference detector output $(Q_u - Q_r)$ calibrated in these same units. Thus $Q_o = Q_r + (Q_u - Q_r) = Q_u$.

Measurement devices and systems vary in the degree to which they depend upon the difference detector and the reference standard outputs in the determination of $Q_u$. In a **direct measurement** device $Q_r$ is zero or constant, and the difference detector output provides the total measurement information. An example of a direct measurement is the meter deflection in an analog multimeter. The **accuracy** and **precision** (see notes 4-1 and 4-2) of a direct measurement are dependent on the accuracy and stability of the difference detector transfer function.

At the other extreme, a **null comparison** measurement is based upon a variation of the reference standard in sufficiently fine increments such that the difference output is adjusted to zero. Since the difference detector output is zero, the value of the reference standard at null is equal to the value of the unknown. The difference detector off-null output need not be calibrated at all in this case. The only requirement for the difference detector is that it have sufficient **sensitivity** (see note 4-3) to indicate a difference between $Q_u$ and $Q_r$. Common examples of null comparison measurements include the measurement of mass with a double-pan analytical balance and the measurement of voltage with a laboratory potentiometer. The precision of a null

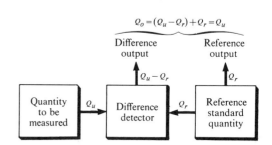

**Fig. 4-1.** Elements of the basic measurement system. The measured value is the reference quantity plus the difference between the reference and unknown quantities.

**Note 4-1. Accuracy.**
The accuracy of a measurement is the degree of agreement between the measured value and the true value.

**Note 4-2. Precision.**
The precision of a measurement is related to its degree of repeatability and its resolution.

**Note 4-3. Sensitivity.**
The sensitivity of a measurement is the smallest change in the quantity to be measured that produces a detectable change in the output. In this case, sensitivity is synonymous with minimum detectability.

comparison measurement depends on the sensitivity of the difference detector; its accuracy depends on both the sensitivity of the difference detector and the accuracy of the reference standard. Because the null comparison technique depends on the calibrated reference standard and not on the transfer function of the difference detector, it is the most accurate measurement method for many quantities.

In many measurement situations, it is inconvenient, too slow, or unnecessary to null the output of the difference detector perfectly. In a third class of measurement systems a close approximation to the null point is made by a coarse adjustment of the reference standard, and the difference detector output gives a readout of the remaining difference between the unknown and the standard. An example is a single-pan balance in which the reference weights are adjustable in 0.1 g increments but the difference scale is readable to 0.1 mg.

## Quantization of Measurements

The total measurement process as conceptualized here is a data-domain conversion in which the unknown quantity is converted to a number of standard units of that quantity. There are several ways to accomplish the quantizing of the unknown quantity. These are considered here in the context of measurements that are aided by electronics.

If the unknown quantity is converted by the input transducer to an electrical signal in one of the analog domains, there are two basic ways in which the resulting signal can be quantized by the null comparison method. Both methods involve the use of a reference standard quantity that is quantized. Figure 4-2 shows the direct analog-to-digital converter (ADC). Here the weighted, digital reference standard is converted to an analog signal by a digital-to-analog converter (DAC). The DAC output is compared with the analog signal input, and the digital reference value is varied until the two analog signals are equal (within the resolution of the reference standard

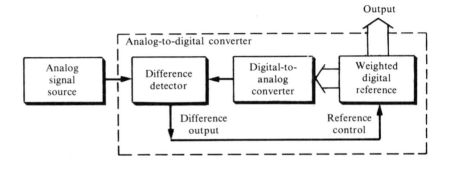

**Fig. 4-2.** Direct analog-to-digital converter. The analog signal is directly compared with an analog quantity derived from the digitally adjustable reference. The digital output is the reference setting that produces the difference detector output nearest null.

and/or difference detector). The output number is then a parallel digital signal related directly to the analog input. Among the common weighting schemes for the reference standard are binary code and binary-coded decimal (BCD) code. A common direct ADC is the popular successive approximation converter described in chapter 13.

A second way of quantizing analog signals is to convert the analog signal to the charge domain and to determine the number of charge increments contained in the signal. This technique is the basis of the dual-slope and voltage-to-frequency converters introduced in chapter 1 and described in more detail in chapter 9. Here the reference standard is unweighted (a source of equal charge increments), and the resulting output is in one of the time domains—that is, a time interval for the dual-slope converter or a frequency for the voltage-to-frequency converter. The final result is obtained by measurement of the time interval or frequency.

There are many measurement situations in which an analog-to-digital conversion is unnecessary because the desired information is the number of discrete events or information about the time of discrete events. For example, if we wish to know how many customers enter a store per day, the quantity to be measured is already in the digital domain. One way to obtain such information is to arrange a light beam and photodetector in such a way that each person entering the store interrupts the beam and causes an electrical pulse to occur at the output of the photodetector. The pulses are then sent to a counter where the total accumulated count is obtained. The electrical signal from the photodetector is a count digital signal, and no data domain conversion is needed to complete the measurement. If the desired information is the rate, or frequency, of customers entering the store, the information in the signal is in the frequency domain and a frequency-to-digital domain conversion is needed to obtain a number for the rate.

We will see in later sections that the output domain of electrical information from an input transducer depends upon the desired information. In many cases the same transducer can produce outputs in all three domain classes.

## 4-2  Counting Measurements

The ability to make high-speed electronic counting measurements has been a central part of the instrumentation revolution of the past few years. In addition to its wide use in counting discrete events, the electronic counting system is the basic element in frequency and period meters and is often used in sequencing and control applications. This section introduces the counting measurement principle and describes the basic functional elements that make up a modern counting system.

## Counting Principles

The purpose of a counting system is to determine the number of events or items $N$ that occur within specified boundary conditions $B$—for example, revolutions per mile, photons per laser pulse, heart beats per minute, seconds required to run 100 m, or, in general, $N/B$.

The functional block diagram of a counting system is shown in figure 4-3. Those electrical signals derived from the events to be measured that meet the criteria set for true events are shaped into logic-level (HI/LO) signals by the input signal shaper and discriminator. The shaped pulses are the input to a **counting gate** (see note 4-4) which either allows pulses through to the counter or prevents them from reaching the counter. The counting gate is opened and closed by start and stop control signals derived from the boundary conditions (miles, laser pulse, time, etc.). When the counting gate is opened by a signal at the start input, shaped pulses enter the counter, and the counter advances by one increment for each pulse. At the end of the counting interval the counter contents are transferred to a digital memory called a **latch** and are held for display while the next count is accumulated. The display provides a visual numerical readout of the counts accumulated while the counting gate was open.

It is important to note that the accuracy of a counting measurement is determined equally by the numerator and denominator of the ratio, events/boundary condition. The counting of events is subject to an inherent uncertainty of $\pm 1$ count when the start and stop signals are not synchronized with the events to be counted. However, even if a million counts are accumulated for the selected boundary conditions, all six digits will not be accurate unless the accuracy of the boundary conditions is at least one part per million.

## Signal Shaper and Discriminator

The signal shaper and discriminator must not only produce logic-level signals from the input pulses, but they must also serve to define an event. The most common type of discriminator in counting systems is the **pulse height discriminator**, or **comparator**, in which pulses above a selectable threshold

**Note 4-4. Counting Gate.**
A counting gate is a circuit that functions like a gate or door. It allows signals to pass through when open and stops signals from passing when closed.

**Fig. 4-3.** An electronic counting system. The events are converted to logic-level transitions which are counted when the gate is open. Start and stop signals open and close the gate at the beginning and end of the appropriate counting boundary. The final count is latched and displayed.

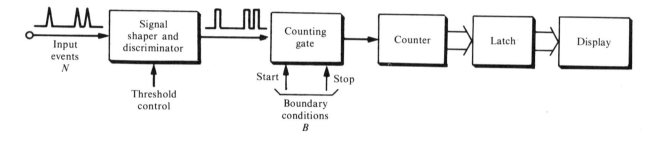

level are shaped into HI logic-level signals and anything below the threshold is considered a LO logic level as illustrated in figure 4-4a. This allows true event pulses to be discriminated from small amounts of noise. The action of another type of discriminator, a **pulse height window discriminator**, is shown in figure 4-4b. Here both an upper and a lower threshold level are present, and only pulses that exceed the lower level without exceeding the upper level are considered true event pulses. There are still other discriminators available in which the pulse width is used as a criterion for selecting true events from undesirable noise pulses.

## Basic Digital Devices

In order to further explore the counting (quantizing) process, two basic digital devices, the gate and the flip-flop, will be introduced. These two devices are combined in different ways to perform a tremendous variety of functions. The gate is used for switching and combining digital signals, and the flip-flop is used for digital data storage and counting. When describing digital functions, the two signal states HI and LO are assigned the values 1 and 0. The most common assignment is that of 1 = HI and 0 = LO. Common examples of these devices and their functions are introduced here; details of their operation and applications are given in chapters 10 and 11.

One kind of gate is called an **AND gate** because its output is 1 only when input $A$ AND input $B$ AND all its other inputs are 1. The symbol and input/output table for a two-input AND gate are shown in figure 4-5a. The AND gate can be used either for its AND logic function or for its action as a gate. When it is used as a gate, one input is the gate control, and the signal to be gated is the other input (see fig. 4-5b). The level at the control input determines whether the gate is open or closed and thus whether $D$ is transmitted to the output or not (see note 4-5).

The second basic gate is called an **OR gate** because its output is 1 when input $A$ OR input $B$ OR any other input is 1. A two-input OR gate symbol

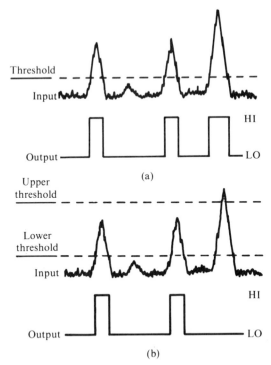

**Fig. 4-4.** Action of discriminators. (a) pulse height discriminator; (b) pulse height window discriminator. The difference is in whether an event is defined as a pulse with an amplitude that exceeds the threshold level or a pulse with an amplitude within a given range.

**Note 4-5. Gate and Switch Terminology.**
A difference should be noted for switch and gate terminology. An open gate allows a signal to pass, while an open switch prevents the signal from passing. Similarly the term closed has opposite meanings for switch and gate circuits.

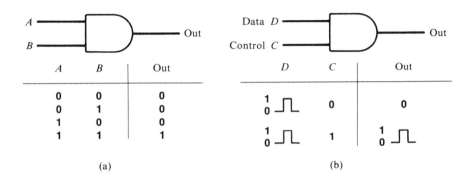

**Fig. 4-5.** The AND gate. The AND gate in (a) is shown by the table to give a 1 output only when inputs $A$ AND $B$ are 1. When used as a gate in (b), control signal $C$ determines whether data signal $D$ is transmitted to the output. When $C$ is 0, the output is 0; when $C$ is 1, $D$ is transmitted.

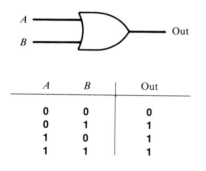

| A | B | Out |
|---|---|-----|
| 0 | 0 | 0 |
| 0 | 1 | 1 |
| 1 | 0 | 1 |
| 1 | 1 | 1 |

**Fig. 4-6.** The OR gate. The OR gate gives a 1 logic-level output when either *A* OR *B* (or both) is 1.

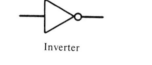

Inverter

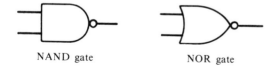

NAND gate          NOR gate

**Fig. 4-7.** Symbols for the inverter and inverting gates. The circle at the output indicates the inverter function in each case.

**Fig. 4-8.** *JK* Flip-flop. A 1-0 transition of the clock signal changes the flip-flop to its alternate state unless that transition is inhibited by a 0 at *J* and/or *K*. The preset and clear functions are achieved by 0 logic levels, and thus these inputs have a circle outside the block to signify inversion.

and its input/output table are shown in figure 4-6. The OR gate can also be used either for its logic function or as a gate; if signal *B* is the control, the output is the same level as *A* when *B* is 0, but it is always 1 when *B* is 1. A circuit that is often used with AND and OR gates is the **inverter.** An inverter converts a 1 logic level into a 0 logic level, and vice versa. If the input signal of an inverter is called *A*, the output is its opposite, NOT *A*, which is written $\bar{A}$. The output is said to be the *complement* of the input. Other gates, such as the NAND, or NOT-AND, gate and the NOR, or NOT-OR, gate are combinations of AND and INVERT and of OR and INVERT, respectively. Symbols for the inverter and the inverting gates are shown in figure 4-7.

The **flip-flop** is the basic digital storage unit. It is a circuit that has only two stable states. Pulses or level changes (edges) can be used to set the flip-flop in either state. The flip-flop performs a memory operation by remaining in that state until commanded by later pulses or edges to change. The state of the flip-flop is given by the level (1 or 0) at its output. The *JK* flip-flop illustrated in block diagram form in figure 4-8 is one of the most versatile. The inputs are the clock input (Ck), the preset input (Pr), the clear input (Clr), and the two inhibit inputs (*J* and *K*). Outputs *Q* and $\bar{Q}$ are complementary, that is, if *Q* is 1, $\bar{Q}$ is 0, and vice versa.

If the two inhibit inputs *J* and *K* are both 1 (or unconnected), the flip-flop is said to **toggle**; that is, the state of the flip-flop alternates on each clock 1-0 transition. Output changes can be inhibited by 0 logic levels at *J* and *K*. A 0 at *J* prevents *Q* from becoming 1 on the next clock pulse, and a 0

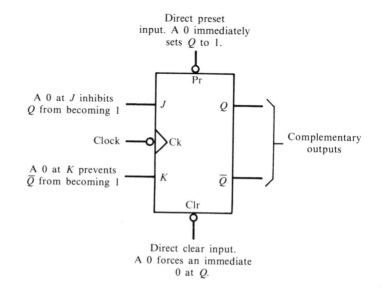

at $K$ prevents $\overline{Q}$ from becoming 1. If both inputs are 0, the clock signal is prohibited from causing any change at the outputs. The direct preset and clear inputs override all other functions. A 0 at preset forces $Q$ to become 1, and a 0 at clear forces $\overline{Q}$ to become 1.

## Counting Gate

The counting gate is like a two-position switch (see note 4-5) with two separate actuators—one to open the gate (close the switch) and the other to close the gate (open the switch). In counting systems it is convenient to label the gate control inputs Start and Stop because an edge applied to Start begins the counting operation, and an edge applied to Stop terminates it. The gating of signals for count measurement requires a gate with separate start and stop command inputs. The stop input should not respond until the start has been activated. The start-stop cycle should not be able to be repeated until the gate is reactivated for the next measurement. Because of these requirements, the counting gate is more complex than the simple gates described above. Its operation is achieved by a combination of flip-flops to sense the start and stop commands and gates to provide the necessary control functions.

Although counting gates can be opened and closed by various logic-level transitions, for this example let us consider that a LO→HI transition at Start is required to begin the counting and a LO→HI transition at Stop is required to end the counting. Figure 4-9 illustrates the action of separate start and stop signals on the gate output. Note that there are many occasions when the start and stop pulses (boundary conditions) might be derived from two different events as in the case of measuring the time a runner requires to pass two different points in a race. There are also many cases in which the start and stop inputs are connected together and signals are allowed to pass for one complete cycle of the boundary condition signal as shown in figure 4-10.

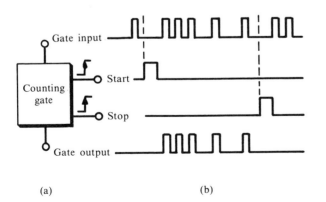

(a)                              (b)

**Fig. 4-9.**  Counting gate with separate start and stop signals. (a) Schematic and (b) waveforms. The symbol ⌐ at the start and stop inputs indicates they are triggered by the rising (LO→HI) edge of the signal.

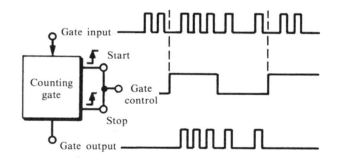

**Fig. 4-10.** Actuating a gate for one complete cycle of a control signal. Triggering at Stop is not enabled until Start has been triggered.

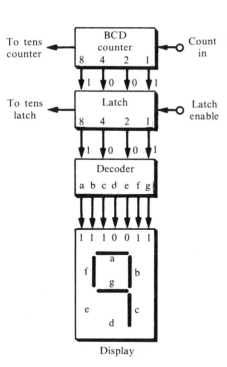

**Fig. 4-11.** Counter, latch, decoder, and display for single decade of counting. A pulse to the latch enable sets the four bits of the latch to the count value. The BCD digit at the latch output is decoded to drive the appropriate segments of the display.

## Counting and Digital Encoding

The counter must advance by one count each time a valid pulse appears at its input. Upon completion of the counting cycle, the counter contents are available as a parallel digital output for storage and display. Some integrated circuit counters encode the count signal as a binary number and some as a binary-coded decimal (BCD) number. Because of the ease in converting BCD code to a decimal number, BCD encoding is used in counting applications when a decimal readout is desired.

In the BCD encoding scheme, each group of four binary bits is allowed to represent the numerals 0 through 9, and each group represents one decimal digit in a number. The 1-2-4-8 BCD code is the same as normal binary coding for the numerals 0 through 9. For example, in the decimal number 47, the four would be represented by the four binary bits 0100 and the seven by the four binary bits 0111 so that 0100 0111 in BCD represents $47_{10}$. Several decimal numbers are given in both binary coding and BCD coding in table 4-1 for comparison purposes. A single **decade counter** is a four-bit counter that is arranged to advance according to the BCD counting sequence

**Table 4-1.** Comparison of binary and BCD codes.

| Decimal number | Binary equivalent | | | | | | | | BCD Equivalent | | |
|---|---|---|---|---|---|---|---|---|---|---|---|
| | $2^7$ | $2^6$ | $2^5$ | $2^4$ | $2^3$ | $2^2$ | $2^1$ | $2^0$ | Hundreds | Tens | Ones |
| | (128) | (64) | (32) | (16) | (8) | (4) | (2) | (1) | | | |
| 47 | 0 | 0 | 1 | 0 | 1 | 1 | 1 | 1 | 0000 | 0100 | 0111 |
| 93 | 0 | 1 | 0 | 1 | 1 | 1 | 0 | 1 | 0000 | 1001 | 0011 |
| 152 | 1 | 0 | 0 | 1 | 1 | 0 | 0 | 0 | 0001 | 0101 | 0010 |
| 231 | 1 | 1 | 1 | 0 | 0 | 1 | 1 | 1 | 0010 | 0011 | 0001 |

and to have one decimal digit of output. Integrated circuits that provide up to eight decades of counting are available in a single package. Internally these packages consist of a number of decade counters connected in series.

## Latch, Decoder, and Display

To provide a decimal readout of the accumulated counts a latch, a decoder, and a decimal display are necessary for a single decade of counting (fig. 4-11).

The latch stores the result of the completed count so that it can be displayed while the counting circuits are reset to zero and the next count is being accumulated. At the end of each counting cycle the new result is transferred to the latch where it is stored to update the count. If the latch is continuously enabled the actual counting sequence is observed instead of just the results of completed counts.

The **decoder** converts the BCD output of the latch to signals that drive the display. Most modern displays are seven-segment light-emitting diodes (LED) or liquid crystal displays (LCD). The decoder therefore functions to convert the BCD signal to signals that light the appropriate segments as shown in figure 4-11 for the decimal numeral 9.

## 4-3    Time and Frequency Measurements

The addition of an internal time base or clock to the general counting system of figure 4-3 allows use of time as one of the parameters in the ratio $N/B$. In this section we shall show how the time between two events, the period of a periodic waveform, and the frequency of a periodic waveform can be measured. By the addition of a few circuits the general counting system can be made to cycle automatically for repetitive measurements.

### Clock

The internal **time base,** or **clock** is derived from a highly precise **crystal oscillator** that has a basic oscillation frequency of 1 MHz or more. In order to obtain a wide range of frequencies from the basic oscillator a **decade frequency divider**, or **scaler**, is used as shown in figure 4-12. Here a selector switch is used to choose the desired clock output frequency. The accuracy of the selected time base is as good as the accuracy of the basic crystal oscillator (often better than one part per million).

### Frequency Meter

Measurement of frequency involves counting the number of input events or cycles per unit time. Hence time is the boundary condition for the measurement, and the time base is connected to the counting gate start and stop

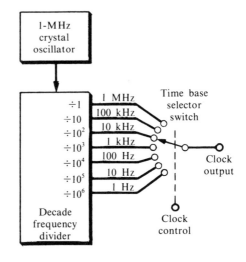

**Fig. 4-12.** Precision clock. With a six-decade divider, seven time base periods from 1 μs (1 MHz) to 1 s (1 Hz) are available from a 1-MHz oscillator.

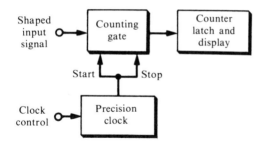

**Fig. 4-13.** Frequency meter. The counting gate is open for exactly one cycle of the clock. The signal cycles that occur in this time unit are counted.

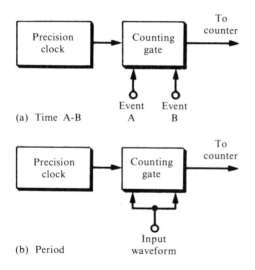

**Fig. 4-14.** Time and period measurements. The counter counts the number of time units from the clock that occur (a) in the time between events A and B or (b) between successive cycles of the input signal.

**Note 4-6.   Convention for Logic-Level Inputs and Outputs.**

The word $\overline{\text{gate}}$ is read "not gate." This implies that the gate is activated on a 0 logic level. An input or output symbol without the bar above it is assumed to be active on a 1 logic level.

inputs as shown in figure 4-13. The start and stop inputs are connected together so that counts are accumulated for one complete period of the clock.

Since time and frequency measurements have an inherent uncertainty of $\pm 1$ count, it is desirable to count as many input cycles as possible without causing the counter to overflow. This is done by changing the time base to allow the accumulation of more counts.

For low-frequency signals ($< 10$ Hz) it would take an inordinately long time to accumulate enough counts for high precision. For a 10-Hz signal the accumulation of 1000 counts (0.1% precision) would require 100 s. For these low-frequency signals the period mode, as described below, can provide higher resolution in a shorter measurement time.

## Time and Period Modes

In time-interval and period measurements the desired units are time per event or time per cycle. Hence pulses from the precision clock are counted for the boundary conditions determined by the input event(s) or waveform. The connections to the counting gate are shown in figure 4-14a and b for time between two events and period, respectively. In the time between events or time interval mode, separate start and stop signals from the events determine the boundary conditions over which clock pulses are counted. In the period mode clock pulses are counted for one complete cycle of the input waveform.

For low-frequency signals the period mode provides higher resolution in a shorter time than does the frequency mode. For example, the period of a 10-Hz input signal can be obtained in 0.1 s. If the clock is set at 10 kHz, 1000 counts would be accumulated in this period for 0.1% precision.

## Automatic Recycling Counter

An **automatic recycling counter** allows for repetitive measurements and thus for regularly updated count information. To accomplish this the following sequence is necessary: (1) the gate is opened and closed to allow the input events to be counted; (2) the information from the completed count is transferred to the latch and displayed; (3) the counter is reset to zero (cleared), and the gate start and stop controls are reactivated to allow another measurement cycle.

A block diagram of an automatic recycling counter is shown in figure 4-15. An internal sequencer generates pulses to operate the gate, the latch, and the counter reset. The internal sequence begins when a rising edge (LO→HI transition) occurs at the start input. This causes a HI→LO transition to occur at the gate control input marked $\overline{\text{Gate}}$ (see note 4-6), and the

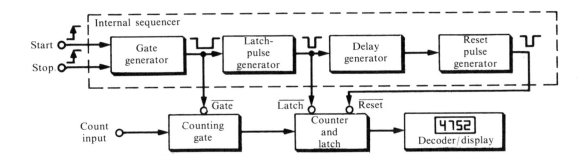

**Fig. 4-15.** Automatic recycling counter. The internal sequencer triggers the latch and reset operations automatically following each closure of the gate. The latch holds the previous count for display during the next gate period.

gate opens. The counting interval stops when a LO→HI transition appears at the stop input. This generates a LO→HI edge at $\overline{Gate}$ and triggers the latch-pulse generator. The latch pulse, a momentary LO of a few microseconds, causes the accumulated count to be stored and displayed. When the latch pulse returns to HI, the counter can be reset without destroying the contents of the latch. The rising edge of the latch pulse triggers a short delay generator which, in turn, triggers the reset-pulse generator to reset the counter. When the counter is reset, the gate generator is rearmed so that another cycle can begin. The entire sequencer, counter, and latch are often present in a single integrated-circuit.

## 4-4  Energy Conversion Transducers

Input transducers can be classified in several ways. For example, one possible classification scheme is to group transducers according to the chemical or physical information being converted to an electrical signal. In this scheme light-input transducers would be grouped together, temperature transducers together, and so on. Another classification scheme is to group transducers according to the electrical data domain at the output—voltage transducers, current transducers, etc. Unfortunately, neither of these popular classification methods gives much insight into the principles by which a transducer functions. Hence in the next three sections of this chapter we will group together transducers that operate by similar physical principles.

We begin with a consideration of energy conversion transducers, devices that convert nonelectrical energy directly to electrical energy. Since these devices generate electrical energy, they require no external power source. An energy conversion converter can produce an output voltage or current related to the phenomenon of interest. Linearity of the transfer function can be optimized by operating in an open-circuit mode (voltage) or a short-circuit mode (current). On the other hand, many of the devices can be used as power sources, and the power output can be optimized by a different

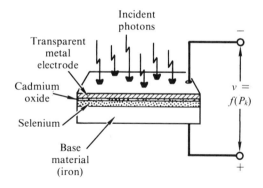

**Fig. 4-16.** Photovoltaic cell. The cadmium oxide insulation layer is a barrier layer across which a potential difference develops when photons are incident on the device.

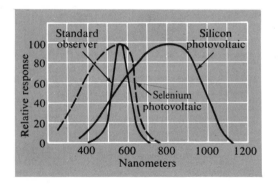

**Fig. 4-17.** Spectral response curves for selenium and silicon photovoltaic cells and for a human eye.

combination. Among the energy conversion transducers are photovoltaic cells, thermocouples, velocity transducers, Hall effect transducers, and several other important devices.

## Photovoltaic Cells

The **photovoltaic cell** or **barrier-layer cell** is a transducer that converts radiant energy (light) into electrical energy. When input photons are incident on the junction of certain dissimilar materials, the energy in the radiation can displace charge carriers and produce a voltage across the junction. A typical photovoltaic cell is illustrated in figure 4-16. The potential difference $v$ is a function of the incident photon flux $P_k$. In many cases the open-circuit voltage as a function of light intensity rapidly saturates, and the short-circuit current shows much better linearity.

Selenium cells are sensitive to radiation in the spectral range 300–700 nm, with maximum sensitivity at about 560 nm (fig. 4-17). It is therefore easy to combine the selenium cell with a filter in such a way that the spectral response of the combination is similar to that of a human eye. Selenium cells are especially useful in camera exposure meters and simple colorimeters.

Another prominent type of photovoltaic cell is made with a silicon semiconductor. Since it can provide relatively large amounts of current, the silicon cell can be used as a power source or solar battery as described in chapter 3.

## Thermocouple

The **thermocouple** is a heat-energy to electrical-energy converter that is widely applicable in temperature measurements. When two dissimilar metals are joined together as shown in figure 4-18a, the voltage $v$ developed between the open ends is a function of the temperature difference between the junctions. For some metal pairs, this thermoelectric effect provides a reproducible relationship between the voltage $v$ and the temperature difference between the two junctions ($T_u - T_r$). The transfer function for a chromel/alumel thermocouple, given graphically in figure 4-18b, can be expressed by the equation

$$v = AT_u + 1/2 \, BT_u^2 + 1/3 \, CT_u^3$$

when $T_r$ is 0°C. Since the coefficients $B$ and $C$ are small in most cases, the transfer function for the thermocouple can be approximated by the simple linear equation $v = A(T_u - T_r)$. The coefficient is approximately $4 \times 10^{-6}$ V/°C for the chromel/alumel thermocouple. Other combinations of materials have considerably different coefficients as shown in figure 4-19. Combinations of thermocouples, called thermopiles, can be used as power sources.

Since the output voltages of thermocouples are relatively small (in the microvolt to millivolt range), they are often either amplified before the voltage measurement step or measured by the null comparison technique with a potentiometer.

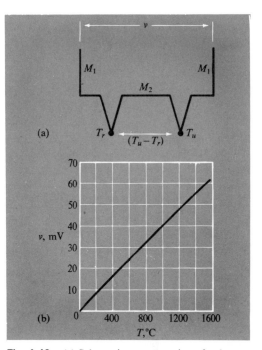

**Fig. 4-18.**   (a) Schematic representation of a thermocouple and (b) graph of the transfer function. To measure the temperature $T_u$, the second junction must be at a known (reference) temperature, $T_r$.

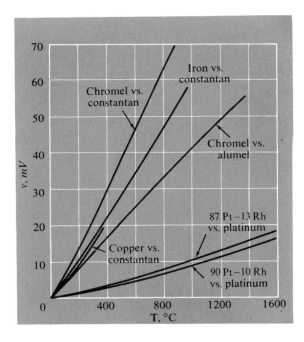

**Fig. 4-19.**   Graph of the transfer functions for several common thermocouples. The voltage for a given temperature difference depends on the composition and the physical treatment of the materials used to make the thermocouple.

## Electromagnetic Transducers

If a conductor is moved in a magnetic field, the voltage induced in the conductor is proportional to the rate at which the conductor traverses the magnetic field lines. This is the basis of the electromagnetic voltage generator as well as of the linear- and angular-velocity transducers described below.

**Linear-velocity transducers.**   A simple form of an electromagnetic linear-velocity transducer is shown in figure 4-20. The induced voltage is proportional to the velocity at which the magnetic core moves in or out of the coil. The object whose velocity is to be measured is attached to the permanent magnet. Alternatively the magnet may be held stationary, and the coil attached to the object.

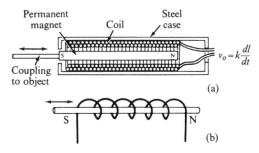

**Fig. 4-20.**   An electromagnetic velocity transducer, (a) pictorial and (b) schematic representation. The motion of the magnetic field relative to the coil induces a voltage in the coil proportional to the rate of motion.

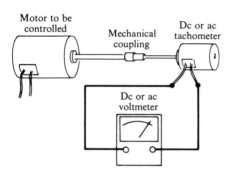

**Fig. 4-21.** Application of a tachometer generator to measuring motor speed. The power to the motor can be adjusted to maintain a constant speed regardless of load.

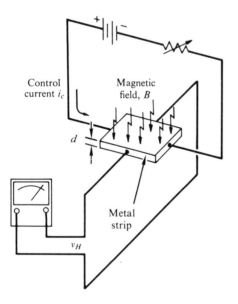

**Fig. 4-22.** Hall effect transducer. The Hall voltage is $v_H = ki_c B/d$, where $B$ is the field strength (gauss), $d$ is the thickness of the strip in centimeters, and $k$ is the Hall coefficient.

**Angular-velocity transducers.** Devices that are designed to determine the rotation rate of mechanical systems by measuring angular velocity are called **tachometers**. The most common types of tachometers are the ac induction tachometer and the dc tachometer. The ac tachometer is made of two sets of coils separated by a rotating cylinder of nonferrous metal. The sets of coils are arranged at right angles to each other so that there is no inductive coupling between them. When the rotating cylinder is at rest, an ac excitation current applied to one set of coils induces no voltage in the other set. The rotation of the metallic cylinder in the magnetic field produced by the excitation coils produces local current loops in the cylinder that induce a voltage in the output coils. The output voltage is of the same frequency as the excitation signal, and its amplitude is proportional to the rate of rotation of the cylinder. The direction of rotation determines whether the output signal is in phase with or 180° out of phase with the excitation signal.

The dc tachometer is a small dc generator. A coil or set of coils is rotated in a steady magnetic field to produce a voltage in the coils as the magnetic field lines are traversed. Wiping contacts (commutators) are used to connect each coil to the output as it moves through the maximum field density. The output voltage has an ac component of some multiple of the rotation frequency, but the average dc level is directly proportional to the rotation velocity. The output polarity depends on the direction of rotation.

An application of a tachometer is illustrated in figure 4-21. The tachometer output voltage can be compared with a reference voltage equal to the tachometer output at the desired speed. The voltage difference is an error signal that can be used to actuate a controller to restore the speed to the desired value.

## Hall Effect Transducers

An important type of interaction known as the **Hall effect** occurs when a current-carrying conductor is introduced into a magnetic field. When a metal strip is fixed in position with its plane perpendicular to a magnetic field and a control current $i_c$ is applied in one direction through the strip (fig. 4-22), then a potential difference called the Hall voltage is developed across the strip at right angles to the current direction and to the magnetic field.

The Hall coefficient $k$ depends on the material and the temperature. It is very large for n-type germanium and for indium arsenide or antimonide, and it is quite small for most metals other than silicon, bismuth, and tellurium. Therefore, most devices are made with n-type semiconductors. The Hall voltage is typically in the range of millivolts per kilogauss at the rated control current. Although the Hall voltage equation indicates that $v_H$ can be increased by increasing the control current $i_c$ or by decreasing the thickness $d$, either change would cause the strip to become hotter and change its characteristics. The internal resistances of typical Hall devices (generators) vary from a few ohms to several hundred ohms.

Since a change in magnetic field strength $B$ causes a proportional change in the Hall voltage $v_H$, Hall effect devices can be used in several configurations to transduce the strength or the position of a magnet into a related electrical voltage.

## Other Energy Conversion Devices

Several other important transducers operate as energy conversion devices. The electrochemical cell converts chemical energy into electrical energy and is the basis for chemical concentration transducers such as ion-selective electrodes and chemical power sources. The piezoelectric effect is used to convert mechanical energy into electrical energy and is the basis of biomedical transducers, phonograph pickups, crystal microphones, and industrial pressure transducers.

## 4-5  Resistive Transducers

Another class of transducers is based upon changes in the resistance of a device as a function of a physical parameter. The resistance of a device is given by $R = \rho l / a$, where $\rho$ is the resistivity and $l$ and $a$ are the length and cross section respectively. Changes in resistance can be effected by changing the dimensions $l$ and $a$ or by changing the concentration of charge carriers, which changes the resistivity $\rho$. The important strain gauge is based upon dimensional changes of a resistor. The thermistor and photoconductive cell are based upon the dependence of charge-carrier concentration in a semiconductor on temperature and light intensity, respectively. In this section these important transducers are described after a brief review of the conductance of materials. The Wheatstone bridge for resistance measurements by the null comparison technique is also introduced.

## Conductance of Materials

The outermost and most easily removed electrons of an atom are frequently called the valence electrons. The interaction of the valence electrons of atoms is primarily responsible for the formation of crystals and molecules. When similar atoms combine to form a crystalline solid, the discrete energy levels of the valence electrons broaden into bands of allowed energy levels. The electrons may have any energy within the band limits. One such band is the **valence band** or **bonding band**. In this band atoms share electrons in covalent bonds. If there is an odd number of electrons in each atom and there are two available energy states per atom in each band, one of the energy bands will

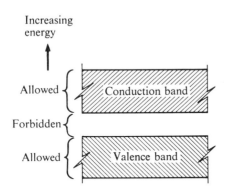

**Fig. 4-23.** Electron energy bands in a solid. If an energy band is only partially filled, the electrons in it are mobile charge carriers.

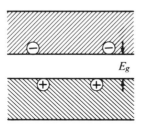

**Fig. 4-24.** Band structure of a semiconductor or insulator with valence electrons excited to the conduction band. Both the valence and conduction bands are now partially filled, and conduction can occur.

be only half filled. The unoccupied energy states in the energy band allow the electrons in that band to be accelerated by an electric field. Thus electrons in a partially filled band are conductive electrons, and a partially filled band is a **conduction band**.

In materials with an even number of electrons in the lowest energy state, the valence bands are filled, and therefore the electrons in them are not free to move. Any allowed energy band higher than the highest valence band will be empty. This is illustrated in figure 4-23. If some of the valence electrons can be promoted into the next higher band, they will be free to move in this partially filled band. The next higher band is the conduction band in this case. The magnitude of the energy gap between the filled valence band and the conduction band determines the conductance type of a given material: If the energy gap is zero, the material is a metallic conductor; if the gap is small, it is a semiconductor; and if the gap is large, it is an insulator.

## Semiconductors

A material with an even number of valence electrons and a forbidden energy gap between the valence and conduction bands cannot conduct electricity at $0°K$. At absolute zero temperature the valence band is filled, and the conduction band is empty. A source of energy is required to promote valence electrons to the conduction band. This energy might be provided by quanta of radiation (photons) as in photoconductors or by thermal excitation at higher temperatures as in thermistors. An energy at least equal to the forbidden gap is needed for excitation to conduction. The result of promoting an occasional valence electron to the conduction band is shown in figure 4-24. For each electron in the conduction band, there is a hole, a lack of one electron, in the valence band. It must be remembered that although the conduction band electrons are free to move throughout the material, a valence electron can move only to an adjacent unoccupied site, a hole, thus creating a hole where it was. The area of a missing valence electron is a localized but movable region of positive charge.

The number of electrons per unit volume of material that are in the conduction band at any given temperature depends on the magnitude of the energy gap. The number of conducting electrons per volume is greater for materials with lower forbidden gap energies $E_g$ and increases with increasing temperature in any material. For some semiconductors the conductivity increases approximately exponentially with increasing temperature. This thermal behavior is the basis of the temperature-dependent resistor (thermistor).

A semiconductor that conducts by equal numbers of holes and electrons created by the excitation of valence electrons is called an **intrinsic semiconductor**. All pure semiconductors and insulators are intrinsic semiconductors. Whether a material is a semiconductor or an insulator depends on the magnitude of the energy gap $E_g$ and on the operating temperatures or energy of excitation. At low enough exciting energies all such materials are insulators, and at high enough exciting energies all such materials are conductors.

## Thermistor

The **thermistor** is made from an intrinsic semiconductor. As we have seen, the concentration of charge carriers in intrinsic semiconductors increases rapidly with increasing temperature. Thus the resistance of the thermistor decreases quickly as the temperature increases. A typical resistance vs. temperature curve is shown in figure 4-25 together with some typical configurations of thermistor probes. The relatively large resistance change per degree makes the thermistor a useful device in temperature measurement or in control devices requiring high accuracy or resolution. Temperature changes as small as $5 \times 10^{-4}\,°C$ can be detected. Thermistors are used in the range of $-100°C$ to $+300°C$. Any current in a resistance does work that appears as heat. Care must therefore be taken that the amount of heat produced by the current in the thermistor does not affect the temperature of the measured system.

Thermistor resistances can be measured in several ways. Differential temperature measurements are most often made with the Wheatstone bridge discussed later in this section.

**Fig. 4-25.** (a) Several practical thermistor probes. (b) The resistance-temperature response of a typical thermistor material compared with that of platinum.

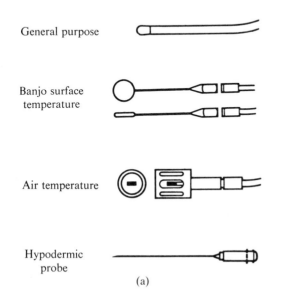

(a)

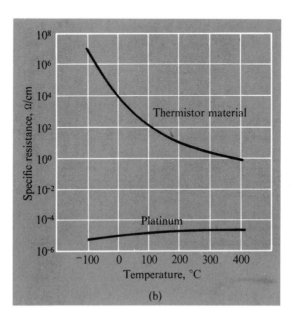

(b)

## Photoconductive Cell

When light falls on a semiconductor, the photon energy may be sufficient to move valence electrons into the conduction band. The resulting increased density of charge carriers due to the light makes the material more conducting. Thus the resistivity decreases as the light intensity increases. The semiconducting sulfide, selenide, and telluride salts of cadmium are most often used for this purpose. Some practical **photoconductive cells** and the spectral response curves of such devices are shown in figure 4-26. In order for each material to photoconduct, the incident light must have a short enough wavelength (high enough energy) for the photons to promote electrons from the valence to conduction states. As the energy of the light increases beyond the minimum required for promotion of valence electrons to the conduction band, the absorption of the light by the material increases rapidly. This results in a marked decrease in the fraction of the light that penetrates to the active region of the device.

## Strain Gauge

If a resistor made of a fine wire is distorted, its resistance changes as a result of dimensional changes. Its resistivity may also change. The result is a resistor for which the resistance is related to the strain. Such a device is called a **strain gauge**. Strain gauges have extensive applications in mechanical and biological measurements. Often the fine wire is bonded to a flexible insulating substrate in such a way that the wire dimensions change as the substrate

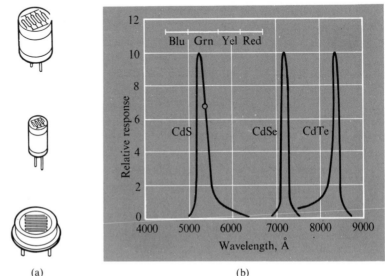

**Fig. 4-26.**  (a) Practical photoconductive cells and (b) the photoconductivity of CdS, CdSe, and CdTe vs. wavelength.

(a)

(b)

is flexed. Some strain gauges are made by depositing a pattern of thin metal film on the substrate material. Several practical strain gauges are shown in figure 4-27. Since the resistance change with strain is a very small fraction of the total resistance, highly sensitive resistance measuring techniques are required. Another difficulty is separating the changes in resistance due to temperature from those attributable to strain. The Wheatstone bridge described below can be used with dual strain gauges kept at the same temperature to minimize the effect of temperature change.

**Fig. 4-27.** Practical strain gauges. Various patterns are used to enhance sensitivity for particular directions of strain or to allow the effects of less desired strains to be balanced out.

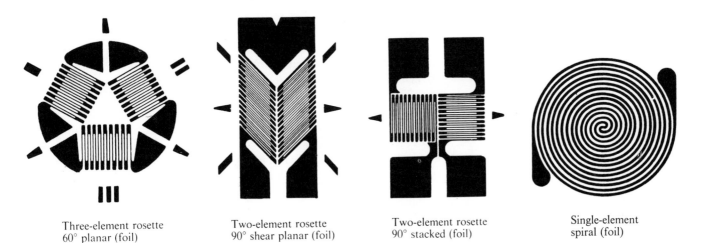

| Three-element rosette 60° planar (foil) | Two-element rosette 90° shear planar (foil) | Two-element rosette 90° stacked (foil) | Single-element spiral (foil) |

## Other Resistive Transducers

Several other resistive transducers are in common usage. The platinum resistance thermometer is an extremely accurate, sensitive, and stable transducer with a resistance that increases with temperature (see fig. 4-25) over the range of –270°C to 1100°C. Linear and angular displacement transducers can be based on the change in some characteristic of an adjustable circuit element such as a potentiometer. The device whose displacement is being measured is connected to the movable contact (wiper) of a potentiometer, and the effective length of the resistance element (proportional to resistance) is changed in proportion to the displacement.

## Wheatstone Bridge

The most direct method of comparing unknown resistances against standard resistances is with a **Wheatstone bridge** (fig. 4-28). In the null operation of the bridge, resistors $R_A$, $R_B$, and $R$ are standards used to determine the unknown resistance $R_u$. Resistance $R$ is made variable and is adjusted until

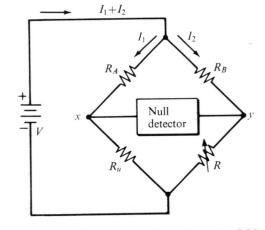

**Fig. 4-28.** Wheatstone bridge. The value of $R$ is adjusted to make the voltage at $x$ and $y$ equal as shown by the null detector. At balance, $R_u = R(R_A/R_B)$.

**Note 4-7.   Linearity of Off-Balance Bridge.**
The output voltage $v_o$ of the bridge is the voltage between points $x$ and $y$ of figure 4-28 and is given by

$$v_o = v_x - v_y = V \left( \frac{R_u}{R_A + R_u} - \frac{R}{R_B + R} \right)$$

*For $v_o$ to be a linear function of $R_u$, the slope $dv_o/dR_u$ must be a constant. Since*

$$\frac{dv_o}{dR_u} = V \left[ \frac{(R_A + R_u) - R_u}{(R_A + R_u)^2} \right] = V \left[ \frac{R_A}{(R_A + R_u)^2} \right]$$

the output will be linear if $V[R_A/(R_A + R_u)^2]$ is a constant. This occurs if $R_A \gg R_u$. Thus,

$$\frac{dv_o}{dR_u} \simeq V \frac{R_A}{R_A{}^2} \simeq \frac{V}{R_A} \text{ (for } R_A \gg R_u)$$

The condition of maximum sensitivity $(dv_o/dR_u)_{max}$, occurs when $R_u = R_A$.

the null detector indicates that the bridge is balanced. At balance there is no current through the null detector and, therefore, the current through $R_A$ and $R_u$ is $I_1$ and the current through $R_B$ and $R$ is $I_2$. Also at balance the voltage between terminals $x$ and $y$ is zero. From the equality of $IR$ drops on each side of the bridge, $I_1 R_u = I_2 R$ and $I_1 R_A = I_2 R_B$. These equations can be solved for $R_u$ to give $R_u = R(R_A/R_B)$.

The unknown resistance $R_u$ is calculated from the values of $R$, $R_A$ and $R_B$. Usually the ratio $R_A/R_B$ is a convenient factor such as $0.01, 0.1, 1, 10,$ or $100$ and is referred to as the *multiplier*.

The major sources of error in a Wheatstone bridge are inaccuracies in the three standard resistors $R$, $R_A$, and $R_B$, but these resistors can be made with errors as low as $0.001\%$. Other factors that could limit the accuracy are null point accuracy, thermal voltages, and changes in resistance values due to heating by too high currents.

The magnitude of the off-balance indication of the null detector is sometimes used as a measure of the change in the resistance $R_u$ (see note 4-7). In this application $R_u$ is usually a resistive transducer. When matched devices are used for $R$ and $R_u$ or for $R_A$ and $R_u$, the off-balance output obtained is related to the *difference* in the resistance of the two devices. A differential resistance measurement with matched transducers tends to cancel the effects of resistance changes caused by environmental effects other than the quantity sought. For example, with strain gauges it is desirable to compensate for the temperature sensitivity of the gauge. If $R_u$ and $R$ are matched strain gauges subjected to the same temperature but different strains, the off-balance output voltage of the bridge will be due only to the difference in strains.

## 4-6   Limiting Current Transducers

The resistive transducers described in the preceding section are all examples of ohmic devices. That is, they obey Ohm's law. There is another class of transducers in which Ohm's law is not obeyed. In these devices the current is not proportional to the voltage but is instead independent of the applied voltage over a relatively wide range of values. These devices are operated in this current-limiting region. In **limiting current transducers** charge carriers arrive at electrode terminals at a rate that is determined by the quantity or phenomenon being measured. The important limiting current transducers discussed in this section include the photodiode, the flame ionization detector, the oxygen electrode, the vacuum phototube, and the photomultiplier tube.

### Photodiode

As we discussed in chapter 3, the current through a reverse-biased pn semiconductor junction arises from the energetic creation of electron-hole pairs

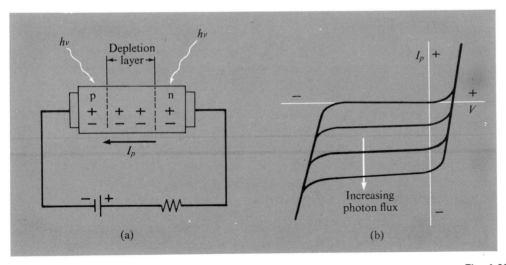

(a)

(b)

in the depletion region at the junction. Like thermal generation of electron-hole pairs, electromagnetic radiation of the proper energy (wavelength) can be absorbed by the material in the depletion region with the creation of electron-hole pairs as illustrated in figure 4-29. If the rate of charge-carrier production by the electromagnetic radiation greatly exceeds the rate of thermal generation, the reverse bias current is directly proportional to the photon flux (light intensity). Under limiting current conditions, the photo-current $I_p$ is given by $-I_p = b_\lambda Q_e P_k$, where $P_k$ is the photon arrival rate (flux), $b_\lambda$ is the quantum efficiency of the material (number of emitted photoelectrons per incident photon), and $Q_e$ is the charge on the electron.

The **photodiode** is an extremely fast responding transducer. Transit times of charge carriers across the junction are often in the sub-nanosecond range. The photodiode is therefore often used for the detection of extremely rapid, high-intensity events such as pulsed laser outputs. Because photodiode output currents are usually in the microampere range, photodiode-amplifier combinations are manufactured in a single unit to preserve the fast response time. Phototransistors, which have internal gain, are discussed in chapter 6, as are several other light-activated switches. In recent years integrated circuit packages containing up to 1024 individual photodiodes (photodiode arrays) have become available. These arrays have found uses as multiwavelength spectroscopic detectors, optical scanners, optical positioners, and as the detection element of solid-state television cameras.

## Flame Ionization Detector

If two metal plates are placed opposite each other and just above a small hydrogen-air flame, as shown in figure 4-30, and if molecules entering the

**Fig. 4-29.** The photodiode. (a) Pictorial representation showing electron-hole pair production; (b) current-voltage characteristics for different photon fluxes. The reverse bias applied in the limiting region is sufficient to collect all the charge carriers generated before they can recombine in the depletion region. The current is then equal to the rate at which the charge carriers are produced by the photon flux.

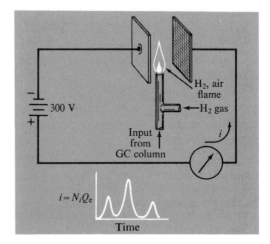

**Fig. 4-30.** Flame ionization detector for gas chromatography. The current $i$ is equal to the number of ions produced by flame ionization that reach the electrode plates each second $N_i$, times the unit charge per charge carrier $Q_e$. Typical currents measured with the flame ionization detector are from about $10^{-6}$ to $10^{-11}$ A.

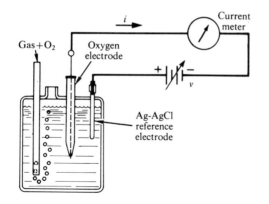

**Fig. 4-31.** Schematic for $O_2$ electrode. A platinum wire is sealed in glass to form a microelectrode that makes contact with the test solution. A silver-silver chloride electrode provides a constant-potential counter electrode. A variable voltage source and a current meter are connected in series with the test ($O_2$) and reference electrodes.

**Fig. 4-32.** (a) Current-voltage curves for $O_2$ electrode. (b) Output current vs. percent oxygen for $v = 0.7$ V. The current-voltage curves are obtained only in order to select an appropriate bias voltage ($\sim$0.7 V in this case). Thereafter, a fixed value of bias is applied, and the $O_2$ concentration is obtained from the measured current and the standard working curve (b).

flame are readily ionizable by the flame energy, then there will be a flow of charge carriers (both electrons and positive ions) between the electrodes. The rate of flow of electrons that occurs in the external circuit depends on the rate of formation of charge carriers by the flame, which in turn is determined by the concentration of ionizable molecules in the flame. The **flame ionization detector** is widely used in gas chromatography. The various constituents in a complex sample are separated as they pass through the chromatographic column, and as each exits from the end of the column, it passes into the flame ionization detector. The peak heights or peak areas of the current-time waveform are a quantitative measure of the concentration of each separated species.

## The Oxygen Electrode

The determination of oxygen concentration levels in lakes, rivers, and oceans, and in many laboratory solutions and biological systems is often based on an electrode system that is a limiting current transducer. The electrode and experimental systems are shown in figure 4-31. If the voltage from the source is varied between 0 and 1.0 V, the current-voltage ($i$-$v$) curves obtained (fig. 4-32a) have the voltage-independent plateau characteristic of limiting current devices. The height of the current plateau is a function of the oxygen concentration as illustrated by the $i$-$v$ curves. Therefore, the current-producing process must be related to the oxygen concentration.

When the platinum microelectrode is made about 0.6 to 0.9 V negative with respect to the reference electrode, the oxygen molecules ($O_2$) that reach the platinum surface immediately accept electrons and become reduced. The

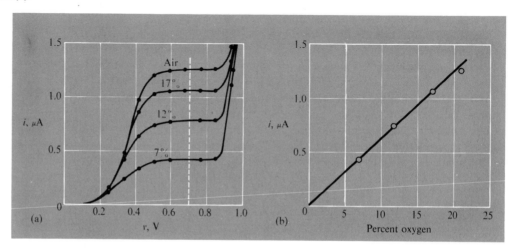

current in the external circuit is then equal to the rate of arrival of oxygen molecules at the platinum surface. This rate is limited by the rate of diffusion of $O_2$ molecules to the electrode surface and is directly proportional to the oxygen concentration as shown in figure 4-32b.

## Vacuum Phototube

The vacuum phototube contains a photosensitive cathode and an anode in an evacuated quartz or glass envelope. A typical **phototube** and its schematic symbol are shown in figure 4-33. Input radiation incident on the photocathode ejects photoelectrons with an efficiency $b_\lambda$ that depends on the energy of the photon ($h\nu$) and the type of cathode surface. If the anode is held at a sufficiently positive potential with respect to the cathode (fig. 4-34a), all the emitted photoelectrons are collected at the anode. The resulting photocurrent $i$ in the external load is shown in figure 4-34b. When the phototube is operated in the limiting current region, the output current is directly proportional to the light intensity (photons incident on the photocathode per second). The proportionally constant is nearly unaffected by the applied voltage.

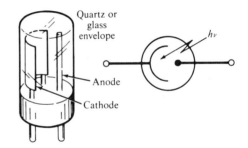

**Fig. 4-33.** Vacuum phototube in pictorial and schematic representations. Light striking the cathode causes the emission of photoelectrons, which are collected at the anode.

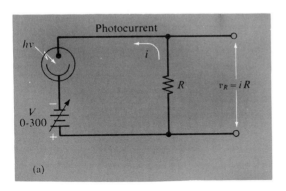

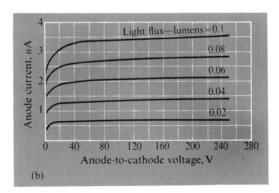

**Fig. 4-34.** Phototube circuit (a) and current-voltage curves (b). When $V$ is sufficient to collect all the photoelectrons, $i$ is proportional to the light flux and independent of $V$.

## Photomultiplier Tube

The photometric transducer commonly used in most types of spectrometers is the **photomultiplier tube (PM)**. It is also used in specialized rapid-scan, submicrosecond time-resolved, T-jump, and chromatogram scanning spectrometers, multichannel spark source direct readers, densitometers, and many other instruments.

The electrode arrangement of a typical PM tube is shown in figure 4-35a. Like the vacuum phototube the PM contains a photosensitive cathode and a

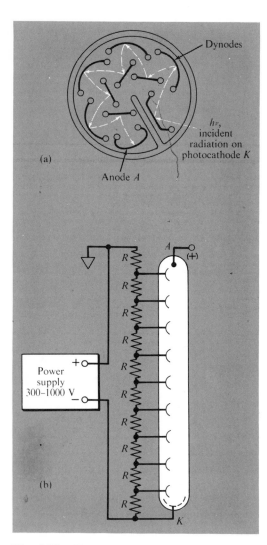

**Fig. 4-35.** Photomultiplier transducer. (a) Cross-sectional view of side-illuminated PM tube; (b) Schematic showing connection to power supply. The voltage divider provides the required successively higher voltage at each dynode.

collection anode. However, the cathode and anode in the PM tube are separated by several electrodes, called **dynodes**, that provide electron multiplication or gain.

Radiation incident on the photocathode gives rise to photoelectrons with an efficiency $b_\lambda$. Most of the photoelectrons emitted by the photocathode are accelerated toward the first dynode by its more positive voltage. Each successive dynode is maintained at a more positive voltage by the voltage divider shown in figure 4-35b. The fraction of photoelectrons collected by the first dynode is the collection efficiency $b_c$, typically about 75%. The energy imparted to the surface of the first dynode by each accelerated photoelectron causes several secondary electrons (e.g., 4 or 5) to be ejected from the dynode. The secondary electrons from the first dynode are attracted to the second dynode, where each electron ejects more secondary electrons, and so on from dynode to dynode until the electrons are collected at the anode. Thus, each photoelectron collected by the first dynode is greatly multiplied within the PM tube by the process of secondary electron emission to form a packet, or pulse, of a relatively large number of electrons at the anode. This pulse is typically only a few nanoseconds long.

The number of anode pulses $N_a$ over certain boundary conditions can be written

$$N_a = b_c b_\lambda N_k$$

where $N_k$ is the number of photons that strike the photocathode within the boundary conditions. Examples of boundary conditions might be a light burst from a spark discharge, or a shutter opening, or simply a unit time interval. If the anode pulses are sufficiently infrequent to avoid overlapping each other, $N_a$ can be determined by counting anode pulses with digital techniques. The coulombic content of the average anode pulse is $\overline{G}Q_e$ where $\overline{G}$ is the effective PM gain (electrons per anode pulse). The gain depends greatly on the power supply voltage, but it is typically $10^5$–$10^7$. For $\overline{G} = 10^6$, the average pulse is $1.6 \times 10^{-13}$ C which gives an average current of 32 $\mu$A over a 5-ns pulse. The counting of anode pulses is the measurement technique that is appropriately called **photon counting**. When used for photon counting, the output of the PM tube is in the digital domain for counts per event or in the frequency domain for counts per time interval. Photon counting measurements are described in chapter 14.

The PM tube is also used as an analog transducer with outputs in the current or charge domains. If the desired measurement is the total number of photons $N_k$ that strike the photocathode over a particular event, the output is the charge delivered to the anode over that event. Thus,

$$Q_a = N_a \overline{G} Q_e = N_k b_c b_\lambda \overline{G} Q_e$$

The desired measurement may instead be the light intensity or photon flux at the photocathode ($N_k$ per second). Now the desired output is the anode current

$$i_a = Q_a/t = (N_k/t) \, b_c b_\lambda \overline{G} Q_e$$

For example, if the photon flux at the photocathode is $10^7$ photons per second, and $b_c$, $b_\lambda$, and $\overline{G}$ are 0.9, 0.1, and $10^6$ respectively, the anode current is $i_a = (10^7)(0.9)(0.1)(10^6)(1.6 \times 10^{-18}) = 1.44 \; \mu\text{A}$. The PM tube owes its great popularity to its versatility as illustrated above, its great sensitivity, and its extremely wide ($10^6$ or greater) dynamic range.

## 4-7 Applications: Detecting Discrete Events

The input transducers discussed in the preceding sections produce a variation in one of the electrical quantities: voltage, current, power, or charge. We often consider such transducers to be analog output transducers. However, if the desired information is the time at which a variation occurs, the number of variations per unit time, the total accumulated number of pulse-shaped variations, or any other time or digitally encoded quantity, the same transducer may produce outputs in any of the electrical domains. It is thus necessary to know specifically which characteristic of the signal variation contains the desired information in order to process the transducer signal appropriately.

In this section a few specific examples illustrate the use of transducers for the detection of discrete events and show how various processing schemes give the desired information.

### Smoke Detector

The optical smoke detector is an excellent example of the use of an *analog* transducer to provide binary (yes or no) information. A typical reflective-type smoke detector is shown in figure 4-36. The light transducer (the photodiode) is the same transducer that would be used in an analog system to measure light intensity, but in this case the photodiode-discriminator system provides a true binary (HI→LO) output. Such considerations as the linearity of the photodiode transfer function are unimportant in this application. The photodiode-discriminator system must merely have enough sensitivity to detect a potentially dangerous level of smoke.

### Opto-Interrupter

A common discrete-events detector with a variety of applications is the optical interrupter or **opto-interrupter**. This device consists of a light source

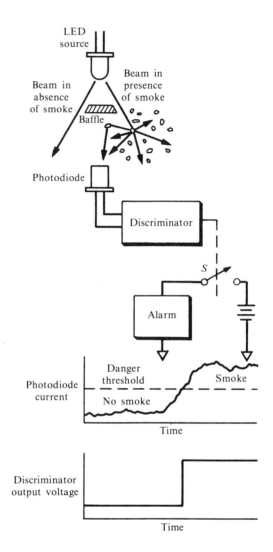

**Fig. 4-36.** Reflective smoke detector. In the absence of smoke the baffle prevents the LED output from striking the photodetector. Smoke, however, diffuses the beam so that light strikes the detector. A discriminator senses when the photodiode output current is above the danger level and closes switch $S$ to sound the alarm.

(usually a LED) and a photodetector (usually a photodiode or phototransistor) in a single package with an open space between them as shown in figure 4-37a. Figure 4-37b shows the opto-interrupter arranged to detect the rotation of a motor shaft. How the electrical signal is processed depends on the information desired. For example, the following questions might be answered from the detector output: (1) Is the motor turning? (2) What is the position of the motor? (3) How many revolutions are occurring per minute? (4) Is the motor speed uniform? To answer the first question the presence or absence of electrical pulses during some predetermined time interval can be used. The presence of pulses during a one-second interval would be sufficient to determine that the motor is on. Again the information is binary, but the discrimination is now in the time domain. Note that this is a more direct way to determine the motor status than a criterion such as whether or not the ac power to the motor is on, since the motor shaft might be frozen.

The motor position can be determined by connecting the opto-interrupter output (after shaping it into logic-level pulses) to the input of a counter and counting the total number of pulses that have occurred since the motor was at a known reference position. The position information is thus directly encoded in the digital domain. The position resolution could be increased by dividing the wheel into several alternating opaque and transparent sections.

The motor speed can be determined by connecting the shaped output of the opto-interrupter to the input of a frequency meter. Speed uniformity can be measured by comparing the results of successive frequency measurements. Note that in the case of motor position and speed, the measurement accuracy depends largely on the quality of the wheel (that is, on how well the wheel is divided into sections); in the case of on/off information, the exact division of the disk into opaque and transparent parts is relatively unimportant.

**Fig. 4-37.** Opto-interrupter. When an opaque object interrupts the light beam, the detector current decreases. The disc in arrangement (b) causes two pulses of detector current for each rotation of the motor shaft.

(a)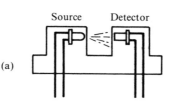

(b)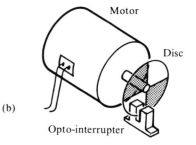

## Summary

The photoelectric detection examples in this section show that the same transducer can provide outputs in several different data domains. The characteristics required of the transducer are significantly influenced by the transducer output domain. In the analog mode the measurement accuracy depends directly upon the reproducibility of the transfer function (electrical output per measured quantity) over the entire measurement range. Frequent recalibrations may be required to compensate for measurement variations due to aging, temperature change, and so on. When used as an event detector with a pulse output, however, the detector system need only maintain the ability to sense the presence or absence of the event reliably. If the event causes a substantial change in the detector output signal, this requirement is easily maintained. Also, the pulse (HI→LO) output signal can be processed with high-reliability, low-cost digital circuitry. For these reasons an increasing number of clever measurement schemes that utilize analog transducers in a pulse output mode are appearing on the market.

## Suggested Experiments

**1. Frequency measurement.**
Connect the function generator (FG) output to both the frequency meter and the scope. Determine the resolution of the measurement every decade from 1 Hz to 1 MHz for both measurement techniques.

**2. Period measurement.**
With the frequency meter in the period mode, determine the resolution of the period measurement for FG frequencies in every decade from 1 Hz to 1 MHz. Determine the frequency of the minimum resolution measurement assuming either frequency or period could be measured.

**3. Digital time base.**
Observe the output of the digital clock circuit with the scope, the frequency meter, and the logic probe over the full range of clock frequencies available.

**4. Basic digital devices.**
Use switch-controlled logic sources and logic-level indicators to obtain the complete table of states for the AND, OR, and INVERT integrated circuit gates. Observe also the response of the *JK* flip-flop integrated circuit to logic signals at the clock, *J*, *K*, preset, and clear inputs. Determine the ratio of clock frequency to output frequency for the *JK* flip-flop in its toggling mode.

**5. Decade counter, latch, and display.**
Connect a decade counter IC, a quad latch IC, a BCD-to-7-segment decoder IC, and a 7-segment display to form a complete counter display circuit for one decade. Operate the counter, and observe the complete count sequence and the action of the latch enable input. Determine the ratio of the frequencies at the *A* input and the *D* output of the decade counter IC.

**6. Events counting.**
Connect the two inputs of a comparator to the FG and VRS sources. Determine what VRS threshold adjustments give reliable detection of the oscillations of the FG for a variety of FG waveshapes, frequencies, and offsets. Use the comparator output as the count input source for the frequency meter in the events-counting mode.

**7. Energy transducers.**
Measure the output voltage of selected energy transducers such as the photovoltaic cell and thermocouple while varying the measured quantity. Apply the output of each transducer to a discriminator, and set the threshold to obtain a greater-than/less-than signal for the measured quantity or alternatively to obtain an event detection based on the quantity.

**8. Resistive transducers.**
Repeat experiment 7 for resistance measurement of selected resistive transducers such as the thermistor and the strain gauge. An ohmmeter and/or a bridge circuit may be used in the measurement. The discriminator application can best be accomplished with the comparator acting as the null detector in a Wheatstone bridge.

**9. Limiting current transducers.**
Repeat experiment 7 for current measurement of selected limiting current transducers. Measure the current through the transducer by measuring the *IR* drop across a series resistor, but keep the *IR* drop small compared to the voltage across the transducer. Determine how independent the current is of the voltage applied for one or more of the transducers.

## Questions and Problems

**1.** Two different voltmeters were used to measure the voltage of a 1.018-V standard cell. The following values were obtained: Meter 1, 1.028, 1.010, 1.006, 1.010, 1.035; meter 2, 1.010, 1.014, 1.009, 1.015, 1.012. (a) Which meter gives the better precision? (b) Which meter is more accurate?

**2.** Give two characteristics of a difference detector and two characteristics of a reference standard that affect the accuracy of a null comparison measurement.

**3.** Measurement devices vary in the degree to which they depend on the difference detector and the reference standard outputs. Classify the following as direct measurement devices, null comparison devices, or partial null devices, and name the difference detector and the reference standard: (a) bathroom scale; (b) ruler for length measurement; (c) tuning fork for tuning a musical instrument; (d) beam balance postage scale.

**4.** How many counts must be accumulated in a counting measurement with asynchronous gating to achieve a precision of 0.05%?

**5.** Sketch a block diagram of the frequency divider circuits required to produce pulses every 0.1 s, every minute, and every hour for a digital clock that operates from the 60-Hz power line as the frequency standard.

**6.** (a) Calculate the relative error due to the ±1 count uncertainty when the frequency meter of figure 4-13 is used to measure the frequencies of 50-Hz, 3-kHz and 250-kHz signals. The clock frequency is exactly 1.000 Hz. (b) Similarly calculate the relative error in the period measurement of the same signals with a standard reference frequency of exactly 1 MHz.

**7.** A frequency/period meter has a 1.000-Hz time base in the frequency mode and a 1.000-MHz time base in the period mode. At what frequency do period measurements and frequency measurements show equal uncertainties? (Assume the time base to be exact.)

**8.** Write the decimal equivalent of the following binary numbers: a) 00101101; b) 11100101; c) 1000011010.

**9.** Express the following decimal numbers in binary and binary-coded decimal: a) 3122; b) 149; c) 215; d) 97; e) 1628. For each number compare the number of bits required in the two codes.

**10.** The OR gate of figure 4-6 is to be used to gate a pulse train to a counter. (a) If input $B$ is used as the gate control, for which state of $B$ is the gate open? (b) When the gate is closed, what is the output logic level?

**11.** By using decoder outputs other than those for the BCD code, a decimal display can produce several alphabetic characters. In the circuit of figure 4-11, which decoder outputs would have to be HI to produce the letters a, b, c, d, e, and f? Do not choose combinations that duplicate those of numerals. Which of the letters are upper case and which lower case?

**12.** (a) From the graph in figure 4-19 estimate the output voltage change per degree celsius for an iron/constantan thermocouple at ~400 °C. (b) If the voltmeter used to measure the thermocouple voltage can resolve 1 mV, what is the minimum temperature change that can be resolved?

**13.** A photomultiplier has an average gain of $5 \times 10^5$, a probability of 0.05 that incident photons will produce a dynode pulse, and an average output current of $1.50 \times 10^{-8}$A. What is the radiant flux at the photocathode in photons per second?

**14.** A photomultiplier tube with an average gain of $1 \times 10^6$ has a collection efficiency of 0.8 and a quantum efficiency of 0.1 at the wavelength of interest. If 5000 photons per second strike the photocathode, (a) what is the anode pulse rate, (b) what is the average anode current?

**15.** (a) What limits the current in a vacuum phototube? (b) Does a phototube under constant illumination follow Ohm's law?

**16.** The resistance of a thermistor is to be measured by applying 5.00 V to a circuit consisting of a thermistor and a current meter in series. A current of 0.99 mA is measured. (a) If the current meter resistance is neglected, what is the thermistor resistance? (b) If the meter is known to have a resistance of 5.0 kΩ on the 50-μA scale, what resistance does the meter have on the 1-mA scale used to measure the thermistor output? (c) What is the true thermistor resistance, and what relative error is caused by the 1-mA full scale meter?

**17.** It is desired to measure the resistance of a thermistor at a temperature near 25°C where the thermistor has a resistance of 10 kΩ. Measurement of the current through the thermistor and of the voltage drop across it are both considered. Only a 1.00-V source is available. Two meters are available: a current meter with a 50-μA basic sensitivity, 5-kΩ resistance, and a 100-μA scale; and a voltmeter with an input resistance of 1 MΩ on the 1-V scale. Which of the two meters will cause less loading error?

**18.** The Wheatstone bridge of figure 4-28 is balanced with $R_A = 909.09$ Ω, $R_B = 90.909$ Ω and the variable standard resistance $R = 628$ Ω. What is the unknown resistance $R_u$?

**19.** The Wheatstone bridge of figure 4-28 is used to measure resistance with $R_A = 30$ Ω, $R_B = 90$ Ω and a current meter as the null detector. If $R_u = 20$ Ω and the current meter resistance is 9Ω, what is the current through the ammeter when the bridge is unbalanced with the standard resistor $R = 10$ Ω? Use Thevenin's theorem.

**20.** Measurements of resistance changes in strain gauges are often made using the variation of the Wheatstone bridge shown in figure 4-38. Resistors $R_1$ and $R_2$ provide a multiplier as usual, and resistor $R_3$ may be only 1–10 Ω for high sensitivity. Resistor $R_u$ is the strain gauge. (a) Consider that $\Delta R_3$ is set to zero and the bridge is balanced with no strain on the gauge, so that $R_u = R_s$. Now consider the gauge to be strained so that $R_u = R_s + \Delta R_s$. Show that when the bridge is rebalanced

$$\Delta R_s = \Delta R_3 (R_2/R_1)$$

(b) The ratio between the relative resistance change and the relative strain change in the strain gauge is called the gauge factor $F$, which is defined as

$$F = \frac{\Delta R_s / R_s}{\Delta l / l}$$

where $l$ is the length of the wire and $\Delta l$ is the change in length upon strain. A 120-$\Omega$ strain gauge with $F = 2$ is used in an equal-arm bridge ($R_1 = R_2$) that is initially balanced with no strain and is rebalanced after straining. Compute the relative strain $\Delta l/l$ that is present if a change of 0.1 $\Omega$ in resistor $\Delta R_3$ is required to rebalance the bridge.

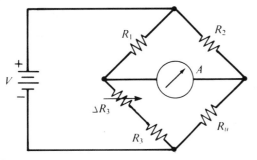

**Fig. 4-38.**

**21.** (a) Discuss whether resistance is an electrical domain or a physical domain. (b) Is capacitance an electrical domain? (c) Inductance?

**22.** If a second opto-interrupter is added to the disc of figure 4-37, it is possible to obtain information concerning the motor direction. Describe how the two opto-interrupter outputs give direction information.

**23.** Two photodiodes are used in the detection and measurement of a laser pulse. Photodiode 1 produces a single pulse for each laser flash and is used to obtain the total number of laser flashes. Photodiode 2 monitors the laser intensity as a function of time. What are the output domains of the two photodiodes? Why are they different?

**24.** The photodiode of figure 4-29 is used to measure light intensity. (a) If the quantum efficiency is 0.3 at the wavelength of interest and the incident light beam has a photon flux of $5 \times 10^{13}$ photons/s, what is the photocurrent $I_p$? (b) The photodiode is reverse biased by a 5.00 V battery and the current is to be calculated from the $IR$ drop across a load resistor. If a 100-k$\Omega$ load resistor is used, what would the $IR$ drop be in the presence of the photon flux given in (a)? (c) What percentage change occurs in the bias voltage in the presence of the above flux? (d) If the load resistor were 1.00 M$\Omega$, what percentage change would occur in the bias voltage?

# Operational Amplifiers and Servo Systems          Chapter 5

The key building block in analog electronic circuits is the operational amplifier. By itself, it is a conceptually simple device—a circuit that greatly amplifies the voltage difference between two input terminals. The operational amplifier and a few passive components can perform many analog functions or operations (hence its name) with great precision. Among these operations are: summing of currents or voltages, signal inversion, impedance buffering, linear amplification, integration, precision control of current or voltage, precision generation of waveforms, and logarithmic amplification. Over the past two decades technology has improved the performance of operational amplifiers by several orders of magnitude while their price has dropped 100-fold. This revolution in analog electronics has been partially overshadowed by the dramatic and concurrent developments in digital devices. However, the operational amplifier is the heart of the analog-to-digital and digital-to-analog converters that provide the interface between the analog world and digital processing and control systems. The operational amplifier often acts as the driving force in error correcting feedback systems (servo systems). Its effectiveness in the servo system configuration gives the operational amplifier high precision in measurement and control applications. In this chapter the operational amplifier is introduced, the servo system concept is described, and applications in the measurement of voltage, current, and charge are explored.

## 5-1  Servo Systems and Automated Null Measurements

The areas of measurement and control are closely related; the measurement of a quantity is prerequisite to its control. In this section we will also see how the control of a quantity can aid in its measurement. The concept of control of a quantity implies the monitoring of the quantity and the application of corrective action in the event the quantity deviates from the desired value. It is this kind of action that keeps the volume of the car radio constant as you drive through regions of varying signal strength and that is operative through you as you steer your car along the road.

## Servo Systems

Any system that detects a difference between the actual and the desired states of a controllable quantity (e.g., voltage, temperature, position) and then feeds the difference information back to a controlling device that causes the difference to become essentially zero falls under the general classification of a **servo system**. If the quantity to be controlled is the position of an object and the controlling device is an electric motor or an electrically controlled hydraulic system, the servo system is an **electromechanical servo system** or **servomechanism**. The position controlling devices in a servomechanism can be powerful enough to position heavy loads in response to command signals. Because of their inherent precision, servomechanisms are often used to position critical parts of delicate instruments in the precise location required for optimum instrument performance.

## Automated Null Measurement

In measurement applications servo systems are based upon the null comparison principle that was introduced in chapter 4. Recall that the unknown quantity $Q_u$ is compared with a reference standard quantity $Q_r$, which is adjusted until a difference detector indicates that $Q_u$ and $Q_r$ are essentially equal. A serious disadvantage of the null comparison measurement in its manual mode is the time and tedium required to adjust the reference standard to produce a true null difference output. But this, of course, is just the kind of task the servo system is conceived to do. A block diagram of the null comparison measurement automated by a servo system is shown in figure 5-1. An electrically activated system is set up to adjust the reference standard. The information from the null detector is used by the reference control system to bring the comparison closer to null. This process repeats until null is achieved. In a continuous null measurement system when the measured quantity changes, the reference readout automatically follows $Q_u$. The automated null measurement concept is developed in the context of precision

**Fig. 5-1.** Block diagram of an automated null comparison measurement system. The off-null signal from the null detector activates the reference standard control system to adjust $Q_r$ to reduce the off-null signal. The reference control system often operates the $Q_r$ indicator as well. At null the $Q_r$ readout displays the measured value of $Q_u$.

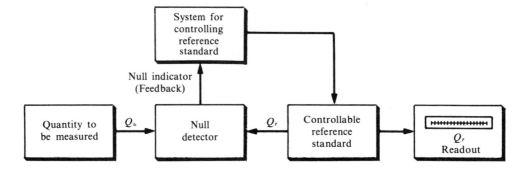

voltage measurement in the next section. In succeeding sections of this chapter, applications of the operational amplifier in precise and idealized measurements of voltage, current, and charge are explored.

## 5-2  Null Comparison Voltage Measurement

The measurement of voltage by a null comparison procedure is represented schematically in figure 5-2. An adjustable reference standard voltage $v_r$ is in opposition to the unknown voltage $v_u$. The reference standard can be adjusted in sufficiently small increments to make $v_r$ equal to $v_u$ within the desired measurement precision. A null detector is used to detect an off-null condition and to aid in adjusting $v_r$ to equal $v_u$. The null detector is a kind of difference detector except that it needs only to indicate the existence of a potential difference at its terminals. When a null measurement is complete, the magnitude of the difference is always zero to within the sensitivity of the null detector.

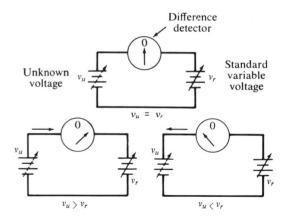

**Fig. 5-2.**  Null comparison voltage measurement. The null detector indicates the sign of the difference between $v_u$ and $v_r$. The value of $v_r$ is then changed in the appropriate direction to bring the difference to zero.

### Voltage- and Current-Sensitive Null Detectors

The difference in voltage between $v_r$ and $v_u$ can be detected by either a sensitive voltmeter with a high resistance or a sensitive current meter with a low resistance. Each has advantages in certain cases. If a null detector measures the current that results from the voltage imbalance between $v_u$ and $v_r$, it should have a low resistance. The sensitivity of a current-type null detector increases as the source resistance decreases, and vice versa. A low-resistance reference standard should be used with such detectors. Voltage null detectors are useful with high-resistance voltage sources as long as the resistance of the null detector is higher than the source resistance.

It should be emphasized that at null there is no measurement error due to loading the unknown or reference voltage sources with either a current-type or voltage-type detector. However, to locate the null point with a low-resistance current-sensitive null detector, the voltage sources must supply current when $v_r$ is off-null. Some voltage sources such as the glass pH electrode do not soon regain a stable output voltage after they have been loaded. Voltage-sensitive null detectors with high input resistance are required for such high-resistance or load-sensitive sources.

### Null Comparison Circuit Variations

The basic null comparison circuit of $v_u$, $v_r$ and the null detector in series is very simple. However, in the choice of devices for practical measurement systems, the placement of the system common is an important consideration (see note 5-1). The three possible locations for the system common connection are shown in figure 5-3. In the first location (figure 5-3a), both $v_u$ and $v_r$ are connected directly to the system common. The null detector connections

**Note 5-1.  Circuit Commons.**
Electronic circuits generally have a very low resistance conductor (a large copper wire or foil) to which one terminal of power supplies, signal sources, and many other circuit components are connected. This conductor, which provides a point of common potential, is frequently called **circuit common**. It is indicated in a circuit diagram by the triangular symbol (▽). Because it is a common point, the voltages at all other points in the circuit are generally measured with respect to circuit common.

When a circuit is linked to other circuits in a system, the commons of those circuits are often connected together to provide the same reference potential for the entire system. The circuit or system common may also be connected to the universal common, earth ground. When a circuit common is not connected to a system common or earth ground, it is said to be **floating**.

**Fig. 5-3.** Locations of system common in the null comparison circuit: (a) floating null detector; (b) floating reference source; (c) floating unknown source. One of the three circuit elements must have neither terminal connected to common. If one terminal of the null detector is common, the other terminal is at a voltage equal to the common when the system is at null.

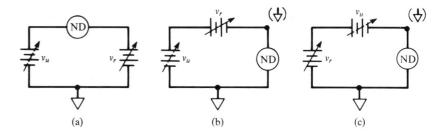

(a)          (b)          (c)

are at voltages $v_u$ and $v_r$ with respect to the common. In this case the null detector must have true **difference inputs** because a difference between $v_u$ and $v_r$ must be detected regardless of the magnitude of either. To accomplish this the null detector inputs must either be completely independent of the system common (floating) or be perfectly balanced with respect to the common point.

The variation shown in figure 5-3b places the common between the unknown source $v_u$ and the null detector. Now one connection to the null detector is always at the common voltage. When the system is at null ($v_u = v_r$), the other terminal of the null detector is also at the common voltage. Thus when null is detected, both null detector connections are at the common voltage. This eliminates any balance requirements of the detector input connections with respect to common. On the other hand, the reference voltage source is now floating; that is, it has no connection to common. Special care is required in the design of a reference source if its output voltage is to remain stable and accurate as the voltage between it and the system common changes. We seem to have simply traded one problem for another, but at null the point between the $v_r$ source and the null detector is at the common voltage. This point is called a **virtual common** (see note 5-2) because when the null condition is fulfilled, its voltage is the same as the common voltage. So even though the reference source is floating, one terminal is always at the common voltage when the reference setting is taken for the measurement value.

A third location for the system common is shown in figure 5-3c. This circuit is the same as that of figure 5-3b except that the position of $v_u$ and $v_r$ have been interchanged. Now one terminal of $v_r$ is connected to common, and the $v_u$ source is floating. The best position for the common connection depends upon which source can best tolerate being disconnected from the common. Neither circuit (fig. 5-3b or c) requires a floating null detector, and both establish a virtual common at the null detector.

## The Comparator Null Detector

For several decades the standard null detector for voltage comparison measurements was the laboratory galvanometer, a moving-coil meter that uses a

**Note 5-2.  Virtual Ground or Common.**
A virtual *common point* in a circuit is a point that is essentially at the circuit common potential but is not connected to the circuit common point. Since the term ground is often loosely used for circuit common even when the common is not connected to earth ground, the term virtual ground is frequently used to identify a point at the circuit common potential.

light beam for a pointer. The galvanometer, which is energized entirely by the input signal, requires no connections to the system common and thus can be used in any of the configurations shown in figure 5-3. More recently, it has become common to amplify the difference signal electronically to reduce the sensitivity requirement of the null indicator itself. When used with sensitive and stable high-gain amplifiers, ordinary meters, indicator lights, earphones, and the like become very effective null indicators.

An amplifier designed specifically for null comparison measurement is the **comparator**. The comparator is a very high-gain amplifier with well-balanced difference inputs and controlled output limits. The symbol for the comparator and a plot of its ideal and practical transfer functions are shown in figure 5-4. Ideally the comparator has just two output states: +limit and −limit. The state of the output indicates whether the condition $v_u < v_r$ or $v_u > v_r$ exists. The inputs are labeled + and − according to the sign of the output voltage when the signal at that input is the more positive. In a practical comparator the region of the transfer function at zero difference has a finite slope. This is the region where the magnitude of the difference times the gain of the amplifier is less than the +limit or the −limit. This slope approaches the vertical ideal as the comparator gain or the smallest difference of interest increases. For example, if the limits are ±12 V and the comparator gain is $10^6$, the output is at limit for any off-null deviation greater than 12 μV. The comparator input circuits are designed for high stability of the null point and good balance with respect to common. However, practical comparators are perfect in neither respect.

Inputs that are perfectly balanced with respect to the common voltage respond only to a difference in voltage. The comparator shown in figure 5-5a is in the floating null detector configuration (fig. 5-3a). To evaluate the balance of the comparator inputs, it is useful to consider that the sources $v_u$ and $v_r$ are a combination of a **difference voltage** $v_d = v_u - v_r$ and a **common mode voltage** $v_{cm}$. The common mode voltage is the average of the two input voltages $(v_u + v_r)/2$. The **difference gain** $A_d$ is the output voltage change $\Delta v_o$ for a given input voltage change $\Delta v_d$, or $A_d = \Delta v_o / \Delta v_d$. The **common mode gain** $A_{cm}$ is similarly $A_{cm} = \Delta v_o / \Delta v_{cm}$. Ideally the common mode response should be zero. The commonly used figure of merit for the quality of the input balance is the **common mode rejection ratio (CMRR)**, which is the ratio of the difference gain to the common mode gain. Therefore

$$\text{CMRR} = A_d / A_{cm} \qquad (5\text{-}1)$$

Of course, the better the balance, the higher the CMRR. A typical comparator CMRR is $10^4$ or more. If such a comparator is used in the floating null detector configuration (fig. 5-3a), the null indication for $v_u = v_r = 1$ V could be different from that for $v_u = v_r = 0$ V by 1 V/CMRR ≈ 100 μV.

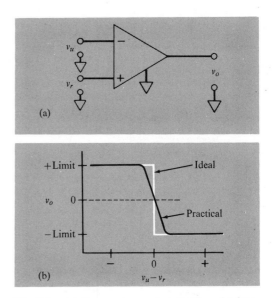

**Fig. 5-4.**  (a) Comparator symbol. (b) Transfer function. The comparator output is +limit or −limit depending on the sign of the difference between $v_u$ and $v_r$.

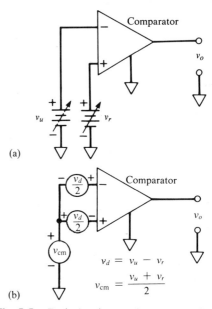

**Fig. 5-5.**  Equivalent input voltage sources that show the difference and common mode voltages. The signal $v_u$ is $v_{cm} + v_d/2$ and $v_r$ is $v_{cm} - v_d/2$. The ideal comparator response is independent of the value of $v_{cm}$.

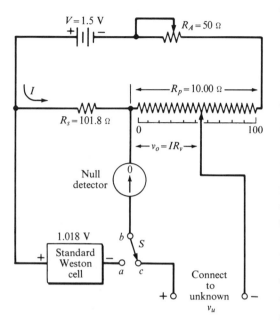

**Fig. 5-6.** Single slidewire potentiometer with span of 0 to 100.0 mV. For calibration switch $S$ is set to close contacts $a$-$b$, and resistance $R_A$ is adjusted until the null detector reads zero. This reading indicates that $IR_s = 1.018$ V. The current $I$ is 10.00 mA. To measure an unknown voltage $v_u$, switch $S$ is set to close contacts $b$-$c$, and the precision divider is then varied until $IR_v = v_u$ as indicated again by the null indication.

This error could be reduced by choosing a comparator with a higher CMRR, or it could be eliminated completely by choosing a configuration in which the null detector is not floating.

## The Laboratory Potentiometer

The standard laboratory potentiometer, the classic precision voltage measurement device, is a combination of a precision adjustable standard voltage source (literally, the potentiometer) and a sensitive null detector in a single instrument. It measures voltage by the null comparison technique described in this section. It also includes a provision for a null comparison standardization of the variable reference voltage source against a standard Weston cell. The circuit of a basic potentiometer with a voltage span of 100.0 mV is shown in figure 5-6.

## The Servomechanical Potentiometer and Recorder

The tedium required to balance manual potentiometers can be eliminated by the use of automated, self-balancing potentiometric systems. Automation is accomplished by combining an electronic amplifier null detector and a motor-driven slidewire as shown in figure 5-7. The motor-driven slidewire is placed in the bridge-type circuit shown. This modification of the basic divider circuit of figure 5-6 provides a control over the position of zero output on the slidewire scale. The output $v_r$ of the bridge circuit is zero volts when the slidewire contact is exactly opposite the zero adjust contact. Slidewire contact positions to the left of this point result in a positive $v_r$ and to the right, a negative $v_r$. The span is determined by the fraction of the bridge drive voltage $V$ that appears across the slidewire.

The output of the slidewire bridge is compared with the input signal voltage $v_u$. The difference is amplified by the comparator, which is usually a voltage amplifier with limited output current capacity. A power booster amplifier is used to provide power to the motor in response to the comparator amplifier output. If $v_r$ is greater than $v_u$, the amplified difference results in the application of power to the motor. The magnitude and polarity of this power causes the motor to move the slidewire contact to the right, thus decreasing $v_r$. This motion continues until the difference between $v_u$ and $v_r$ is too small to produce enough power to turn the motor shaft.

If the servo amplifier had the ideal response function shown in figure 5-4, the slightest difference voltage between $v_u$ and $v_r$ would provide a sufficient servomotor signal to correct the difference. However, there is always a minimum null detector input signal that will cause servomotor rotation, and there is also a small zero drift or input imbalance in the null detector. The combination of these effects results in a small uncertainty, or error, $v_s$,

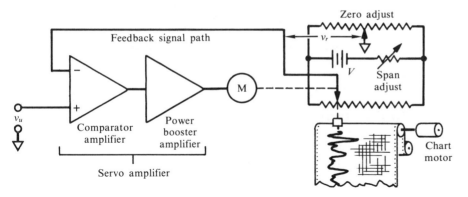

**Fig. 5-7.** Servomechanical potentiometer and recorder. The slidewire bridge circuit generates a voltage $v_r$ that is directly related to the slidewire pen position. Any difference between $v_r$ and $v_u$ is amplified to produce a signal that causes the motor to move the slidewire to reduce the difference to zero. The pen position then follows the value of $v_u$.

so that $v_u = v_r \pm v_s$. The amplification (gain) of the servo amplifier should, of course, be made very high so that $v_s$ is negligible. As long as it remains high, the gain can vary over rather wide limits (from components aging, temperature variation, etc.) without affecting the measurement accuracy. For many servo systems used in potentiometric measurements, the error is less than a few microvolts for full-scale spans of 1 mV or more. For certain specialized instruments the error may be only a new nanovolts. In most cases the error is negligible. Note that $v_r$ is a low-impedance voltage source equal in magnitude to $v_u$. A load, such as a readout indicator, can be connected to $v_r$ without loading the input signal $v_u$ at all. Thus, the generation of the signal $v_r$ which follows $v_u$ is extremely useful in isolating the $v_u$ source from undesirable loads. In this mode, the servomechanical potentiometer acts as a voltage follower or buffer amplifier.

In some servomechanisms advantage is taken of the mechanical motion to provide a recording function as shown in figure 5-7. The pen-positioning mechanism, the servomotor, and the $v_r$ adjustment are all mechanically coupled.

In a **strip chart recorder** a long strip of graph paper is moved at a constant rate at right angles to the pen motion. Since the pen position is linearly related to $v_r$ and since $v_r$ follows $v_u$ exactly, the variation of $v_u$ with time is recorded on the chart paper. An **x-y recorder** uses two complete servomechanisms: one to position the pen along the $x$-axis of a graph paper in response to $v_x$, and another to position the pen along the $y$-axis in response to $v_y$. As $v_x$ and $v_y$ vary, a plot of $v_x$ vs $v_y$ is produced. For all servo systems a certain amount of time is required for the motor to adjust $v_r$ in response to a change of the value of $v_u$. For servo recorders the response time typically allows full-scale travel of the pen in 0.5 to 2 s.

Servomechanical components in this same configuration also find extensive application in the precise mechanical positioning of instrument components (slits, gratings, detectors, probes, etc.) in response to an electrical control signal ($v_u$). When the electromechanical servo system is used only to generate the signal following voltage $v_r$ rather than for its mechanical positioning operation, the same operation can be accomplished in the all-electronic operational amplifier system described in the next section.

## 5-3  The Voltage Follower

The electronic **voltage follower** differs from the electromechanical servo potentiometer of the previous section in that the null detector amplifier output signal is used to provide the reference comparison voltage directly rather than to drive a motor that adjusts the reference voltage value. The amplifier used in all electronic servo systems is the **operational amplifier**. In this section some of the characteristics of the operational amplifier are introduced, and its application in the voltage follower amplifier circuit is described.

## The Operational Amplifier

The operational amplifier (OA or op amp) combines the null detector amplifier and feedback signal source in a single unit. As a null detector amplifier, the op amp has a well-balanced difference input and a very high gain. As a feedback signal source, it has a low-impedance output that can produce voltages within a few volts of the power supplies used to operate it at output current levels from 2 to 50 mA or more. The symbol for the op amp and its transfer function are shown in figure 5-8.

The op amp symbol and transfer function are very similar to those shown for the comparator amplifier in figure 5-4. There is a high-gain region in which a small value for the input signal voltage difference $v_s$ produces a larger output voltage. Values of $v_s$ larger than a fraction of a millivolt result in a limited output voltage that at each extreme approaches the values of the power supplies used. The primary difference between op amps and comparators is that *the op amp is designed to operate in the linear region through feedback control that keeps $v_s$ very small* whereas the comparator is designed to produce well-defined limit output voltages that indicate the sign of $v_s$ and to respond quickly to changes in the sign of $v_s$. Similarities in their characteristics allow op amps to serve as comparators in some applications. However, the focus in this chapter will remain on op amp applications in analog or linear servo systems.

In the linear operating region the output voltage $v_o$ is equal to the input difference voltage $v_s$ times the gain. Thus,

$$v_o = A(v_+ - v_-) = -Av_s \qquad (5\text{-}2)$$

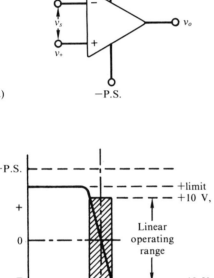

(a)

(b)

**Fig. 5-8.** The op amp (a) symbol and (b) transfer function. The op amp has a very high gain so that its output limit is reached with only a very small input voltage. Although the transfer function is similar to that of the comparator, the op amp is designed to operate in the linear region, not at limit.

where $v_+$ and $v_-$ are the signal voltages connected to the amplifier inputs labeled $+$ and $-$, respectively. These inputs are labeled $+$ and $-$ to indicate the direction of output voltage change for a positive change in signal voltage at that input. That is, a positive voltage change at the $+$input produces a positive change in $v_o$, and a positive voltage change at the $-$input produces a negative change in $v_o$. The $-$input terminal is therefore called the **inverting input**, and the $+$input terminal, the **noninverting input**.

## The Op Amp Voltage Follower

The circuit for the op amp voltage follower is shown in figure 5-9. Comparison of this circuit with the servomechanical potentiometer of figure 5-7 shows them to be similar in that an amplified form of the difference between $v_r$ and $v_u$ is used to adjust the feedback signal so that the difference between $v_r$ and $v_u$ is decreased. The elimination of the mechanical elements in the servo system has the advantages of simplicity and the possibility of more rapid response to changes in voltage.

The design of the voltage follower circuit can be rationalized as follows: Since $v_r$ is to be obtained from $v_o$ and since $v_r$ is to follow $v_u$, the output voltage should change in the same way as $v_u$. For this reason $v_u$ is connected to the noninverting input of the amplifier. The signal connected to the inverting input, the other input of the null detector, is $v_r$, the signal that is to follow $v_u$. This completes the path for the feedback information. The control of $v_o$ by the difference in $v_u$ and $v_r$ at the inputs is seen to have the correct sense. If $v_u$ becomes more positive than $v_r$ ($v_s$ is negative), $v_o$ and $v_r$ also become more positive, thus reducing the difference signal $v_s$.

The approximate operating characteristics of op amp servo systems can be analyzed by beginning with the assumption that the amplifier is operating in its linear region. This assumption can be stated thus: *If the output voltage $v_o$ is not at limit, the input difference voltage $v_s$ is negligibly small.* Applied to the voltage follower circuit of figure 5-9, this assumption can be stated, $v_r \simeq v_u$. Since the amplifier output and $-$input are directly connected, $v_o = v_r$ and thus, $v_o \simeq v_u$. For an op amp with a gain $A$ of $10^5$, $v_s = v_o \div 10^5$, and thus the assumption that $v_s$ is negligible introduces an error of one part in 100 000. The error arises because according to equation 5-2 there must be some difference input $v_s$ in order to maintain a finite output voltage. The exact relation between $v_o$ and $v_u$ (see note 5-3) is:

$$v_o = v_u A / (A+1) \tag{5-3}$$

This equation confirms our approximate analysis that $v_o \simeq v_u$ with an error of one part in $A$. Note again that the actual value of $A$ is relatively unimportant as long as it is large.

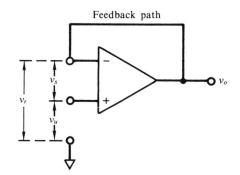

**Fig. 5-9.** Op amp voltage follower. Because of the feedback path, $v_o = v_u + v_s$. For all values of $v_o$ less than limit, $v_s$ is very small. Therefore $v_o \simeq v_u$.

**Note 5-3.  Voltage Follower Response.**
The exact response function of the follower is derived as follows: From equation 5-2, $v_o = A(v_u - v_r)$. Since $v_r = v_o$, $v_o = A(v_u - v_o)$. Collecting terms gives $v_o(1+A) = v_u A$. Solving for $v_o$ gives $v_o = v_u A/(1+A)$. This equation can be rearranged as $v_o = v_u - v_o/A$, in which the term $-v_o/A$ is the error. That is, it is the difference between $v_o$ and $v_u$. As expected, the error voltage is proportional to $v_o$ and inversely proportional to $A$. For an amplifier with a maximum $v_o$ of 10 V and an amplification $A$ of $10^6$, the maximum error is 10 $\mu$V.

The op amp voltage follower is not a complete voltage measurement system in that its output voltage must still be measured to obtain the value of $v_u$. In the domain sense the amplifier output information is still in the voltage domain and must be converted to the number domain to complete the measurement. The great merits of the potentiometric amplifier are that the inputs of the amplifier have the very high resistance and balance (CMRR) characteristic of the operational amplifier and that the output is a low-resistance source of voltage equal to $v_u$ and referenced to common. This allows almost any convenient voltage-measuring device to be connected to $v_o$ to convert the voltage to the scale position or digital domains.

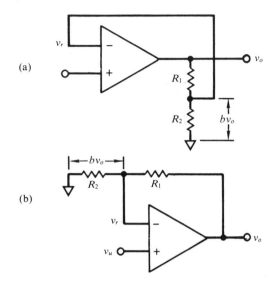

**Fig. 5-10.** Follower with gain, (a) drawn to empha-size voltage divider on $v_o$ and (b) conventionally drawn. The fraction $b$ of $v_o$ is essentially equal to $v_u$ as long as $v_o$ is not at limit. Therefore, $v_o \simeq v_u/b \simeq v_u[(R_1 + R_2)/R_2]$.

## The Follower with Gain

When the use requires greater sensitivity or sensitivity that is adjustable, the op amp voltage follower can be connected to provide an output voltage that is a multiple of the measured voltage. This is done by using only a fraction of the output voltage $v_o$ for the comparison voltage $v_r$. Thus, $v_o$ must be greater than $v_u$ to produce a value of $v_r$ equal to $v_u$. The resulting circuit is shown in figure 5-10. A voltage divider composed of $R_1$ and $R_2$ is used to produce a fraction $b$ of $v_o$. The signal $bv_o$ is used as $v_r$ at the inverting input of the op amp.

The gain of the resulting amplifier can be determined approximately by again starting with the assumption that $v_u \simeq v_r$ as long as $v_o$ is not at limit. The signal $v_r$ is $bv_o$ so that $v_o = v_r/b \simeq v_u/b$. From the voltage divider principle, $b = R_2/(R_1 + R_2)$ and thus,

$$v_o \simeq v_u \left(\frac{R_1 + R_2}{R_2}\right) \tag{5-4}$$

If $R_1$ and $R_2$ are 99 kΩ and 1 kΩ respectively, $b = 1/100$. Then $v_o = 100 v_u$, and the amplifier has a gain of 100. For an adjustable gain amplifier, a potentiometer can be used to vary the value of $b$. The precision and constancy of this gain depend upon the values of resistors $R_1$ and $R_2$ and not upon the amplification $A$, within the limits of the approximation. The exact gain equation (see note 5-4) is:

$$v_o = v_u A/(bA + 1) \tag{5-5}$$

Equation 5-5 indicates that the error is one part in $bA$. If $b$ were $1/100$ (gain of 100) and the op amp amplification $A$ were $10^5$, the error would be one part in 1000. As the desired follower gain increases, one must use an op amp with higher amplification or sacrifice some accuracy.

Because of its near-ideal input characteristics and its precise gain, the follower with gain is used as an input signal amplifier with a variety of

**Note 5-4.   Follower with Gain Response.**
The exact response function of the follower with gain is derived as follows: From equation 5-2, $v_o = A(v_u - v_r)$. Since $v_r = bv_o$, $v_o = A(v_u - bv_o)$. Collecting terms and solving for $v_o$ gives $v_o = v_u A/(bA + 1)$. This equation can be rearranged as $v_o = v_u/b - v_o/bA$ in which the term $-v_o/bA$ is the error. If the maximum $v_o$ is 10 V, $A$ is $10^6$, and $1/b$ is 100, the maximum error voltage is 1 mV.

readout devices such as recorders, pH meters, and digital voltmeters. Frequently the readout device has a single sensitivity or full-scale deflection. In these cases, a range of measurement sensitivities is obtained by selecting values of $b$ from a precision multitap voltage divider.

## 5-4  The Current Follower and Summing Amplifier

The concept of null comparison can be applied to the measurement of current as well as voltage. A variable reference current source is used to offset the unknown current source. The reference current is then adjusted to bring the null detector to a zero difference indication. The same null detectors that are used for null comparison voltage measurements can be used for null comparison measurements of current, and the servo systems that are used to automate potentiometric measurements can be used to automate the null comparison current measurements. Automation is achieved through a current source that is continuously adjusted to follow the measured current. The op amp servo current follower system described in this section is widely used in the precision measurements of small currents, and the related summing amplifier is the basic circuit used in analog computers.

### Null Comparison Current Measurement

The basic current null comparison circuit is shown in figure 5-11. The null detector is connected as a current meter would be for measuring the magnitude of the unknown current $i_u$. A reference current source is then connected to the same null detector in such a way that its current through the null detector is opposite that of the unknown current. The null detector current $i_{nd}$ is thus equal to the difference between the unknown and reference source currents. The sign of the null detector current indicates whether $i_r$ is larger or smaller than $i_u$.

When the current through the null detector is zero, the voltage across the null detector must also be zero. Therefore, at null the potential difference between $J_1$ and $J_2$ is zero. This means that *at balance the output voltage of both current sources is zero.* Both sources are under the ideal condition of a short circuited output and are thus delivering the full values of $i_u$ and $i_r$ to the comparison circuit. The current comparison at null idealizes the load of the current sources just as the null voltage comparison unloads the voltage sources. If point $J_2$ is connected to the circuit common, point $J_1$ is also at the common potential at null; that is, $J_1$ is a virtual common. As in the case of the null comparison voltage measurement, either a high-resistance voltage-sensitive or a low-resistance current-sensitive null detector can be used.

For the basic null voltage comparison circuit of figure 5-3, three positions for the connection of the circuit common are possible because of the

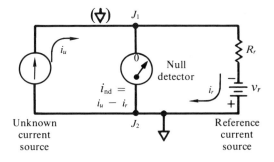

**Fig. 5-11.** Basic current comparison circuit. Any difference in the currents $i_u$ and $i_r$ must pass through the null detector. Either $v_r$ or $R_r$ may be adjusted so that $i_u = i_r$. At null, the voltage across the null detector is zero, as though there were a short circuit between $J_1$ and $J_2$. Thus the comparison circuit presents an ideal load to the unknown current source.

series arrangement of the two sources and the null detector. Note in figure 5-11 that in the basic current comparison circuit all components of the circuit are connected in parallel. There are only two possible positions for the connection of the circuit common: $J_1$ or $J_2$. Brief inspection shows that these two positions are equivalent. Therefore, there is only one common point configuration for the current comparison circuit, and all components have one connection to the circuit common. Note also that the other connection to each of the components is at the common voltage when the system is at null. This is indicated by the virtual common symbol ($\triangledown$) next to $J_1$. This eliminates the need for floating sources and the concern about common mode error at the input to the null detector.

The null comparison technique for current measurement is far superior to the measurement of $IR$ drop with a voltmeter since it requires no assumption about the ideality of the unknown current source. Even though manual null current comparison measurements are rarely employed, a number of circuits that are very commonly used with servomechanical systems and operational amplifiers are automatic null current comparison systems. Several of these circuits are described in the remainder of this section. The current comparison concept helps greatly in understanding the operation and characteristics of this family of circuits.

## Op Amp Current Follower

The null comparison measurement of current shown in figure 5-11 is automated by providing a reference current source that is controlled by the null detector output in such a way as to maintain a null condition. When an op amp is used as the servo system, its differential inputs are the null detector, and its output voltage produces a proportional reference current through a series resistance. The resulting circuit is shown in figure 5-12. Comparison with figure 5-11 reveals the analogous components.

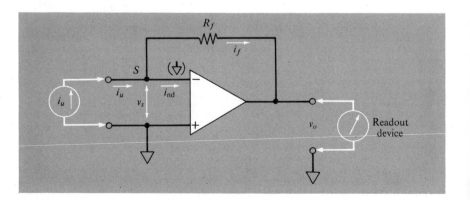

**Fig. 5-12.** Op amp current follower. If $i_{nd}$ is negligible, $i_f \simeq i_u$, and if $v_s$ is negligible, $v_o \simeq -i_f R_f$. Therefore $v_o \simeq -i_u R_f$.

The unknown current source $i_u$ is connected directly to the op amp input terminals. The output voltage $v_o$ generates the comparison current $i_f$ through the resistor $R_f$. The subscript $f$ is used for signals and components in the feedback path. The inverting input is used for the feedback current information because it provides the proper corrective direction at the output. If $i_u$ increases, point $S$ becomes more positive, making $v_o$ more negative and thus increasing $i_f$ as desired. The connection of the +input to common is required because the other terminal of the op amp output signal is common.

An approximate analysis can again be made by assuming that $v_s$ is negligible as long as $v_o$ is not in limit. Since the +input is connected to common, the voltage at the −input is $v_s$ and $v_s \simeq 0$. The voltage $v_o$ thus appears across $R_f$, and $i_f \simeq -v_o/R_f$ from Ohm's law. The null detector current $i_{nd}$ for the op amp, called the **bias current** $i_b$, is generally very small because of the high impedance inputs of the op amp and the virtual ground at point $S$. The summing of the currents at point $S$ yields $i_u = i_f + i_b$. If $i_b$ is negligible compared to $i_f$, the currents $i_u$ and $i_f$ are essentially equal. That is, *all the current to point S from the signal source appears in the feedback path through $R_f$.* Substitution of $i_u$ for $i_f$ in the equation $i_f \simeq -v_o/R_f$ yields

$$v_o \simeq -i_u R_f \qquad (5\text{-}6)$$

Thus $i_f$ follows $i_u$ and produces an output voltage proportional to $i_u$. If a value of 1 MΩ were chosen for $R_f$, the output voltage would be one volt per microampere of input current.

The current follower circuit is in many ways analogous to the voltage follower. The output voltage produced is the same as the one that would occur if $R_f$ were simply a load resistor across the current source $i_u$. However, the current follower presents a nearly ideal load to the current source and provides an output voltage suitable for driving a variety of readout devices.

The limitations in the application of the current follower are seen in the exact relation between $v_o$ and $i_u$ (see note 5-5).

$$v_o = -R_f\,(i_u - i_b)\left(\frac{A}{1+A}\right) \qquad v_o = -i_u R_f + i_b R_f - \frac{v_o}{A} \qquad (5\text{-}7)$$

From either form of equation 5-7 it can be seen that when $A$ is very large and $i_b$ is much less than $i_u$, the output voltage $v_o$ is proportional to $i_u$ as expected from equation 5-6. The range of currents that can be measured by a current follower and readout is limited on the low end by the input current of the op amp $i_b$ and on the high end by the op amp's output current capability. Op amp input bias currents vary from $10^{-9}$ to $10^{-15}$ A, making the current follower useful into the picoampere current range with careful choice of the op amp. The output current capability of an op amp is generally 2 – 100 mA. The op amp output must supply both the feedback current $i_f$ and the readout device current.

---

**Note 5-5. Current Follower Response.**
The exact relationship between the input current and the output voltage for the current follower is obtained by summing the currents at point $S$ to obtain $i_u = i_f + i_b$ and by summing the voltages in the loop containing $v_o$, $R_f$, and $v_s$ to obtain $v_o = v_s - i_f R_f$. Combining these equations to eliminate $i_f$ yields $v_o = v_s - (i_u - i_b)R_f$. From equation 5-2 $v_s = -v_o/A$, and thus $v_o = -i_u R_f + i_b R_f - v_o/A$. Combining terms in $v_o$ yields the other form of equation 5-7.

The effective input resistance $R_{in}$ of the current follower circuit as seen by the unknown current source is the input voltage $v_s$ divided by the input current $i_u$; that is, $R_{in} = v_s/i_u$. Since $v_s = -v_o/A$ and from equation 5-6 $i_u \simeq -v_o/R_f$,

$$R_{in} \simeq R_f/A$$

The loading effect of $R_f$ on the current source $i_u$ is thus improved $A$ times by using the current follower. Because $A$ is generally $10^5$ to $10^7$, the use of the current follower can and does provide a dramatic improvement in current measurements.

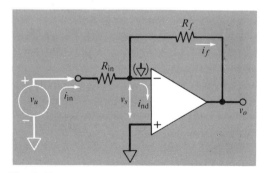

**Fig. 5-13.**  Op amp inverting amplifier. The current $i_u$ is approximately equal to $v_u / R_{in}$ since $v_s \simeq 0$. From the current follower equation, $v_o = -v_u (R_f / R_{in})$.

**Note 5-6.   Inverting Amplifier Response.**
The exact relationship between the output and input voltages of the inverting amplifier is obtained as follows: Sum the voltages in the input and feedback circuits to obtain $v_u = i_u R_{in} + v_s$ and $v_o = v_s - i_f R_f$. Substitute $-v_o / A$ for $v_s$ in both equations. For $i_f$ in the $v_o$ equation, substitute $i_u - i_b$ (from the sum of currents at point S) to obtain $v_o = -v_o / A - i_u R_f + i_b R_f$. Into this equation, substitute $i_u = v_u / R_{in} + v_o / R_{in} A$ (obtained from the $v_u$ equation) to obtain

$$v_o = -v_u \frac{R_f}{R_{in}} - \frac{v_o}{A}\left(1 + \frac{R_f}{R_{in}}\right) + i_b R_f$$

From this equation, the errors due to the less than infinite value of $A$ and the greater than zero value of $i_b$ are clear. In the ideal case, equation 5-8 is verified.

## Op Amp Inverting Amplifier

The **inverting amplifier** is a simple variation of the current follower circuit in which the unknown current source is a voltage source and resistor in series. The resulting circuit is shown in figure 5-13. Since $v_s \simeq 0$, the source current $i_u \simeq v_u / R_{in}$. If $v_u / R_{in}$ is used for $i_u$ in equation 5-6 for the current follower,

$$v_o \simeq -v_u \frac{R_f}{R_{in}} \tag{5-8}$$

Thus the output voltage is a constant times the input voltage, and the constant $-R_f / R_{in}$ is dependent only upon the values of the resistors $R_f$ and $R_{in}$. This provides another possibility for precision amplification of a voltage signal. In contrast to the voltage follower precision amplifier of figure 5-10, the inverting amplifier inverts the input signal as indicated by the minus sign in equation 5-8. The inverting amplifier also differs from the voltage follower in that it has one input connected to the circuit common, eliminating the common-mode rejection error, and it has a simpler relationship between gain and resistance, allowing whole decade values of resistors to produce whole decade values of system gain. For example, if $R_f$ is 1 M$\Omega$, and $R_{in}$ is 10 k$\Omega$, a gain of 100 is obtained. On the other hand, the input source $v_u$ is loaded by the input resistor $R_{in}$ in the inverting amplifier, but it is measured potentiometrically in the follower with gain. In both inverting and follower amplifiers, higher gain is achieved with some loss in gain accuracy (see note 5-6). Each amplifier circuit has merits depending upon the desirability of inversion and the problems of source loading.

## Summing Amplifier

Since the current follower amplifier of figure 5-12 provides an input that is maintained at the common voltage, a current source connected to point $S$ has the same output that it would have if it were connected to the common. Therefore, multiple current sources connected to point $S$ do not interfere with each other. The result is the **current summing amplifier** shown in figure 5-14. Each current source applies its current to the virtual common point $S$. Since the sum of all currents to point $S$ must be equal to $i_f$, $i_f \simeq i_1 + i_2 + i_3$. For this reason, point $S$ is often called the **summing point** of the op amp in current follower applications. From equation 5-6 we see that

$$v_o \simeq -R_f(i_1 + i_2 + i_3) \tag{5-9}$$

Therefore the output voltage is proportional to the sum of the input currents. Equation 5-7 can be applied to determine the limitations of this circuit.

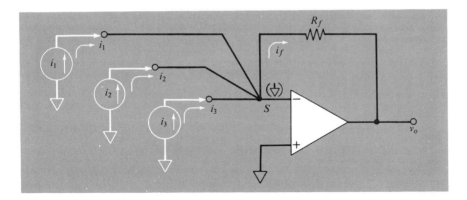

**Fig. 5-14.** Summing current follower. All current sources to point $S$ are ideally loaded and do not affect each other. The current $i_f$ is the simple sum of $i_1$, $i_2$ and $i_3$, and $v_o$ is proportional to that sum.

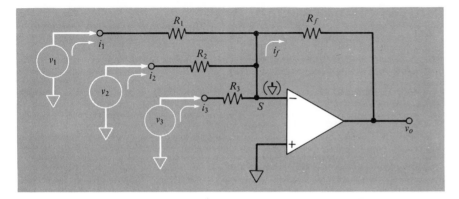

**Fig. 5-15.** Summing amplifier. This voltage summing amplifier is the summing current follower in which input resistors are used to obtain input currents proportional to the input voltages. The separate input resistors allow a different gain for each input connection.

The current sources in a current summing amplifier can be voltages with series resistors as shown in figure 5-15. Assuming point $S$ to be at the common potential, $i_1 \simeq v_1/R_1$, $i_2 \simeq v_2/R_2$, and $i_3 \simeq v_3/R_3$. Substituting these relations for the current in equation 5-9, we obtain

$$v_o \simeq - \left( v_1 \, \frac{R_f}{R_1} + v_2 \, \frac{R_f}{R_2} + v_3 \, \frac{R_f}{R_3} \right) \qquad (5\text{-}10)$$

Equation 5-10 shows that the output voltage is the sum of the input voltages, each multiplied by its own $R_f/R_{in}$ ratio. This circuit is called the **weighted summing amplifier** because the contribution of each input voltage to the sum is weighted by its individual gain factor. Frequently a simple sum is desired, so that $R_1$, $R_2$, and $R_3$ are all equal to $R_{in}$. Then

$$v_o \simeq \frac{-R_f}{R_{in}} (v_1 + v_2 + v_3) \qquad (5\text{-}11)$$

This circuit has the same output and gain error characteristics as the inverting amplifier. Summing amplifiers are used in instruments and in analog computers to perform the addition function on data in the voltage or current domains. A summing amplifier can also be used to add a constant to a signal voltage or current and thus to introduce (or eliminate) an offset in the transfer function. The input resistance of the summing amplifier is $R_{in}$ at each input. If this presents an excessive load on the voltage source, a voltage follower circuit (fig. 5-9) is used between the voltage source and the summing amplifier input.

## Linear Resistance-to-Voltage Converters

According to Ohm's law, $R = V/I$, the voltage across a resistor is proportional to the resistance for a constant value of current. Similarly the conductance, $G = I/V$, is proportional to the current through a resistor for a constant applied voltage. To take advantage of these linear relationships requires a constant current source in the first case and a current measurement with negligible input resistance in the second case. The current follower circuit introduced in this section provides a convenient and accurate means to achieve either of these requirements.

Recall that the current in the feedback resistor follows the current applied to the summing point and that the output voltage is proportional to $R_f$ and the input current. In the circuit of figure 5-16a, $V$ and $R_c$ produce a constant current $I_c$ which appears in the unknown resistor $R_u$ in the feedback path. The output voltage is then

$$v_o \simeq -VR_u/R_c \quad \text{or} \quad v_o \simeq R_u(-V/R_c) \tag{5-12}$$

showing that $v_o$ is proportional to $R_u$.

To obtain an output proportional to conductance, the positions of $R_u$ and $R_c$ are reversed as in figure 5-16b. The combination of $V$ and $R_u$ produce a current proportional to the conductance $G_u$ of $R_u$. Because the output voltage $v_o = -i_u R_f$,

$$v_o \simeq -VR_c/R_u \quad \text{or} \quad v_o \simeq G_u(-VR_c) \tag{5-13}$$

**Fig. 5-16.** Resistance-to-voltage converters. In circuit (a), $v_o \simeq -I_c R_u$ based on the current follower equation. Similarly, in circuit (b), $v_o \simeq -i_u R_c \simeq -VR_c/R_u \simeq -G_u VR_c$.

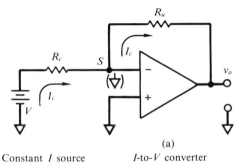

(a)

Constant $I$ source      $I$-to-$V$ converter

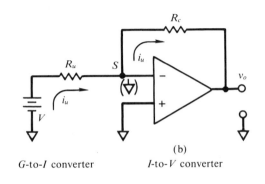

(b)

$G$-to-$I$ converter      $I$-to-$V$ converter

Of course, both circuits are simply the op amp inverting amplifier, for which $v_o = -v_{in}R_f/R_{in}$ with $v_{in} = V$ and $R_u$ taking the place of $R_{in}$ in the first case and of $R_f$ in the second. This simple circuit is an extremely convenient method for converting the resistance or conductance of various resistive transducers to a proportional voltage. It is also the circuit used for the resistance scales in many digital multimeters.

## 5-5  The Integrator (Charge Follower)

The use of a resistor in the feedback path of the current follower amplifier results in an output voltage proportional to the feedback current. The linear current-voltage relationship is due to the ohmic nature of the feedback resistor. Devices with other current-to-voltage or charge-to-voltage relationships (transfer functions) could be used in the feedback or input paths of the basic current follower to produce a circuit that implements the transfer function of the components used. In this section we shall see how a combination of capacitors and resistors can be used with an operational amplifier to produce a charge-to-voltage converter (integrator), a rate-of-charge-to-voltage converter (differentiator), and a charge amplifier. These operations on information in the charge domain are very useful with certain types of transducers—especially biomedical ones.

### Charge-to-Voltage Conversion

Since the voltage across a capacitor is directly proportional to the charge on it ($V = Q/C$) the capacitor is the basic device used in **charge-to-voltage converter** circuits. The measurement of charge and current are related because current is the rate of charge transfer. The basic operational amplifier charge-to-voltage converter is the integrator shown in figure 5-17. A positive charge applied to the input accumulates on capacitor $C$ since there is no other path for it to take. The op amp establishes an output voltage $v_o$ which maintains point $S$ at virtual common. Since the voltage at point $S$ is negligible, the total voltage across the capacitor appears at the output. Neglecting $v_s$, the voltage $v_o$ is thus

$$v_o \simeq -q_{in}/C \qquad (5\text{-}14)$$

The output voltage $v_o$ is thus proportional to $q_{in}$. The value of $C$ is chosen for the range of charge values to be measured. Equation 5-14 would be perfectly accurate for any amount of charge over any time period if it were not for the leakage of charge to point $S$ from sources other than the input. The major source of leakage current is often the amplifier input bias current $i_b$. The total charge measured should be larger than the charge leaked by $i_b$ during the measurement time.

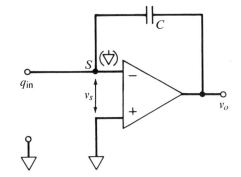

**Fig. 5-17.**  Charge-to-voltage converter. Charge applied to the input appears on the capacitor, $C$ so that $v_o = -q_{in}/C$.

**Note 5-7.   Integration.**
The integral sign $\int$ is a convenient symbol for the total accumulation of a quantity over a particular interval. For instance, equation (5-15) could be read, "$q_{in}$ is equal to the total accumulation of $i_{in}$ for each increment of time $dt$ from time 0 to time $t$." By analogy, the balance in your bankbook is the integral of the positive and negative cash flow (deposits and withdrawals) from the beginning of the account to the present.

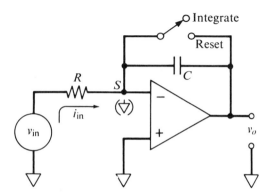

**Fig. 5-18.**   Op amp integrator. The resistor $R$ makes the rate of charge flow to point $S$ proportional to $v_{in}$. The output voltage $v_o$ is equal to $-1/C$ times the integral of the input current from the time of the opening of the integrate switch to the present.

**Note 5-8.   The Derivative.**
The derivative $dq/dt$ is the increment of charge per increment of time for vanishingly small increments. It is therefore exactly equal to the rate of change of charge. The derivative $dv/dt$ is similarly the rate of change of voltage. Equation 5-18 is obtained by expressing the equation $q = Cv$ in terms of the rates of change of $q$ and $v$.

**Integration.**   Since the total charge passing a point over a given time is the integral or summation of the instantaneous currents during that time, the charge-to-voltage converter can be used to perform the mathematical function of integration (see note 5-7) on a signal in the current domain. If $q_{in}$ is applied to the **integrator** of figure 5-18 as an input current $i_{in}$ over time $t$,

$$q_{in} = \int_0^t i_{in}\, dt \qquad (5\text{-}15)$$

Combining equations 5-14 and 5-15,

$$v_o \simeq -\frac{1}{C} \int_0^t i_{in}\, dt \qquad (5\text{-}16)$$

A signal encoded in the voltage domain can be integrated if a resistor is used to convert the input voltage to a proportional current. Since point $S$ is maintained at the common voltage, $i_{in} = v_{in}/R$. Substituting for $i_{in}$ in equation 5-16

$$v_o \simeq -\frac{1}{RC} \int_0^t v_{in}\, dt \qquad (5\text{-}17)$$

A switch (often electronic) is used to discharge the capacitor and reset the integrator to begin another measurement. Such switched circuits are discussed in the next chapter. An important application of the integrator is as the generator of a linear sweep voltage. If $v_{in}$ in figure 5-18 is constant, $v_o$ increases at a constant rate determined by the values of $R$ and $C$.

**Differentiation.**   The time derivative of a signal is proportional to its rate of change. The rate of change of charge on a capacitor, $dq/dt$ (see note 5-8), is related to the rate of change of voltage across the capacitor by the following variation of the basic $Q = CV$ relationship:

$$\frac{dq}{dt} = C\,\frac{dv}{dt} \qquad (5\text{-}18)$$

where $dq/dt$ is the rate at which charge is brought to the capacitor; in other words, $dq/dt$ is the current $i$. Thus, imposing a change in voltage on a capacitor produces a current proportional to the rate of change. The measurement of the charging current is then a measurement of the time derivative of the voltage signal. A derivative-measuring circuit using an op amp current follower for the current measurement is shown in figure 5-19. Since $i_{in} = C(dv_{in}/dt)$ and $v_o \simeq -Ri_{in}$,

$$v_o \simeq -RC\,\frac{dv_{in}}{dt} \qquad (5\text{-}19)$$

Derivative circuits sometimes called **differentiators** are useful for sharpening signal transitions for easier identification of regions of change as well as for measurements of rate. Because the output voltage is proportional to the rate of input voltage change, a small but very rapid pulse could produce a larger output signal than a slower but larger change in signal level. The response of practical derivative circuits to noise spikes is reduced by introducing either a small resistor in series with the input capacitor to limit its rate of charge, or a small capacitor in parallel with the feedback resistor to limit the response rate of the amplifier, or both.

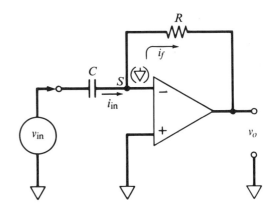

**Fig. 5-19.** Op amp differentiator. The current $i_{in}$, and thus the output voltage $v_o$, are proportional to the rate of change in the voltage across $C$. Because $S$ is a virtual common, the voltage across $C$ is $v_{in}$.

## Charge-Coupled Amplifier

From equation 5-14 the integrator circuit is clearly very useful for the measurement of charge, or more precisely, for the conversion of data from the charge domain to the voltage domain. Thus it is particularly useful with transducers having an output in, or easily converted to, the charge domain. Several charge-output transducers were described in chapter 4. Since the charge on a capacitor is a function of the voltage across the capacitor and the capacitance value, both voltage and capacitance variations are easily converted into charge variations.

A common charge generating and measuring circuit is shown in figure 5-20. Since one terminal of $C_{in}$ is at voltage $v_{in}$ and the other terminal is held at the common voltage by the operational amplifier, the charge on the input capacitor is $q = C_{in}v_{in}$. If $v_{in}$ varies by an amount $\Delta v_{in}$, the charge on $C_{in}$ varies by

$$\Delta q = C_{in}\Delta v_{in} \qquad (5\text{-}20)$$

This same change in charge will appear across $C_f$, so that

$$\Delta v_o = \Delta q / C_f = -\Delta v_{in} C_{in} / C_f \qquad (5\text{-}21)$$

Thus the input voltage variations appear at the output multiplied by the factor $C_{in}/C_f$. This **charge-coupled amplifier** is often used with biological transducers that produce voltage variations but have a limited charge supply capability. It is necessary to short out $C_f$ occasionally or to bypass it with a large resistor to keep the output voltage from drifting off as a result of the long-term integration of low-level leakage currents.

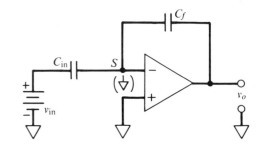

**Fig. 5-20.** Charge-coupled amplifier. A change in the charge across $C_{in}$ appears as a proportional change in the voltage $v_o$. This circuit is analogous to the inverting amplifier, but the use of capacitors for input and feedback elements makes it sensitive to charge rather than to voltage.

## Capacitive Transducers

The geometry of the conductors in a capacitor determine its capacitance according to the equation $C = \epsilon_0 K_d a / d$ where $a$ is the common area between the conductors, $d$ is the distance between the conductors, $K_d$ is the dielectric constant of the insulator between the conductors, and $\epsilon_0$ is the permittivity of free space. This relationship is the basis of several capacitive transducers. Because the capacitor is so sensitive to dimensional changes, it is often used where the parameter to be measured can cause a physical motion. This motion can cause a variation in the distance between the plates, a change in the common area of the plates, or a change in the dielectric constant of the insulator. Thus, there are capacitive transducers for displacement, velocity, pressure, fluid level, and so on. Capacitive transducers can also be used to measure the dielectric constant of materials.

If the capacitive transducer is used as either $C_{in}$ or $C_f$ in the circuit of figure 5-20, the output voltage changes as the capacitance varies. Equation 5-20 can be rewritten to show that the change in $q$ that results from a change in $C_{in}$ is $\Delta q = v_{in} \Delta C_{in}$. In this case the output voltage is $\Delta v_o = \Delta q/C_f = -\Delta C_{in}v_{in}/C_f$, proving the circuit to be a linear **capacitance-to-voltage converter**.

## 5-6 Practical Considerations in Amplifier Circuits

An **amplifier** is a device that increases the amplitude of an electrical signal. The amplifier is used when a load or readout device requires more current, voltage, or power than is available from the signal source directly. The signal amplification is accomplished by using the input signal to control the delivery of power from the amplifier's power supply to its output terminals as shown in figure 5-21. Amplification is thus seen to be not so much an operation *on* the input signal as the control of current, voltage, or power in response *to* the input signal. The power control elements used in amplifiers are generally solid-state devices such as the bipolar transistors or field-effect transistors described in chapter 7. The amplification is a result of the small voltage or current required by the control element in order to control larger voltages or currents from the power supply. The controlled output voltage (or current) ideally follows some known and desired function related to the input signal.

The many amplifiers used in practice are aimed at meeting certain design requirements. Some possible design requirements are high-frequency response, the ability to provide the necessary control of power supply voltage and current even when the input signal varies at rates as high as 100 MHz or more; the ability to respond to a floating signal source, as with the differential amplifier; the ability to provide an unusually large output voltage or current, as with power or driver amplifiers; or the ability to operate with very low power, as for portable equipment.

No amplifier is either ideal or universal although the versatility of the modern op amp is very great. One must choose among the available options to meet the basic application requirements and then determine that the

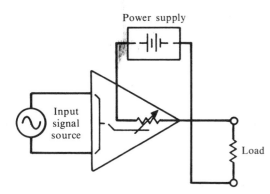

**Fig. 5-21.** An amplifier as a signal-responsive control system. The input signal is used to control the delivery of power from the power supply to the load. If the voltage or the current in the load is greater than that of the input signal, amplification has occurred.

deviations from ideality are within acceptable limits. The primary considerations with respect to the nonideality of op amps for several applications are illustrated in this section.

## Amplifier Response Function

A typical amplifier transfer function is shown in figure 5-22a. Ideally the change in $v_o$ resulting from a given change in $v_{in}$ (the gain) is absolutely constant throughout the linear region. For amplification of signals in the voltage domain, it is important to know the gain and the limits of the linear range. The effect of changing gain on the transfer function is shown in figure 5-22b.

One difficulty in designing a linear amplifier is that the control elements do not have completely linear transfer functions. Approximately linear response is obtained in some amplifiers by using just a small portion of their response function. In the op amp, as we have seen in this chapter, the approach is to make the amplification very large. The amplification may vary with the signal level (that is, it is nonlinear), but as long as it remains very large, the op amp circuit gain is determined predominantly by the feedback circuit components and thus remains constant. In this way the feedback idealizes the gain linearity as well as the input-loading and output-driving characteristics of the amplifier.

The transfer functions shown in figure 5-22a and b are those of amplifiers with zero **offset voltage**; i.e., those in which $v_o$ is zero when $v_{in}$ is zero. Figure 5-22c shows the transfer function for several positive and negative values of offset. Most amplifiers have some offset, either unintentional—caused by circuit characteristics or component instability—or intentional—to compensate for an offset inherent in the voltage source. A "zero" or "balance" adjustment is often provided to reduce offset to within the offset stability of the amplifier.

The voltage offset characteristic of an op amp is given in terms of the input signal amplitude required to produce a zero output voltage. The offset voltage contributes directly to the error in the null point maintained by the op amp through the feedback circuit. Many op amps have less than 1 mV offset with no balance control required. With a balance adjustment the offset can be reduced to the level of the offset instability, often about 10 $\mu$V and in some cases as low as nanovolts. Other contributions to input errors are the **input resistance** and the **input bias current**. The input resistance of a modern op amp is typically $10^{12}$ to $10^{15}$ $\Omega$. With the very low input voltages maintained by the null comparison feedback, the current through the input resistance would be almost infinitesimal. However, all transistors and other control elements have some current leakage into the control input. This

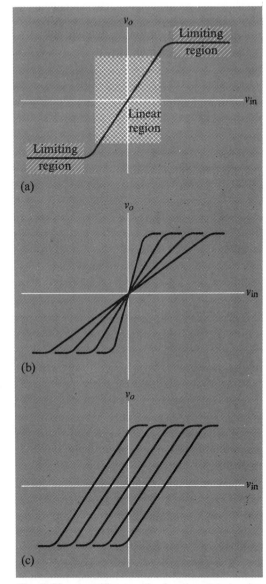

**Fig. 5-22.** Amplifier transfer functions (a) showing linear and limiting regions, (b) for various gains, and (c) for various offsets. In the linear region, this change in $v_o$ for a given change in $v_{in}$ (the gain) is constant. A change in gain affects the slope of the input/output curve. The offset is equal to the value of $v_{in}$ that produces a $v_o$ of zero.

results in a current at the op amp input terminals even when the input voltage is zero. The input bias current is of the order of $10^{-8}$ A for inexpensive op amps and as low as $10^{-15}$ A for special low-current types. This current creates an $IR$ drop in the resistors connected to the op amp input, and this voltage appears at the amplifier input. A different $IR$ drop in the $+$ and $-$ input circuits results in an additional source of input offset voltage. If the resistances to the $+$ and $-$ op amp inputs are made equal, the $IR$ drops approximately balance, and the important characteristic is then the *difference* in input currents called the **input offset current**. In any application where the op amp input voltages are not at very nearly the common (or at least constant) voltage, the common mode response of the amplifier can contribute a significant error. The evaluation of the errors due to offset, bias current, and CMRR is illustrated below for several of the op amp applications discussed in this chapter.

## Voltage Follower Amplifiers

Ideally, the voltage follower output voltage $v_o$ exactly equals the input voltage $v_u$. As indicated in figure 5-23a the error, or difference $v_o - v_u$ is equal to $v_s$. The maximum value for the error is the sum of (1) the input difference required to produce $v_o$ volts of output ($v_o/A$ from eq. 5-2), (2) the common mode error, and (3) the offset voltage. Assuming the maximum output voltage of 10 V, an amplification of $10^6$, a CMRR of $10^5$, and an offset voltage $v_{\text{off}}$ of 100 $\mu$V, the maximum difference between $v_o$ and $v_u$ would be $10/10^6 + 10/10^5 + 10^{-4} = 2.1 \times 10^{-4}$ V. Since the common mode voltage is equal to $v_u$, the CMRR is a critical characteristic for the voltage follower application. Furthermore, the CMRR rating of some op amps decreases as the common mode voltage increases. The figure used for $v_{\text{off}}$ should include the voltage error caused by the input bias current.

The follower with gain circuit is shown in figure 5-23b. The amplifier gain $(1/b)$ is $(R_1 + R_2)/R_2$. Again, the difference between $bv_o$ and $v_u$, or the **input error**, is equal to the sum of the amplification error $v_o/A$, the common mode error $bv_o/$CMRR, and the offset $v_{\text{off}}$. The **output error** is equal to the input error times the amplifier gain $1/b$ since the total input signal is amplified by this factor. For the same amplifier used in the follower error illustration, and with a $1/b$ gain of 100, the maximum input error is $10/10^6 + 10^{-1}/10^5 + 10^{-4} = 1.1 \times 10^{-4}$ V, and the maximum output error is $1.1 \times 10^{-2}$ V.

The CMRR error is peculiar to the floating null detector configuration for voltage null comparison circuits (see fig. 5-3). It is possible to use the op amp in a floating unknown source configuration as shown in figure 5-24a. Note that $v_{\text{off}}$ now contributes the major source of error. Since the common mode input voltage does not change, there can be no error due to a finite

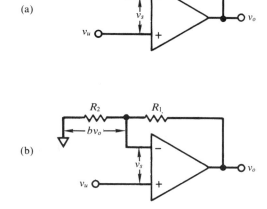

(a)

(b)

**Fig. 5-23.** (a) Operational amplifier voltage follower. (b) Follower with gain. Errors in $v_o$ result from the magnitude of $v_s$ required by the gain, the common mode rejection error, and the input offset error.

CMRR. Because both op amp inputs are essentially at the circuit common, there is negligible voltage applied to the amplifier input resistance, and the loading of the signal source is greatly reduced. The floating signal source configuration is superior to the more common voltage follower circuit when the signal source does not require a direct connection to the op amp circuit common. The same considerations apply to the floating signal source follower with gain circut shown in figure 5-24b.

## Current Follower Amplifiers

The range of currents that can be converted to voltages by the op amp current follower is limited on the high end by the output current limit of the op amp, which must supply both $i_f$ and the current to the output load. On the low end it is limited by the op amp input bias current $i_b$ as shown in equation 5-7, by $v_s$ as shown in note 5-5, or by the maximum practical value for $R_f$. The best way to reduce $i_b$ is by the proper choice of op amp, but a steady value of $i_b$ can be compensated for by applying a constant source of current opposite to $i_b$ to point $S$. The value of $v_s$ compared to $i_f R_f$ is sufficient to cause appreciable error only when a low-gain amplifier is used or when the output voltage $v_o$ is very low.

The maximum practical value of $R_f$ is limited by the state of the art in producing and using stable, accurate, high-value resistors. A circuit that has the effect of a high $R_f$ but is made of lower-value resistors is shown in figure 5-25. The output voltage is divided by $R_2$ and $R_3$, and the resulting fraction of $v_o$ is applied to $R_1$ to provide $i_f$. The resistance $R_1$ is a load on the $R_2$-$R_3$ divider. If point $S$ is assumed to be at the common voltage, $R_1$ and $R_3$ are effectively in parallel and the equivalent resistance $R_{f(eq)}$ between the output and point $S$ is derived to be

$$R_{f(eq)} = \frac{R_1 R_2 + R_2 R_3 + R_1 R_3}{R_3} \tag{5-22}$$

To produce a larger equivalent resistance, $R_3$ must be smaller than either $R_1$ or $R_2$. For instance, for $R_1 = R_2 = 10^6 \ \Omega$ and $R_3 = 10^3 \ \Omega$, $R_{f(eq)} = 10^9 \ \Omega$. This circuit should be used with the caution that the maximum value of $v_s$ due to op amp gain, offset, and drift limitations must remain much smaller than the voltage at the junction of the three resistors. In the example given, this voltage is only $v_o/1000$.

Error considerations in other current-follower-based op amp circuits are similar since all input circuits are current sources to the basic current follower circuit. For the inverter and voltage summing amplifiers, there is an error in $i_u$ if $v_s$ is not negligible compared to $v_u$. In the integrator the bias current is also integrated and is often the limiting factor in the length of time over which an integration is accurate.

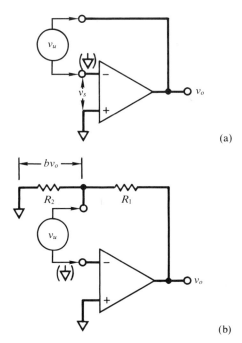

**Fig. 5-24.** Floating signal source comparison measurement. (a) Voltage follower. (b) Follower with gain. This configuration eliminates common mode error and decreases the loading of $v_u$.

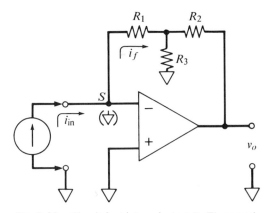

**Fig. 5-25.** Circuit for high equivalent $R_f$. Three moderate resistance resistors respond as a feedback resistance of much higher value. $R_3$ must be smaller than $R_1$ or $R_2$.

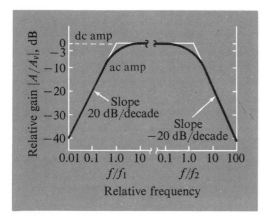

**Fig. 5-26.** Amplifier frequency response.

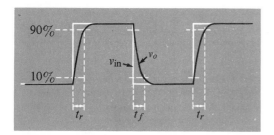

**Fig. 5-27.** Distortion of square wave due to high-frequency response limit.

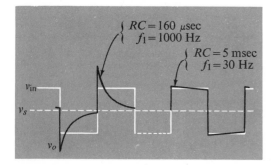

**Fig. 5-28.** Response of a high pass circuit to a square-wave signal.

## Amplifier Bandwidth

The combined high, low, and midrange frequency dependence of an amplifier can be summarized by the log-log plot of figure 5-26. Here the magnitude of the relative gain $A/A_v$ in decibels is plotted against the log of frequency. In the midrange of frequency, where $A = A_v$, the relative gain is 0 dB. At $f = f_1$ and $f = f_2$ the relative gain is $-3$ dB. Beyond the 3 dB points the gain rolls off with frequency with an asymptotic slope of 6 dB/octave or 20 dB/decade. Such rolloff characteristics are merely those of the low and high pass networks. The **bandwidth** of an ac amplifier is usually taken to be $f_2 - f_1$, and although there is appreciable gain outside this frequency range, it should also be remembered that the attenuation at $f_1$ and $f_2$ is already 30%. Note that there is no low-frequency rolloff with dc amplifiers such as an op amp.

It seems reasonable to assume that if the frequency components of a signal fall within the bandwidth of the amplifier, it will be amplified without distortion. This assumption can be tested by the square wave for which the frequency spectrum is shown in figure 2-7. It has been demonstrated that the discontinuities of the square wave are made up of the highest frequency components. Since the Fourier expansion of the square wave is a series that extends indefinitely to the high harmonics, a square waveform with an instantaneous transition from one voltage level to the other contains frequency components to infinite frequency. A practical amplifier with a finite high-frequency limit distorts the signal as shown in figure 5-27. The ability of an amplifier to respond to instantaneous signal changes is measured in terms of the **rise time** $t_r$, the time required to go from 10% to 90% of the applied change. It is a useful rule of thumb that the rise time of an amplifier is approximately related to $f_2$ by

$$t_r \simeq 1/(3f_2)$$

An even more dramatic example of square wave distortion is provided by the low-frequency limitations of the amplifier. The lowest frequency term in the Fourier series expansion for a 1000-Hz square wave is the 1000-Hz fundamental. A high pass filter with an $f_1$ of 1000 Hz has an $RC$ time constant of $1/(2\pi f_1) = 160\ \mu s$ (from eq. 2-3). However, a half-cycle is 500 $\mu s$ long, and the amplitude of a square wave subjected to such a filter would drop to a small fraction of the initial step as shown in figure 5-28. Even when $f_1 = 30$ Hz and $RC = 5$ ms, there is a definite slope, or a "droop," to the square wave. The percentage droop $D$ can be determined from the expression

$$D = 100\pi\ \frac{f_1}{f}$$

where $f$ is the frequency of the square wave. For a 10% droop on a 1000-Hz square wave, $f_1 = 32$ Hz. If only 1% droop can be tolerated, the low-frequency response of the circuit must be extended to 3.2 Hz. This droop in the square wave can be detected on a scope when the ac position of the input switch is used. When quantitative information is contained in the wave shape, careful attention must be paid to distortion due to bandwidth limitations.

## 5-7    Application: Polarography

An instrument that measures the current-voltage relationship for an electro-chemical cell is called a **polarograph**. Ions in solution gain or lose electrons at an electrode if the electrode voltage is sufficient to oxidize or reduce the ionic species. When the electrode voltage is scanned, a current due to the transfer of electrons between the electrode and the ion appears at the characteristic voltage for that ion. As the electrode voltage increases, the current increases until it is limited by the rate that the ions can diffuse through the solution to the electrode. The diffusion limited current is proportional to the ion concentration in solution. Thus the voltage at which the current begins and the limiting current value contain information about the species of ion and its concentration. A recording of the electrode current vs. the scanned electrode voltage, called a **polarogram**, is similar to the current-voltage curve of the $O_2$ electrode in figure 4-28a.

A very effective polarograph can be made from four op amps as shown in figure 5-29. The current at the working electrode W is converted to a proportional voltage by op amp 1 which also maintains the electrode at the common potential. The voltage at electrode W with respect to the solution in the electrochemical cell is measured by the reference electrode R which has a constant electrode/solution voltage difference. The output of the voltage follower (op amp 2) is thus the voltage difference between the reference and the working electrode. A voltage follower is used here so that there is negligible current in the reference electrode. The working electrode current is supplied to the cell through the counter electrode C by op amp 3. The feedback loop for op amp 3 includes the cell and the op amp 2 follower. Op amp 3 acts as a summing amplifier at its input so that the current through the cell is kept at the value for which the sum of $v_2$, $V_{init}$, and $v_{sweep}$ is zero. With the start/reset switch of the sweep generator of op amp 4 in the reset position, $v_{sweep}$ is zero and op amp 3 controls the current $i$ so that $v_2 + V_{init} = 0$. That is, $v_W - v_R = V_{init}$. Therefore, the working electrode potential before the scan starts is determined by setting $V_{init}$. The direction of the scan can then be selected by the switch $S_D$ and the rate of scan by setting $R_{sw}$. The scan begins when the sweep generator control is set to start. The polarogram is obtained by plotting $v_1$ vs. $v_2$ on an $x$-$y$ recorder or oscilloscope.

The versatility of the op amp is amply demonstrated in this application which includes sweep generation, summing, feedback control, high-impedance voltage measurement, and current-to-voltage conversion.

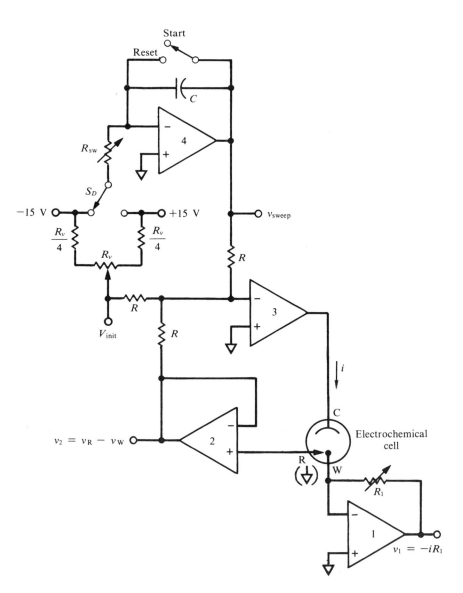

**Fig. 5-29.** A polarograph. The voltage between the working electrode W and the reference electrode R is controlled to be equal to $V_{init} + v_{sweep}$. The current at the working electrode is measured as a proportional voltage $v_1$ as the electrode potential is scanned from $V_{init}$ by the sweep generator. A polarogram is obtained by plotting $v_1$ vs $v_2$.

## Suggested Experiments

**1. Null voltage measurement.**
Use the comparator with logic-level indicator and the VRS to make a voltage comparison measurement of several voltage sources. For the circuit of figure 5-5a, measure the difference between $v_u$ and $v_r$ for several values of $v_r$ including 0, and calculate the CMRR for the comparator. Operate an op amp without feedback as shown in figure 5-8 to determine the $+$ and $-$ output voltage limits, and estimate the gain and the input voltage offset.

**2. Voltage follower.**
Connect a unity gain voltage follower, and determine its input/output function. Use the follower amplifier to eliminate loading of a high-resistance voltage source during measurement with a DMM or scope. Observe the change in output voltage of a follower amplifier as the output load is varied. Determine the follower output resistance and the linear output current limit for the op amp. Connect and test a follower with gain circuit for several gains between 10 and 1000.

**3. Current follower.**
Wire a current follower circuit, and test its response for several values of $R_f$ between $10^5$ and $10^7$ $\Omega$. Use the current follower to determine the reverse bias current of a signal diode. Connect the current comparison circuit of figure 5-11 to measure the current from the same source. Use the comparator for the null detector and the VRS with a large resistor for the reference current source.

**4. Inverting and summing amplifier.**
Connect an inverting amplifier, and determine its gain for several combinations of $R_f$ and $R_{in}$. Demonstrate the operation of a current summing amplifier. Design and wire a voltage summing amplifier to produce an output voltage that is proportional to the sum of $-v_1$, $-2v_2$, and $-4v_3$. Arrange an inverting amplifier with a thermistor as one resistor in such a way that the output voltage becomes more positive as the temperature increases. Set the gain so that the output changes 10 mV/$^\circ$C around room temperature. Connect a current offset to the summing input so the output voltage is 250 mV when the temperature is 25$^\circ$C.

**5. Integrator.**
Wire an integrator, and confirm that the output voltage changes linearly for a constant input current. Use the integrator to obtain the integral of several waveforms from the FG. Explain the resulting waveshapes. Integrate the charge from an unknown capacitor that was charged to a known voltage, and calculate the capacitor value.

**6. Differentiator.**
Connect a differentiator, and observe the input/output relationship for signals of several waveshapes and frequencies from the FG. Explain the waveforms observed.

**7. Op amp characteristics.**
Determine several characteristics for two or three different op amp types. The input offset voltage can be obtained from the voltage follower output error. The input bias current can be measured from the output voltage change of a voltage follower when a large resistor is in series with the noninverting input connection. The CMRR can be calculated from the change in the difference between $v_{in}$ and $v_o$ for a voltage follower as $v_{in}$ is changed from 0 to 5 V.

## Questions and Problems

**1.** In a voltage comparison measurement, the current from the source may not be exactly zero even when the null detector is as close to null as it can be set. (a) What is the worst-case current (after nulling) for the measurement in which the null indicator sensitivity is 0.5 $\mu$V, the resolution of the reference source is one part in $10^5$ for 1.000 00 V full scale, and the signal source and null detector resistances are 1 k$\Omega$ and 10 k$\Omega$ respectively? (b) What is the worst-case voltage error due to loading for the above measurement?

**2.** A comparator is to be used as a null detector in a voltage measurement. The output voltage changes 10 V for an input difference voltage of 500 $\mu$V, and the output voltage changes by 1 V for a common mode input voltage of 500 mV. What is the common mode rejection ratio (CMRR) of the comparator?

**3.** A comparator with a CMRR of 83 dB has been zeroed for 0 V input. (a) What is the voltage error in the threshold when the reference input is set at 2.5 V? (b) How do the circuits of figure 5-3b or c avoid the common mode error? (c) What are the conditions needed for the virtual common to be at exactly the same voltage as common?

**4.** For the potentiometer of figure 5-6, what value of $R_s$ should be substituted if the span is to be calibrated for 500 mV full scale?

**5.** For a servomechanical recorder like that of figure 5-7, the slidewire bridge has a voltage source $V = 1.34$ V and zero, span, and slidewire potentiometers of 10 kΩ, 5 kΩ, and 10 kΩ respectively. The slidewire moves from 10% to 90% of its full span as the pen moves from 0 to full scale on the chart paper. (a) What are the minimum and maximum full-scale sensitivities possible for this recorder? (b) A challenge question: To obtain greater sensitivity a voltage divider is often used between the slidewire bridge output and the feedback input of the comparator amplifier. What should the total resistance of such a divider be in order to keep the error due to loading the slidewire bridge output less than 0.25%?

**6.** An operational amplifier has a voltage gain of 200 000 and output voltage limits of +13 V and −14 V when used with a ±15-V power supply. The linear amplification range is guaranteed to be at least ±10 V. Sketch the transfer function for this amplifier as in figure 5-8, and label the axes with actual voltage values.

**7.** An operational amplifier with a voltage gain of $10^5$ and an input resistance of $10^{12}\Omega$ is used in the voltage follower circuit of figure 5-9 to measure a voltage source that has a value of about 2 V and a source resistance of 10 kΩ. What are the output voltage errors due to (a) the finite gain of the op amp, and (b) the loading of the voltage source?

**8.** (a) Design a follower with gain amplifier such as figure 5-10 to have a gain of exactly 200. Choose values for the divider resistors that are high enough to use only a small fraction of the op amp's output current capacity of 5 mA but are low enough to minimize the noise and instability that occur in high resistance circuits. (b) If the op amp has linear output voltage limits of ±10 V, what is the maximum range of input voltages usable with the follower with gain? (c) What is the maximum output voltage error at the maximum input voltage? ($A = 10^5$)

**9.** Compare the output resistance of an ideal voltage source and an ideal current source. Compare also the input resistance of ideal voltage and current measurement devices.

**10.** (a) In the current comparison circuit of figure 5-11, indicate the location of the virtual common point of the circuit if $J_2$ is connected to common (assuming the null condition). (b) Which, if any, of the circuit components (unknown source, null detector, and reference source) must be able to float? (c) What are the common mode rejection requirements of the null detector? (d) What characteristic of the circuit makes it an ideal load on the unknown current source?

**11.** (a) Design a current follower that will produce a 1-V output change for a 10-$\mu$A change in the input current. The op amp to be used has a voltage gain of $2 \times 10^5$ and an input bias current of 20 nA. (b) What is the effective input resistance of this circuit? (c) What is the percentage output error for an input current of 25 $\mu$A?

**12.** (a) Design an inverting amplifier that has a gain of −50. Choose resistance values that satisfy the criteria in question 8. The op amp to be used has a gain of $10^5$, an input bias current of 50 nA, a linear output voltage range of ±10 V, and an input resistance of $10^{12}\Omega$. (b) What is the range of usable input voltages? (c) What is the amplifier input resistance? (d) Calculate the output voltage errors due to finite gain and input bias current when the input voltage is 1 mV.

**13.** An inverting amplifier is to be designed. What is the highest voltage amplification accurate to 0.1% obtainable with an op amp whose gain is (a) $10^4$, (b) $10^6$, and (c) $10^8$?

**14.** Design a voltage summing amplifier that produces an output voltage that is $-50v_1 - 3v_2 - 15v_3$. Use reasonable values of resistance.

**15.** Challenge question: Design a circuit for which the output voltage is $v_o = 2 \times 10^4 i_1 - 10^6 i_2 + 18v_3 - v_4$. Present ideal loads to all input voltages and currents, and use reasonable valued components. More than one op amp will be required.

**16.** The circuit of figure 5-16a is used as an $R$-to-$V$ converter in some digital multimeters. If the digital voltmeter sensitivity is 200 mV full scale and the internal reference voltage is 1.30 V, what value of $R_c$ should be chosen so that the voltmeter output reads directly in kilo-ohms when the decimal point of the display is in the position, 00.00?

**17.** In the op amp current measurement circuit of figure 5-25 the values of $R_1$, $R_2$, and $R_3$ are 5 MΩ, 1 MΩ, and 5 kΩ. What is the equivalent feedback resistance $R_{f(eq)}$?

**18.** Choose the capacitor value for an op amp integrator to produce a 5.0-V output from the integration of 0.050 mC.

**19.** The op amp integrator of figure 5-18 has an output voltage limit of ±12 V. The values of $R$ and $C$ are 100 kΩ and 0.1 $\mu$F, and the inverting input has an offset voltage of 50 $\mu$V with respect to the circuit common. The capacitor is initially shorted with a switch of 1 Ω resistance. (a) What is the initial output voltage when the switch is closed? (b) How long does it take the op amp to reach its voltage limit after the switch is opened if $v_{in} = 0.00$ V? (c) What average input voltage is required to give an output integral accurate to 1%? (d) How long can the integration of the input voltage found in part (c) proceed before the op amp reaches its output voltage limit? (e) What is the maximum tolerable offset at the inverting input for a 1000 s integration accurate to 1% if the output voltage at 1000 s is to be 10 V?

**20.** (a) Design an op amp differentiator that gives an output of $-5.0$ V for an input rate of change of $+8$ V/s. (b) Design a differentiator to give an output of $+1.0$ V for an input rate of change of $+15$ V/s.

**21.** (a) A capacitor is held at a constant voltage of 5.0 V while its capacitance changes from 25 to 35 pF as a result of the relative motion of its plates. Calculate the corresponding change in the charge on the capacitor. (b) If this capacitor is $C_{in}$ in the charge-coupled amplifier of figure 5-20 and if $C_f$ has a value of 100 pF, what change in $v_o$ results from the change in $C_{in}$?

**22.** Challenge question: (a) Show that the charge-coupled amplifier is an example of a general form of inverting amplifier for which $v_o = -Z_f/Z_{in}$ where $Z_f$ and $Z_{in}$ are the complex impedances of the feedback and input components. (b) Comment on the principal causes of the limitations of the charge-coupled amplifier for low-frequency variations of $v_{in}$.

**23.** A voltage integrator with $R_{in} = 100$ k$\Omega$ and $C_f = 1.00$ $\mu$F is used to integrate a signal for 10 s between resets. The input offset voltage of the op amp is 1.0 mV, and the input bias current is 50 nA. Calculate the errors in the output voltage due to the input offset voltage and the input bias curent.

**24.** A unity gain voltage follower is made with an op amp that has a gain of $5 \times 10^4$, a CMRR of 75 dB, and an input offset (at 0 V) of 75 $\mu$V. Calculate the output errors at $+5$ V due to (a) the finite gain, (b) the CMRR, and (c) the input offset voltages.

# Chapter 6

# Programmable Analog Switching

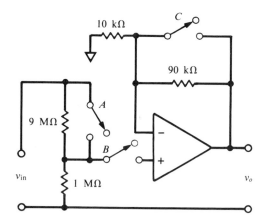

**Fig. 6-1.** Switches in an analog circuit. Switch *A* controls signal attenuation, switch *B* controls the amplifier input connection, and switch *C* controls the amplifier gain.

An ideal switch has only two states: it is either a perfect conductor or a perfect insulator. In its conducting state it directly connects two parts of a circuit, and in its insulating state that connection is broken. The change in the circuit caused by the change in state of a switch affects the analog quantities (voltage, current, charge, and power) in the circuit. Thus, a switch can be used to alter or direct the encoded quantity in an analog circuit. For example, a switch may determine whether or not a signal is connected to an amplifier input, whether or not a signal is attenuated, or whether the gain of the amplifier is 1 or 10 as shown in figure 6-1.

Note that each switch (contact pair) exerts a two-state, or binary, influence on the circuit. When multiple contact pairs are used in combination to produce a greater number of switched states, the transition from one state to another is still step-wise, or quantized. The switch is thus a crucial element that interfaces the digital and analog domains by providing digital control over the states of analog circuits.

In this chapter the principles and characteristics of switches and switched circuits are described. No switch is ideal, but the various switch types have special features as well as characteristic limitations. Applications in the areas of the control of the flow of analog information, waveform generation, and data sampling are discussed and illustrated in this chapter. Throughout the remainder of this book, the switch will be encountered as a critical element in many circuits.

## 6-1 Switching Principles

The accuracy with which analog information is transmitted from one point to another in modern electronic systems depends largely on the ideality of the switches used to direct the signals. In particular, the open and closed resistance of the switch can have a large effect on the accuracy of analog data transmission. This section explains how the nonideal resistance characteristics of switches influence data transmission. Then the basic configurations of switches in analog transmission gates are discussed. The general considerations of nonideal switch behavior discussed in this section are extended to mechanical and solid-state switches in the next sections.

## Switch Resistance

An ordinary mechanical switch (e.g., light switch) has two electrical states: open and closed (or OFF and ON). The ideal switch has zero resistance between its contacts when closed and zero conductance (infinite resistance) when open. The schematic symbols that are often used for open and closed switches are shown in figure 6-2a. However, in considering a real switch in an electronic circuit it is often important to represent the nonideal open and closed switch resistances $R_{so}$ and $R_{sc}$ as illustrated by the schematic for a nonideal switch in figure 6-2b.

In the generalized switching circuit shown in figure 6-3, the current in the circuit (through the load) is controlled by the switch. Because of the resistance of the switch, the source voltage $v_s$ is divided between the resistance of the load and the resistance of the switch. For purely resistive switches, when $R_{so}$, $R_{sc}$, and $R_L$ are known, the effectiveness of a given switch in a desired application can be evaluated with the circuit model of figure 6-3.

## Analog Switch Configurations

A basic **analog transmission gate** is shown in figure 6-4. An analog gate is designed to transmit from its input to its output an exact reproduction of the input waveform during the selected interval when it is open and to have a zero output when it is closed. Other terms used for such a device are **analog switch**, **sampling gate**, **transmission gate**, and **linear gate**. The simplest analog gate could be a manual switch, but the term gate generally refers to an automatically actuated switch.

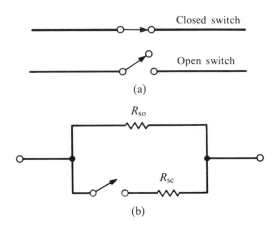

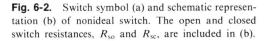

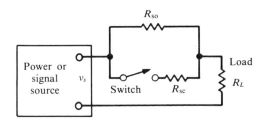

**Fig. 6-2.**  Switch symbol (a) and schematic representation (b) of nonideal switch. The open and closed switch resistances, $R_{so}$ and $R_{sc}$, are included in (b).

**Fig. 6-3.**  Generalized switching circuit. In order for the maximum signal to be transmitted to the load, the closed switch resistance $R_{sc}$ should be very small compared to $R_L$. Conversely, $R_{so}$ should be much greater than $R_L$ so that the current is effectively turned off by the switch.

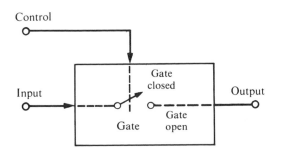

**Figure. 6-4.**  Basic gate. When the switch is ON (open gate), the signal at the gate input appears at the output. When the gate is closed, the output signal is zero. The gate control signal level determines whether the gate is open or closed.

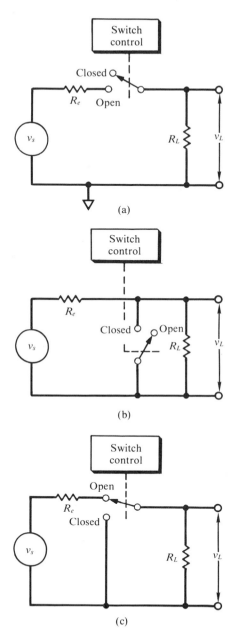

**Fig. 6-5.** Basic types of analog switch gates. In (a) the switch is in series with the load. A shunt switch is shown in (b), and a series-shunt voltage switch in (c). The state of the gate corresponding to each switch position is indicated.

The actual switching devices can be arranged in the gate circuit in several different ways. The three most common arrangements are shown with the signal sources and loads in figure 6-5. The choice among these possibilities depends on the switch characteristics and the gate application requirements. The simple **series switch** gate is shown in figure 6-5a, where $R_e$ is the equivalent source resistance. If the switch were ideal, the output voltage would be $v_s R_L / (R_e + R_L)$ when the gate is open (switch closed) and zero volts when the gate is closed. In order to approach this ideal, the switch's ON resistance $R_{ON}$ must be much less than $R_L$, and its OFF resistance $R_{OFF}$ very much greater than $R_L$. To obtain an accuracy of 0.1% for both transmission and rejection, a switch with an $R_{OFF}/R_{ON}$ ratio of $10^6$ or more would be required. Note that in the closed gate the only connection to common at the gate output terminal is through the load. This is an advantage if other signal sources are to be gated to this same load when this gate is closed, but it is a disadvantage if a well-defined zero voltage output is required for the closed-gate state.

The gate circuit of figure 6-5b employs a **shunt switch**. The gate is open when the switch is open, producing the same ideal output voltage as the series switch above. When the switch is closed, the gate input and output terminals are shorted. A true zero output in the closed-gate state is provided when $R_{ON} \ll R_e$. Practical signal sources and switches generally require a resistor in series with the source to increase the effective $R_e$. At the same time, $R_{OFF}$ must be much larger than $R_L$ to avoid distorting the signal voltage.

Better switching efficiency with fewer demands on the source and load resistance values is obtained with a combination of series and shunt switches like that in figure 6-5c. When the gate is open, the circuit acts like the series switch; and when the gate is closed, the source is disconnected (as in the series switch), but the load is shorted (as in the shunt switch). Accurate transmission therefore requires that $R_{ON} \ll R_L$ and, for a good zero output, that $R_{OFF} \gg R_{ON}$. Thus, the $R_{OFF}/R_{ON}$ ratio requirement for 0.1% accuracy is 1000 times less than that for the series switch, and the closed output voltage level is well defined at zero. One common application of the **series-shunt switch** is as a **chopper**. A chopper is a device that is used to alternately transmit and interrupt a signal at regular intervals. Generally the purpose of chopping an input signal is to compare its level accurately with a known reference level (often the common). For this reason it is important that the gate circuit used for chopping have a well-defined closed-gate output voltage.

## 6-2   Mechanical Switches

No switch is ideal, but different kinds of switches have various features and limitations. Mechanical switches have a pair of metal contacts that touch when the switch is closed and separate when the switch is open. All mechanical switches require some form of pressure to close (make) and open (break) the switch contacts. Mechanical switch contacts that can be actuated by the pressure of a human hand are known as **manual switches**. Contacts can also be actuated by mechanical pressure, such as the pressure exerted by a cam rotated by a motor. These switches are **mechanically actuated switches**. Finally, switch contacts can be actuated by electromagnets or permanent magnets. Such remotely controlled mechanical switches are known as **relays**.

## Manual Switches

Manual switches are used where complete operator control over opening and closing switch contacts is desired and where switching speed is relatively unimportant. For example, the on/off power switches on instruments are manually operated. Likewise manual switches are used in instruments for function selection, range changing, and many other control functions.

There are many different arrangements of switch contacts for manual switches. Some of these arrangements and circuit symbols are illustrated in figure 6-6. The simple two-contact switch is called a **single-pole single-throw (SPST)** switch. A combination of two stationary contacts and one movable contact is called a **single-pole double-throw (SPDT)** switch. When more than one movable contact is actuated by the same mechanism, the switch is called a **multi-pole switch.** An example is the double-pole double-throw switch illustrated in figure 6-6. Two different types of multiple contact switches exist: those that change states by breaking the existing connection before making the new connection are called **break-before-make (B-M)** and those that make the new connection before breaking the existing connection are called **make-before-break (M-B)**. Specific circuit applications usually dictate the most appropriate arrangements of contacts and connection mechanism. Break-before-make contacts are used when it is permissible to have a momentary open circuit and undesirable to have two circuit components connected together even momentarily. Make-before-break contacts are used when such a momentary open circuit would be undesirable, as in the switch selection of op amp feedback resistors for a current follower, where a momentary open circuit would drive the op amp to limit.

## Electrical Actuation of Switches

When an electrical signal is used to actuate a switch, there are generally two electrical circuits involved: the actuating circuit and the switched circuit as shown in figure 6-7. The degree of interaction between the actuating and

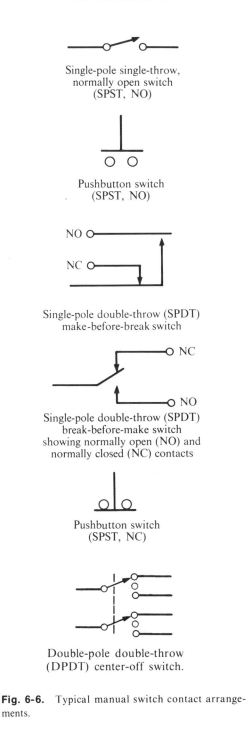

Single-pole single-throw, normally open switch (SPST, NO)

Pushbutton switch (SPST, NO)

Single-pole double-throw (SPDT) make-before-break switch

Single-pole double-throw (SPDT) break-before-make switch showing normally open (NO) and normally closed (NC) contacts

Pushbutton switch (SPST, NC)

Double-pole double-throw (DPDT) center-off switch.

**Fig. 6-6.** Typical manual switch contact arrangements.

**Fig. 6-7.** Electronically actuated switching devices. A signal applied to the actuating element controls the state of the switch and thus the signal or power level applied to the load.

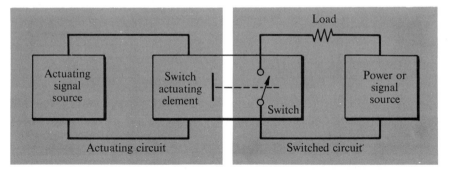

switched circuits depends on the type of switch and the circuit. Some switching devices have one or two connections inherently in common between the switch and the actuating element. The three possible degrees of interconnection for simple switches are shown in figure 6-8. The actuating element for electrically actuated switches is called a **switch driver**. The switch driver converts logic-level (HI/LO) signals into appropriate voltage levels or drive currents to actuate the switch.

**Fig. 6-8.** Switching devices. With the four-terminal switch (a), the actuating circuit is electrically isolated from the switched circuit. The three-terminal switch (b) has one connection in common, and the two-terminal switch (c) has both connections in common.

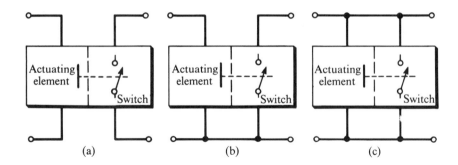

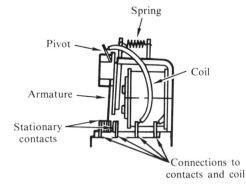

**Fig. 6-9.** Pictorial diagram of single-pole double-throw, (SPDT) relay. If the current in the coil (and, therefore the magnetic force) exceeds a certain minimum value, the armature moves the movable contact until it touches the stationary contact on the right.

## Electromagnetic Relays

Relays are remotely controlled mechanical switches. Electromagnetic relays utilize a current through a coil to provide a magnetic field that moves the switch contacts, as illustrated in figure 6-9 for an **armature relay**. The minimum current required to move the armature is called the **pull-in current**, because at or above that current the armature "pulls in" to close the normally open (NO) contact. At a somewhat lower current the armature "drops out," and the NO contacts open. Switching circuits are generally designed to exceed the pull-in current by several times the minimum to ensure operation of the relay. The electromagnetic relay is a four-terminal device in which the actuating terminals are electrically isolated from the switched signal terminals. They have activation response times of a few milliseconds, very high open-circuit resistance, very low contact resistance, and often, the ability to switch high currents and/or voltages.

The **reed relay** contains two or more metal reeds enclosed in a hermetically sealed glass capsule. A normally open SPST reed relay is shown in figure 6-10. The overlapping reeds can be closed or opened by positioning a permanent magnet near or away from the reed contacts.

Relays for various applications differ in the number of contacts and contact arrangements. The nomenclature and symbolism for the four most common contact forms are given in figure 6-11. Many other contact forms are available. These may be combined in a variety of multiple-form arrangements.

Most relays require several milliseconds to complete the transition from one contact state to the other. A circuit for the observation of relay response times is shown in figure 6-12a. A typical relay response for the coil drive signal is shown in figure 6-12b. After the application of coil drive current

Reed switch relay

Dry reed in coil
Switch operates in any position

**Fig. 6-10.** Reed relay. The reed contacts are switched by actuating an electromagnet.

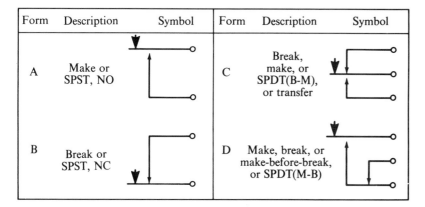

| Form | Description | Symbol | Form | Description | Symbol |
|---|---|---|---|---|---|
| A | Make or SPST, NO | | C | Break, make, or SPDT(B-M), or transfer | |
| B | Break or SPST, NC | | D | Make, break, or make-before-break, or SPDT(M-B) | |

**Fig. 6-11.** Four common forms of relay contacts with designations. The heavy arrow indicates the position and direction of the force from the coil when energized.

**Fig. 6-12.** Observation of operate, transfer, bounce, and release times. The circuit in (a) provides outputs of 0, 2.5 and 5 V for the relay contact being normally closed, in between, and normally open, respectively. The time response of the relay is shown at the top of (b) in relation to the drive signal shown below.

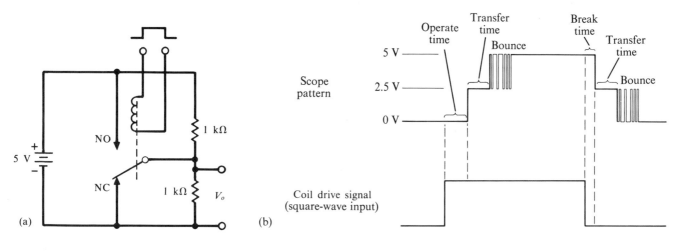

there is a finite time delay, called the **operate time** for normally closed (NC) switches, before the movable contact breaks away from the NC contact. After the NC contacts break, there is a finite **transfer time** before the movable contact reaches the normally open (NO) contact. When the NO contacts first touch the contacts bounce apart and together for a time before a firm connection occurs. This **contact bounce** can seriously distort the switched signal and severely limit switching speeds. When the drive signal is removed, the movable contact requires a finite **break** or **release time** before it disengages the NO contact. Again there is contact bounce when the movable contact first strikes the NC contact. Contact bounce is also characteristic of manual switches. Mercury-wetted relays and switches are sometimes used to overcome the contact bounce limitation of normal switches.

## 6-3    Solid-State Switches

In a mechanical switch the switch state changes from conducting or nonconducting by actual motion of a metallic contact. In the solid-state switch, there is no physical motion of contacts; rather, the material between the contacts is made either conducting or nonconducting in response to an external signal. Because of the speed at which a solid-state device can change states, such devices are imperative for the short time (micro- to nanosecond) on-off control operations that many modern circuits and devices require.

The widespread availability of analog switches in integrated circuit form has made elegant high-speed analog switching applications possible at relatively low cost. This section describes IC analog switch types and characteristics and the increasingly useful optically coupled analog switch. The reader is referred to chapter 7 for a description of the specific bipolar and field-effect transistor devices used in integrated circuit switching packages.

### Integrated Circuit Analog Switches

Both solid-state switching elements and the associated switch drivers are integrated in a single package in modern IC analog switches. Symbolism for such switches and switch drivers is shown in figure 6-13. Because the actuating signal and the switch contacts have no terminals in common, the IC analog switch is an example of a four-terminal device. However, the switched circuit is not completely isolated from the actuating circuit as in the relay. Most IC analog switch packages contain two or more sets of switches and drivers in the same unit. They are available with a variety of "contact" arrangements (SPST, SPDT, etc.) and a variety of switching characteristics.

IC analog switches are of two types: voltage switches and current switches. A **voltage switch** is used to transmit an analog voltage from input

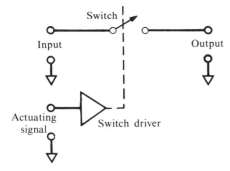

**Fig. 6-13.** Normally open SPST analog switch. The switch driver converts logic-level signals to the appropriate voltage or current levels to actuate the solid-state switch.

to output when the switch is closed. The switch state is determined by a logic-level actuating signal. Figure 6-14 illustrates typical waveforms for the analog voltage switch of figure 6-13.

Analog **current switches** are made to operate with one of the switch contacts connected to the system common or to the virtual common of an operational amplifier. Figure 6-15 shows the use of a typical current switch to switch a current source input to an op amp current follower. The small voltage drop across the switch (limited by the diode) assures fast switching speeds and allows the switch to be opened and closed with relatively small drive voltages, such as the HI/LO logic levels of the TTL family. Analog current switches can often switch states several times faster than comparable voltage switches. The input signal is shunted to ground through the diode when the switch is open. Therefore, voltage sources that should not be shorted should be connected to a current switch through a resistor.

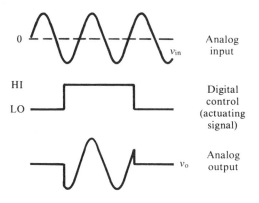

**Fig. 6-14.** Analog voltage switch waveform. The analog signal $v_{in}$ appears at the output when the drive signal is at the HI logic level and the switch is closed.

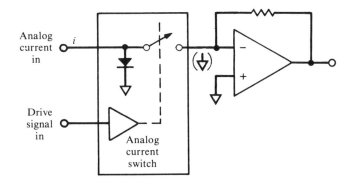

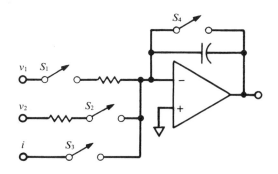

**Fig. 6-15.** Analog current switch used to switch current signal $i$ to an op amp current follower. The input diode limits the voltage drop across the switch to 0.6 V in the OFF state by shunting the input signal to ground.

Figure 6-16 shows an op amp integrator with three switched input sources and a shorting switch for the integrating capacitor. Switch $S_1$ controls the application of voltage $v_1$ and should be a voltage switch. Switches $S_2$ and $S_3$ can be current switches since in each case one contact is connected to the op amp summing point and the other contact is not directly connected to a voltage output. Switch $S_4$ should be a voltage switch since it is connected to the op amp output.

Integrated circuit analog switches are characterized by several important parameters. First the ON resistance should be very low for accurate voltage or current switching. Typical ON resistance values vary from less than ten ohms to several hundred ohms. The OFF resistance of the switch should be very high for excellent isolation of input and output in the OFF state. Typical OFF resistance values are in the $10^9$ to $10^{12}$ $\Omega$ range. Another important parameter is the change in $R_{ON}$ with applied analog voltage,

**Fig. 6-16.** Op amp integrator with multiple switched sources. Switches $S_1$ and $S_4$ are voltage switches; switches $S_2$ and $S_3$ can be current switches.

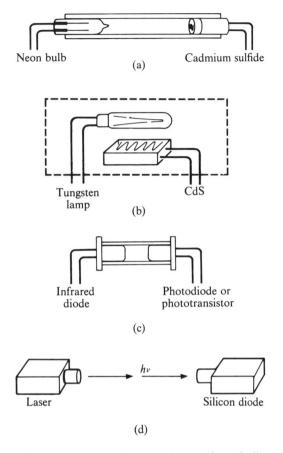

Neon bulb          Cadmium sulfide
(a)

Tungsten          CdS
lamp
(b)

Infrared          Photodiode or
diode             phototransistor
(c)

Laser             $h\nu$          Silicon diode
(d)

**Fig. 6-17.** Source-detector pairs used in optically coupled analog switches.

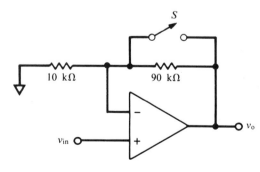

$S$

$10\ \text{k}\Omega$     $90\ \text{k}\Omega$

$v_{\text{in}}$          $v_o$

**Fig. 6-18.** A follower with gain circuit. The gain is 10 when the switch is open and 1 when it is closed.

sometimes called **ON resistance modulation**. For good switches $R_{\text{ON}}$ changes by only a few percent over the full range of allowed analog signal levels. When several switches and drivers are contained in a single IC, an important measure of the isolation of one switch from another is the switch **crosstalk**. Analog signals applied to the input of one switch can appear at the output of another switch unless the crosstalk is very low. Because the drive signal is not totally isolated from the analog signal, the drive signal can sometimes **feed-through** to the analog output. The optically coupled switches described next completely prevent feedthrough distortion. Finally, the switch capacitance should be very low since capacitance can limit the switching speed. Integrated circuit analog switches have switching speeds in the nanosecond to microsecond range. Capacitance effects are discussed in section 6-4.

## Optically Coupled Analog Switches

The combination of a light source and a photodetector inside an opaque enclosure can produce a totally isolated analog switch often called an **opto-isolater**. The actuating circuit is the light source, and the photosensitive device acts as the switch. No electrical coupling is needed between the actuating circuit and the switch: Light is the coupling link. Because the light source terminals and photodetector terminals are electrically isolated, the pair forms an almost ideal four-terminal device in which there is essentially no interaction between switch and driver circuits.

Some typical optical links between matched source and detector pairs are illustrated in figure 6-17. In recent years completely encapsulated **optical couplers** have become available commercially. Many of these use light-emitting diodes (LEDs) as sources and phototransistors as detectors. The LED-phototransistor optical coupler is described in chapter 7. In addition to providing electrical isolation of switch and actuator, optical coupling eliminates ground loops and isolates noise sources.

## 6-4   Transient Behavior of Switched Circuits

The opening or closing of a switch causes the currents and voltages in the switched circuit to undergo a transition between those of the open-switch condition and those of the closed-switch condition. The voltages and currents for the two states of the switch can be determined independently. For example, the circuit of figure 6-18 is a voltage follower with a gain of ten when switch $S$ is open and a gain of one when switch $S$ is closed. For the open-switch condition, the voltage at the $-$input of the op amp is $(v_o/10)$ volts, $v_o$ is $10\ v_{\text{in}}$ volts, and the current through the two resistors is $(v_o/10^5)$ amperes. In the closed-switch condition, $v_{\text{in}}$, $v_o$, and the $-$input are all of equal voltage, and the current through the 10 k$\Omega$ resistor is $(v_o/10^4)$ amperes.

The time required for the signal levels to change from one condition to another depends on factors inherent in the switch as well as on the ability of the switched circuit to respond to rapid changes. The response of the switched circuit will be explored first. We shall see that, even if the switch were to open or close instantaneously, the transition of the signal values between the closed and open levels would not be instantaneous. The finite time required is due to inductive and capacitive reactances in the circuit that resist sudden changes in current and voltage respectively. Every component in a circuit has some inductance or capacitance, and these reactances, along with the resistances, limit the response time of the circuit. The influence of the capacitance in the switch itself on the maximum switching speed must also be considered.

## Series *RC* Circuit

Suppose that in the circuit of figure 6-19 $V = 10$ V, $R = 100$ Ω, $C = 1$ μF, and the switch has been at position $B$ for a long time. Any charge on the capacitor has been discharged through $R$ and the switch; the charge $q$ on the capacitor is therefore now zero, and the current is zero. Since $v_R = iR$, $v_R = 0$, and since $v_C = q/C$, $v_C = 0$. Kirchoff's voltage law is satisfied since $v_C + v_R = 0$ V, the voltage applied in position $B$. Now the switch is turned to position $A$, and the applied voltage is $V$, i.e., 10 V. At the instant of closing, the charge on the capacitor is still zero so $v_C = 0$. Because $v_C + v_R$ is now 10 V, the entire 10-V drop appears across $R$. The current $i$ at this instant is $v_R/R = 10/100 = 0.1$ A.

The current immediately starts to charge the capacitor. At any time,

$$V = v_R + v_C = iR + \frac{q}{C} \qquad (6\text{-}1)$$

Thus, as $q$ and $v_C$ increase, $i$ and $v_R$ decrease. All four of these quantities change exponentially with time as shown in figure 6-20. Note that the time scale is calibrated in units of $RC$. The product $RC$ has the units of seconds

$$RC = \frac{v}{i} \times \frac{q}{v} = \frac{\text{volts}}{\text{coulombs/seconds}} \times \frac{\text{coulombs}}{\text{volts}} = \text{seconds}$$

and is called the **time constant**. The current $i$ at any time $t$ after turning the switch to $A$ is given by equation 6-2 (see note 6-1).

$$i = \frac{V}{R} e^{-t/RC} \qquad (6\text{-}2)$$

After a time $t = RC$, the current $i_t = (V/R)e^{-RC/RC} = (V/R)e^{-1} = (V/R) \times 0.368$. In figure 6-19 at time $t = RC = 100 \times 1 \times 10^{-6} = 10^{-4}$ s, the

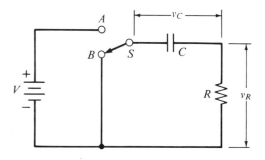

**Fig. 6-19.** A series *RC* circuit and step signal source. In switch position $A$ the capacitor charges towards the applied voltage $V$. In switch position $B$ the capacitor discharges through resistor $R$.

**Note 6-1.  Derivation of Charging Equation.**
Equation 6-2 can be derived by substituting $dq/dt$ for $i$ in equation 6-1 and solving the resulting differential equation. The result is

$$v_C = V(1 - e^{-t/RC})$$

To obtain the $i$ vs. $t$ relation, we substitute the above equation into equation 6-1:

$$v_R = V - v_C = Ve^{-t/RC}$$

Since $v_R = iR$, it follows that

$$i = \frac{V}{R} e^{-t/RC}$$

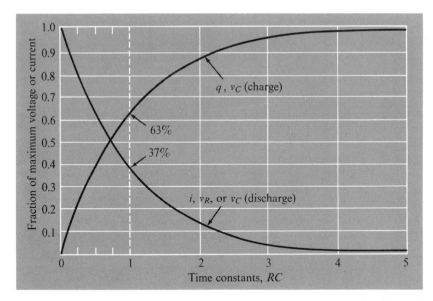

**Fig. 6-20.** Charge and discharge curves in the *RC* circuit. On charging, the capacitor voltage $v_C$ reaches 63% of the impressed voltage in one time constant. On discharging, the capacitor voltage falls to 37% of its value during one time constant.

current is 36.8% of the current at the instant of impressing voltage *V*. This means that at a time equal to one time constant the voltage across the resistor is only 36.8% of its initial value, and the capacitor is charged to 63.2% of the impressed voltage. The time constant of the *RC* circuit, is given the symbol $\tau$. Table 6-1 gives values of $v_C$ and $v_R$ for different multiples of $\tau$, both for the charging of the capacitor by impressing voltage *V* and for discharging a capacitor that had been charged to a voltage *V*. Note that when $t = 4.6\ \tau$, the capacitor is charged to 99.0% of the impressed voltage

**Table 6-1.**  Output voltages across capacitor and resistor in series *RC* circuit.

| Time | Capacitor charging | | Capacitor discharging | |
|---|---|---|---|---|
| | $v_C$ % *V* applied | $v_R$ % *V* applied | $v_C$ % *V* initial | $v_R$ % *V* initial |
| $RC = \tau$ | 63.2 | 36.8 | 36.8 | 36.8 |
| $2\tau$ | 86.5 | 13.5 | 13.5 | 13.5 |
| $2.3\tau$ | 90.0 | 10.0 | 10.0 | 10.0 |
| $3\tau$ | 95.0 | 5.0 | 5.0 | 5.0 |
| $4\tau$ | 98.2 | 1.8 | 1.8 | 1.8 |
| $4.6\tau$ | 99.0 | 1.0 | 1.0 | 1.0 |

and $v_R$ is only 1% of its initial value. For practical purposes the capacitor is often considered to be fully charged when $t \simeq 5\tau$. It is important to keep in mind that the voltage across a capacitor cannot change instantly; instead, it changes exponentially with time.

When an oscilloscope is connected to the series $RC$ circuit first across the resistor and then across the capacitor, various waveshapes similar to those in figure 6-21 can be observed when the switch of figure 6-19 is turned on and off, or when a rectangular pulse source is used as the input. The output waveform depends on the ratio of the $RC$ time constant $\tau$ to the pulse width $T_p$. It is interesting to observe that the leading edge of the output across the resistor is always steep as long as the input voltage has a steep leading edge. In contrast, the leading edge of the capacitor output always changes exponentially. Note that the sum of voltages across the capacitor and resistor equals the input voltage at each instant for a given $RC$ time constant. This can be observed by comparing the pairs of curves in figure 6-21c, d, and e.

Sharp positive and negative pulses can be obtained across the resistor when the $RC$ time constant is much shorter than the pulse width as in figure 6-21e. This type of response finds application in many circuits. When the time constant is greater than the pulse width (fig. 6-21c), the voltage across the capacitor is a rather linear sawtooth voltage. In this case, the $RC$ circuit is often referred to as an integrator because the rate of capacitor charging is nearly proportional to $v_i$ as long as $v_C \ll v_i$. This condition is met only when the $RC$ time constant is much longer than the pulse width. In the op amp integrator in chapter 5, the servo system idealizes the $RC$ integrator by keeping $v_R$ always equal to $v_i$. This equality keeps the rate of change of the capacitor charge proportional to $v_i$.

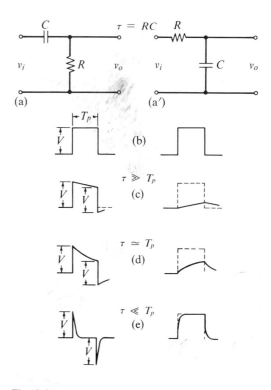

**Fig. 6-21.** Waveforms in the $RC$ series circuits for the output taken cross the resistor (a) and the output taken across capacitor (a'). The input voltage $v_i$ is shown in (b). The waveforms in (c), (d) and (e) result from various ratios of the time constant $\tau$ to the pulse width $T_p$.

## Parallel *RC* Circuit

To determine the shape and time constant of voltage changes in a circuit, it is necessary to determine only (1) whether the output signal is across the capacitance or across the resistance, and (2) what components determine the $RC$ time constant in the case of each change. To take another example, consider the parallel $RC$ circuit of figure 6-22. When switch $S$ is thrown to the ON position, the capacitor begins to charge. The charging rate is determined by $C$ and by the output resistance of the charging circuit, which consists of the source $V$ and the voltage divider. Thevenin's theorem tells us that the equivalent resistance of a voltage divider is the parallel combination of the divider resistors (see note 6-2), or $R_1R_2/(R_1 + R_2)$. Thus, the charging time constant is $CR_1R_2/(R_1 + R_2)$. When switch $S$ is turned OFF, the capacitor discharges toward zero with a discharge time constant of $CR_1R_3/(R_1 + R_3)$.

**Note 6-2. Equivalent Resistance of a Voltage Divider.**

Thevenin's theorem states that the voltage divider in figure 6-22 ($V$, $R_1$ and $R_2$) can be replaced by a single equivalent voltage $V_e$ in series with a single series equivalent resistance $R_e$, as shown in figure 1-23 in chapter 1. The equivalent resistance of the divider is $R_e = R_1R_2/(R_1 + R_2)$.

**Fig. 6-22.** Parallel *RC* switching circuit. When switch *S* is thrown to the ON position, the capacitor *C* charges towards the voltage established by the voltage divider $R_1$ and $R_2$. When *S* is turned off, the capacitor discharges through the parallel paths $R_1$ and $R_3$.

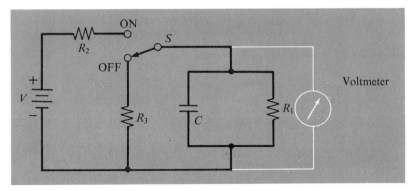

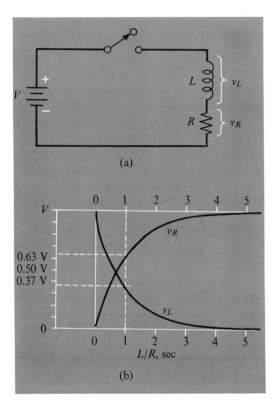

(a)

(b)

**Fig. 6-23.** Series *LR* circuit and response. When the switch is closed, the voltage $v_L$ decays exponentially from its value of *V* at the instant of switch closure to zero. The voltage $v_R$ increases exponentially toward *V*.

**Note 6-3.  Series *LR* Response.**

When the switch in figure 6-23 is closed, $v_R = iR$ and $v_L = L\, di/dt$. From Kirchoff's voltage law $V = v_R + v_L = iR + L\, di/dt$. Solving this differential equation yields $v_L = Ve^{-tR/L}$ and $v_R = V(1 - e^{-tR/L})$.

Since it is the voltage across the capacitor that is being observed, the waveforms resemble those in the right-hand column of figure 6-21.

## Inductive Circuits

The presence of inductance in a circuit element makes it impossible to change the current value instantaneously just as capacitance makes it impossible to change the voltage instantaneously. In either case there is a time lag between the attempt to change the voltage or current level in the switched circuit and the attainment of that change.

A series inductance-resistance (*LR*) circuit is shown in figure 6-23a. Here the inductance *L* might be an actual inductor intentionally placed in the circuit or, more often, an undesirable **stray inductance** associated with the circuit connections or with the circuit elements. When the switch is closed, the sum of $v_L$ and $v_R$ changes from 0 to *V*. However, the inductor prevents an instantaneous change in current. Thus, immediately after the contact is made, *i* and $v_R$ are zero, and $v_L$ equals *V*. Since $v_L = L(di/dt)$, the current is increasing at a rate of $V/L$ amperes per second. As the current increases, some of the voltage *V* appears across *R*. This decreases both $v_L$ and the rate of increase in the current. The result is an exponential decrease in $v_L$ according to $v_L = Ve^{-t/RL}$ and an exponential increase in $v_R$ according to $v_R = V(1 - e^{-tR/L})$ (see note 6-3). The quantity $R/L$ in the exponent is the reciprocal of the time constant for *LR* circuits. Thus $L/R = \tau$ and has the units of seconds. The curves for $v_L$ and $v_R$ as functions of $L/R$ units are shown in figure 6-23b. At large values of $L/R$, $v_L$ approaches zero, while $v_R$ approaches the steady-state value *V* and the current through the inductor and resistor is $V/R$. When the switch is opened, the current suddenly becomes zero. The inductor reacts against this sudden change in current by developing a large negative value of $v_L$. Since $di/dt$ is very large when the switch is opened, the value of $v_L$ when the current is interrupted in real inductors can be thousands of volts. This can cause arcing across the open switch contacts and eventual destruction of the switch. This same effect is used to generate high voltage sparks for internal combustion engines and other applications.

## Switch Capacitance

The preceding discussion has included the effects of the switched circuit capacitance and inductance on the attainment of steady-state current and voltage levels. Factors inherent in the switching device itself also affect switching speed. One such factor is capacitance, which is unavoidable in the construction of any switch.

To account for switch capacitance the schematic of the nonideal switch given in figure 6-2 can be modified to include an **equivalent switch capacitance** $C$ as shown in the boxed area of figure 6-24. The nonideal switch is shown connected in a series circuit in figure 6-24, and the effect of the equivalent switch capacitance will now be noted. Assume that the switch $S$ is initially closed. The voltage $V_{sc}$ across the switch is $V_{sc} = VR_{sc}/(R_L + R_{sc})$ as determined by the voltage divider $R_L$ and $R_{sc}$. When the switch is opened, the voltage across the open switch rises exponentially as the equivalent capacitance $C$ is charged to the maximum open-switch voltage $V_{so}$ (fig. 6-25). The value of $V_{so}$ is $VR_{so}/(R_L + R_{so})$ as determined by the voltage divider $R_L$ and $R_{so}$. The charging time constant $\tau_0$ is determined by $C$ and by the output resistance of the voltage divider $R_L$ and $R_{so}$. Thus,

$$\tau_0 = CR_LR_{so}/(R_L + R_{so}) \qquad (6\text{-}3)$$

If $R_{so} \gg R_L$, as is usually the case, then $\tau_0 \simeq CR_L$.

The discharge time constant $\tau_c$ is determined by $C$ and by the parallel discharge paths $R_{so}$, $R_{sc}$, and $R_L$. Since we assumed that for an effective switch $R_{so} \gg R_{sc}$,

$$\tau_c = CR_LR_{sc}/(R_L + R_{sc}) \qquad (6\text{-}4)$$

In most cases $R_{sc} \ll R_L$, so that $\tau_c \simeq CR_{sc}$.

In the circuit of figure 6-24, if $C = 100 \text{ pF}$, $R_L = 10 \text{ k}\Omega$, $R_{so} = 10^9 \ \Omega$, and $R_{sc} = 100 \ \Omega$, then

$$\tau_0 = CR_L = 100 \times 10^{-12} \times 10^4 = 10^{-6} \text{ s} = 1 \ \mu\text{s}$$

and

$$\tau_c = CR_{sc} = 100 \times 10^{-12} \times 10^2 = 0.01 \ \mu\text{s}.$$

Note that the value for $\tau_c$ is much less than that for $\tau_0$, as we would expect because $R_{sc} \ll R_L$. Therefore, a circuit with significant switch capacitance reaches a steady state more rapidly when the switch closed than when it is

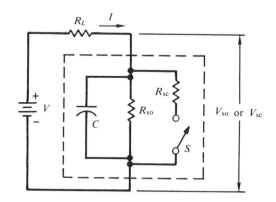

**Fig. 6-24.** Switch circuit with nonideal switch. The resistances $R_{so}$ and $R_{sc}$ are the open and closed values for switch $S$, and $R_L$ is the load resistance of the circuit. The equivalent switch capacitance is $C$.

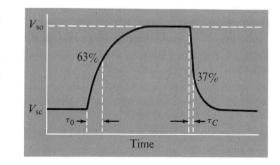

**Fig. 6-25.** Effect of capacitance on transition between voltage levels. The charging time constant $\tau_0$ is given by equation 6-3 and is $\sim CR_L$. The discharge time constant $\tau_c$ is $\sim CR_{sc}$, which is usually much smaller than $\tau_0$.

opened. There are other factors inherent in certain devices such as diodes and transistors that determine switching speed; these are discussed in chapter 7.

## 6-5   Voltage-Programmed Switching and Timing Circuits

For every electrically actuated switch there is some minimum level of drive signal required to ensure actuation. For example, a relay has a minimum pull-in current, and a diode has a minimum forward conduction bias. The state of the switch is determined by whether the drive signal is above or below a certain level. The voltage required for actuation of a switch can be made precise and adjustable by means of a comparator switch drive. The result is a **voltage-programmed switch**. In this section a number of practical applications of voltage-programmed switches are illustrated. These include decision thresholds for analog transducer outputs, simple overvoltage protection circuitry, timing and delay circuits that use an $RC$ charging waveform, and function generators.

### Voltage-Actuated Switches

The actuation level of most electrically actuated switches is not sufficiently reproducible to use as a threshold detector. Instead a comparator is used as, or in conjunction with, a switch driver (fig. 6-26). Because the comparator output is either HI or LO, it gives an unambiguous signal to the switch driver. In some cases the comparator output may be an appropriate switch drive signal directly. The $v_{in}$ threshold level for the HI/LO transition of the comparator is adjustable within a few millivolts or less. An obvious application for such a switch is as an event indicator or **critical level indicator**. A transducer is arranged to monitor the level or event, and the threshold is adjusted so that the switch causes the appropriate action (sound alarm, trip counter, remove power, etc.) when the event occurs or when the critical level is exceeded. A voltage-actuated switch is part of a **zero-crossing switch**, an ac power switch that turns on or off only when the supply voltage is at zero. Such switches, described in more detail in chapter 7, are useful in reducing noise caused by ac power switching.

The simplest voltage-actuated switch of all is the pn junction diode. Since it is a two-terminal switch (fig. 6-8c) the actuating and switched signals are the same. The actuation threshold for conduction is a forward bias of 0.6 V for the silicon diode. Despite these limitations the diode is a remarkably useful switch. Diode switches can be used to transmit selectively or to clip off any part of a signal that exceeds the threshold voltage. Figure 6-27 illustrates a **diode clipping circuit** to clip a signal at $V_1$ on the positive half-cycle and $V_2$ on the negative half-cycle. Zener diodes with breakdown voltages of $V_1$ and $V_2$ could replace the diodes and voltages in figure 6-27.

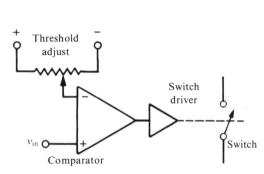

**Fig. 6-26.** A comparator/driver voltage programmed switch. The comparator provides a precise threshold for $v_{in}$ and a clean HI/LO input to the switch driver. The switch can be either mechanical or solid state.

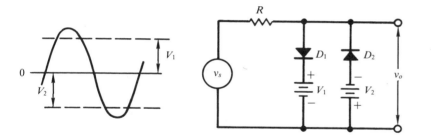

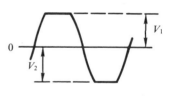

Practical applications of the diode clipping circuit are shown in figure 6-28. Most op amps would be destroyed by a voltage at an input greater than either power-supply voltage. The diode clipper keeps such destruction from happening for the follower of figure 6-28a. Some op amps have such protection diodes built in to their circuitry. The input of the inverting amplifier of figure 6-28b is similarly protected against overload by the diodes. As long as the summing point remains a virtual common (within a few millivolts), the diodes do not affect the amplifier's operation. Low leakage diodes should be used.

## *RC* Timers and Monostable Multivibrators

The time required for a capacitor that is charging through a resistor to reach a particular level of charge is the basis for **RC timers**. The elements of a timer circuit are shown in figure 6-29. Before the start of a timed interval, the capacitor remains discharged by the closed condition of switch S. When the timed interval is to begin, the logic level at the trigger input changes from HI to LO. This causes a change from LO to HI at the flip-flop output, which in turn causes the switch driver to open switch S. The capacitor is now free to charge from voltage source $V_1$ through R. The capacitor voltage is monitored by the comparator. The comparator output changes state when the capacitor voltage exceeds the fraction of $V_2$ selected by the comparator input divider.

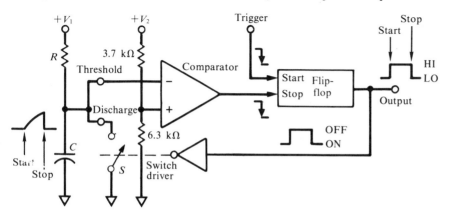

**Fig. 6-27.** Diode clipping circuit. Diode $D_1$ conducts whenever $v_s$ is larger than $V_1 + 0.6$ V. The difference between $v_s$ and $V_1$ appears as an *IR* drop across R, and only $V_1 + 0.6$ V appears at the output. When $v_s$ is less than $V_1$, $D_1$ is reverse biased and does not affect the output. Diode $D_2$ provides a similar clipping action at a maximum negative limit set by $V_2$.

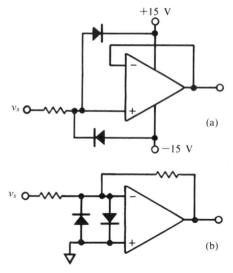

**Fig. 6-28.** Clipping circuit applications. In (a) $v_s$ is prevented from exceeding the supply voltages, in (b) the summing point is kept near the common voltage.

**Fig. 6-29.** *RC* timer and monostable multivibrator. A HI→LO trigger signal causes the flip-flop output to go HI opening switch S. Capacitor C charges until the comparator threshold is reached. The HI→LO comparator output causes the flip-flop output to go LO closing switch S. The HI→LO transition at the output appears *RC* seconds after the trigger signal is applied. The popular 555-type of timer operates in this way.

The HI→LO transition at the comparator output stops the timing by causing the flip-flop output to go LO, closing switch $S$. The net result of this cycle is to produce an output pulse of a duration determined by the values of $R$, $C$, $V_1$, $V_2$, and the comparator divider. Typically the $RC$ charging and divider circuits are connected to the same voltage, and the divider is set at 63% of $V$. The capacitor charges to $0.63V$ in one time constant $\tau = RC$ (recall fig. 6-20). Thus, the HI→LO transition at the output appears $RC$ seconds after the trigger transition is applied and is independent of $V$.

The $RC$ timer output can be used to cause some operation to occur or to begin a controlled time after the trigger signal. The extensive need for controlled intervals and delays in electronic circuits has made $RC$ timers a common element. Integrated circuit timers contain one to four $RC$ timers that are complete except for $R$ and $C$. The 555 timer can produce pulse durations of from a few microseconds to hours. A 1% reproducibility of the pulse duration is attainable but it is generally limited by the stability of $R$ and $C$. The **monostable multivibrator (MS)** is a controlled duration pulse generator designed to operate within a particular family of logic circuits (see chapter 11). The MS is characterized by higher speed (10 ns—100 ms) and lower accuracy, but it operates on the same principle as the $RC$ timer.

## Function Generators

A signal source for which the output voltage varies in a specific way as a function of time is a **function generator**. The waveforms produced by function generators are useful for a variety of testing and control applications. One of the most useful waveforms is the sweep or ramp signal. A signal that changes amplitude at a constant rate is generally obtained by accurately integrating a constant signal. This operation is performed by the op amp integrator circuit in figure 6-30. The constant voltage $V_{in}$ produces a constant current $I_{in}$ at the summing point of the integrator. This current charges the capacitor at the rate of $I_{in}$ coulombs per second. This capacitor voltage $v_C$ then changes at the rate of $I_{in}/C$ volts per second. Since $I_{in} = V_{in}/R$, the rate of change of $v_C$ equals $V_{in}/RC$. Since $v_o = v_C$, the selection of $V_{in}$, $R$, and $C$ can provide almost any desired sweep rate.

For example, if $V_{in} = +5$ V, $R_{in} = 100$ kΩ, and $C = 0.1$ μF, the sweep rate would be $-5/(10^5 \times 10^{-7}) = -5 \times 10^2$ or $-500$ V/s. If the op amp has a maximum output voltage of ±10 V, the sweep is linear for $10/500 = 1/50$ s before limiting at $-10$ V. When the switch is closed, the capacitor discharges through the switch, and $v_o$ quickly returns to zero. This simple sweep circuit is useful in many laboratory applications for sweep times in the range of microseconds to minutes.

To produce a repetitive periodic sweep signal requires an automatic periodic reset for the integrator. This can be provided in two ways. One is to use the output of an oscillator to close the reset switch momentarily at

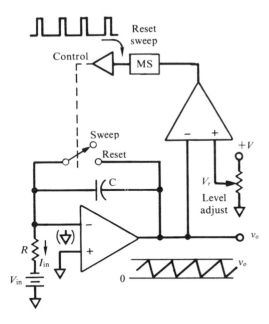

**Fig. 6-30.** Sweep generator with automatic level switch reset. This op amp integrator produces a linear sweep until the output voltage $v_o$ exceeds the comparator reference level $V_r$ at which time the reset closes momentarily and shorts the integrating capacitor. The cycle then repeats.

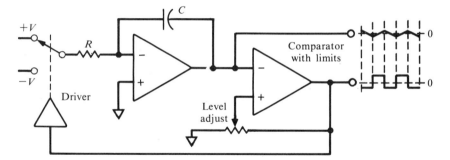

**Fig. 6-31.** Level-controlled square- and triangular-wave generator. The input switch is used to reverse the polarity of the integrating current when the comparator threshold is reached. A fraction of the comparator output is used to obtain the threshold level.

regular intervals. With this method the values of $R$ and $C$ affect the sweep rate and the maximum amplitude, but they do not affect the frequency because that is controlled by the oscillator. The other method is to use a voltage-programmed switch to close the reset switch momentarily when the desired maximum sweep voltage is reached. This is the technique illustrated in figure 6-30, which shows a generalized comparator-operated switch. As the sweep output $v_o$ increases toward the comparison voltage $V_r$, the comparator output is positive. When $V_r$ is reached, the comparator output becomes negative, triggering the monostable multivibrator to generate a reset pulse to the reset switch driver. The reset pulse is set to a duration that allows the integrating capacitor to be discharged completely. In the level-control method of automatic resetting, the values of $V_{in}$, $R$, and $C$ affect both the sweep rate and the frequency. They do not affect the amplitude because that is controlled by the comparator. A change in the level control changes the amplitude and frequency but not the slope.

A variation on the sweep generator of figure 6-30 is used to generate both square and triangular waveforms. In this circuit the switch is used to reverse the polarity of the integrating current as shown in figure 6-31. The integrator output has a negative slope when the integrating current is switched to the positive source. The comparator output is a negative voltage in this state. Because a fraction of the comparator output is used for the comparator reference source, the reference is also a negative voltage. When the integrator output crosses the negative reference level, the comparator output becomes positive, the switch driver connects the integrator to the negative source, and the integrator output has a positive slope. The new positive comparator reference level that is established will reverse the levels again when it is reached. This generator can operate in many control modes. The amplitude can be changed at a constant slope, or the frequency can be varied at a constant amplitude. Choice of unequal values for $+V$ and $-V$ results in different positive- and negative-going slopes. A generator with voltage-controlled frequency is obtained by varying $+V$ and $-V$. This general circuit is the basis for the popular laboratory function generator. A sine-wave output is sometimes obtained by shaping the triangular waveform as described in the next section.

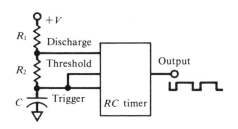

**Fig. 6-32.** *RC* timer connections for astable multi-vibrator operation. When the capacitor charges to the threshold value, the discharge switch closes. The discharge time constant is $R_2C$. When the capacitor voltage crosses the trigger level, the discharge switch opens, and $C$ charges with a time constant of $(R_1 + R_2)C$. The frequency is determined by the values of $R$ and $C$.

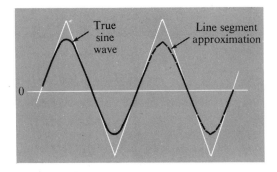

**Fig. 6-33.** Synthesis of a sine wave from a triangular wave by line-segment approximation.

An *RC* timer like that of figure 6-29 can be used in a self-triggering mode to generate a repetitive pulse or rectangular wave. This **astable multivibrator** mode is achieved by changing the connections between the timer and the *RC* circuit as shown in figure 6-32.

## Diode Waveshaping

In the function generator, the basic waveform is the triangular wave; the sine wave must somehow be derived from it. Obtaining the sine wave from the triangular wave requires an attenuator that diminishes the larger input voltages much more than the smaller ones as shown in figure 6-33. A **diode waveshaping** circuit to serve this purpose can be made with diodes, switches, and resistors as shown in figure 6-34. In the single diode circuit of figure 6-34, for $v_{in} < v_b$ the output signal is the unattenuated input signal. When $v_{in}$ is greater than $v_b$, the diode conducts a current proportional to $v_{in} - v_b$. This current causes an *IR* drop in $R_s$ proportional to $v_{in} - v_b$. Thus the slope of the output signal changes at $v_b$. The point of the slope change is set by the **breakpoint** $v_b$ adjustment. The amount of slope change depends on the value of the slope-adjust resistor. Additional diodes and resistors can be added to provide additional breakpoints and slope changes until the desired waveform is approximated by a series of line segments. The line-segment approximation is shown in figure 6-33, and a multiple-breakpoint circuit with four positive and four negative breakpoints is shown in figure 6-34b.

The sine-wave output produced by diode waveshaping in a good function generator is generally of high quality. The line segments are not observable with an oscilloscope. The function generator approach is the best way to produce low-frequency (below 20 Hz) sine waves; they are difficult to produce satisfactorily by standard harmonic oscillator circuits.

## 6-6  Sampling Measurements

In an analog sampling measurement the signal amplitude is acquired or sampled at specific instants or intervals. This is done in order to operate on data acquired at times of particular significance, to digitize and store data points at regular intervals, to share a readout among a number of signal sources connected successively, or to perform other similar operations. This section begins with a discussion of multiplexing, the time sharing of a data channel among a number of data sources. Then the sample-and-hold circuit is introduced. This conceptually simple application of op amps and switches is capable of accurately holding the signal level acquired at a precisely controlled instant. Several elegant applications of the sampling concept are illustrated in later chapters.

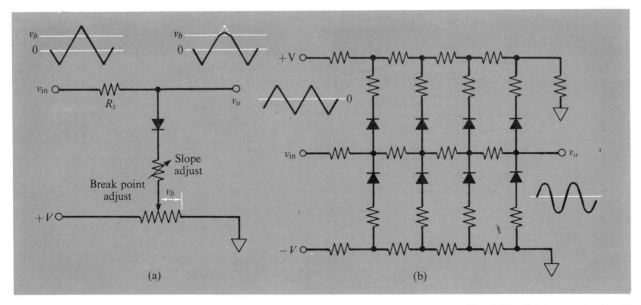

**Fig. 6-34.** Diode waveshaping. The single diode in (a) is connected as a shunt clipping circuit. The output signal is the unattenuated input below the breakpoint voltage $v_b$. In (b) additional diodes and resistors are added to provide multiple breakpoints.

## Multiplexing of Input Sources

The sharing of a single analog-input measurement and recording system by more than one input signal is accomplished by a signal switching technique called **multiplexing**. The analog multiplexer is simply a controlled selector switch like that in figure 6-35. The switch position is controlled by a channel control circuit that can be set to any channel or instructed to change to the next channel by external signals. A digital output is generally provided to indicate which channel is active.

The multiplexer switch can be an electromagnetically operated rotary selector switch or a set of relays or solid-state switches controlled in such a way that only one channel is connected at a time. Low-level and/or high-impedance signal sources require careful choice of switches to avoid noise and error. The multiplexer in figure 6-35 has two connections per channel to allow the use of floating signal sources when a differential input is available on the measurement system. Multiplexing is most frequently used with groups of similar signal sources such as identical thermocouples that monitor the temperatures in different locations. When disparate sources that require different range or offset adjustments of the measurement system are multiplexed, they must be conditioned to a common range and offset before they are multiplexed. Analog multiplexing is frequently used with recorders, oscilloscopes, printing digital voltmeters, and analog-to-digital converter inputs for computers and data loggers.

**Fig. 6-35.** Four-channel analog multiplexer. In the switch position shown channel 2 is connected to the output. As the switch position changes, the other input channels are connected in turn to the output.

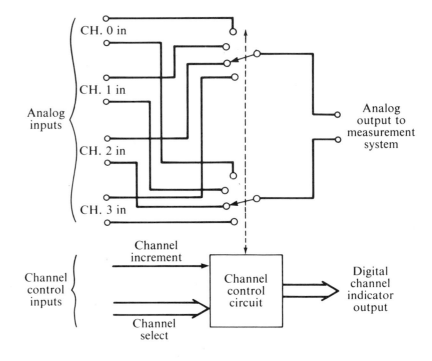

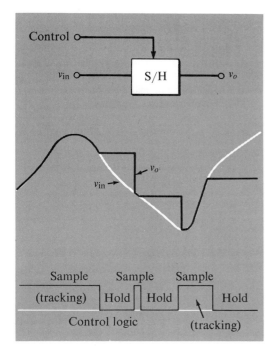

**Fig. 6-36.** Typical sample-and-hold (S/H) waveforms. When a hold command is received, the voltage level at the sample-and-hold input is stored and maintained at the output. After the held value is measured or used in some way, the circuit is switched to the sample condition to respond again to the input voltage.

## High-Speed Analog Sampling

One major application of sampling measurements is the determination of the instantaneous amplitude of a signal that can vary significantly during the response time of the measurement system. For this application a **sample-and-hold** circuit, a kind of analog memory, is used (fig. 6-36). In the hold condition, the steady voltage can be measured by a relatively slow device. The sample-and-hold function is generally accomplished by charging a capacitor with the signal value during the sample interval, then measuring the voltage across the capacitor with a high input impedance amplifier during the hold period.

A reliable sample-and-hold circuit that uses the voltage follower amplifier is shown in figure 6-37. The input signal charges the capacitor $C$ through resistance $R$ when the circuit is in the sample mode. The time constant $RC$ and the response time of the amplifier must be short compared to the rate of change of the input signal, so that the follower amplifier input and output follow the input signal variations. When the switch is changed to the hold position, the voltage on the capacitor is maintained (held) at the amplifier noninverting ($+$) input (and consequently, at the output) until the circuit is returned to the sample mode. The time of the transition from sample to hold determines the instant at which the input signal is sampled. The deviations

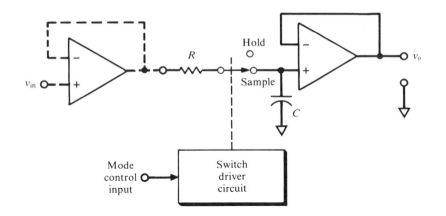

**Fig. 6-37.** Voltage follower sample-and-hold circuit. At the desired sampling instant the switch is changed to the hold position. This isolates the input signal and leaves the voltage across capacitor $C$ at that instant at the amplifier input. The buffer amplifier, shown in dashed lines is used, when necessary, to supply the capacitor charging current.

from ideal behavior of practical sample-and-hold circuits are discussed in chapter 13 where we shall discuss the indispensable role that sample-and-hold circuits play at the interface between the analog and digital domains.

## Equivalent Time Conversion of Repetitive Signals

Not infrequently, one wishes to measure a signal waveform that has a rate of change faster than the response speed of the measuring device. If the signal has a repetitive waveform, data points sampled from many repetitions of the waveform can be combined to produce a reasonable data point recording of the waveform. The relatively simple device that accomplishes this is shown in figure 6-38.

**Fig. 6-38.** Equivalent time converter. A sample is acquired whenever the slow and fast sweep generators are of equal voltage. One sample is thus obtained each time the fast sweep is triggered. Since the sampling instant occurs a little later in the fast sweep each time, the sampled point is taken at increasing times in the signal waveform. The waveforms show that plotting the sampled amplitude against the sampling delay time gives points on the original waveform.

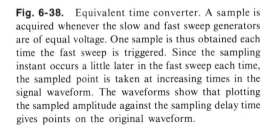

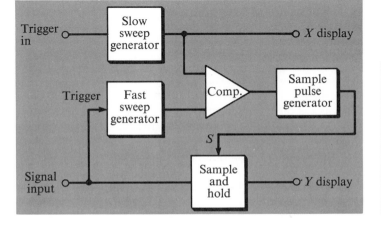

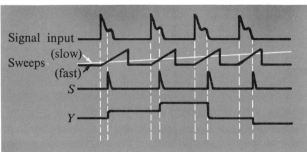

The repetitive input signal is applied to the input of a sample-and-hold circuit and to the trigger input of a fast sweep generator. The generator is adjusted to be triggered by the feature of the signal that is to be recorded and is set to complete its sweep in the period of interest after the trigger. A sweep generator that is slow enough for the *x-y* measurement system to follow is also started. The time between the fast sweep trigger and the sampling instant is called the **sampling delay time**. The amplitude of the slow sweep is proportional to the sampling delay time, and thus an equivalent time sweep signal for a slow measurement system is provided. The ratio of equivalent time to real time is equal to the ratio of the slow sweep rate to the fast sweep rate. The number of data points recorded in a single slow sweep is equal to the number of fast sweep triggers that occur in that time.

Equivalent time conversion is used with both oscilloscopes and *x-y* recorders. An equivalent time oscilloscope is called a **sampling oscilloscope**. Equivalent time sampling is used to extend oscilloscopic observations into the picosecond time range. Much slower versions of the equivalent time converter are available for *x-y* recorders. One application for these devices is plotting the steady displays of repetitive signals from an oscilloscope for a permanent record. Such a device is very simple because the fast and slow sweep signals can be provided by the oscilloscope sweep generator and the *x-y* recorder time base respectively.

## 6-7   Application: Programmable Gain Amplifiers

The gain of an amplifier that uses an op amp depends on the values of the input and feedback resistors. A **selectable gain amplifier** would allow for several resistance combinations to be switched into the circuit. If the switches were programmable analog switches, the amplifier would be capable of having its gain programmed by remotely generated logic-level signals. In this section, the basic **programmable gain amplifier** is developed, and the similar digital-to-analog converter circuit is introduced.

### Amplifier Gain Control

There are two basic op amp voltage amplifier configurations: the follower with gain and the inverting amplifier. Both can be gain-programmed by using switches to change the gain-determining resistance values. A programmable follower with gain amplifier is shown in figure 6-39. The follower has an advantage over the programmable inverting amplifier in that the switches are not in series with the gain-determining resistors. This avoids the problem of the amplifier gain being affected by the switch ON resistance. In figure 6-39, the ON switch carries negligible current and is used only to connect the appropriate feedback voltage to the inverting input.

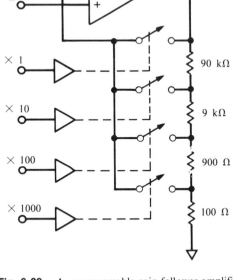

**Fig. 6-39.** A programmable gain follower amplifier. Signals are applied to the switch drivers so that only one switch is closed at a time. The closed switch determines the fraction of the output signal fed back and thus the amplifier gain.

The programmable gain amplifier is useful if the amplifier and its gain controls cannot be very near each other. The connection of the amplifier input and feedback signals to a remote gain switch would add noise to the system, but long switch driver control connections do not produce amplifier noise. The gain of a programmable gain amplifier can be controlled automatically. Comparators connected to a logic circuit can sense a $v_o$ value of $>10$ V (overrange) or $<1$ V (underrange) and decrease or increase the gain as appropriate.

## Digital-to-Analog Converter

The circuit for a programmable gain inverting amplifier is shown in figure 6-40. If the input resistors of this amplifier follow the binary or BCD weights for digital data encoding, and if a digital data word is used to drive the gain-control switches, a **digital-to-analog converter (DAC)** is produced. Each resistor produces at the amplifier summing point a current that is proportional to the coded weight of the bit when the switch for that bit is closed. The simultaneous closure of all the switches for which the digital input signal is HI results in an output voltage proportional to the digital signal value. For the resistor values in the figure and a $-10$-V analog reference source, a BCD input signal of 0110 0101 ($65_{10}$) produces an output of 6.5V. The value of the input resistances shown in figure 6-40 must include the ON resistance of the series switch. Errors due to variation or unpredictability in switch resistance are greatest for the most significant bits of the digital signal.

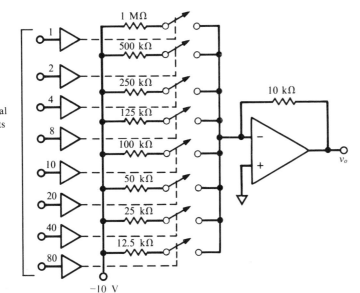

**Fig. 6-40.** A two-digit BCD digital-to-analog converter. The DAC function is obtained by the appropriate choice of summing resistors and the use of a reference voltage as the analog input. The eight bits of digital input are applied to the switch driver inputs.

The basic DAC circuit of figure 6-40 has another interesting application. If the $-10$-V analog reference input is replaced with a variable input voltage $v_{in}$, the output voltage would be $v_o = -v_{in} (nn/100)$, where $nn$ is the two-digit BCD input value. This is a programmable gain amplifier with the gain adjustable in integer steps from 1% to 99%. In this application the circuit is called a **multiplying DAC (MDAC)**. DAC circuits and their characteristics are further discussed in chapter 13.

# Suggested Experiments

**1.  Relay switches.**
Connect a relay with a switch driver to a TTL drive signal. Use an oscilloscope to measure the relay operate time, transfer time, and bounce time on making and the break time, transfer time, and bounce time on breaking. Determine the frequency at which the movable contact fails to make connection with the normally open contact.

**2.  Solid-state switches.**
Apply a constant current through a closed IC switch, measure the voltage drop across the switch with a voltage follower, and calculate the ON resistance. Measure the leakage current of the open switch with an op amp current-to-voltage converter, and calculate the OFF resistance. Investigate an optically coupled switch.

**3.  Transient response of RC circuits.**
Connect a series $RC$ circuit to a square-wave generator. Observe the output across $R$ and $C$ for several $RC$ time constants and several input frequencies. Investigate a parallel $RC$ circuit.

**4.  Inductive circuits.**
Observe the response of $LR$ and $LC$ circuits to a square-wave stimulation.

**5.  Switching speed.**
Connect an IC switch to a square-wave drive, and switch a constant voltage to a load. Measure the turn ON and turn OFF times for the switch.

**6.  Level-actuated switch.**
Use the comparator output as the switch driver input of an IC switch. Show that the switch can be made to turn on and off at a controllable level. Construct a diode clipping circuit to clip a waveform above and below a preset level.

**7.  Monostable multivibrator.**
Connect a 555 type IC timer as a monostable multivibrator, and observe the output time delay as a function of $R$ and $C$.

**8.  Astable multivibrator.**
Connect the 555 timer as an astable multivibrator. Measure the oscillation frequency for several values of $R$ and $C$.

**9.  Function generator.**
Make a sawtooth generator by connecting a linear sweep generator (integrator) to a comparator-actuated switch that shorts the integrating capacitor at a preset voltage. Make a triangular-wave generator.

**10.  Analog multiplexer.**
Connect analog switches to switch two different signal sources to an op amp amplifier. Show that the two sources can be automatically connected to the op amp under control of a digital signal.

**11.  Sample-and-hold circuit.**
Connect a sample-and-hold circuit and confirm its operation. Measure its leakage rate during the hold period. Sample a varying signal, and use the sample-and-hold to obtain the peak amplitude.

**12.  Programmable gain amplifier.**
Use three switches to connect three different input resistors to an op amp inverting amplifier. Show that a digital signal can control the gain. Construct a four-bit BCD digital-to-analog converter, and measure the output voltage as a function of the digital input.

# Questions and Problems

**1.**  In the nonideal switch circuit of figure 6-3, the switch has an open resistance $R_{so} = 900$ k$\Omega$ and a closed resistance $R_{sc} = 80$ $\Omega$. It is desired to switch a 5-V source $v_s$ to a 50-k$\Omega$ load $R_L$. What are voltage drops across the switch and across the load for each state of the switch?

**2.**  In the nonideal switch circuit of figure 6-3, $R_{so} = 10^{10}$ $\Omega$ and $R_{sc} = 100$ $\Omega$. Find the total range of load resistance values for which the error in switching $v_s$ to the load $R_L$ is less than 0.1%.

**3.**  In the series switch circuit of figure 6-5a, a 5-V source has an equivalent resistance $R_e$ of 500 $\Omega$. It is desired to switch the source to a load $R_L = 25$ k$\Omega$. The switch is the nonideal switch of figure 6-3 with $R_{so} = 10$ M$\Omega$ and $R_{sc} = 500$ $\Omega$. Find the error in switching for each state of the switch.

**4.**  The same source, switch, and load as in problem 3 are to be used in the series-shunt switching circuit of figure 6-5c. Find the error in switching for each state of the switch. Compare these errors to those of the series switch of problem 3.

**5.**  In the series $RC$ circuit of figure 6-19, $V = 1.0$ V, $R = 100$ k$\Omega$, and $C = 0.1$ $\mu$F. Switch $S$ is an ideal switch. Find the voltage across the capacitor $v_C$ and the voltage across the resistor $v_R$ at 0.5 ms, 1.0 ms, 5.0 ms, 10.0 ms, 50.0 ms and 100 ms after switch has been turned to position $A$.

**6.** A 5.0-V rectangular pulse with a pulse width of 1.0 mA is applied to a series $RC$ circuit at time $t_0$. (a) If $R = 100$ k$\Omega$ and $C = 1$ nF, calculate the voltages $v_R$ and $v_C$ across $R$ and $C$, at $t_0$, $t_0 + 0.1$ ms, $t_0 + 0.3$ ms, $t_0 + 1.0$ ms, $t_0 + 1.3$ ms, and $t_0 + 1.5$ ms. (b) Repeat part (a) for $R = 1$ M$\Omega$ and for times $t_0$, $t_0 + 1.0$ ms, and $t_0 + 1.5$ ms.

**7.** In the series $LR$ circuit of figure 6-23a, $L = 200$ mH, $R = 100$ $\Omega$, and $V = 4.5$ V. Find the voltage across the inductor and that across the resistor at 0.5 ms, 1.0 ms, 2.0 ms, 10 ms, and 100 ms after the switch has been closed.

**8.** In the parallel $RC$ circuit of figure 6-22, $V = 5.0$ V, $R_1 = 25$ k$\Omega$, $R_2 = 50$ k$\Omega$, and $C = 0.22$ $\mu$F. (a) How soon after switch $S$ is turned to the ON position is capacitor $C$ charged to within 1% of its final value? (b) If $R_3 = 50$ $\Omega$, how long does it take for $C$ to discharge to within 1% of its final value?

**9.** In the nonideal switch shown in figure 6-24, $R_{so} = 100$ M$\Omega$, $R_{sc} = 250$ $\Omega$, and the equivalent switch capacitance $C = 500$ pF. If a 5-V source $V$ is connected to a 100-k$\Omega$ load $R_L$, (a) what are the voltage drops across the switch and load for both states of the switch? (b) What are the two time constants $\tau_0$ and $\tau_c$ for opening the switch and closing the switch?

**10.** Describe in words the major reason that a switch with significant capacitance gives steady-state circuit conditions more rapidly on closing than on opening.

**11.** For a series $RC$ circuit find the fraction of the applied voltage $V$ that appears across $C$ and $R$ at 0.5 $RC$, 1.5 $RC$, 4.5 $RC$ and 5.0 $RC$ (a) for the capacitor charging, and (b) for the capacitor discharging.

**12.** The series $RC$ circuit of figure 6-21 is considered for converting a rectangular pulse with a peak voltage of 5.0 V and a pulse width of 100 $\mu$s into a positive-going spike that occurs during the rising edge of the pulse. (a) What value of $RC$ should be used in order that the spike decays to one-half the peak voltage in 1.5 $\mu$s? (b) How can the negative-going spike (see fig. 6-21c) be eliminated?

**13.** It is desired to add another four bits to the BCD DAC of figure 6-40 to make a three-digit DAC. (a) What resistance values are required? (b) What are the ON and OFF resistance requirements of the switches in such a three-digit DAC if 0.1% accuracy is required?

**14.** In the optically coupled analog switches shown in figure 6-17, to what element is the switch driver signal connected? To what element is the analog signal connected?

**15.** Design a circuit in which two optically coupled switches could be used in a series-shunt arrangement to make a chopper. What advantages would such a chopper have over one made with relays?

**16.** Design diode clipping circuits that restrict the output voltage $v_o$ to the following limits regardless of the range of input voltages: (a) $0 < v_o < +5$ V, (b) $-10$ V $< v_o < +10$ V, (c) $+50$ V $< v_o < +75$ V.

**17.** The op amp sweep generator of figure 6-30 has $V_{in} = -1.5$ V, $R = 250$ k$\Omega$, and $C = 0.5$ $\mu$F. (a) What is the sweep rate of the generator? (b) If the maximum output voltage of the op amp is $\pm 12$ V, for how many seconds does the sweep remain linear? (c) If the comparator reference level in figure 6-30 is $+3$ V, what is the output frequency of the sawtooth waveform?

**18.** Challenge question: Derive an equation that predicts the output frequency of the sawtooth generator of figure 6-30 as a function of $V_{in}$, $R$, $C$, and $V_r$.

**19.** For the $RC$ timer of figure 6-29 derive an equation to show how the output time delay is related to the voltage applied to the $RC$ circuit $V_{RC}$, the voltage applied to the comparator divider $V_{com}$, and the $RC$ time constant.

**20.** For the astable multivibrator shown in figure 6-32, draw the waveform at the discharge input during operation. (Hint: refer to figure 6-29 for details of the circuit.)

**21.** The sample-and-hold circuit of figure 6-37 has $R = 1$ k$\Omega$, $C = 0.01$ $\mu$F and a follower with $R_{in} = 10^{10}\Omega$. (a) If $v_{in} = 0.5$ V, what is the minimum time the switch should be in the sample position for accurate sampling (0.1% error)? (b) If the signal is held for 1 s, how much leakage occurs? (c) How much leakage occurs when the signal is held for 10 s and 20 s?

**22.** The input signal to the sample-and-hold circuit of figure 6-37 is a 1-V pulse, 10 ns long. (a) If $R = 100$ $\Omega$, what is the maximum value of $C$ that gives an output of 1 V in the hold position? (b) If the follower has an input resistance of $10^{10}$ $\Omega$, what is the output voltage after 0.01 s, 0.1 s, 1 s, and 5 s? (c) If $R = 100$ $\Omega$ and $C = 0.01$ $\mu$F, what is the maximum output voltage of the follower in the hold mode?

**23.** The sample-and-hold circuit of figure 6-37 in the tracking mode is a low pass filter for an ac input signal. If $R = 1$ k$\Omega$, $C = 0.01$ $\mu$F, what is the upper 3 dB point of the filter?

**24.** The switch for a sample-and-hold circuit must have excellent resistance characteristics. Consider the pulse signal of problem 22. (a) For the value of $C$ chosen to give a 1-V output, what is the maximum switch ON resistance for 0.1% accuracy? (b) What value of $C$ would be needed for a 1-V output if $R_{ON}$ were 100 $\Omega$? (c) How long would be required to fully charge the capacitor for the values chosen in (b)?

# Chapter 7

# Solid-State Switches and Amplifiers

The advent of solid-state devices for switching and amplification has been an essential part of the revolution in electronics and modern instrumentation. Solid-state switches enable digital (ON/OFF) control functions to be implemented on a nanosecond to microsecond time scale. These switches are the basic building blocks of digital circuits. Solid-state amplifiers have been an integral part of the recent development of extremely high quality, inexpensive operational amplifiers, instrumentation amplifiers, and automatic feedback control systems.

Despite dramatic differences in their application, switching and amplification circuits have much in common. In each case the active device controls the current in one part of the circuit in response to a controlling signal applied through another part. In switch applications, the active device is clearly either ON or OFF; in amplifier applications, the active device operates throughout the region between the maximum conducting and nonconducting limits.

Two main classes of transistors are used as active devices for both switching and amplification. They are the bipolar junction transistor (BJT) and the field-effect transistor (FET). In bipolar devices the controlled current path includes one or more pn junctions; in field-effect devices the current through a single type of semiconductor (n or p) is controlled by the influence of an electric field. This chapter begins with a description of both types of transistors.

The analog switch packages considered functionally in chapter 6 are opened up in this chapter and described in terms of their internal elements and circuits. The principles and power switching applications of four-layer devices such as silicon-controlled rectifiers and triacs are also considered. Basic bipolar and field-effect transistor amplifiers are introduced, and the use of feedback to improve amplifier performance is discussed. Several of the basic amplifier circuits are then combined to form a representative example of the versatile operational amplifier.

## 7-1 Bipolar Devices

Most solid-state switches and amplifiers depend upon the characteristics of the pn junction for their control capabilities. For this reason pn junctions and diodes are reviewed briefly before the bipolar transistor is introduced.

A pn junction is formed in a semiconductor at every interface between p-doped and n-doped regions. Under conditions of zero bias, electrons from the n region and bonding electron vacancies (holes) from the p region neutralize each other in the junction region. The concentration profiles for holes and electrons are shown in figure 7-1. In the example shown the dopant concentration is slightly less in the n region than in the p region. As the figure shows, the total concentration of charge carriers is depleted in the junction region by many orders of magnitude. This region of reduced charge-carrier concentration is the **depletion region**, and the contact junction potential appears across this insulating region as shown in figure 7-1c. Recall from chapter 3 that the contact potential is about 0.3 V for germanium pn junction and about 0.6 V for silicon.

The concentration of charge carriers in the depletion region is essentially that of the intrinsic semiconductor. These charge carriers are electron-hole pairs generated by thermal excitation. The contact potential difference across the depletion region causes the generated holes to move to the p region and the electrons to move to the n region. The current due to carriers in the intrinsic region is thus from right to left as shown in figure 7-1d. The magnitude of the intrinsic carrier current $I_i$ depends on the rate of carrier generation in the depletion region, which is a function of temperature only. The net flow of positive charge from right to left due to the intrinsic carrier current would tend to reduce the contact potential across the junction. However, as the potential barrier is decreased, the number of majority carriers on each side with enough energy to cross the junction increases. The current due to majority carriers $I_m$ is from left to right and increases exponentially with increasing temperature and decreasing junction potential:

$$I_m = Ke^{-Q_e V_{np}/kT} \tag{7-1}$$

where $k$ is the Boltzmann constant in joules per degree Kelvin and $T$ is the temperature in degrees Kelvin. When no external current source is applied to the pn junction, the net current across the junction must be zero; that is, the intrinsic and majority currents must cancel exactly. For any given temperature there is only one value of the junction potential for which this is true. One can see qualitatively how the higher intrinsic carrier generation rate of germanium over silicon results in the lower junction potential for germanium.

When an external bias is applied, essentially all of the bias voltage appears as a change in the voltage across the pn junction as discussed in

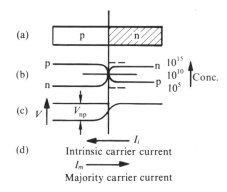

**Fig. 7-1.** pn junction. (a) The block representation; (b) the concentration profile of charge carriers in carriers per cubic centimeter; (c) the potential profile; and (d) the junction current directions.

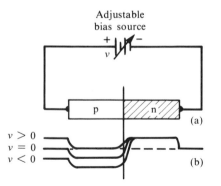

**Fig. 7-2.** pn junction with bias voltage source connected (a) and potential profiles (b) showing the effect of bias voltage on the pn junction potential.

chapter 3. This is shown by the family of potential profiles in figure 7-2. The change in $V_{np}$ has a great effect on the majority carrier current $I_m$ but little effect on the intrinsic current $I_i$. At zero applied bias $I_m = I_i$; so that

$$I_i = I_m^0 = Ke^{-Q_e V_{np}^0/kT} \qquad (7\text{-}2)$$

where $I_m^0$ and $V_{np}^0$ are the values for the majority current and junction potential respectively at zero bias. At nonzero bias the junction potential is $V_{np}^0 - v$, and a net forward current $i$ can cross the junction. The net current is the difference between $I_m$ and $I_i$:

$$i = I_m - I_i = Ke^{-Q_e(V_{np}^0 - v)/kT} - I_i \qquad (7\text{-}3)$$

Rearranging equation 7-3, we obtain the Shockley equation,

$$i = I_i(e^{Q_e v/kT} - 1) \qquad (7\text{-}4)$$

The Shockley equation describes the familiar exponential current-voltage behavior of the pn junction under forward bias conditions (see figure 3-10) and shows that under reverse bias conditions the current is essentially $I_i$, often called the **reverse bias saturation current**. The reverse bias current depends on temperature as shown in equation 7-2 and upon other sources of energy at the junction (e.g., light), but it is essentially independent of the reverse bias voltage applied.

## The Bipolar Transistor

Seldom has a device had such a dramatic impact on our leisure and work as has the transistor. This remarkable device has brought about a complete revolution in electronic circuitry and has made commonplace products that were science fiction visions only a few years ago.

The **bipolar junction transistor (BJT)** is composed of two n- or p-type semiconductor regions separated by a thin region of the opposite type semiconductor. This arrangement is illustrated in figure 7-3 with the standard symbols for BJTs. Two types of transistors result, the npn and pnp, named for the order of the semiconductor types used in their construction. The bipolar transistor is a combination of two pn junctions. The thin center region is called the **base**, and the other two regions are called the **emitter** and the **collector**. In three terminal switch operation an actuating signal that is applied between the base and emitter controls conduction between the collector and emitter.

An npn transistor in the ON state is diagrammed in figure 7-4. The base-emitter junction is forward biased by voltage $V_{BE}$. Under the influence of the forward bias, emitter electrons are accelerated into the base region. Some of these electrons encounter holes in the base and produce the base current $I_B$. However, the base region is so thin and so lightly doped that there

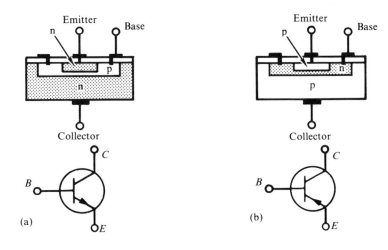

**Fig. 7-3.** Basic transistor types. In (a) the structure and symbol of a npn transistor are shown; (b) illustrates the pnp transistor.

is a high probability that the electrons from the emitter will traverse the base region without being neutralized by a hole. These electrons are collected by placing a positive voltage on the collector, and the collector current $I_C$ results. Since only a small fraction of the electrons from the emitter combine with holes in the base, the collector current is much larger than the base current. Therefore, a small base current controls a much larger collector current. When the transistor conducts, the collector voltage $V_{CE}$ is larger than the base voltage $V_{BE}$, and the collector-base junction is reverse biased. When the base-emitter junction is reverse biased, both pn junctions are reverse biased and the collector-emitter conduction path is turned off. The operation of the pnp transistor is completely analogous: holes are emitted and collected rather than electrons, and the current arrows and battery polarities are reversed.

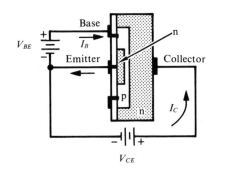

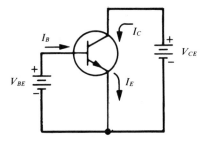

**Fig. 7-4.** The npn transistor biased in the ON state. The transistor is in the common-emitter configuration.

**Fig. 7-5.** Collector characteristic curves for the common-emitter connection of an npn transistor. The dashed load line connects $V_{CC}$ and $V_{CC}/R_L$. The collector voltage and current are given by the point on the load line that is intersected by the characteristic curve for the applied base current. For $I_B = 68$ $\mu$A, $V_{CE}$ and $I_C$ are given by the point in the middle of the load line.

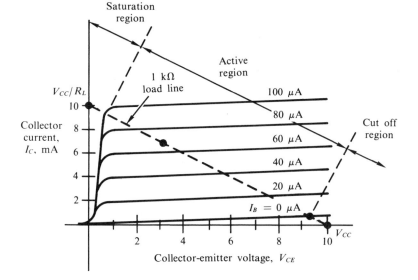

**Current-voltage characteristics and the load line.** A current-voltage curve for a representative transistor is shown in figure 7-5. There are actually a whole family of current-voltage curves, one for each value of base current. For values of $V_{CE}$ greater than 1 V, the collector current is relatively constant and is approximately proportional to the base current. The **forward current gain characteristic** $\beta$, is defined as the ratio of the collector current to the base current, $\beta = I_C/I_B$. The value of $\beta$ is approximately constant in the linear amplification region and is about 100 for the transistor type shown in figure 7-5. The value for $\beta$ is roughly the probability of an emitted charge carrier passing through the base region and arriving at the collector (100:1 in this case). Values of $\beta$ for commercially available transistors vary from ~5 to > 1000. From the flatness of the current-voltage curve, it can be seen that once $V_{CE}$ is high enough to collect the emitted charge carriers, a further increase in $V_{CE}$ does not give a significantly higher current.

A simple npn transistor switch circuit is shown in figure 7-6. The actuating circuit provides the base current $I_B$ that controls the conducting state of the transistor. The switched circuit consists of the collector supply voltage $V_{CC}$ and the load $R_L$. When the base current is high enough to force the transistor into conduction, the much larger collector current $I_C$ exists in the load. When the transistor is turned off by the actuating signal, the current through $R_L$ is greatly reduced.

The switching circuit of figure 7-6 can be analyzed to find $I_C$ for the two conduction states of the transistor by a simple graphical method using the characteristic curves of figure 7-5 and a linear equation called the **load line**

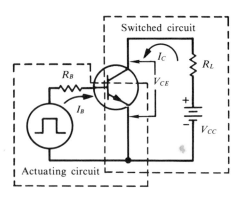

**Fig. 7-6.** A common-emitter npn transistor switching circuit. The actuating circuit provides base current, which controls the conducting state of the transistor. For the analysis of figure 7-5, the actuating signal is assumed to vary from 0 V to +4.0 V and $R_B = 50$ k$\Omega$, $R_L = 1$ k$\Omega$ and $V_{CC} = 10$ V.

for the switched circuit. The switched circuit is a complete loop, and Kirchoff's voltage law can therefore be applied. The result is $V_{CC} = I_C R_L + V_{CE}$. This equation is the load line; it is the set of all the possible collector current and collector-emitter voltage combinations that the switched circuit can provide. Two points on the line are obtained very easily. When $V_{CE} = 0$ V, $I_C = V_{CC}/R_L$. For the specific case of figure 7-5, $I_C = 10$ V$/1$ k$\Omega = 10.0$ mA. Thus, $I_C = 10.0$ mA, $V_{CE} = 0$ V, is one point on the line as shown in figure 7-5. A second point on the line is obtained when $I_C = 0$ mA. At this point $V_{CE} = V_{CC} = 10$ V. These two points establish the complete load line shown by the dotted line in figure 7-5.

To find the collector current $I_C$ for any state of the switch, it is necessary only to know the base current $I_B$ that results from applying the drive signal. When the actuating signal is 0 V, $I_B \simeq 0$ $\mu$A, and the transistor is OFF. The OFF operating point is the intersection of the $I_B = 0$ $\mu$A curve with the load line. Thus the OFF operating point is ($V_{CE} \simeq +9.5$ V, $I_C \simeq 0.4$ mA). When the actuating signal is $+4.0$ V, the transistor is ON, and the voltage across $R_B$ is 3.4 V (4 V less the 0.6-V forward bias of the base-emitter junction). For $R_B = 50$ k$\Omega$, $I_B$ is thus 3.4 V$/50$ k$\Omega = 68$ $\mu$A. The collector current can then be estimated from the characteristic curve or calculated from the value of $\beta$. For the transistor of figure 7-5, $\beta = 100$, and $I_C \simeq 100 \times 68$ $\mu$A $= 6.8$ mA. This ON operating point is also confirmed by noting that the intersection of $I_B = 68$ $\mu$A and the load line in figure 7-5 occurs at $I_C \simeq 6.8$ mA and $V_{CE} \simeq 3.2$ V. From this analysis, we see that the transistor switch is not ideally ON in that the voltage across the switch is not 0 V. In fact, this operating point is in the linear amplification region for the transistor where $I_C$ is essentially proportional to $I_B$.

For better operation as a switch, $I_B$ should have values that allow only for the maximum or minimum values for $I_C$. For the load line given the maximum value of $I_C$ is $V_{CC}/R_L = 10$ V$/1$ k$\Omega = 10$ mA. The value of $\beta$ can be used to estimate the value of $I_B$ required: $I_{B(ON)} \simeq I_{C(max)}/\beta \simeq 10$ mA$/100 \simeq 100$ $\mu$A. The operating point on the load line for $I_B = 100$ $\mu$A is readily identified in figure 7-5. For this point $V_{CE} \simeq 0.2$ V. Since $V_{BE} = 0.6$ V, the base-collector voltage is $\sim +0.4$ V. In other words, the base-collector junction is actually forward biased by 0.4 V. In this condition the transistor is said to be in **saturation**. The saturation, or maximum ON condition, is a state achieved by applying enough base current to control a collector current that is greater than the circuit allows.

## Light-Activated Switches

In a **light-activated switch**, the conductivity or the current between the two switch contacts depends upon the amount of light falling on a photosensitive device. These devices are used as light detectors as described in chapter 4.

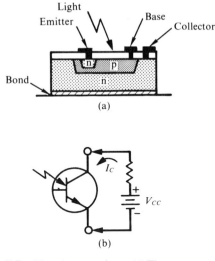

**Fig. 7-7.** The phototransistor. (a) The structure of a phototransistor and (b) the circuit with a typical load and bias.

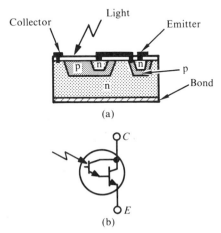

**Fig. 7-8.** The photodarlington structure (a) and circuit symbol (b).

The light can also be used as the actuating signal in switching applications of the photodetectors. This application has found extensive use in position indication where mechanical motion is used to interrupt the light path from source to detector. The combination of light and detector used as actuator and switch has the advantage of excellent electrical isolation between the actuating and controlled circuits as described in chapter 6. This allows transmission of data between systems that are highly noise sensitive or that may differ in voltage by thousands of volts.

The photoconductive cell and the photodiode described in chapter 4 have some application as light-activated switches. These applications are generally limited to low signal levels. Although the OFF (no-light) conductivity or current can be very low, the ON (full-light) conductivity or current is often not very high (that is, $R_{ON}$ is hundreds of ohms or $I_{ON}$ is tens of microamperes).

A very popular photodetector for switch applications is the **phototransistor** shown in figure 7-7. As in normal transistors, the collector-base junction is reverse biased. Light strikes this region through a window and creates electron-hole pairs that give rise to a positive "leakage" current to the base. The positive charge, collecting in the doped base region, forward biases the base-emitter junction to produce a collector current $I_C$. The collector current is $\beta$ times the light-induced base current, where $\beta$ is the forward current gain of the transistor. Collector ON currents of several milliamperes are obtained with this type of device. (Note that the thermal collector-base leakage current that occurs in ordinary junction transistors is amplified in this same way if the base is connected to a large resistance or is left open.)

Still larger current capacity is obtained with a **photodarlington**, an integrated two-stage amplifier device shown in figure 7-8. The emitter of a phototransistor is connected directly to the base of another transistor that shares a common collector with the first. The collector-emitter current of the phototransistor is the base current for the second transistor. Thus the light-induced current is effectively amplified by $\beta^2$. With this device the second transistor can be driven into saturation with moderate levels of actuating light.

## 7-2  Field-Effect Devices

Field-effect transistors (FETs) find a variety of applications in switching and in linear circuits. As switches, they have OFF/ON resistance ratios superior to those of bipolar transistors. The high input impedance of FET amplifiers makes them nearly ideal as voltage amplification elements. This section considers the basic operating principles of the junction field-effect transistor and the metal oxide semiconductor field-effect transistor.

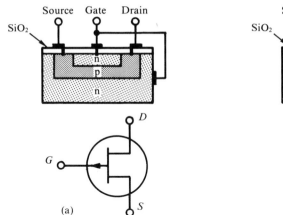

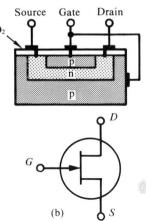

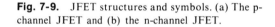

**Fig. 7-9.** JFET structures and symbols. (a) The p-channel JFET and (b) the n-channel JFET.

## The Junction Field-Effect Transistor

The **junction field-effect transistor** (**JFET**) is made of a channel of unipolar (only one type of charge carrier) semiconductor with a contact at each end (the **source** and the **drain**). Conductivity between the source and drain depends upon the charge-carrier density and the dimensions of the channel, which are, in turn, affected by the depletion layer thickness of one or more reverse biased pn junctions placed along the channel. This is shown in figure 7-9 for JFETs made of both p-type and n-type channels. The channel dimension control region is called the **gate**.

The bias on an n-channel JFET is shown in figure 7-10. The junction of gate and channel is reverse biased by the gate-source voltage to create a depletion region in the channel. The drain-source voltage $V_{DS}$ must then be of opposite polarity to the gate to maintain this reverse bias. Increasing the reverse bias on the gate narrows the effective source-drain channel and causes a decrease in $I_D$. For the p-channel JFET, the voltage polarity and current directions are, of course, reversed. In switch and amplifier applications, the control or actuating signal is connected to the gate, and the source and drain connections are placed in the circuit to be controlled.

The current-voltage characteristic curves are shown in figure 7-11 for a typical n-channel JFET. As expected, the drain current is smaller for increasingly negative values of $V_{GS}$. In the ohmic region ($V_{DS} < V_p$), an increase in $V_D$ increases $I_D$ proportionally. The effect of the gate bias voltage is just to change the channel resistance. At higher values of drain voltage, the drain current becomes self-limiting at a value determined by $V_{GS}$. This is called the "pinch-off" region because the $IR$ drop in the channel affects the bias on the junction. The result is to keep the $IR$ drop constant. From the

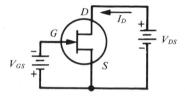

**Fig. 7-10.** Voltages and current in an n-channel JFET.

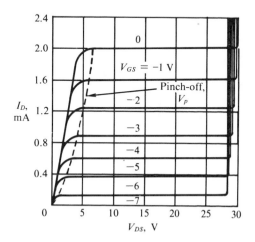

**Fig. 7-11.** Characteristic curves for the n-channel JFET.

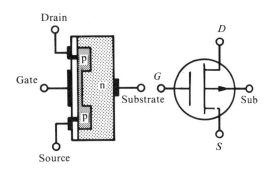

**Fig. 7-12.** Basic structure of and symbol for a p-channel depletion mode MOSFET.

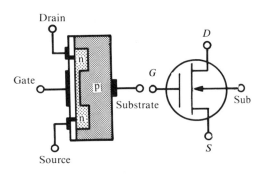

**Fig. 7-13.** Structure of and symbol for an n-channel depletion mode MOSFET.

current-voltage curves the JFET can be seen to make an excellent switch. The ON condition is achieved at $V_{GS} = 0$ V and the OFF condition occurs when $V_{GS}$ is a large negative value. The ON resistance is typically 25 to 100 $\Omega$ and the OFF resistance is $10^9$ $\Omega$ or greater.

There are several significant differences between the bipolar junction transistor (BJT) and the JFET. The BJT is actuated by a current signal at the base; the JFET responds to a voltage signal at the gate. The base connection of the BJT is of low impedance since the base-emitter junction is forward biased. The gate connection of the JFET is of high impedance ($\sim 10^9$ $\Omega$) because the gate-channel junction is reverse biased. The BJT is bipolar and introduces two opposing but unequal junction potentials in the emitter-collector circuit path. The JFET is unipolar and acts simply as an electrically controlled resistor, especially in the ohmic region.

## The MOS Field-Effect Transistor

In the JFET the gate-channel junction serves primarily to impose a reverse bias electric field on the channel. Since the junction is rarely used in forward-bias conduction, a normal pn junction is not required. In the **metal oxide semiconductor field-effect transistor** (**MOSFET**) an insulating layer of metal oxide is interposed between the gate electrode and the channel (fig. 7-12). The channel is an extremely thin layer of semiconductor of type opposite to that of the **substrate** (the semiconductor piece on which the device is fabricated). On the surface of the channel are successively deposited a few micrometers of metal oxide insulator and metallic gate connection. In figure 7-12 an n-type substrate is shown; the channel and the drain and source contacts are therefore p-type semiconductor. Just as in the JFET, a reverse bias (in this case, positive) applied to the gate reduces the number of charge carriers in the channel. This is called the **depletion mode** because increasing the gate bias depletes the channel conductivity. The structures and symbols for n-channel and p-channel MOSFETs are shown in figures 7-12 and 7-13. The current-voltage curve has the same features as that of the JFET shown in figure 7-11. An advantage of the MOSFET is its extremely high gate input impedance of $10^{12}$ $\Omega$.

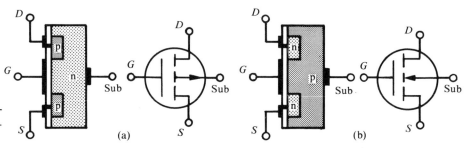

**Fig. 7-14.** Structures of and symbols for enhancement mode MOSFETs. (a) The p-channel MOSFET. (b) The n-channel MOSFET.

The presence of the insulating layer prevents conduction between the gate and channel even for what would have been a forward bias in the JFET. Because of this feature it has been possible to build MOSFETs in which the gate bias is used to enhance the channel conductivity. The structures and symbols of **enhancement mode** MOSFETs are shown in figure 7-14. In these FETs, no channel is diffused at all. The application of a forward bias actually creates a channel of the opposite type on the surface of the substrate. This surface channel provides the conducting link between the source and drain contacts. Figure 7-15 shows the bias voltage polarities and the characteristic current-voltage curves for an n-channel enhancement mode MOSFET. Note that the characteristic curves are not as flat as those of the depletion mode FETs. An advantage of the enhancement mode FET is that the gate signal is of the same polarity as the supply for the drain circuit.

All types of FETs are used extensively as analog switches. They have low noise, no distortion due to junction characteristics, high impedance in the actuating circuit, and fast response. The relative simplicity and low power requirement of MOSFETs have fostered today's "one chip microcomputer" revolution. As loose components, MOSFETs must be handled with care, for the gate insulation can be permanently damaged by relatively small static charges.

## 7-3  Analog Switches

The integrated circuit analog switch packages discussed in chapter 6 are made from bipolar and field-effect devices. In many switches combinations of devices are used to provide the various driver and switch functions. In this section several analog switching circuits that use the BJT, the JFET and the MOSFET are described.

## Transistor Analog Switches

Several BJT analog switch circuits are illustrated in figure 7-16. In the series switch circuit (fig. 7-16a) the gate is closed when the transistor base is positive with respect to $v_s$. The reverse biased base-collector and base-emitter junctions make an open circuit output when the gate is closed. A negative pulse is used to open the gate (and turn the transistor on). The output voltage is 0.1 to 0.2 V less than the input voltage because of $V_{CE(sat)}$ and follows any input voltage that is more than 0.6 V positive with respect to the base voltage. When negative signal levels are to be switched, an npn transistor is used with a positive-going turn-on signal. Note in figure 7-16a that the base of the transistor is connected to the common point of the actuating and controlled signals. Accordingly, this arrangement is called the **common-base configuration**.

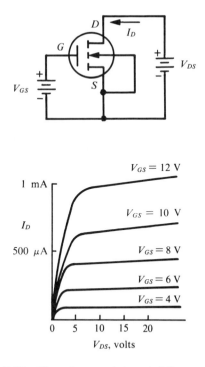

**Fig. 7-15.** Bias voltages and characteristic curves for n-channel enhancement mode MOSFET.

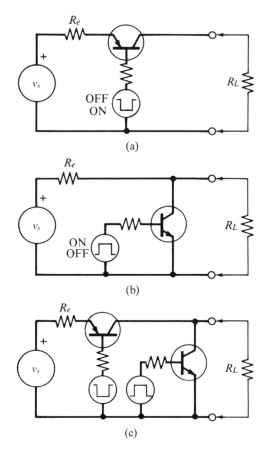

**Fig. 7-16.** Transistor analog switch circuits. (a) The series switch, (b) the shunt switch, and (c) the series-shunt switch.

The shunt switch shown in figure 7-16b uses an npn transistor to short the output terminals for positive signals when the switch base is driven positive. The current-limiting resistor $R_e$ avoids shortcircuiting the source when the transistor is on. This switch gives better transmission accuracy for the open gate and has a closed-gate output voltage of $V_{CE(sat)}$. The shunt switch is in the familiar **common-emitter configuration**. The series-shunt combination shown in figure 7-16c is used for precision analog switching. BJTs are frequently used in an inverted or **common-collector mode** with the emitter and collector connections reversed from those of figure 7-16b. The voltage offset is smaller when a BJT is operated in the inverse mode, the ON resistance can be as low as 2 Ω, and the switching speed can be as short as 10 ns.

## Transistor Switching Speeds

Consider the basic common-emitter transistor switch of figure 7-17a. Figure 7-17b shows the response of the circuit to an activating signal $V_B$. When the activating signal changes from 0 V to $V_B$, there is no immediate increase in collector current because of capacitance inherent in the base-emitter junction. This input capacitance must charge through $R_B$ before the base-emitter junction is actually forward biased. In addition there is a finite transit time for the first minority carriers to cross the base region and a finite time for the collector current to reach 10% of its final value. These effects give rise to the **delay time** $t_d$, the time that elapses between application of the base voltage and the collector current's reaching 10% of its final value.

Next a finite time called the **rise time** $t_r$ is needed to establish a sufficient flux of minority carriers in the base region (i.e., electrons in the p-type base of the npn transistor) to carry 90% of the final value of $I_C$. The combination

**Fig. 7-17.** Transistor switching times. Measurement of the collector current $I_C$ and input voltage $V_B$ as functions of time in the circuit shown in (a) produces the curves shown in (b).

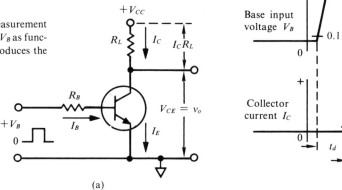

of $t_d$ and $t_r$ is the **turn-on time** $t_{ON}$. The rise time, and thus $t_{ON}$, can be substantially reduced if the base current is larger than the minimum needed for saturation.

When the activating signal is removed, a finite time called the **storage delay time** $t_s$ is required to clear the collector-base region of excess minority carriers. The **fall time** $t_f$, the time required for $I_C$ to decay from 90% to 10% of its maximum value, depends on the collector-emitter capacitance and the time required for the minority carriers in the base to be collected. The sum of the storage time and the fall time is called the **turn-off time** $t_{OFF}$. Thus, forcing the transistor hard into saturation increases the minority carrier concentration in the base; this decreases $t_{ON}$ but substantially increases $t_{OFF}$.

## Field-Effect Transistor Analog Switches

As might be expected, the field-effect transistor family (JFET and MOSFET) make excellent analog switches. The FET switch offers high analog transmission accuracy because it does not introduce a junction potential in series with the signal. It also offers the versatility of being able to conduct current in either direction. Its disadvantages with respect to the BJT switch are higher on resistance and generally slower switching speed.

When an FET is used as a voltage switch (i.e., to switch a signal that might have an amplitude of several volts), a switch driver circuit is needed to convert HI/LO logic signals into signals suitable for turning the FET on and off. Figure 7-18 illustrates a FET voltage switch and its associated two-transistor driver. The driver circuit must provide a gate voltage that is enough more positive than the switched voltage in one state and enough more negative than the switched voltage in the other state to turn the FET completely on and off. A voltage swing larger than the maximum switched voltage is generally required. Values of $R_1$ and $R_2$ are chosen so that for the more positive value of the control signal (say +4 V), transistor $Q_1$ will be OFF, and for the more negative value (say +0.5 V), it will be ON. When $Q_1$ is on, $Q_2$ is turned on and produces $\sim -15$ V at the gate of the FET switch. For the more positive control input level, $Q_1$ and $Q_2$ are OFF, and the gate voltage of the FET is $\sim +15$ V. This $\pm 15$ V gate voltage swing is enough to switch signal levels of $\pm 10$ V with most FET devices.

When a JFET is used as a current switch (i.e., to switch a current source to a point at common), it can be turned on and off by ordinary TTL logic-level signals. Figure 7-19 illustrates use of a JFET current switch at the input of an op amp current follower. When the JFET is OFF, the input diode limits the drain-to-source voltage to 0.6 V. Since the drain and source voltages are always near zero, the driver voltage needs to vary only between zero and the pinch-off voltage ($< 5$ V). The small voltage drop across the JFET and the corresponding small driver voltage makes the current switch faster than a voltage switch.

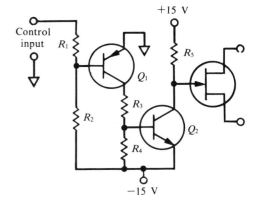

**Fig. 7-18.** FET voltage switch and its associated two-transistor switch driver. The drive circuit provides gate voltages sufficient to turn the FET completely on in one state and completely off in the other.

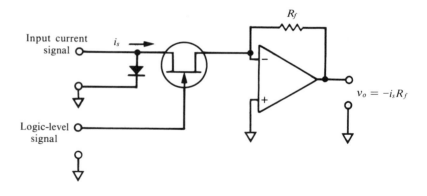

**Fig. 7-19.** A JFET current switch. When used as a current switch, the JFET can be turned on and off by ordinary logic-level signals.

$$v_o = -i_s R_f$$

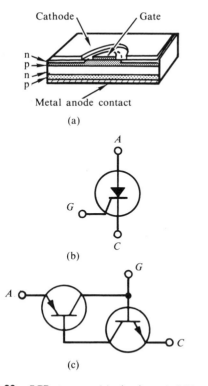

**Fig. 7-20.** SCR structure (a), circuit symbol (b), and two-transistor equivalent (c).

## 7-4    Power Control Switches

There are several semiconductor switches called **thyristors** that have at least four layers such as pnpn. Thyristors can be triggered into conduction, but they do not require any control current to maintain conduction. Because of this the circuitry for controlling a thyristor switch is usually simple and consumes very little power. Thyristors find many applications in the control of ac and dc power including the zero-crossing switch described at the end of this section.

### Silicon-Controlled Rectifier (SCR)

The SCR is a four-layer pnpn device. Its structure and symbol are illustrated in figure 7-20a and b. The three adjacent layers starting at the anode form a pnp transistor, and the three starting at the cathode constitute an npn transistor. In the two-transistor representation of the SCR shown in figure 7-20c, the pnpn structure is considered a complementary npn-pnp transistor pair.

If the anode of the SCR is connected to the positive terminal and the cathode to the negative terminal of the power supply, the center np junction is reverse biased, both transistors are off, and the SCR does not conduct. Now if base current is supplied to the npn transistor by way of the gate terminal, the npn transistor turns on. The collector of the npn transistor draws current from the base of the pnp transistor and turns it on. The on pnp transistor provides current through its collector to the base of the npn transistor; this current keeps the npn transistor on even after the original gate current is removed. Thus, all that is needed to initiate conduction is a pulse of gate current and to maintain conduction, a minimum anode-cathode current called the **holding current**. To stop conduction in the SCR the current is reduced below the holding current by removing the anode-to-cathode supply voltage, or, if the supply is ac, the SCR turns off when the

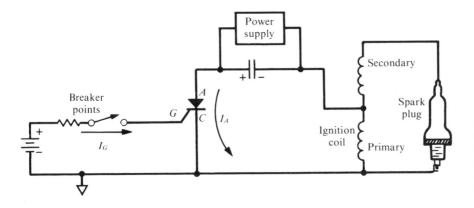

**Fig. 7-21.** Solid-state automobile ignition system. A small gate current switched to the SCR by the breaker points causes the SCR to conduct and discharges the capacitor through the primary turns of the ignition coil. The high voltage produced in the secondary causes the spark.

supply reverses polarity. A few thyristors known as **gate turn-off switches** are turned off by withdrawing current from the gate, but for normal SCRs this is an inefficient method of stopping conduction. Because the SCR conducts in only one direction, it is classified as a **reverse-blocking triode thyristor**.

Figure 7-21 illustrates the use of an SCR in a solid-state automobile ignition system. The capacitor is charged from the power supply. When the spark plug is to fire, the breaker points close momentarily, and a small gate current is applied to the SCR. The SCR turns on, and the charged capacitor discharges through the few primary turns of the ignition coil. The sudden surge of current in the ignition primary generates the very high voltage in the secondary of the coil that produces the spark. As soon as the capacitor has discharged, the current through the SCR falls below the holding current, the SCR turns off, and the charging cycle begins again. In an ordinary ignition system the points wear out because they are required to switch the full ignition coil current.

## The Triac

A **bidirectional triode thyristor**, unlike the SCR, can conduct in both directions. The **triac**, an example of such a device, behaves much like two SCRs connected in a head-to-toe manner as the triac circuit symbol and circuit model in figure 7-22 show. Because the triac can conduct in both directions, the terminals are labeled MT1 and MT2 (main terminals 1 and 2) rather than anode and cathode. The gate signal is connected to both SCRs to allow conduction to be triggered by either positive or negative gate currents.

The current-voltage curve of the triac is illustrated in figure 7-23. In quadrant I, the voltage at MT2 is positive with respect to MT1. The triac can be triggered to conduct forward current at voltages less than the breakover voltage $V_{BO}$ by a control current at the gate. In quadrant III, the voltage at MT2 is negative with respect to MT1, and conduction can occur in the

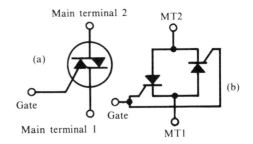

**Fig. 7-22.** Triac circuit symbol (a) and circuit model (b).

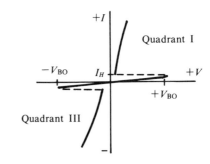

**Fig. 7-23.** Current-voltage curve of the triac.

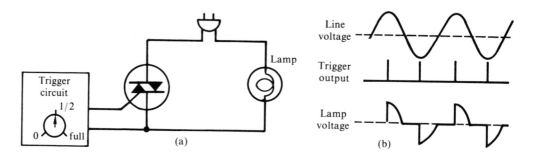

**Fig. 7-24.**  Triac lamp dimmer.

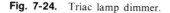

reverse direction when the triac is triggered. In normal triac use the ac voltage to be switched should always be less than $V_{BO}$ since a voltage that exceeds $V_{BO}$ will cause conduction that is not controlled by the gate. The triac remains in its conducting state until the main terminal current falls below the holding current $I_H$.

The triac is an extremely versatile switch. It can be triggered by either positive or negative dc currents, by ac currents, or by pulses. Triacs are used in power control applications where full-wave control is desirable. Figure 7-24 shows a triac used as the control element of a solid-state lamp dimmer. A trigger circuit gates the triac into conduction at a time during each half-cycle determined by the setting of the control dial. This method of power control is known as **phase control** because the triac is triggered at a specific point in the cycle (phase) of the ac waveform. Here for the half-power setting the triac is triggered into conduction in the middle of each cycle. This method of regulation is called **full-wave phase control** switching. Phase control regulation is much more efficient than current control by resistive devices that convert the unused power ($I^2R$) to heat.

## Zero Voltage Switches

Whenever a significant current is switched on or off very suddenly as it can be with solid-state switches, the inductance of the conducting path acts as an antenna generating a broad-band, high-frequency electromagnetic signal called **rf noise** or radio-frequency noise. The generation of rf noise in ac power switching can be minimized by turning the switch on or off when the ac current is zero. The combination of a circuit to sense when the switched ac waveform crosses zero and an SCR or triac power switch can provide an almost ideal zero-crossing switch. A typical IC zero-crossing triac controller is shown in figure 7-25. In addition to the zero voltage sensing and trigger circuits, it contains an input comparator and buffer. Here a voltage signal from the device being controlled is compared to a reference signal. When the

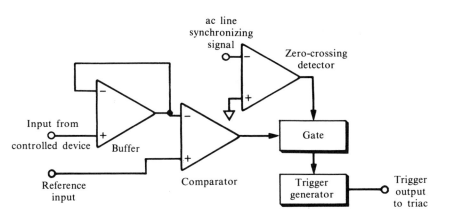

**Fig. 7-25.** Zero-crossing switch. The trigger generator produces a triac trigger pulse the first time that the
ac line voltage crosses zero after the control signal falls
below the reference level.

control input falls below the reference input, the comparator output changes
state and enables the gate to pass the zero-crossing detector output to the
trigger generator. Thus the triac is triggered into conduction on the first zero
crossing after the control signal falls below the desired level. Note that the
zero-crossing detector is a voltage comparator that senses when the ac *voltage* crosses zero volts. In order for this zero crossing to correspond to the
time of zero current, the load controlled by the triac must be resistive, i.e.,
nonreactive. Encapsulated modules that include a zero voltage switch controller and a triac (often optically isolated) are commercially available. These
**solid-state ac relays** are rugged, easy to use, and respond to standard logic
levels.

## 7-5  Transistor Amplifiers

In amplification an active electronic device such as a BJT or FET is used to
control the voltages and currents from the power supply that are applied to
the load, as shown in figure 7-26. The BJT or FET control element is
actuated by the input signal. The amplifier gain is a result of the small
voltage or current required by the control element to control larger voltages
and currents from the power supply.

## Basic JFET Amplifiers

In figure 7-27 a silicon n-channel JFET is shown in a basic common-source
amplifier configuration. The term **common-source** indicates that the source
terminal of the active element is common to both the input and output
terminals of the amplifier.

Majority carriers (electrons for the n-channel silicon FET) flow from
the source $S$ to the drain $D$ because of the drain supply voltage $V_{DD}$. Because
the gate-channel junction is reverse biased (in this example by a voltage

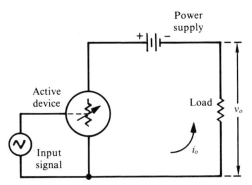

**Fig. 7-26.** Principle of amplification. The active
device (BJT or FET) controls a large voltage or current
from a power supply in response to the small input
signal.

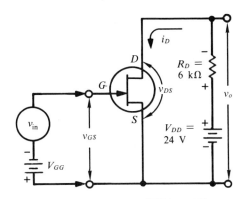

**Fig. 7-27.** Common-source JFET amplifier.

source $V_{GG}$), $i_G$ is very small ($10^{-8}$ to $10^{-13}$ A). Therefore, the gate-source voltage $v_{GS}$ is equal to $v_{in} + V_{GG}$. The current $i_D$ through the channel of the JFET is controlled by $v_{GS}$, and thus $v_{GS}$ also controls the voltage drop $i_D R_D$ across the drain load resistor $R_D$.

The output voltage $v_o$ is equal to the difference between the supply voltage $V_{DD}$ and the voltage drop across $R_D$ caused by the drain current $i_D$. Thus,

$$v_{DS} = v_o = V_{DD} - i_D R_D \qquad (7\text{-}5)$$

The linear equation 7-5 relates the output voltage and device current and can be drawn as a load line on the JFET characteristic curve as shown in figure 7-28. This provides a graphic illustration of how an input signal is amplified by a specific JFET. It should be noted that the input signal varies about an average value of $v_{GS}$. Thus, when $v_{in} = 0$, there is an average drain current $I_D$ that causes an average drain-source voltage $V_{DS}$. These average values, which are the dc components of the varying signals, are known as the **quiescent voltage** or **quiescent current**, and they define a **quiescent operating point** about which the signal varies. The quiescent values are designated by capital letters.

For the JFET amplifier, the application of an input signal causes the instantaneous circuit values of $v_{GS}$, $i_D$, and $v_{DS}$ (fig. 7-27) to differ from the corresponding quiescent values $V_{GS}$, $I_D$, and $V_{DS}$. The differences between the actual values and the quiescent values, called the **signal values**, are given the symbols $v_{gs}$, $i_d$, and $v_{ds}$. In other words, $v_{gs} = v_{GS} - V_{GS}$, where $v_{gs}$ is the change in gate-to-source voltage that causes a change in drain current $i_d = i_D - I_D$ and an output signal $v_{ds} = v_{DS} - V_{DS}$.

In analyzing the operation of practical amplifier circuits, it is convenient to use a circuit model that assumes that a linear equivalent circuit is a valid approximation over small portions of the operating range. From the analysis

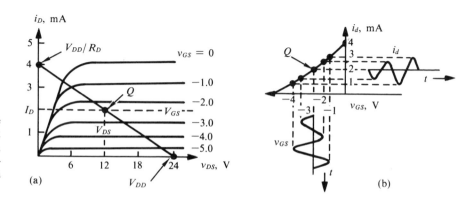

**Fig. 7-28.** JFET characteristic curve with load line (a) and transfer characteristic (b). The quiescent point ($Q$) in (a) is shown as $I_D = 2$ mA, $V_{GS} = -2.4$ V, $V_{DS} = 12$ V. In (b) the transfer characteristic illustrates how an input signal $v_{GS}$ produces a given drain current $i_d$.

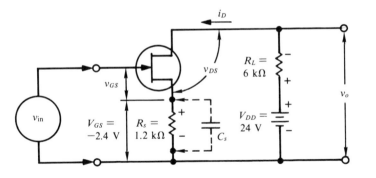

**Fig. 7-29.** A JFET self-biased amplifier. The quiescent drain current $I_D$ provides a voltage drop across source resistor $R_s$ to establish the quiescent gate-source voltage $V_{GS}$.

**Note 7-1. A Small-Signal Circuit Model for the JFET.**
The actual circuit quantities $v_{GS}$, $i_D$, and $v_{DS}$ are interdependent variables. If $i_D$ is chosen as the dependent variable, it is a function of $v_{GS}$ and $v_{DS}$ as shown below:

$$i_D = f(v_{GS}, v_{DS})$$

If the differential of $i_D$ is taken, equation 7-6 results:

$$di_D = \left(\frac{\delta i_D}{\delta v_{GS}}\right)_{v_{DS}} dv_{GS} + \left(\frac{\delta i_D}{\delta v_{DS}}\right)_{v_{GS}} dv_{DS} \tag{7-6}$$

This simply indicates that the change is equal to the change in $i_D$ due to the change in $v_{GS}$ if $v_{DS}$ were constant plus the change in $i_D$ due to the change in $v_{DS}$ if $v_{GS}$ were constant when all the changes are small. The quantity $(\delta i_D/\delta v_{DS})_{v_{GS}}$ is the **drain conductance**, $g_d$ or $1/r_{ds}$. The quantity $(\delta i_D/\delta v_{GS})_{v_{DS}}$ is called the **mutual conductance** or **transconductance** and is given the symbol $g_m$. It has the units 1/ohms ($\Omega^{-1}$), or mhos. The amplification capability of the JFET is due to its transconductance. Typical values of $g_m$ are in the range 0.5–20 m$\Omega^{-1}$.

Since the quantities $di_D$, $dv_{GS}$, and $dv_{DS}$ are the small-parameter changes from quiescent values of drain current, gate-to-source voltage, and drain-to-source voltage, respectively, they can be represented by $i_d$, $v_{gs}$, and $v_{ds}$. Therefore, equation 7-6 can be written as

$$i_d = g_m v_{gs} + \frac{v_{ds}}{r_{ds}} \tag{7-7}$$

Equation 7-7 suggests that the drain current variation $i_d$ is the sum of the main current variation $g_m v_{gs}$ and a small current variation caused by a slight change in channel resistance due to a change in drain voltage. The resistance $r_{ds}$ is known as the **drain resistance**. It is the reciprocal of the slope of the $i_D$ vs. $v_{DS}$ curve shown in figure 7-28. Since the drain characteristic curve is nearly horizontal in the region of pinch-off, $r_{ds}$ is quite high. Typical values are in the 20 k$\Omega$ to several M$\Omega$ range. The usual circuit model of the JFET amplifier is shown in the figure on the left.

in note 7-1 and the accompanying figure, we see that the small signal variation in drain current $i_d$ is approximately equal to the transconductance of the FET times the small signal variation in $v_{gs}$.

**Self-bias of JFET amplifier.** In the JFET amplifier of figure 7-27, the operating point was established with a separate bias voltage $V_{GG}$. The JFET amplifier is more often self-biased by inserting a resistance $R_s$ in the source circuit as shown in figure 7-29. When $v_{in} = 0$, the current through $R_L$ is $I_D$, and the voltage drop $I_D R_s$ provides the quiescent bias voltage $V_{GS}$ between gate and source. For the circuit of figure 7-27, for example, the desired quiescent point might be as shown in figure 7-28 ($I_D = 2$ mA, $V_{GS} = -2.4$ V). The source resistance $R_s$ can be found as the ratio of $V_{GS}$ to the desired $I_D$ or, in this case, $R_s = 2.4$ V$/2$ mA $= 1.2$ k$\Omega$. In order to prevent the bias voltage from changing with a variation in the input signal, a bypass capacitor $C_s$ is placed in parallel with $R_s$. This provides a low-impedance path around $R_s$ for $i_d$. The gain of this practical amplifier circuit can be obtained through the small-signal model (see note 7-2). The concept of the small-signal model is used extensively in analyzing amplifier circuits. It must be realized, however, that such models are restricted to a limited range of operating conditions.

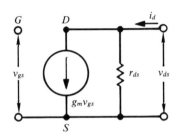

**Note 7-2.    Derivation of the JFET Amplifier Gain Equation.**

For the amplifier circuit of figure 7-29, a circuit model can be drawn as shown in the accompanying figure. This model includes the input source $v_{in}$ and the output load resistance $R_L$. The resistor $R_s$ is omitted because to an ac signal the bypass capacitor offers negligible impedance. If $r_{ds} \gg R_L$, $v_o \simeq -g_m v_{gs} R_L$. The minus sign means an inversion or phase reversal of the output signal compared to the input. Since $v_{gs} \simeq v_{in}$, the voltage gain $A_v$ is

$$A_v = \frac{v_o}{v_{in}} = -g_m R_L$$

For a typical JFET amplifier with $g_m = 4\ m\Omega^{-1}$ and $R_L = 6\ k\Omega$, $A_v = -4 \times 10^{-3} \times 6 \times 10^{-3} = -24$.

**Note 7-3.    Characteristics of the Source Follower.**

The ac model of figure 7-30 is shown below. The voltage gain of the source follower is

$$A_v = \frac{v_o}{v_{in}} = \frac{g_m}{g_m + (1/r_{ds}) + (1/R_s)}$$

If $r_{ds}$ and $R_s$ are both very large, $A_v \simeq 1$, and the source voltage closely follows the input voltage. In addition to almost unity gain, the source follower presents a very high input resistance to the voltage source because $v_{gs}$ changes much less than $v_{in}$. The output resistance $R_o$ is the parallel combination of $1/g_m$, $R_s$, and $r_{ds}$. If $R_s$ and $r_{ds}$ are both large compared to $1/g_m$, $R_o \simeq 1/g_m$.

The source follower is thus seen to possess high input resistance, nearly unity gain, and low output resistance. These characteristics make it a popular input or output stage in multistage amplifiers.

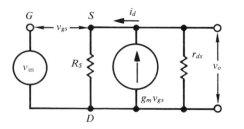

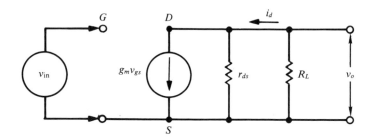

**A JFET source follower.**    In the source follower circuit shown in figure 7-30, the load resistor is connected to the source rather than to the drain. The output voltage, $v_o = i_D R_s$, and the input voltage $v_{in} = v_{GS} + i_D R_s$. Since $i_D R_s$ is much larger than $v_{GS}$ for reasonable values of $R_s$ and the JFET transconductance, most of the input variation appears across $R_s$. Application of the small-signal circuit model confirms the prediction that the amplifier is noninverting and has a gain of slightly less than unity (see note 7-3).

**Fig. 7-30.**    A JFET source follower. The amplitude is noninverting and provides nearly unity gain.

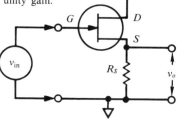

## Basic BJT Amplifiers

A basic BJT common-emitter amplifier is shown in figure 7-31. The principle of amplification is analogous to that in the JFET amplifier, but one great difference between the BJT and the FET amplifiers is the appreciable base current required by the BJT for control of the emitter-collector current. The base current $I_B$ and base-emitter voltage $V_{BE}$ influence the charge distribution and thus the operating characteristics of the BJT. As a result the characteristics of the output circuit (emitter to collector) depend on certain parameters of the input circuit. Similarly conditions in the output circuit have a small effect on the characteristics of the input circuit. To a first approximation this latter complication can be neglected (see note 7-4). According to the simple circuit model that results, the BJT amplifier can be considered as a current generator controlled by the input current.

The dc analysis of the common-emitter BJT amplifier is very similar to that of the JFET. Consider the circuit of figure 7-31 with $V_{CC} = 12$ V, $R_C = 1.5\ k\Omega$, $V_{BB} = 1.6$ V, and $R_B = 10\ k\Omega$. The load line can be determined

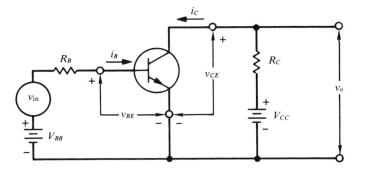

graphically by drawing a line connecting $V_{CC}$ on the voltage axis with $V_{CC}/R_C$ on the current axis of the collector characteristic curve as shown in figure 7-32. The values of $V_{CC}$ and $R_C$ should be chosen so that the load line falls well within the maximum collector power dissipation curve, which is a characteristic of the particular transistor. In this case a choice of $V_{CC} = 12$ V and $R_C = 1.5$ kΩ is safe.

The quiescent base current is chosen to put the transistor in the middle of its linear amplification region for its quiescent state ($v_{in} = 0$). For the transistor whose characteristics are shown in figure 7-32, a quiescent base current of 100 μA is a reasonable choice. To develop 100 μA of base current, $V_{BB}$ must supply the forward bias voltage of the base-emitter junction ($\sim$0.6 V for silicon transistors) and the $IR$ drop across $R_B$. If $R_B = 10$ kΩ, $V_{BB}$ should be $0.6 + (10^{-4}\text{A} \times 10^4\ \Omega) = 1.6$ V for $I_B = 100$ μA. There are several self-biasing methods that can be used with BJT amplifiers to avoid having a separate bias source $V_{BB}$. One clever method is illustrated in figure 7-33. Here the divider $R_1$ and $R_2$ brings the base to the proper voltage to forward bias the base-emitter junction. This circuit also makes the quiescent point less dependent on $\beta$.

**Fig. 7-31.** A BJT common-emitter amplifier.

**Note 7-4.  Small-Signal Model for the BJT Amplifier.**
A simple two-parameter circuit model for the BJT consists of an input resistance $r_\pi$ and an output current generator $\beta i_b$. The input resistance $r_\pi$ is the slope of a plot of $v_{BE}$ vs. $i_B$ and is given by

$$r_\pi = \left(\frac{\delta v_{BE}}{\delta i_B}\right)_{V_{CE}}$$

The ac current gain $\beta$ is the slope of the $i_C$ vs. $i_B$ characteristic curve,

$$\beta = \left(\frac{\delta i_c}{\delta i_B}\right)_{V_{CE}}$$

For most purposes this ac $\beta$ is approximately equal to the dc $\beta$ given in section 7-1. If, as before, we define signal values as differences between actual values and quiescent values, we can write $v_{be} = v_{BE} - V_{BE}$, $i_b = i_B - I_B$, $i_c = i_C - I_C$, $v_{ce} = v_{CE} - V_{CE}$. Here a lower-case variable with upper-case subscript refers to the actual circuit value; an upper case variable refers to the quiescent value. The simple two-parameter model for the BJT amplifier is then given by the accompanying figure. Note that the collector current $i_c$ is derived from the current generator, which multiplies the base current $i_b$ by $\beta$. In contrast to the FET, a finite base current is required to control the output current generator.

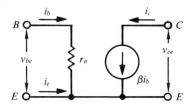

This simple model is adequate for low-frequency operation and for small values of external load resistance. More accurate four-parameter models consider the nonideality of the current generator and the influence of the output circuit on the input. These models often make use of the generalized **hybrid parameters** $h_{ie}$ (*hybrid input* resistance for common *emitter* configuration), $h_{re}$ (*reverse voltage amplification parameter*), $h_{fe}$ (*forward current gain*) and $h_{oe}$ (*output conductance*).

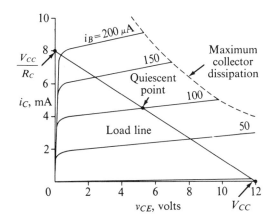

**Fig. 7-32.** Direct-current analysis for the BJT amplifier of figure 7-31.

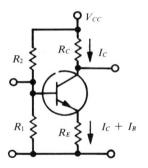

**Fig. 7-33.**  Self-biased BJT amplifier. This circuit can provide the appropriate base bias and stabilize the quiescent point. An attempted increase in $I_C$ gives a larger current through $R_E$ and raises the emitter voltage. This reduces $V_{BE}$, the base-emitter forward bias, which lowers $I_B$ and brings $I_C$ back to its original value.

Once the quiescent point has been chosen and the parameters $r_\pi$ and $\beta$ have been determined, the amplifier gain characteristics and its input and output resistances can be calculated. Table 7-1 shows the resulting ac circuit model of a common-emitter amplifier and the approximate equations for the amplifier characteristics. More accurate equations derived from more complicated models can be found in several of the references for this chapter in the bibliography at the end of the book.

The equations in table 7-1 are idealized in that the source resistance is not included in the calculation. The voltage gain can be substantially reduced if the source resistance is comparable to $r_\pi$ because the source resistance and $r_\pi$ form an input divider.

For a typical transistor with $\beta = 50$ and $r_\pi = 2.5\ \text{k}\Omega$, the amplifier characteristics for $R_L = 1.5\ \text{k}\Omega$ are $A_i = -50$, $A_v = 30$, $A_p = 1500$, $R_{\text{in}} = 2.5\ \text{k}\Omega$, and $R_o = 1.5\ \text{k}\Omega$.

The common-emitter BJT amplifier is thus seen to have voltage gain, current gain, phase inversion of the input signal, and relatively low input resistance.

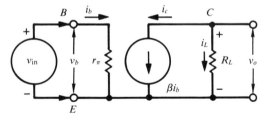

**Table 7-1.**  Common-emitter amplifier characteristics.

| | |
|---|---|
| Current Gain: | $A_i = \dfrac{i_L}{i_b} = \dfrac{-i_c}{i_b} = \dfrac{-\beta i_b}{i_b} = -\beta$ |
| Input Resistance: | $R_{\text{in}} = \dfrac{v_b}{i_b} = r_\pi$ |
| Voltage Gain: | $A_v = \dfrac{v_o}{v_{\text{in}}} = \dfrac{i_L R_L}{i_b r_\pi} = \dfrac{-\beta R_L}{r_\pi}$ |
| Output Resistance: | $R_o = R_L$ |
| Power Gain: | $A_p = A_i A_v = \dfrac{\beta^2 R_L}{r_\pi}$ |

**Other BJT configurations.**  The other BJT configurations are the common-collector and the common-base. The common-base configuration provides low input impedance, very high output impedance, and is non-inverting. It has almost unity current gain, a potentially high voltage gain, and finds use as a constant current source and as an impedance-matching stage.

The common-collector amplifier shown in figure 7-34 is usually called an **emitter-follower** amplifier. Its operation is very similar to that of the

JFET source follower. The emitter-follower circuit has the desirable characteristics of relatively high input impedance, low output impedance, and nearly unity voltage gain. It is a noninverting configuration and is commonly used as a buffer amplifier or an output stage.

The transistor pair connected as shown in figure 7-35 is known as a **Darlington pair**. In this configuration the collectors of the pair are connected together, and the emitter of the input transistor $Q_1$ is connected to the base of the second transistor $Q_2$. The Darlington pair can function either as an emitter-follower (fig. 7-35a) or as a voltage amplifier (fig. 7-35b). The Darlington emitter-follower provides a higher input impedance, lower output impedance and more nearly unity voltage gain than its single transistor equivalent. The overall gain of the pair is nearly $\beta_{Q_1} \times \beta_{Q_2}$. In the voltage amplifier configuration (fig. 7-35b), the circuit can be viewed as an emitter-follower ($Q_1$) driving a common-emitter stage ($Q_2$). It provides the voltage gain of the common-emitter amplifier while retaining the high input impedance of the emitter-follower. Darlington pairs are commercially available in single packages.

## Difference Amplifier

An amplifier whose output is a function of the difference between two input voltages is a **difference amplifier**. Ideally the output of a difference amplifier is not responsive to common mode voltages, temperature variations, and supply voltage fluctuations.

A basic FET difference amplifier is shown in figure 7-36. The sources of the two FETs are connected through source resistors $R_{s1}$ and $R_{s2}$ to a constant current source made from resistor $R_{cm}$ and the negative supply voltage $-V$. The differential output $v_{od}$ is taken from drain-to-drain as illustrated. An effective differential amplifier should have high gain to a difference signal but low gain to any common mode signal (high common mode rejection ratio). The common mode response is determined largely by the quality of the constant current source. A common mode signal should produce a negligible change in the currents through the drain resistors and thus produce a correspondingly low change in output voltage. To obtain a bias current that is independent of the common mode signal, $R_{cm}$ should be a very large resistance. Alternatively, the current source can be controlled to be constant by an active circuit such as the transistor circuit of note 9-2.

When a difference signal $v_d$ is applied between the two inputs, one-half of $v_d$ appears across each stage when the resistances and FETs are matched (see fig. 7-36). This equal division of the differential input signal produces equal and opposite current changes in the two source resistors and no change in the overall current supplied by the constant current source. The resulting

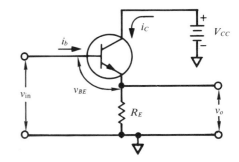

**Fig. 7-34.** Emitter-follower circuit. The amplifier provides nearly unity voltage gain, high input impedance, and low output impedance.

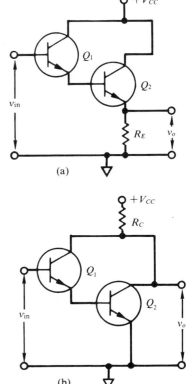

**Fig. 7-35.** Darlington pairs. (a) The emitter-follower Darlington amplifier and (b) the Darlington voltage amplifier.

**Fig. 7-36.** A JFET difference amplifier. The constant current source composed of $R_{cm}$ and $-V$ makes the output insensitive to a common mode signal. The difference voltage $v_d$ is divided equally across each stage and produces equal and opposite current changes in the drain load resistors. This produces equal and opposite output voltages from each FET.

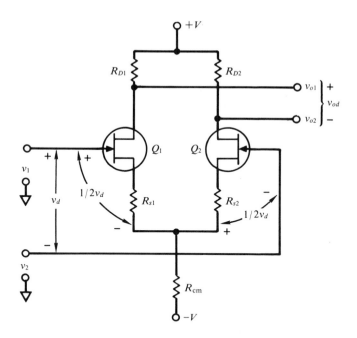

current changes in the drain load resistors produce equal and opposite output voltages. The differential output voltage is then

$$v_{od} = v_{o1} - v_{o2} = A_1 \left(\frac{v_d}{2}\right) - A_2 \left(\frac{v_d}{2}\right)$$

where $A_1$ and $A_2$ are the voltage gains of FETs $Q_1$ and $Q_2$ respectively. If $Q_1$ and $Q_2$ are well matched, $A_1 = A_2 = A$, and $v_{od} = A v_d$.

The difference amplifier is widely used as an input stage because of its ability to reject common mode signals, its relative insensitivity to environmental changes, and its amplification of true difference signals. A difference amplifier is illustrated in section 7-7 as the input stage for a general purpose operational amplifier. For actual measurement of the difference between two voltage signals, operational amplifier circuits are available to perform the difference function as described in chapter 8.

## 7-6  Amplifiers with Feedback

An amplifier is said to have **feedback** when part of its output signal is returned to its input. **Negative feedback** occurs when the magnitude of the input signal is decreased, and **positive feedback**, when the magnitude is increased. In general, negative feedback has a stabilizing influence on a system and positive feedback has the opposite effect. The presence of feedback affects almost every electrical characteristic of an amplifier. Feedback can be used to control the gain of an amplifier, improve its stability and impedance characteristics, and reduce noise and distortion. Feedback is also very useful in controlling and modifying the frequency response characteristics of an amplifier. In this section some of the basic characteristics of amplifiers with feedback are discussed.

## Gain

A quantitative relation for the effect of feedback on the gain of an amplifier can be derived through reference to figure 7-37. A fraction $\beta$ of the output signal is fed back and added to the input signal. The fraction $\beta$ is determined by the voltage divider circuit in the feedback loop. The input voltage to the amplifier $v_{in}$ is the sum of the signal input voltage $v_{sig}$ and the feedback voltage $\beta v_o$

$$v_{in} = v_{sig} + \beta v_o \tag{7-8}$$

From the amplification $A$ of the amplifier,

$$v_o = A v_{in} \tag{7-9}$$

The gain of the amplifier with feedback $A_f$ is the ratio of $v_o$ to $v_{sig}$. From equations 7-8 and 7-9 $A_f$ is expressed by:

$$A_f = \frac{v_o}{v_{sig}} = \frac{A}{1 - \beta A} \tag{7-10}$$

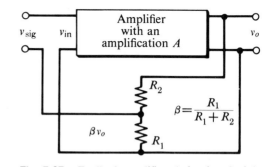

**Fig. 7-37.** Feedback amplifier. A fraction $\beta$ of the output voltage is fed back to the input.

The gain of the feedback amplifier $A_f$ or $K$ is often called the **closed-loop** gain and $A$ is called the **open-loop** gain.

If $\beta A$ is negative, $1 - \beta A$ is greater than 1, and the gain of the amplifier is less than $A$. In this case the feedback is said to be negative. An important limiting case of equation 7-10 occurs when $\beta A \gg 1$, that is, when the open-loop gain is very large. In this case the gain of the feedback amplifier $A_f$ reduces to $-1/\beta$. This result means that the closed-loop gain of the feedback amplifier can be made virtually independent of the amplifier's open-loop characteristics. We have encountered that characteristic with op amps in chapter 5. Note from figure 7-37 that the feedback fraction $\beta$ can be accurate and easily varied. Note, too, that relatively large changes in $A$ influence $A_f$ only slightly. For example, if $A$ is $-10^5$ and $\beta = 0.01$, the closed-loop gain $A_f$ is $-99.9$. If the amplifier gain were to decrease 10%, the new closed-loop gain $A_f$ would be $-99.89$, a change of less than 0.1%.

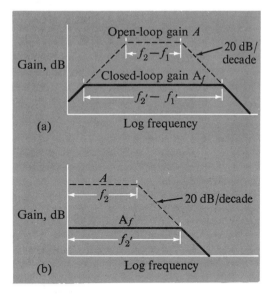

**Fig. 7-38.** Frequency response of feedback amplifiers. (a) An ac amplifier and (b) a dc amplifier. Note the increased bandwidth with negative feedback.

## Frequency Response

It was shown above that as the amount of negative feedback is increased, the gain of the amplifier becomes less dependent on the amplifier and more a function of the feedback network. If the impedance of the feedback network is nonreactive, and the fraction $\beta$, therefore, independent of frequency, the gain $A_f$ of the feedback amplifier is less dependent on frequency, and the bandpass of the amplifier is extended. Positive feedback, on the other hand, tends to decrease the bandpass of the amplifier.

The relationship between the increase in frequency response and feedback can be visualized on the plots of gain against frequency in figure 7-38. The dotted lines of figure 7-38a show the open-loop gain vs. log $f$ for an ac amplifier. The midfrequency gain is $A$, and the amplification rolls off at 20 dB/decade. The original bandwidth is $f_2 - f_1$, as shown. If a fraction $\beta$ is now fed back in such a way that closed-loop gain $A_f$ results, it is apparent that the bandwidth has been extended to $f_2' - f_1'$. The same improvement in bandwidth for a dc amplifier such as an operational amplifier is shown in figure 7-38b. Here the bandwidth has been extended from $f_2$ for the amplifier without feedback to $f_2'$ for the feedback amplifier. The product $A_f f_2'$ is constant and is often called the **gain-bandwidth product**.

## Noise

Nonlinearity of amplification and noise from the amplifier are also reduced by negative feedback. A noise voltage at the amplifier output that is fed back attenuated and out of phase to the input causes the amplifier to counteract the spurious signal at the output. Suppose that a noise voltage $N$ is generated in the amplifier. The result of this noise that appears at the output will be designated $N'$. A voltage $\beta N'$ is fed back to the amplifier input. The actual output noise $N'$ is then the sum of $N$ and $A\beta N'$, the noise fed back and amplified. Solving for $N'$,

$$N' = N + A\beta N' = \frac{N}{1 - \beta A}$$

Noise generated in the amplifier is reduced by the same factor as the gain.

## The Possibility of Oscillation

The gain of the voltage feedback amplifier was given in equation 7-10 as $A_f = A/(1 - \beta A)$. When $\beta A$ is positive and the quantity $1 - \beta A$ is less than 1, $A_f$ is greater than $A$. In other words, there is positive, or regenerative, feedback. As $\beta A$ approaches 1, the gain $A_f$ approaches infinity. This generally causes the amplifier to "peg" at its positive or negative limit. When the feedback is frequency selective and $\beta A$ equals 1 for a particular frequency, the feedback voltage is sufficient to maintain an output signal without any input signal. Such an amplifier is acting as an **oscillator**.

Unless an oscillator is desired, it is essential that the positive feedback conditions necessary for oscillation do not exist. Even when the feedback is negative at normal frequencies, the possibility of positive feedback exists for frequencies on the fringe of the amplifier bandpass where considerable phase shift can occur. Feedback oscillators made with operational amplifiers are discussed in chapter 8.

## 7-7  The Operational Amplifier

The operational amplifier introduced in chapter 5 is used in hundreds of different ways to provide elegant solutions to measurement and control problems. It is, of course, not necessary to know what is inside the op amp block (represented simply by a triangular symbol) in order to use an op amp profitably. However, most users feel more comfortable when they have some idea of how op amps are designed, and certainly the user can appreciate specifications and limitations better after investigating the internal operation of a typical op amp. Therefore, the first part of this section is devoted to a look inside the triangular symbol. The section concludes with a summary of specifications for operational amplifiers.

### Integrated Circuit Operational Amplifier

The op amp that we have chosen to investigate is constructed on a semiconductor microchip. This integration of several basic amplifier circuits into a single chip has resulted in units no larger than, and costing no more than, a single transistor. The general purpose op amp is to function as a gain block with dc integrity, high input impedance, low output impedance, and wideband frequency response with negative feedback. The ways in which these characteristics are obtained can be understood by studying figure 7-39, a simplified circuit diagram of an IC op amp.

The input circuit is a differential amplifier using matched p-channel JFETs as source followers. An external 10-k$\Omega$ potentiometer can be used to balance the differential amplifier and compensate for any offset. The differential input provides for the inverting and noninverting inputs and for excellent CMRR. The JFETs also provide a very high input impedance and a very low input bias current.

BJT transistors (shown as a single transistor in fig. 7-39) are coupled together to produce a second stage with high current gain. This stage is usually a Darlington pair with a current gain approximately equal to the product of the $\beta$ values of the two transistors.

The output stage is a cascaded complementary emitter-follower circuit. It has the necessary low output impedance ($<$ 1 $\Omega$ with negative feedback) for driving external loads.

## Summary of Op Amp Specifications

Every commercial op amp has a set of specifications. Although some of these have been described earlier, the most important specifications are summarized here as a useful reference.

**Open-loop gain.**   The open-loop gain is the gain without feedback in decibels. Since gain (dB) $= 20 \log (v_o / v_{in})$, an open-loop gain of 100 dB means that $A = 100\,000$.

**Frequency response.**   The frequency response characteristic reflects the variation in the open-loop gain with frequency (see fig. 7-38). It is normally given as the 3 dB point or the unity-gain bandwidth.

**Input offset voltage.**   Because the input differential amplifier is not perfectly balanced, there is a small, relatively constant, but temperature dependent offset voltage between the input terminals. The offset voltage causes an output voltage when both inputs are at 0 V. An external balance potentiometer may be required to null the offset voltage and reduce offset voltage errors. The offset voltage changes with temperature, supply voltage, and time.

**Input bias current.**   Even when the input voltage is zero, there is an input current in each input terminal which, for JFET input stages, results from the gate currents of the JFETs and any leakage currents within the amplifier. Typical bias currents for good op amps are in the range 0.1–100 pA at 25°C. The bias current is quite sensitive to temperature changes; it typically doubles for each 10°C change in temperature. The difference between the two input bias currents is called the input offset current.

**Input resistance.**   Modern FET amplifiers have extremely high input impedances, typically in the range $10^9$–$10^{14}$ $\Omega$. An amplifier with $10^{14}$ $\Omega$ input resistance can keep the input current in the subpicoampere range for full common mode voltage swings of $\pm$ 10 V.

**Input noise.**   The input noise of the amplifier is the factor that limits signal resolution. It varies as a function of source impedance and frequency. Graphs of typical changes of input noise with source resistance are usually available from the manufacturer.

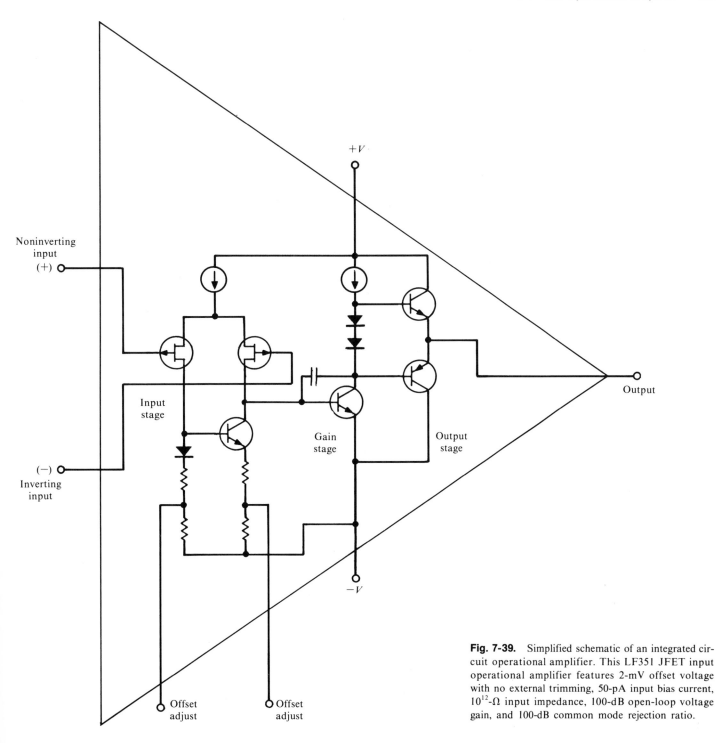

**Fig. 7-39.** Simplified schematic of an integrated circuit operational amplifier. This LF351 JFET input operational amplifier features 2-mV offset voltage with no external trimming, 50-pA input bias current, $10^{12}$-$\Omega$ input impedance, 100-dB open-loop voltage gain, and 100-dB common mode rejection ratio.

**CMRR.**   The common mode rejection ratio of op amps should be very high, particularly for configurations that require a differential input. Typical CMRR values for good op amps are in the range 80–100 dB.

## 7-8   Application: Regulated and Programmable Power Supplies

The principles of power supply regulation were introduced in chapter 3 where the ideal power supply was seen to produce a constant output voltage independent of load and line variations. Modern power supplies use IC voltage regulators to achieve high stability and very low ripple voltages. These regulators are based on the series regulator diagrammed in figure 3-21. Internally the three-terminal regulator contains a difference amplifier, a feedback network, a series control element, and a reference voltage source as figure 7-40 illustrates. The power transistor $Q$, called the **pass transistor**, acts as the variable resistance control element. It provides current gain and the variable voltage drop which compensates for any fluctuations in $V_{in}$ or $I_L$. An internal Zener diode provides the reference voltage $V_r$.

Fixed voltage IC regulators are available with positive or negative output voltages preset to industry standard values (5, 6, 8, 12, 15, 18, and 24 V). Many allow external pass transistors to be added to increase the output current. Adjustable output regulators enable the user to set the output voltage.

**Fig. 7-40.**   Regulated power supply. This voltage feedback circuit controls $V_o$ to be approximately $V_r/\beta$. A reference voltage $V_r$ and a fraction $\beta = R_2/(R_1 + R_2)$ of the output voltage $V_o$ are compared by the op amp. The amplifier output voltage $V_A$ controls the pass transistor $Q$, an emitter follower. The pass transistor output voltage $V_o \simeq V_A$, and $V_o = V_r A/(1 + \beta A)$, where $A$ is the open loop gain of the op amp.

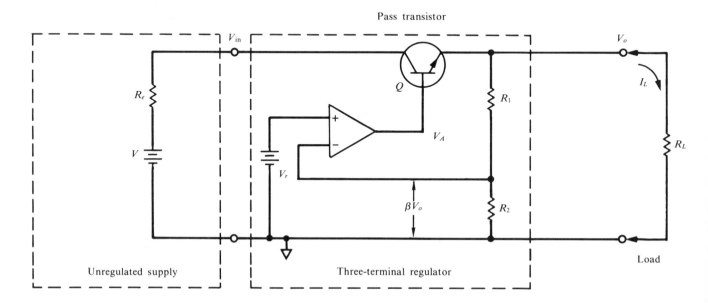

A power supply in which the output voltage varies in response to a remote command is known as a **programmable power supply**. With the follower configuration of figure 7-40, external control over $V_o$ could be achieved by a variation in $V_r$ or by allowing divider resistor $R_1$ to be externally controlled. Some commercial power supplies, known as **operational power supplies** are based on the inverting amplifier configuration of figure 7-41. They can be programmed by varying $v_r$ or the feedback resistor $R_f$. Many provide user access to the summing junction so that typical op amp operations can determine the voltage output. Operational power supplies can also be readily configured so that the controlled quantity is the load current rather than the output voltage. Such a **current-regulated power supply** is highly useful in many control applications.

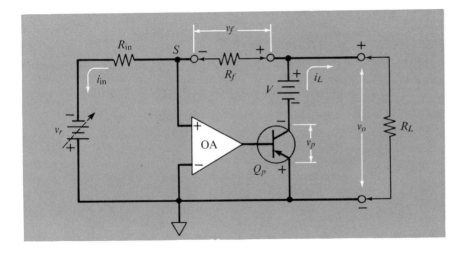

**Fig. 7-41.** Operational power supply. The output voltage $v_o$ is given by $v_o = -v_r(R_f/R_{in})$. The combination of the op amp, the pass transistor and the raw supply can be considered as a high-power op amp and used for adding, subtracting, integrating and other operations. For a bipolar output, an npn transistor and oppositely connected raw supply must be added to the above circuit so that the output polarity depends on the direction of the input current.

## Suggested Experiments

**1.  Bipolar junction transistor characteristics.**
Obtain the collector characteristic curves for an npn transistor in the common-emitter configuration. Measure $V_{CE}$ in saturation and in cutoff. Obtain $\beta$ from the characteristic curves.

**2.  FET characteristics.**
Obtain characteristic curves for a JFET from a curve tracer. From the curves obtain the ON/OFF characteristics of the FET.

**3.  Transistor switching speeds.**
Wire a basic BJT switching circuit, and observe its characteristics. Determine the ON and OFF switching times with an oscilloscope.

**4.  The SCR and the triac.**
Determine the current-voltage curves for the SCR and the triac. Measure the turn-on time as a function of gate current.

**5.  SCR and triac applications.**
Wire either an SCR overvoltage protection circuit for a power supply or a triac lamp controller to keep the intensity constant by controlling the power applied.

**6.  Basic transistor amplifiers.**
Connect either a JFET amplifier in the common-source configuration or a BJT amplifier in the common-emitter configuration. Choose an appropriate quiescent point from the characteristic curves. Connect a self-biased JFET or BJT amplifier, measure its gain, and compare with expected values. Obtain the amplifier gain as a function of frequency.

**7.  Source follower.**
Connect an FET as a source follower or a BJT as an emitter follower. Determine the input and output resistance and voltage gain.

**8.  Difference amplifiers.**
Construct a difference amplifier from matched JFETs. Determine the gain and the common mode rejection ratio.

**9.  Feedback amplifiers.**
Connect an operational amplifier follower with gain. Observe its frequency response for several different gains. Determine the upper 3 dB point, the slope of the open-loop roll off, and the unity-gain bandwidth.

## Questions and Problems

**1.**  (a)  For the transistor that is characterized by the current-voltage curves in figure 7-5, draw the load line for $V_{CC} = 5$ V and $R_L = 1$ k$\Omega$. Determine the values of $V_{CE}$ and $I_C$ at the OFF operating point. (b) What is the minimum value of $I_B$ needed to ensure that saturation is achieved? (c) If the input signal $V_B$ is 4 V, what is the maximum value of $R_B$ that can be used and still achieve saturation?

**2.**  Determine the transconductance of the JFET whose characteristic curves are given in figure 7-11. Assume that $V_{DS} = 10$ V. Is the transconductance independent of $V_{GS}$?

**3.**  (a) A reverse biased photodiode has a quantum efficiency of 0.3 at the wavelength of an incident photon flux of $10^{12}$ photons/s. What is the photocurrent $I_P$? (b) A phototransistor has a dc current gain of 50. If it is operated under the same conditions as the photodiode in part (a), what is the emitter current $I_E$?

**4.**  In the transistor switching circuit of figure 7-17, $\beta = 100$, $V_{CE(sat)} = 0.2$ V and $V_{BE(on)} = 0.6$ V. The input signal varies between 0 and +1 V, $R_B = 1$ k$\Omega$ and $V_{CC} = 5$ V. (a) When the input signal is +1 V, is the transistor in saturation? (b) Calculate the collector current $I_C$ when the transistor is on, and calculate the ON resistance.

**5.**  The transistor switch circuit shown in figure 7-42 is used to control the current through a relay coil. The relay coil has a resistance of 200 $\Omega$ and requires 20 mA to operate. (a) Does the transistor have to be in saturation to energize the relay coil? Explain. (b) If a +4-V signal at the control input is to turn the relay on, what current is drawn from the source ($V_{BE(on)} = 0.6$ V)? (c) What is the minimum value of $\beta$ that the transistor must have to operate the relay?

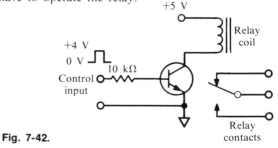

**Fig. 7-42.**

**6.** The storage time caused by allowing a transistor to saturate can significantly limit the switching speed. One method of preventing saturation in a common-emitter switch is to put a diode between the base and collector to clamp the collector. (a) Which end of the diode (anode or cathode) should be connected to the collector? (b) Describe how the diode prevents saturation.

**7.** A BJT analog switch is connected in the series-shunt configuration of figure 7-16c. The ON and OFF resistances of the transistors are 150 $\Omega$ and 100 k$\Omega$, respectively. If the source $v_s$ has an output voltage of 5 V, an equivalent resistance $R_e$ of 100 $\Omega$, and $R_L = 1$ k$\Omega$, calculate the voltage drop $v_L$ across $R_L$ for the two states of the switch.

**8.** Describe the effect on the lamp voltage if the lamp dimmer of figure 7-24 is operated by an SCR instead of a triac. Would this still be full-wave phase control switching? Describe the lamp voltage waveform.

**9.** In the JFET amplifier of figure 7-29, circuit conditions are changed to $R_s = 1$ k$\Omega$, $R_L = 10$ k$\Omega$, $C_s = 20$ $\mu$F, and $V_{DD} = 30$ V. For the particular JFET used, the transconductance $g_m = 2$ m$\Omega^{-1}$ and the drain-source resistance $r_{ds} = 100$ k$\Omega$. (a) Assume that the current $v_{ds}/r_{ds}$ is negligible, and calculate the voltage gain. (b) For $v_{gs} = 1$ V, determine whether $v_{ds}/r_{ds}$ is significant compared to the current $g_m v_{gs}$.

**10.** (a) Draw the small-signal model of the JFET amplifier of problem 9. Give actual values for the small-signal components. (b) What would the small-signal model of an ideal voltage amplifier be? (c) Compare the JFET model found in (a) to the model of an ideal voltage amplifier.

**11.** For a particular BJT, $\beta = 75$, $r_\pi = 1$ k$\Omega$. The transistor is to be used in the circuit of figure 7-31 with $V_{BB} = 1.20$ V, $R_C = 1.5$ k$\Omega$, $V_{CC} = 12$ V, and $R_B = 2$ k$\Omega$. (a) Calculate the current gain, the voltage gain, and the power gain of the amplifier. (b) Calculate the input and output resistances (assume $R_C = R_L$).

**12.** (a) Draw the small-signal model of the common-emitter amplifier of problem 11. Include specific values for the small-signal components. (b) What would the model of an ideal current amplifier be? (c) Compare the BJT model developed in (a) with the ideal current amplifier. Describe how the BJT falls short of ideality.

**13.** In the JFET source follower of figure 7-30, the FET used has a transconductance $g_m$ of 15 m$\Omega^{-1}$ and a drain-source resistance $r_{ds}$ of 10C k$\Omega$. The source resistor $R_s = 500$ k$\Omega$. Find the voltage gain $A_v$ and the output resistance $R_o$.

**14.** The output voltage of a difference amplifier changes by 5 V for an input difference signal of 250 $\mu$V; the output voltage changes by 1 V for a common mode input signal of 500 mV. (a) What is the common mode rejection ratio (CMRR) of the difference amplifier? (b) Express the CMRR in decibels.

**15.** It is desired to measure a 50-mV signal to an accuracy of 1% using an oscilloscope with a differential input. If the CMRR of the oscilloscope amplifier is $10^4$, what is the maximum common mode input signal that can be tolerated?

**16.** One circuit model for a difference amplifier is shown in the figure 7-43. The parameters are $\mu = 1000$, $R_{in} = 5$ k$\Omega$, $R_o = 200$ $\Omega$, and $R_L = 1$ k$\Omega$. If the signal $v_2 = 0$ V, calculate the voltage gain, the current gain, and the power gain.

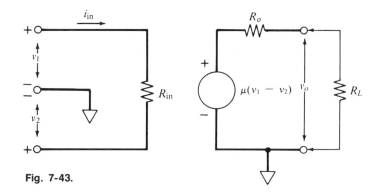

**Fig. 7-43.**

**17.** The same amplifier as in problem 16 is operated with a sine-wave input at $v_1$ of 5-mV rms amplitude and $v_2 = 0$ V. Consider the load resistor $R_L$ to be variable. (a) What is the maximum output power that can be drawn from the amplifier by adjustment of $R_L$? (b) What is the power gain at the value of $R_L$ that gives maximum power?

**18.** Feedback affects both the input and output impedances of an amplifier. (a) If the $IR$ drop across the output impedance $R_o$ of an amplifier is considered, equation 7-9 is modified to

$$v_o = A v_{in} - iR_o$$

Beginning with this equation, calculate the effect of voltage feedback on the output impedance of the amplifier. (b) Beginning with equation 7-9 and the assumption that the output load is small (no $IR$ drop across $R_o$), calculate the effect of voltage feedback on the input impedance of an amplifier.

**19.**  It is desired that the gain of a feedback amplifier be stable to better than 0.01% for changes in the open-loop gain as high as 10%. (a) If the open-loop gain is $10^5$, what is the maximum closed-loop gain? (b) If the closed-loop gain must be 100, what is the minimum necessary open-loop gain?

**20.**  The gain of an amplifier behaves much like a low pass filter in its frequency response. At high frequencies the gain is given by $A = X_C/R$. (a) Show that $A(\text{dB}) = -20 \log f - 20 \log (2\pi RC)$. (b) Show that changing the frequency by a factor of 2 (one octave) gives a 6-dB change in gain in this region. (c) What change in linear gain results from an octave change in frequency?

# Linear and Nonlinear Op Amp Applications

# Chapter 8

The versatile op amp introduced in chapter 5 has become the universal "gain block" for analog circuits of all kinds. In this chapter, the applications of op amps are developed beyond the basic circuits introduced earlier. Precision linear difference amplifiers called instrumentation amplifiers are studied first. Special amplifiers that accurately perform a nonlinear operation (such as taking the absolute value or logarithm) on the input amplitude are studied next. The analog multiplier, a very versatile nonlinear device, allows products, ratios, squares, and square roots of analog signals to be readily obtained.

An underlying theme of op amp applications is the use of high gain and feedback to improve the quality of an operation over that obtainable with passive circuits. This theme is particularly clear in the use of op amps for active filters, band-limited amplifiers, and oscillators. Amplifiers have had a key part in the dramatic advances of the last decade in the recovery of very low level signals from relatively larger amounts of noise. The lock-in amplifier and the chopper-input amplifier accomplish this signal improvement by relocating the measured amplitude variations in a narrow band of the frequency spectrum well away from the major noise frequencies. The lock-in amplifier is a particular application of modulation, the technique by which a high-frequency signal is used to carry lower-frequency information. Amplitude modulation (AM) and demodulation are described in this chapter. Frequency modulation is discussed in chapter 9.

## 8-1   Difference Amplifiers

As the name implies, the difference amplifier produces an output voltage that is proportional to the difference between two input voltages. That is, $v_o = K(v_2 - v_1)$. The difference amplifier is very convenient where the desired measurement is the difference between two transducer outputs, the off-balance signal of a Wheatstone bridge, or the voltage difference between any two points, neither of which is at the common voltage. It is also the type of amplifier most often used to eliminate noise that occurs between the common of the signal source and the common of the amplifier as shown in figure 8-1.

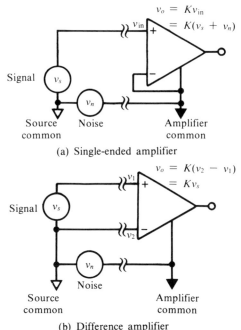

(a) Single-ended amplifier

(b) Difference amplifier

**Fig. 8-1.** The effect of noise in amplifier input connections. The single-ended amplifier (a) amplifies signal $v_s$ and the noise $v_n$, which is the voltage difference between the source and amplifier common. The difference amplifier (b) can be used to exclude this noise voltage from the signal to be amplified.

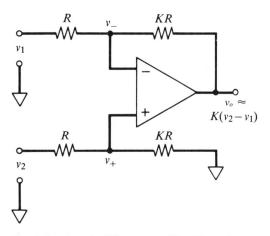

**Fig. 8-2.** A basic difference amplifier. The resistance ratios must be carefully matched for good CMRR. A trimmer adjustment on any of the resistors would allow precise balance adjustment.

**Note 8-1. Derivation of Exact Gain Equation for the Basic Difference Amplifier.**
The relation between the output voltage and the two input voltages can be obtained by substituting the expressions for $v_+$ and $v_-$ into $v_o = A(v_+ - v_-)$ as follows:

$$v_+ = v_2 \frac{KR}{R + KR}$$

$$v_- = v_1 + (v_o - v_1) \frac{R}{R + KR}$$

$$v_o = \frac{AK(v_2 - v_1)}{1 + K + A} \simeq K(v_2 - v_1) \text{ for } A \gg K + 1$$

The relationship $v_o = K(v_2 - v_1)$ for the ideal difference amplifier indicates that the gain is the same for signal $v_2$ as for signal $v_1$ and that this is true for all values of $v_o$. If the amplifier is perfectly balanced for the two inputs, an identical voltage change applied to both inputs cancels and has no effect on $v_o$. Thus the output responds only to the *differences* in the input voltages. A measure of the degree of balance in a difference amplifier is the ratio between the response of the amplifier to a signal applied between the difference inputs and its response to a signal applied between both inputs and common. This ratio is called the common mode rejection ratio (CMRR) since the portion of the input signal that is applied identically to both inputs is called the common mode signal. For example, if $v_o = 1000(v_2 - v_1)$ for a difference signal and $v_o = 0.1v_2$ when $v_2 = v_1$, for a common signal, the CMRR $= 1000/0.1$ or 10 000, which is 80 dB.

## Basic Difference Amplifier

The circuit shown in figure 8-2 is that of a basic difference amplifier. It combines the inverting amplifier and the follower with gain. A fraction of the input voltage $v_2$ is applied to the noninverting op amp input, and the same fraction of the difference between $v_1$ and $v_o$ is applied to the inverting op amp input. The difference between $v_-$ and $v_+$, the – and + op amp input voltages, is $v_o / A$, where $A$ is the open-loop gain of the op amp. For most op amps, $A$ is $10^5$ or more, and $v_o / A$ is negligibly small with respect to the input voltages. When this assumption is made,

$$v_o \simeq K(v_2 - v_1) \tag{8-1}$$

As equation (8-1) shows the output voltage is equal to the difference voltage multiplied by $K$ (see note 8-1). With the selection of a high CMRR op amp and carefully matched resistance ratios, the amplifier of figure 8-2 can be very useful. The common mode response of a difference amplifier can be checked by connecting the output of a signal generator to both inputs in parallel and observing the amplifier output. If the amplifier has a balance control, it should then be adjusted for a minimum output.

## Instrumentation Amplifier

A complete **instrumentation amplifier** combines the advantages of the difference input with the high input resistance of the voltage follower. This is readily achieved by simply putting a voltage follower amplifier before each input of the difference amplifier of figure 8-2. The use of follower with gain circuits to achieve higher gain is not recommended, however, because the follower amplifier gains would have to be well matched in order to achieve a high CMRR. A very clever circuit that cross-couples the two follower with

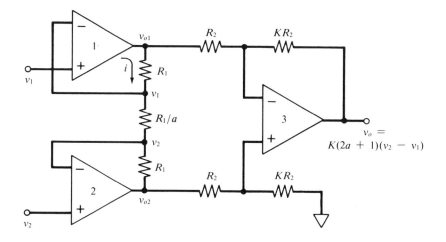

**Fig. 8-3.** An instrumentation amplifier using a cross-coupled differential follower input. The gain can be adjusted with the single resistor $R_1/a$. The difference gain of the input stage is $1 + 2a$.

gain circuits so that they track each other is shown in figure 8-3 with the difference amplifier. The gain and cross-coupling are provided by the three resistors between the two follower outputs. The follower amplifiers 1 and 2 keep the feedback points equal to $v_1$ and $v_2$ respectively. This results in a current through the three resistors of $i = a(v_1 - v_2)/R_1$. The follower output voltages are then equal to the sum of the feedback voltage and the $IR$ drop through the feedback resistor. Thus, $v_{o1} = v_1 + a(v_1 - v_2)$ and $v_{o2} = v_2 - a(v_1 - v_2)$. These expressions show that each follower amplifies its input signal by 1 and the difference signal by $+a$ or $-a$. The difference gain $A_d$ of the cross-coupled followers is

$$A_d = \frac{v_{o1} - v_{o2}}{v_1 - v_2} = 1 + 2a$$

and the common mode gain, $A_{cm}$ is

$$A_{cm} = \frac{(v_{o1} + v_{o2})/2}{(v_1 + v_2)/2} = 1$$

The above gain equations do not depend upon precision matching of $R_1$ (see note 8-2). Since the common mode gain is one, the CMRR of this stage is equal to the difference gain, which is generally between 10 and 1000. The gain of this stage can be adjusted by changing the single resistor $R_1/a$. In general, a resistor with $a = 4.5$ for a difference gain of 10 is wired into the circuit, and other resistors are switched in parallel with this resistor when higher gains are desired. The CMRR of the cross-coupled input amplifiers is then multiplied by the CMRR of the difference amplifier to produce an instrumentation amplifier with excellent CMRR, high input impedance, and stable, easily adjustable gain. For instance, if the gain of the input stage is set

**Note 8-2.  Effect of Unmatched Values of $R_1$ in the Circuit of Figure 8-3.**
The gain equations for the cross-coupled follower amplifier can be derived using $R_1'$ and $R_1''$ for the feedback resistors of op amps 1 and 2, respectively. The output voltages are $v_{o1} = v_1 + a(v_1 - v_2) R_1'/R_1$ and $v_{o2} = v_2 - a(v_1 - v_2) R_1''/R_1$. From these, $A_d = 1 + a(R_1' + R_1'')/R_1'$, and $A_{cm} = 1 + a[(v_1 - v_2)/(v_1 + v_2)] (R_1' - R_1'')/R_1$. These equations show that both $R_1'$ and $R_1''$ affect the difference gain and that for reasonably close values of $R_1'$ and $R_1''$ the common mode gain is essentially unity.

at 100 and the gain and CMRR of the difference stage are 10 and $10^4$ respectively, the resulting amplifier would have a gain of 1000 and a CMRR of $10^6$.

Instrumentation amplifiers are extremely versatile general purpose amplifiers for low-level signals. They are frequently used to amplify the microvolt and millivolt outputs from biological transducers and as the input amplifiers for sensitive recorders. Because of their common use, they are now available in convenient IC form from several manufacturers.

## 8-2    Waveshaping

There are circumstances in which perfectly linear amplification of a signal is not the desired operation. In some cases the signal voltages should be limited to safe or useful values. In other cases the logarithm, square, or some other function of the signal voltage is desired. The circuits in this section illustrate how operational amplifiers in conjunction with other components can be used to perform such operations with great accuracy.

### Precision Limiter

The ideal limiter circuit provides a linear transfer function between the input and the output over a limited range. Beyond this range the output voltage responds much less or not at all to changes in the input voltage. This circuit has many applications in keeping signals within a safe range as illustrated in the voltage-programmed switching section of chapter 6. In its precision form the limiter function is also very useful in nonlinear analog signal conditioning. The heart of the limiter circuit is the pn junction diode, which is not, by itself, a suitable component for precision operations on low-level signals because of the large ($\sim$ 0.6 V) forward bias needed for conduction and the variability of the required bias both among devices and with temperature. However, the diode function (unidirectional conduction) can be included in an op amp circuit in such a way that the resulting function is very precise.

A precise limiter circuit that performs well even at signal levels in the millivolt range or less is shown in figure 8-4. The circuit is very similar to the summing amplifier except that it includes two feedback paths, one for each direction of the feedback current. If the sum of the currents from $V_R$ and $v_{in}$ is positive, $D_1$ conducts the feedback current required to keep point $S$ at virtual common. From the forward bias across $D_1$, $v_A$ is $-0.6$ V. The output voltage $v_o$ is exactly zero in this case since the output is connected to common through $R_L$ and to virtual common through $R_f$, and diode $D_2$ is reversed biased. The voltage required to make $D_1$ conduct does not then appear at the output. In the case when the sum of the currents through $R_R$ and $R_{in}$ is negative, the only feedback path is through diode $D_2$ and $R_f$. The amplifier

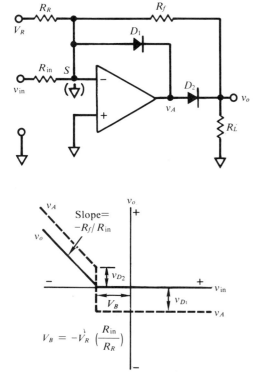

**Fig. 8-4.**  Precision inverting limiter. This circuit eliminates the forward bias errors of the diodes by including them in the feedback control loop. The op amp controls $v_o$ in an inverting amplifier mode when the net input current from $v_{in} + V_R$ is negative. For net positive currents, diode $D_2$ is reverse biased and $v_o = 0$. Note that the op amp supplies the extra voltage at $v_A$ to forward bias the diodes in each case. The breakpoint voltage $V_B$ is set by adjusting $V_R$ or $R_R$.

output $v_A$ must be sufficiently positive to provide the required forward bias for $D_2$ plus the $IR$ drop across $R_f$ as shown by the dashed line in figure 8-4. Since point $S$ is still maintained at virtual common, $v_o$ is simply $-i_f R_f$, where $i_f$ is equal to the total input current $(v_{in}/R_{in}) + (V_R/R_R)$. Thus,

$$v_o = -R_f \left( \frac{v_{in}}{R_{in}} + \frac{V_R}{R_R} \right) \text{ for } v_{in} < -V_R R_{in}/R_R \qquad (8\text{-}2)$$

It is important to include an actual load resistor $R_L$ such that significant current ($\sim 1$ mA) is drawn through $D_2$ when $v_o$ is positive. This ensures good control of $v_o$. It should also be pointed out that when $v_o$ is negative, the op amp is not in control of $v_o$. Therefore, $v_o$ should only be connected to an amplifier input or through a resistive load to common. Both the slope and the breakpoint of this circuit are easily adjusted. When $V_R$ is zero or unconnected and when $R_{in} = R_f$, the precision limiter acts as an ideal diode that provides unity transfer for one polarity of $v_{in}$ and zero transfer for the other. Reversing both diodes in the circuit makes zero the upper bound of $v_o$ and changes the condition of validity of equation 8-2 to $v_{in} > -V_R R_{in}/R_R$.

A similar limiter circuit based on the noninverting voltage follower is shown in figure 8-5. When $v_{in}$ is positive, the feedback path through $D_2$ and the 10-k$\Omega$ isolating resistor enable the op amp to establish a voltage essentially equal to $v_{in}$ at the $-$ input (within the error $v_o/A$). The voltage difference between $v_o$ and the $-$ input is the $IR$ drop across the feedback resistor due to the op amp input current. Since this is negligibly small, $v_o = v_{in}$ for all positive values of $v_{in}$. For negative values of $v_{in}$, the feedback path is $D_1$, and $D_2$ is reverse biased. This would at first appear to disconnect the circuit from $R_L$ as in the case of the inverting limiter of figure 8-4. However, in this case, the op amp inverting input is not held at virtual common; it follows $v_{in}$ negative, and the voltage $v_{in}$ is divided between $R_L$ and the 10-k$\Omega$ feedback resistor. Thus this circuit should be used in conjunction with some other circuit that controls $v_o$ at zero volts should it tend to become negative. The 10-k$\Omega$ feedback resistor in this circuit limits the feedback current that must be supplied by the lower bound control circuit. The actual resistance value is not critical.

These basic limiter circuits have been used in a variety of ingenious combinations to produce many useful nonlinear functions, two of which are described below.

## Absolute Value Function

As its name suggests, this circuit produces a positive output voltage equal in magnitude to the input voltage regardless of the sign of the input voltage. A

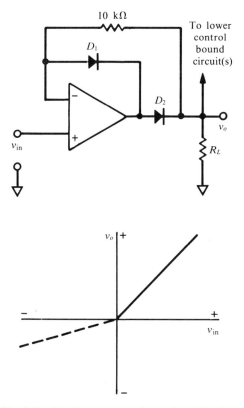

**Fig. 8-5.** Precision noninverting limiter. This circuit is based on the voltage follower. The output voltage $v_o$ follows $v_{in}$ precisely for all positive values. For negative values, $v_o$ follows $v_{in}$ loosely with a gain of less than one (dashed line) unless it is controlled by another circuit.

simple implementation of this function combines the inverting and noninverting limiter circuits of figures 8-4 and 8-5 as shown in figure 8-6. The upper, noninverting limiter circuit establishes the voltage $v_{in}$ at the output for all positive values of $v_{in}$. Similarly, the lower, inverting limiter establishes a voltage of $-v_{in}$ at the output for all negative values of $v_{in}$. The two resistors in the lower circuit should be carefully matched to ensure unity gain for both polarities of $v_{in}$. If all the diodes are reversed, the same circuit produces a negative $v_o$ for either polarity of $v_{in}$.

The absolute value circuit is actually a precision full-wave rectifier circuit for use with analog signals. When the measurement information is encoded as the magnitude of voltage variations of both polarities, the absolute value circuit converts the signal variations to a proportional unipolar signal suitable for a dc voltage measurement system.

## Multiple-Line-Segment Functions

The absolute value circuit above is a simple two line-segment function as its input-output function curve shows. Precision multiple line-segment functions can be obtained by combining separate limiter circuits for each segment desired. This is most easily done by summing the outputs of inverting limiter circuits (fig. 8-4) at the inputs of an op amp summing amplifier (fig. 5-15). Each limiter circuit provides an independently adjustable breakpoint and slope change. The initial signal itself can also be summed with this combination. In this way a great variety of functions can be implemented by precision line-segment approximation.

## Logarithmic Function

The logarithmic transfer function is one that has many uses, including compensation for logarithmic function transducers, compression of especially wide dynamic range signals, and implementation of nonlinear arithmetic operations such as multiplication and division. The logarithmic transfer function, $v_o = \log v_{in}$, is generally achieved in one of two ways; by approximation with a multiple line-segment function generator or by taking advantage of the approximately logarithmic relationship between current and voltage in a semiconductor pn junction. The latter technique, which can provide good accuracy and a smooth transfer function over a wide dynamic range, is described below.

The current-voltage relationship of a forward biased pn junction (eq. 7-4) can be solved for the voltage across the diode (see note 8-3) to yield

$$v = \frac{0.059\eta T}{300}(\log i - \log I_i) \qquad (8\text{-}3)$$

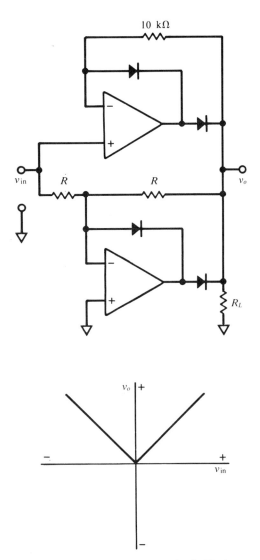

**Fig. 8-6.** Precision absolute value circuit. The inverting and noninverting limiter circuits are combined in this circuit. It is useful where signals must be unipolar (as at the input of a voltage-to-frequency converter) or when precision full-wave rectification of a small signal is required.

where $I_i$ is a temperature-dependent term related to the reverse bias current, $T$ is the temperature in degrees Kelvin and $\eta$ is an empirical parameter ($\sim 2$ for silicon, 1 for germanium). As shown in equation 8-3, the voltage across the diode changes by $59\eta$ mV for each tenfold change in current through it. A simple circuit that utilizes this concept is shown in figure 8-7. The voltage $v_{in}$ is converted to a proportional current by resistor $R_{in}$ connected to the virtual common at the summing point of the operational amplifier. The output voltage establishes the exact diode bias voltage required to pass the input current through the diode. Thus, the output voltage $v_o$ is $-v$, the forward bias voltage of the diode.

Substituting $v_o = -v$ and $i = v_{in}/R_{in}$ in equation 8-3, we obtain

$$v_o = -\frac{0.059\eta T}{300}\left(\log\frac{v_{in}}{R_{in}} - \log I_i\right) \tag{8-4}$$

For most diodes, equation 8-4 is valid over from two to five orders of magnitude change in current. The trend recently is to use the base-emitter junction of a transistor in place of the diode as shown in figure 8-7b. For a transistor the relationship between the collector current and the emitter-base junction voltage also follows equation 8-3. The op amp applies the exact

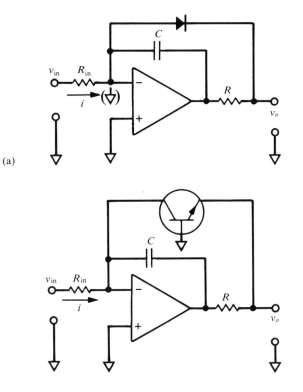

(a)

**Note 8-3. Derivation of pn Junction Voltage as a Function of Current.**

In equation 7-4

$$i = I_i(e^{vQ_e/kT} - 1)$$

the exponential term is greater than 100 for forward bias values greater than 0.25 V for silicon and 0.12 V for germanium. For all but very small forward bias voltages,

$$i \simeq I_i e^{vQ_e/\eta kT} \quad \text{for } i > 100\ I_i$$

where $\eta$ is an empirical parameter about 2 for silicon devices and 1 for germanium. Taking the $\log_{10}$ of both sides of this equation and solving for $v$, we obtain

$$v = \frac{2.3\eta kT}{Q_e}(\log i - \log I_i)$$

The factor $2.3\ kT/Q_e = 0.059$ V at 300 K; so that

$$v = \frac{0.059\eta T}{300}(\log i - \log I_i)$$

The range of currents over which this equation is valid is limited on the low end by the minimum bias voltage required for the approximation of $i > I_i$ to be valid and on the high end by the assumption that the $IR$ drop across the semiconductor, contacts, and leads is negligible compared to $v$.

**Fig. 8-7.** Logarithmic amplifier. An output voltage related to the logarithm of the input voltage is obtained in these circuits as a result of the logarithmic relationship between current and voltage in the diode (a) and in the transistor (b). The circuits are basically current followers in which the input current appears in the diode or transistor in the feedback loop. Components $R$ and $C$ (typically 1 k$\Omega$ and 0.01 $\mu$F) stabilize the amplifier but do not affect the output value. These simple circuits are useful, but they have a large temperature dependence of slope and offset.

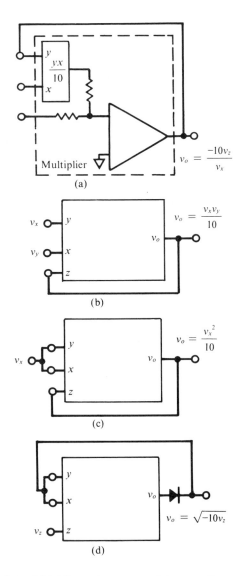

(a)

(b)

(c)

(d)

**Fig. 8-8.** Multiplier and op amp module connections to perform (a) division, (b) multiplication, (c) square, and (d) square root. In the divider the product $v_o v_x / 10$ is made equal to $v_z$. In the multiplier and square circuits the op amp serves as an inverting buffer output amplifier. The square root and divider circuits are related. In the square root circuit, the diode prevents $v_o$ from becoming negative. The value of $v_z$ must be negative.

emitter-base bias voltage required to maintain a collector current of $i$. In practice the transistor has been found to have a better logarithmic accuracy, a wider dynamic range, and a value of $\eta$ very near unity.

One of the greatest difficulties with logarithmic function circuits has been maintaining reasonable accuracy in the face of large variations of the output voltage with temperature. Recently, use of effective temperature compensation circuits has replaced temperature-controlled ovens. These circuits involve a matched transistor on the same silicon chip to provide a compensating current to offset $I_i$ and a temperature-dependent gain amplifier (using a thermistor) to offset the variation of $0.059 \eta T / 300$ with temperature. A useful dynamic range of $10^7$ has been reported for a circuit of this type. Logarithmic function circuits with temperature compensation are now available as complete function modules.

## 8-3   Analog Multipliers

An analog multiplier circuit produces an output voltage that is proportional to the product of two input voltages. As analog multiplier function modules have improved in accuracy, speed, versatility, and economy, their use has spread rapidly. A few basic computational applications are illustrated in this section. Other applications such as modulation and demodulation appear in later sections and chapters.

### Analog Multiplier Function Modules

The basic input-output relationship for an analog multiplier is $v_o = v_x v_y / 10$. The input voltages $v_x$ and $v_y$ have maximum values of 10 V, and the output voltage is scaled down by a factor of 10 to stay within the 10-V maximum output voltage. Some analog multipliers will not accept all combinations of polarities of $v_x$ and $v_y$. There are four possible combinations of input polarities: $++$, $+-$, $-+$, and $--$, corresponding to the four quadrants defined by the $x = 0$ and $y = 0$ coordinates in an ordinary Euclidean graph of $x$ vs. $y$. A multiplier that accepts only one of those combinations (usually $++$) is called a **single-quadrant multiplier**. More versatile multipliers accept signals in two or all four quadrants.

Several approaches to the design of analog multipliers have been used in recent years. These include the summing of logarithmic functions and then using an antilog function. However, the **transconductance multiplier** in its integrated circuit form is the type most responsible for the great increase in multiplier applications. It is based on the proportionality between the current through a transistor and its transconductance (ratio of output current to input voltage). One input signal is used to control the voltage applied to a matched pair of transistors in a differential amplifier circuit, and the other input signal is used to control the combined collector currents. The difference in the collector currents is then approximately proportional to the product of the two input voltages. The technique depends on the transistors being carefully matched for temperature stability and on operating over a very narrow range of input voltages and currents. Therefore the input signals are scaled down before application to the transconductance circuit, and the differential output signal is amplified afterward. A more sophisticated circuit based on the same principle is called the **current-ratioing multiplier**. It incorporates a differential current control circuit that provides better stability and accuracy. The transconductance multiplier also has very good response speed and can be made into a four-quadrant multiplier.

Applications of multipliers for basic mathematical operations are shown in figure 8-8. Division is accomplished by using an op amp to adjust one of the multiplier inputs so that the product is equal to the numerator. The resulting circuit is shown in figure 8-8a. The op amp controls $v_o$ so that $v_x v_y / 10 = -v_z$ at the summing point. Since $v_o = v_y$, $v_x v_o / 10 = -v_z$ and $v_o = -10 v_z / v_x$. As this equation shows, $v_o$ becomes indeterminately large as $v_x$ approaches zero; low values of the denominator must therefore be used with caution. With the circuit of figure 8-8a negative values of $v_o$ must be avoided because for $v_o < 0$ the multiplier inverts the sense of the $y$ signal, and the feedback control loop becomes positive and immediately goes to limit. As figure 8-8 demonstrates, the op amp is useful for so many applications that it is frequently included as part of the multiplier circuit module.

## Operational Transconductance Amplifier

A special, but versatile, form of the op amp has a current source output for which the output current (rather than the voltage) is a function of the differential input voltage. This is called the **operational transconductance amplifier (OTA)** because the transfer function ($\Delta i_o / \Delta v_{in}$) is a transconductance and has the units mhos ($\Omega^{-1}$). The functional equivalent diagram of the OTA is shown in figure 8-9. The OTA circuit is basically a transistor difference amplifier fed from a constant current source controlled by the input current connection $I_{ABC}$. The currents through the two legs of the difference amplifier drive two current generators (one of them inverting), which are summed to produce the output current. Thus the current applied to $I_{ABC}$ determines both the maximum output current according to the relationship $I_{o(max)} \simeq \pm I_{ABC}$ and the transconductance $g_m$ according to the relationship $g_m = F I_{ABC}$, where $F$ is a temperature-dependent constant with a value of 15 to 20 $V^{-1}$ at room temperature (see note 8-4). Because of the action of this gain control input, the OTA is also called a **variable operational amplifier** or **programmable operational amplifier**. The OTA thus operates as a multiplier since the output current magnitude is proportional to the product of $v_{in}$ and

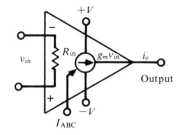

**Fig. 8-9.** The operational transconductance amplifier. The output is a near-ideal source of current proportional to the input voltage $v_{in}$. The proportionality constant is $g_m$, the amplifier transconductance, which can be controlled by the current applied to the $I_{ABC}$ connection.

**Note 8-4. Relationship between $g_m$ and $I_{ABC}$.** The transconductance difference amplifier has a basic gain of $Q_e \alpha I_C / 2kT$ where $\alpha$ is the collector-to-emitter current ratio for the transistors and $I_C$ is the amplifier collector current. For the OTA, $I_C \simeq I_{ABC}$ and $\alpha$ has a maximum value that approaches 1. Therefore, for the input difference stage, $g_m \geq 19.32 I_{ABC}$ at room temperature ($T = 300$ K). The $g_m / I_{ABC}$ ratio F, for the overall OTA, may be somewhat less, depending upon losses in the input or output circuits.

$I_{ABC}$ ($i_o = g_m v_{in}$ and $g_m = FI_{ABC}$, or $i_o = v_{in} FI_{ABC}$). The OTA is a two-quadrant multiplier since $v_{in}$ can be either polarity but $I_{ABC}$ must be positive.

The basic OTA response is linear only for an input voltage range of $\pm 10$ mV although there are advanced designs with a diode circuit at the input that linearizes the response for signals of $\pm 100$ mV or more. The input resistance is relatively low and decreases with increasing $I_{ABC}$. Loading of the input signal should therefore be carefully considered. The output resistance is very large as it should be for a current source. It is over 60 M$\Omega$ for $I_{ABC}$ of 100 $\mu$A and increases with decreasing $I_{ABC}$. The output current remains essentially independent of the output voltage to within a volt or so of the supply voltage values. The output voltage is equal to the output current times the output load resistance. Thus, for a given load resistance $R_L$, the voltage gain is $g_m R_L$.

The circuit for a basic programmable gain amplifier is shown in figure 8-10. The output is often buffered by a voltage follower. Alternately, resistor $R_L$ can be used as the feedback resistor in a current follower. If a current follower is used, the outputs of several OTAs can be summed at its input to provide mixing or multiplexing of the input signals. When used in multiplexing, $I_{ABC}$ can be turned on and off by a logic-driven analog switch circuit. (For $I_{ABC} = 0$, $g_m = 0$, and the current output is essentially an open circuit.) The OTA is also convenient for applications in voltage-controlled filters, oscillators, demodulators, function generators, and many other circuits.

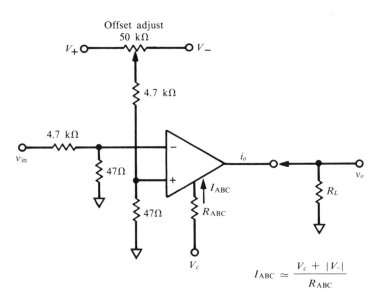

**Fig. 8-10.**   The OTA as a variable gain amplifier. The input voltage is divided to operate the OTA in its linear range. The offset adjust is required to place the variation of $v_{in}$ in the linear operating range. The control current is supplied by a voltage $V_c$ referenced to the negative supply voltage, $V_-$. The output voltage is $v_o = g_m R_L v_{in} = FI_{ABC} R_L v_{in}$.

## 8-4  Active Filters and Tuned Amplifiers

A filter is a circuit that favors the transmission of signals of certain frequencies over signals of other frequencies. Filters generally fall into three categories: high pass filters, which reject low-frequency signals, low pass filters, which reject high-frequency signals and band pass filters, which reject signals having frequencies both higher and lower than the desired band of frequencies. Filters are made by using devices with a frequency dependent impedance in voltage divider or amplifier gain determining circuits. Filter circuits vary in the sharpness of the transition between the transmitted and rejected frequencies, accuracy of transmission in the accepted band, ease of frequency adjustment, independence of load, and availability of gain. The ability to select and reject certain frequencies of a signal has made radio communication possible. It is an essential part of many techniques that help separate the informational part of a signal from the noise (see chapter 14).

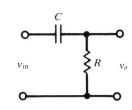

**Fig. 8-11.** A high pass $RC$ filter. This circuit is a frequency-dependent voltage divider in which an increasing fraction of $v_{in}$ appears across $C$ as the frequency decreases.

### First Order High Pass Filters

The simplest high pass filter is the $RC$ voltage divider introduced in section 2-3 and shown in figure 8-11. The signal $v_{in}$ is divided by $C$ and $R$, and the output $v_o$ is the fraction of $v_{in}$ that appears across $R$. This fraction is the ratio of $R$ to the total impedance of $C$ and $R$. The impedance of the resistor is its resistance $R$ and the impedance of the capacitor is its reactance, $X_C = 1/(2\pi fC)$, where $f$ is the frequency of $v_{in}$. The resistance and reactance cannot simply be added together because they are different kinds of impedance. In a resistance, the current is proportional to the voltage at every instant; if the signal is a sine wave, the current and voltage are in phase. In a capacitive reactance, the current is proportional to the rate of change of the voltage; for sinusoidal signals, the current waveform is the cosine of the signal values, 90° ahead of the voltage waveform.

Resistance and reactance can be represented as vector quantities at right angles as shown in figure 8-12. The capacitive reactance is drawn as a length along the $j$ (vertical) axis in the negative direction (see note 8-5). The resulting impedance $\mathbf{Z}$, the vector sum of $R$ and $-jX_C$, can be written $R - jX_C$ because the $-j$ is a reminder of the vector nature of these quantities. The magnitude of $\mathbf{Z}$ is $(R^2 + X_C^2)^{1/2}$ according to the formula for the length of the hypotenuse of a right triangle, and the angle $\theta$ can be obtained by $\tan \theta = X_C/R$. The impedance $\mathbf{Z}$ is the load imposed on the sine-wave source by $R$ and $C$, in which the current leads the voltage by the angle $\theta$. The voltage across $R$ is in phase with the current, and thus $v_R$ will be $\theta$ degrees ahead of $v_{in}$. The capacitor voltage will be $90 - \theta$ degrees behind $v_{in}$.

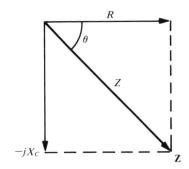

**Fig. 8-12.** Impedance vectors for a series $RC$ circuit. The magnitude of $\mathbf{Z}$ is $Z = \sqrt{R^2 + X_C^2}$, and the tangent of $\theta = X_C/R$. The voltage across $R$ leads the source voltage by $\theta$ degrees. The boldface $\mathbf{Z}$ is a reminder that impedance is a complex quantity.

**Note 8-5.  Operations on $j$.**
The quantity $j$ can be treated as though it had the value $\sqrt{-1}$. Therefore, $j^2 = -1$, $1/j = -j$, and so on.

The fraction of $v_{in}$ that appears at the output of the filter of figure 8-11 can now be calculated. This fraction ($v_o/v_{in}$) is called the **transfer function** $H(j\omega)$. The $j\omega$ is a reminder that the transfer function depends on frequency ($\omega = 2\pi f$) and that there may be a phase difference between $v_o$ and $v_{in}$.

$$H(j\omega) = \frac{R}{R - jX_C} \quad \text{and} \quad |H(j\omega)| = \frac{R}{\sqrt{R^2 + X_C^2}} \quad (8\text{-}5)$$

The frequency dependence of the transfer function is clearer when $1/j\omega C$ is substituted for $-jX_C$ in equation 8-5 to give

$$H(j\omega) = \frac{R}{R + \dfrac{1}{j\omega C}} = \frac{j\omega RC}{1 + j\omega RC} \quad (8\text{-}6)$$

The Bode plot for the response of this filter is shown in figure 8-13. Three points can be identified from equation 8-6. At very high frequencies, the $j\omega RC$ term is much greater than one, and $H(j\omega)$ is unity. At very low frequencies, the $j\omega RC$ term is much less than one and $H(j\omega)$ is very small. When $\omega = 1/RC$ [$f = 1/(2\pi RC)$], $|H(j\omega)| = 1/\sqrt{1+1} = 1/\sqrt{2} = 0.707$. This last point is called the **cutoff frequency** of the filter. The rolloff portion of the response function has a slope of 20 dB per decade change in frequency.

**Fig. 8-13.** Bode and phase shift plots for a first order high pass filter. The attenuation in decibels or phase shift in degrees is plotted against the logarithm of the frequency relative to the cutoff frequency.

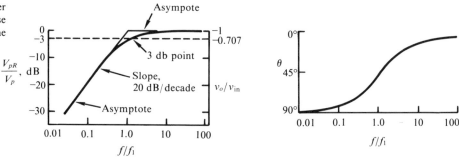

A high pass filter can also be realized with an $LR$ combination as shown in figure 8-14. The inductor is a reactance in which the current lags behind the voltage by $90°$. Its reactance is $X_L = 2\pi f L = \omega L$, and its vector is plotted along the $j$ axis in a positive direction. The transfer function for this filter is

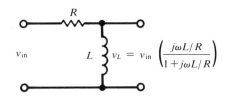

**Fig. 8-14.** An $LR$ high pass filter. The fraction of $v_{in}$ that appears across $L$ increases as the frequency increases.

$$H(j\omega) = \frac{jX_L}{R + jX_L} = \frac{j\omega L/R}{1 + j\omega L/R} \quad \text{and} \quad |H(j\omega)| = \frac{X_L}{\sqrt{R^2 + X_L^2}} \quad (8\text{-}7)$$

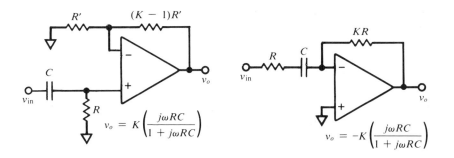

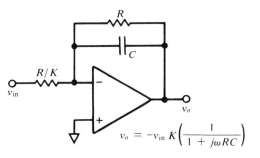

**Fig. 8-15.** Active high pass filters. The noninverting circuit combines the passive $RC$ circuit with a follower with gain buffer amplifier. The inverting circuit uses a frequency-dependent input impedance to produce a frequency-dependent gain.

The $LR$ filter has the same transfer function as the $RC$ filter except that the time constant is $L/R$ instead of $RC$. The phase angle $\theta = \tan^{-1}(X_L/R)$, and the voltage across $R$ lags $v_{in}$ by $\theta$ degrees.

The passive $RC$ and $LR$ filters shown in figures 8-11 and 8-14 provide no gain for the selected frequencies. They must be followed by very high impedance loads to avoid a loading effect on their efficiency and their cutoff frequency. The inclusion of an amplifier to form an **active filter** can avoid these problems. Figure 8-15 shows two forms of active high pass filter. In the inverting filter, the signal $v_{in}$ causes a current of $v_{in}/(R + 1/j\omega C)$ at the summing point, which gives an output signal of $-KR$ times the input current. The transfer function is thus $-KR/(R + 1/j\omega C)$, or $-K$ times the transfer function given in equation 8-6.

## First Order Low Pass Filters

Passive and active low pass filters are shown in figure 8-16. In the passive circuits the components have been interchanged from the high pass filter reversing the divider fraction. The filter transfer functions are given in the figure. The output voltage of the low pass filter lags the input by a phase

**Fig. 8-16.** Passive $RC$ and $LR$ and active $RC$ low pass filters. The passive circuits are frequency-dependent voltage dividers. A buffer amplifier is added for the noninverting active filter. The inverting active filter has a frequency-dependent feedback impedance.

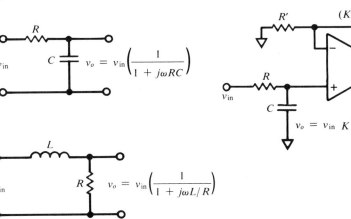

angle of $90 - \theta$ degrees. The Bode plot for the low pass filter shows unity gain at low frequencies and begins to roll off at 20 dB per decade at the cutoff frequency of $1/(2\pi RC)$ or $L/(2\pi R)$.

To analyze the active inverting low pass filter of figure 8-16, it is necessary to obtain the impedance of a parallel resistor and capacitor. The **admittance** of a parallel circuit is the vector sum of the admittances of the parallel components (see note 8-6). The admittance of a parallel $RC$ circuit is thus $j\omega C + 1/R$. The impedance is the reciprocal of the admittance. The gain of the inverting amplifier is the negative ratio of the feedback and input impedances. Thus

**Note 8-6.  Admittance.**
The admittance of a component is the reciprocal of the impedance of that component. It is thus a measure of the component's ability to conduct a signal. The term admittance, like impedance, is generally used with frequency-dependent (reactive) circuits.

$$H(j\omega) = -\left(\frac{1}{j\omega C + 1/R}\right)\left(\frac{1}{R/K}\right) = -K\left(\frac{1}{1 + j\omega RC}\right) \qquad (8\text{-}8)$$

Equation 8-8 is the simple low pass filter transfer function with inverted gain.

## Higher Order Filters

The simple filter's relatively gentle rolloff characteristic of 20 dB per decade change in frequency is not adequate for many purposes. Filter circuits with more than one reactive component have been designed to produce sharper rolloffs. A second order filter has a rolloff of 40 dB per decade; a third order, 60 dB per decade, and so on. Filters with orders of six or more are practical.

A simple second order filter is the combination of the $RC$ and $LC$ low pass filters shown in figure 8-17. The transfer function for this filter is the impedance of $C$ over the combined impedance of $R$, $L$, and $C$.

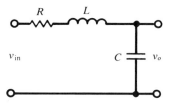

**Fig. 8-17.**  *RLC* second order low pass filter circuit. The frequency-dependence of both $L$ and $C$ produces a sharper rolloff than when $L$ or $C$ are used alone.

$$H(j\omega) = \frac{1/j\omega C}{R + j\omega L + 1/j\omega C} = \frac{1}{j^2\omega^2 LC + j\omega RC + 1} = \frac{1}{1 - \omega^2 LC + j\omega RC}$$

It is the $\omega^2$ term in the denominator that produces the second order rolloff. The relationship between the magnitude of $H(j\omega)$ and $\omega$ is best seen when the vector quantity in the denominator is evaluated:

$$|H(j\omega)| = \frac{1}{\sqrt{(1 - \omega^2 LC)^2 + \omega^2 R^2 C^2}} \qquad (8\text{-}9)$$

A special case occurs when $\omega^2 = 1/LC$ because the term $(1 - \omega^2 LC)$ is zero and the transfer function is equal to $1/\omega RC$. If $R$ is very small, for example, the transfer function can become much greater than 1. The frequency $1/\sqrt{LC}$ is called the **resonant frequency** for an $LC$ circuit and is given the symbol $\omega_0$. Equation 8-9 can be rewritten with $\omega_0$ substituted for $1/\sqrt{LC}$.

$$|H(j\omega)| = \frac{1}{\sqrt{(1 - \omega^2/\omega_0^2)^2 + R^2(C/L)\omega^2/\omega_0^2}} = \frac{1}{\sqrt{(1 - \omega^2/\omega_0^2)^2 + 4d^2\omega^2/\omega_0^2}} \quad (8\text{-}10)$$

where $d = \frac{1}{2}R\sqrt{C/L}$ and is called the **damping factor**.

The response of the $RLC$ filter is shown graphically in figure 8-18. The filter is **underdamped** for values of $d$ less than 1 and **overdamped** when $d$ is greater than 1. At high frequencies ($\omega \gg \omega_0$) the $(1-\omega^2/\omega_0^2)^2$ term is dominant in the determination of $|H(j\omega)|$, and this produces the 40-dB per decade rolloff. For a practical filter the component values are selected to give the desired values of $\omega_0$ and $d$.

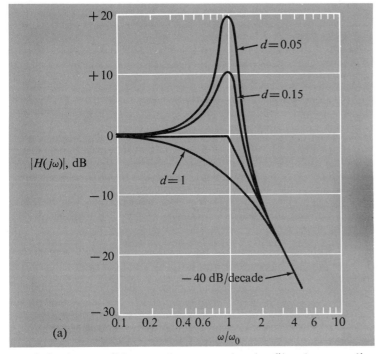

(a)

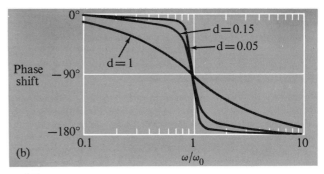

(b)

**Fig. 8-18.** Bode plot and phase shift plot for an $RLC$ high pass filter. The effect of $d$ on the response function near the resonant frequency is shown. The flattest response occurs when $d = 1$ and the transition to the 40 dB/decade rolloff is smooth. As $d$ approaches 0, the response becomes more and more peaked. At $d = 0$, the circuit becomes an oscillator.

It is also possible to make a second order filter by cascading two of the simple $RC$ filters shown in figure 8-16. This is not recommended for passive filters because the second filter loads the first filter and degrades its performance. Second order active filters can be made with a single op amp as shown in figure 8-19. A comparison of the transfer function with equation 8-10 shows that the response has the same characteristic as the $RLC$ filter. The action of the cascaded $RC$ filter section is enhanced by the positive feedback

**Fig. 8-19.** Active second order filters of the Sallen-Key type. The cutoff frequency $\omega_c$ is $1/RC$, and the damping factor $d$ is $(3-K)/2$. The gain is restricted to an upper limit less than three. For $d=1$, $K=1$. The Bode plots and phase shift curves are the same as those for the $RLC$ filter shown in figure 8-18.

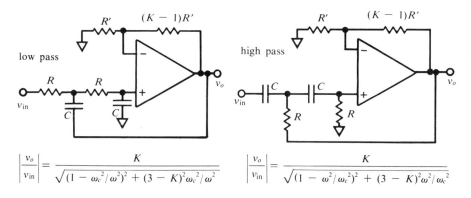

low pass

high pass

$$\left|\frac{v_o}{v_{in}}\right| = \frac{K}{\sqrt{(1-\omega_c^2/\omega^2)^2 + (3-K)^2\omega_c^2/\omega^2}}$$

$$\left|\frac{v_o}{v_{in}}\right| = \frac{K}{\sqrt{(1-\omega^2/\omega_c^2)^2 + (3-K)^2\omega^2/\omega_c^2}}$$

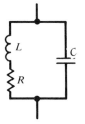

**Fig. 8-20.** The parallel $LC$ tuned circuit. It has a resonant frequency of $\omega_0 = 1/\sqrt{LC}$. At $\omega_0$ its impedance reaches a maximum value of about $\omega_0^2 L^2/R$.

**Note 8-7. Impedance of an $LC$ Tuned Circuit.** The impedance is the reciprocal of the sum of the admittances of the $C$ and the $LR$ combination.

$$\frac{1}{Z} = j\omega C + \frac{1}{R + j\omega L}$$

and

$$Z = \frac{\frac{R}{j\omega C} + \frac{j\omega L}{j\omega C}}{R + \frac{1}{j\omega C} + j\omega L} = \frac{\frac{R}{j\omega C} + \frac{L}{C}}{R + \frac{1}{j\omega C} + j\omega L}$$

At resonance, $j\omega L = 1/j\omega C$ and $\omega = \omega_0$

$$Z = \frac{Rj\omega_0 L + \omega_0^2 L^2}{R} = \omega_0^2 L^2/R + j\omega_0 L$$

For $R \ll \omega_0 L$, $Z \simeq \omega_0^2 L^2/R = Q\omega_0 L$

where $Q = \omega_0 L/R$. Since $\omega_0 = 1/(LC)^{1/2}$,

$$Q = \frac{L}{R\sqrt{LC}} = \frac{1}{R}\sqrt{\frac{L}{C}} = \frac{1}{2d}$$

to the first section. The filter can be tuned using ganged variable resistors for $R$ and switched identical capacitors for $C$. This filter is convenient, can be tuned over a wide range, and avoids the expense and inconvenience of an inductor. Third and still higher order filters can be made by cascading first and second order active filters.

## Band Pass Filters and Tuned Amplifiers

A band pass filter is designed to transmit a relatively narrow band of frequencies and reject signals of higher or lower frequency. A traditional form of band pass filter is the parallel $LC$ tuned circuit shown in figure 8-20. The resistance, even when not intentional, is an unavoidable part of the inductor wires. The impedance of the $LC$ tuned circuit is (see note 8-7)

$$Z = \frac{R/j\omega C + L/C}{R + 1/j\omega C + j\omega L} \tag{8-11}$$

Equation 8-11 shows that $Z$ is a maximum at the resonant frequency $\omega_0 = (1/\sqrt{LC})$ where $1/j\omega C = -jwL$. At this frequency the denominator of equation 8-11 contains only $R$, and the impedance is equal to $\omega_0^2 L^2/R + j\omega_0 L$. If $R$ is small compared to the inductive reactance, the resonant impedance equals $Q\omega_0 L$ where $Q$ is $\omega_0 L/R$, the **quality factor** of the resonant circuit. When this circuit is used as a gain-determining element in an amplifier, a circuit that is selective for the resonant frequency results. The higher the value of $Q$, the sharper the frequency selectivity of the circuit. The factors $Q$ and $d$ are related as shown in note 8-7. The $LC$ filter is convenient at high frequencies ($> 100$ kHz) where the inductors are compact, reasonably inexpensive, and nearly ideal.

For frequencies in the range below 100 kHz, a variety of op amp based active filter circuits are available. One that is especially versatile and easy to use is the **state variable filter** shown in figure 8-21. The input signal is

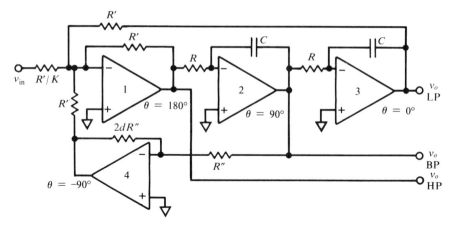

**Fig. 8-21.** State variable second order filter. This circuit provides three outputs simultaneously: high pass, low pass, and band pass. The cutoff frequency $\omega_c = 1/RC$. The gain is adjusted by the input resistor to op amp 1, and the damping is determined by the gain of op amp 4.

inverted by amplifier 1 and then successively integrated by amplifiers 2 and 3. The transfer function for each integrator, $1/j\omega RC$, indicates that the amplifier has a phase shift of 90° and a unity gain when $\omega = 1/RC$. For lower frequencies the gain is greater. The signal at the LP output has been shifted 180° + 90° + 90° and is thus in phase with the input. The feedback path at the top of the diagram is positive (reinforcing) feedback, but it is tempered by the feedback of the 90° component of the BP output through amplifier 4. The greater the gain of amplifier 4, the greater the damping (and the smaller the $Q$). The cutoff and resonant frequency can be changed by varying $R$, $C$, or both, but the $RC$ product must remain the same for both amplifiers. The input voltage must be small enough so that the quantity $v_{in}KQ$ (the BP output voltage at the resonant frequency) does not exceed the op amp output voltage limit. The state variable filter can serve as a tuned amplifier with practical $Q$ values as high as 500 when good quality op amps are used.

Band pass filters can be cascaded to increase the sharpness of the roll off in the frequency response function. Two second order filters in cascade provide a fourth order rolloff (80 dB/decade). If a wider band is wanted in conjunction with the steeper rolloff, the center frequencies of the cascaded filters can be offset from each other somewhat.

A variable tuning control can be used with the filters of figure 8-19 and 8-21 if the two $R$s are ganged potentiometers. The state variable filter also lends itself to relatively easy electrical control of the tuning frequency. The two $R$s can be replaced by multiplying DACs if digital control is desired or by transconductance amplifiers (the two-quadrant analog multiplier) for analog control of the tuned frequency.

A useful variation on the band pass filter is its complement, the **notch filter.** The notch filter specifically rejects a particular band of frequencies and accepts those higher and lower than the rejection band. Again, the versatile state variable filter can be used in this application. The low and high pass outputs are summed to provide a signal that is the complement of and has essentially the same $Q$ as the BP output.

Several manufacturers of analog circuits (National Semiconductor, Datel/Intersil, Burr Brown, General Instrument Corp., etc.) now offer state variable filters in integrated circuit form. Only a few external components are required to adapt these devices to specific applications.

## 8-5    Regenerative Oscillators

An oscillator is a signal generator with a repetitive output waveform. The waveform produced may be a sine wave, square wave, sawtooth wave, pulse, or any other basic shape. Oscillators are used as precision time generators, sources of signals for synchronizing operations, test-signal generators, carrier-frequency sources for transmission or recording by modulated carrier, to name only a few applications. There are two basic oscillator forms: the relaxation oscillator and the regenerative feedback oscillator. The relaxation oscillator is based on alternately charging and discharging a capacitor; the reversal occurs at particular charge values. Several oscillators of this type are described in section 6-5. Examples of relaxation oscillators are the astable multivibrator, the sweep generator, and the function generator. The regenerative feedback oscillator is basically a sine-wave generator. It is based on the principle of a signal loop that is regenerative (self-sustaining) but only for signals of a single frequency. In this section several examples of regenerative feedback oscillators are described.

## Regenerative Feedback

An amplifier has feedback when its output signal has an effect on the signal at its input. As described in section 7-6, this feedback is positive if a change in the output signal causes the input signal to change further in the same sense. The input signal is thus augmented, or "regenerated" by the output signal. An amplifier with positive feedback has higher gain than the same amplifier with no feedback. It is also possible to feedback the output signal so as to oppose or decrease the effective signal change at the input. This negative feedback has the effect of decreasing the overall gain.

As derived in section 7-6, the gain of an amplifier with feedback is

$$A_f = \frac{A}{1 - \beta A}$$

where $A$ is the amplifier gain without feedback and $\beta A$ is the fraction of the output signal added to the input. The value of $\beta$ is between $-1$ and $+1$, the sign indicating negative or positive feedback. When $\beta A$ is negative, $A_f$ is less than $A$ as expected for negative feedback. An important limiting case occurs when $-\beta A$ is much larger than 1. This is most readily achieved by a very large value for $A$. In this case $A_f \simeq 1/\beta$; that is, the gain depends almost entirely on the feedback components and very little on $A$. We have seen this principle at work in op amp applications where negative feedback components produce precisely controlled results (see note 8-8).

If $\beta A$ is positive and $\beta A < 1$, $A_f$ is greater than $A$. This is not generally a desirable way to obtain more gain as all other amplifier characteristics deteriorate under these conditions (see chapter 7). The special case in which $\beta A = 1$ is of interest in this section, for in that case the gain is infinite; the output is present without any input signal. Thermal noise within the amplifier is sufficient to start the regeneration process at all frequencies. If the feedback loop is frequency selective so that the condition $\beta A = 1$ is true for only one frequency, then a signal of that frequency will be amplified to the limit by regeneration. The system is then said to be "in oscillation." Frequency-selective circuits useful for practical op amp oscillators include the Wien bridge and the state variable oscillators as shown by the following examples.

## Wien Bridge Oscillator

As its name suggests, this oscillator circuit contains an ac impedance-measuring bridge called the **Wien bridge**. The circuit of figure 8-22 shows the components in the standard bridge configuration. The amplifier input is the imbalance signal between the negative feedback provided by the resistive divider on the right side of the bridge and the positive feedback from the reactive divider on the left side. As the frequency increases the phase angle of the voltage across the series $RC$ arm increases and that across the parallel $RC$ arm decreases. For maximum positive feedback the signal at the positive

**Note 8-8. Operational Amplifier Feedback Circuits.**
Operational amplifiers can be used with either positive or negative feedback as shown in the accompanying figures. In op amps, negative feedback occurs when the feedback loop is connected to the inverting input. In the accompanying figure, the follower with gain (a), employs negative feedback, and

$$\frac{v_o}{v_{in}} = \frac{A}{1 + \beta A} \simeq \frac{R_1 + R_2}{R_1} \text{ for } \beta A \gg 1$$

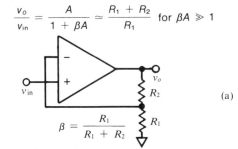

$$\beta = \frac{R_1}{R_1 + R_2}$$

(a)

Positive feedback occurs when the loop is connected to the noninverting input as in (b). Here the closed loop gain is

$$\frac{v_o}{v_{in}} = -\left(\frac{A}{1 - \beta A}\right)$$

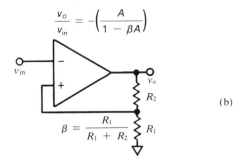

$$\beta = \frac{R_1}{R_1 + R_2}$$

(b)

The minus sign occurs because the signal is connected to the inverting input. Other combinations can be used as in the difference amplifier of figure 8-2 where feedback occurs to both inputs.

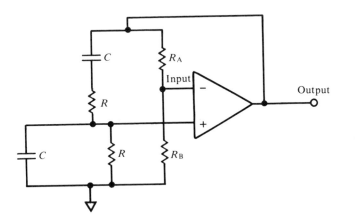

**Fig. 8-22.** A Wien bridge oscillator circuit. The frequency of balance, $\omega_0 = 1/RC$, is the condition of zero phase difference between the reactive arms on the left side of the bridge and is one condition of balance of the bridge. The negative feedback from arms $R_A$ and $R_B$ must be low enough to allow oscillation to be maintained but not so low that excessive distortion occurs. The optimum feedback can be automatically maintained if $R_B$ is a dynamic resistance that increases as the output amplitude increases. In the simplest implementation a #344 or #1869 lamp is used as a dynamic resistance for $R_B$, and $R_A$ is a 1 k$\Omega$ adjustable resistor.

### Note 8-9.    Regenerative Frequency of the Wien Bridge Oscillator.

The reactive arms form a divider for which the divider ratio is $Z_p/(Z_s + Z_p)$ where $Z_s = R + 1/jwC$ and $Z_p = R/(1 + j\omega RC)$. Solving for the divider ratio gives

$$\frac{Z_p}{Z_s + Z_p} = \frac{R}{3R + j[\omega R^2 C - 1/(\omega C)]}$$

In order for the phase angle of the divider to be zero, the $j$ term in the divider ratio must be zero. Therefore, $\omega_0 R^2 C = 1/(\omega_0 C)$ and $\omega_0 = 1/RC$. At $\omega_0$ the divider fraction is $R/3R = 1/3$.

amplifier input should be the same as the amplifier's output. This condition is attained at the frequency $\omega_0 = 1/RC$, at which frequency $\beta = 1/3$ (see note 8-9).

The bridge is in balance for a signal frequency of $\omega_0$ when $R_B = R_A/2$, but the amplifier input is zero. To produce oscillation the negative feedback is decreased to achieve the desired amount of net positive feedback. The purest sine wave will be produced when $\beta A = 1$ exactly. However, this is not practical. To build up the oscillation initially, $\beta A$ must be greater than one, and variations in $A$ or $\beta$ can cause wide fluctuations in the output amplitude. Where a low-distortion waveform is not important, the negative feedback is reduced enough to run the amplifier from limit to limit while maintaining stable, though distorted, oscillation. For pure sine-wave regenerative oscillators some means of automatically adjusting the negative feedback ratio to maintain constant amplitude must be added. A simple but effective amplitude control for the Wien bridge oscillator is to use a lamp or thermistor for $R_B$. As the output amplitude increases, the current through $R_B$ increases. This increases the temperature and resistance of $R_B$. Adjusting $R_A$ affects both the amplitude and the distortion. A more elaborate technique is to derive a dc signal proportional to the peak-to-peak output voltage. This signal is multiplied with the signal from the resistive arms to provide the negative feedback signal. Such an arrangement is necessary to obtain distortion (harmonic content) less than 1%.

## State Variable, or Quadrature, Oscillator

A great many oscillators that use the regenerative feedback principle can be designed around different active filter circuits. The **state variable oscillator** has been chosen as an example because of its unique features and the great versatility of the filter. The state variable oscillator is a variation on the state variable filter shown in figure 8-21. In the filter, however, damping (negative feedback) provided by op amp 4 prevents oscillation. The state variable oscillator of figure 8-23 omits the damping amplifier. Slight regeneration

**Fig. 8-23.**    State variable oscillator. At

$$\omega_0 = \frac{V_c}{10\,RC}$$

the loop gain through $A_1$, $A_2$, and $A_3$ is unity. The multipliers provide voltage control of the oscillator frequency. The 100 $R'$ resistor provides regenerative feedback for start-up and control response; the degenerative loop through the Zener diodes gives dynamic amplitude control. Two outputs 90° out of phase (quadrature outputs) are available. With $R = 10$ k$\Omega$ and $C = 0.015$ $\mu$F, the output frequency can be tuned from 100 Hz to 1 kHz with a 1–10 V range of $V_c$. A convenient value for $R'$ is 50 k$\Omega$. 9-V Zener diodes are used for ±10-V op amp circuits.

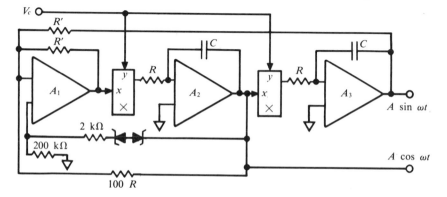

above unity loop gain is provided from the $A_2$ output through the $100R'$ resistor. The Zener diode degenerative loop becomes active at an output amplitude less than the amplifier saturation levels and controls the output at that level. The somewhat clipped waveform that results at the $A_1$ output is smoothed once by $A_2$ and again by $A_3$. Distortion is less than 1%. The range of frequency adjustment depends on the quality of the multipliers used. A ratio of 100:1 or more is practical.

## 8-6 Amplitude Modulation

**Modulation** is the alteration of some property of a **carrier wave** by a signal in such a way that the carrier wave can be used to convey signal information. Carrier waves may be either sinusoidal or pulse train waveforms. If the carrier is sinusoidal, either its amplitude or its frequency may be altered (modulated) by the signal. Pulsed waveforms may be modulated in pulse amplitude or pulse width.

Modulation is used for three principal purposes. One is to use the carrier signal as a conveyor of the modulation signal through a medium unsuited to the modulation signal itself. Examples of this are radio and TV transmissions and the use of frequency modulation for recording low-frequency signals on magnetic tape. A second purpose is to make it possible to convey many channels of information via a common medium by using carriers of different frequencies. The frequency assignments of radio transmitters and the 1000-channel fiber optic communication links are examples of this application. The third common use of modulation is in electrical measurement and data processing. Modulation can be used to move the bandwidth of information of a signal to a portion of the spectrum where it is less subject to noise and more distinguishable from noise by its unique modulation pattern. Some interdomain converters such as the voltage-to-frequency converter used in measurement systems are actually modulation devices.

To be useful, modulation must be reversible. The process of recovering the signal from the modulated carrier wave is called **demodulation**. This section discusses the amplitude modulation and demodulation of sinusoidal carrier waves. Frequency modulation techniques are described in chapter 9.

**Amplitude modulation** involves the alteration of the amplitude of a sinusoidal carrier wave by a signal. The frequency of the carrier wave is generally much greater than the frequencies that make up the signal. Amplitude modulation translates the signal information upward in frequency. This provides a means of moving dc and low-frequency signal information out of the $1/f$ noise region (see note 8-10) and into a region of lower noise. Double sideband modulation and AM modulation are described here as examples of modulation techniques.

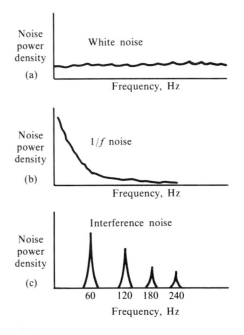

**Note 8-10. Sources of Noise.**
Three common sources of noise are **white noise**, **one-over-F (1/f)-noise** and **interference noise**. The power density spectrum (in watts/Hz) vs. frequency for each of these is shown in the figure. White noise (a) is a mixture of signals of all frequencies with random amplitudes and phase angles. The power density of $1/f$ noise (b) increases approximately with the reciprocal of frequency at low frequency. Such a spectrum is typical of low-frequency drifts that are common in transducers, amplifiers, and measurement systems. Interference noise generally occurs at specific frequencies such as at the power-line frequency and its harmonics, as shown in (c). In a real system the overall noise is likely to be the sum of all three types of noise. At low frequencies, the $1/f$ noise generally predominates; white noise is the major contributor at high frequencies.

## Double Sideband Modulation

For simplicity assume that the carrier is a sinusoidal wave $A_c \cos \omega_c t$ and that the signal is a lower frequency sinusoidal wave $A_s \cos \omega_s t$. Double sideband modulation is carried out by multiplying these two together as shown in figure 8-24. This figure and its equations are greatly simplified by assuming a sinusoidal modulating signal. If the signal contains many frequencies or a continuous range of frequencies, then a sum and a difference frequency are generated for each signal frequency. The modulated carrier frequency spectrum shows the signal frequency spectrum reflected on each side of the carrier frequency. The group of frequencies greater than the carrier frequency is called the **upper sideband** and that group below the carrier frequency, the **lower sideband**. Note that the bandwidth of the modulated carrier is double that of the signal alone. The complete spectrum is called a double sideband, and thus this particular type of modulation is called **double sideband modulation (DSB)**. It is also referred to as double sideband suppressed-carrier modulation, because the carrier wave is not present in the spectrum of the modulated waveform.

**Fig. 8-24.**    Double sideband modulation. The signal and carrier are applied to two inputs of a four-quadrant multiplier to produce the modulated carrier. The mathematical product of the waveforms indicates a modulated waveform of two frequencies—one higher and one lower than the carrier frequency by an amount equal to the signal frequency. This is shown in the frequency spectrum of the modulated carrier.

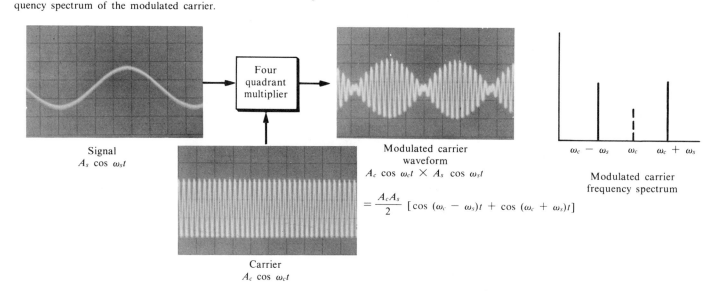

Signal
$A_s \cos \omega_s t$

Four quadrant multiplier

Carrier
$A_c \cos \omega_c t$

Modulated carrier waveform
$A_c \cos \omega_c t \times A_s \cos \omega_s t$

$$= \frac{A_c A_s}{2} \left[ \cos (\omega_c - \omega_s)t + \cos (\omega_c + \omega_s)t \right]$$

$\omega_c - \omega_s \quad \omega_c \quad \omega_c + \omega_s$

Modulated carrier frequency spectrum

## Amplitude Modulation

The amplitude modulation utilized for AM radio differs from DSB in that the carrier wave is specifically added to the modulated carrier wave (see figure 8-25). DSB is thus the same as AM except for the suppression of the carrier wave. This seemingly minor difference has important consequences with respect to the demodulation of these carriers. The significant difference

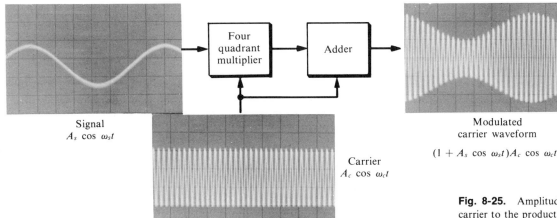

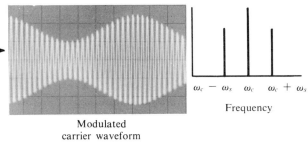

Signal
$A_s \cos \omega_s t$

Four quadrant multiplier

Adder

Carrier
$A_c \cos \omega_c t$

Modulated carrier waveform

$(1 + A_s \cos \omega_s t)A_c \cos \omega_c t$

$\omega_c - \omega_s \quad \omega_c \quad \omega_c + \omega_s$

Frequency

**Fig. 8-25.** Amplitude modulation. The addition of carrier to the product of carrier and signal produces a modulated carrier waveform for which the envelope (the line joining the peak values) is the same as the modulating signal waveform. This type of modulation, used in AM radio, can be demodulated with a simple rectifier and filter.

between the DSB and AM modulated waveforms can be seen by comparing figures 8-24 and 8-25. The envelope of the AM carrier is the same as the signal waveform; in general, this is not true for the DSB carrier. If the signal goes negative, that information is encoded in the DSB modulated carrier as a phase reversal. Recovery of only the envelope of a DSB signal would not faithfully reproduce the original signal. Because circuits for envelope recovery can be quite simple, an AM carrier is considerably easier to demodulate than a DSB carrier.

## Demodulation

The general method for demodulation of amplitude modulated carriers (both DSB and AM) is synchronous multiplication by a sinusoidal signal exactly equal to the carrier in frequency and phase, followed by low pass filtering. This translates the signal information downward in frequency to its original location and generates some higher-frequency components. Demodulation is shown in figure 8-26.

Synchronization of this multiplication with the carrier wave is not always easy. Often a small amount of the original carrier is added to the

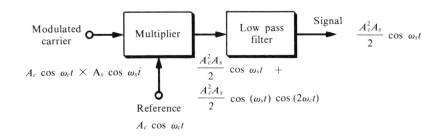

Modulated carrier

Multiplier

Low pass filter

Signal

$\dfrac{A_c^2 A_s}{2} \cos \omega_s t$

$A_c \cos \omega_c t \times A_s \cos \omega_s t$

Reference

$A_c \cos \omega_c t$

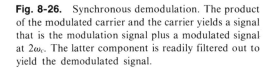

$\dfrac{A_c^2 A_s}{2} \cos \omega_s t \; +$

$\dfrac{A_c^2 A_s}{2} \cos (\omega_s t) \cos (2\omega_c t)$

**Fig. 8-26.** Synchronous demodulation. The product of the modulated carrier and the carrier yields a signal that is the modulation signal plus a modulated signal at $2\omega_c$. The latter component is readily filtered out to yield the demodulated signal.

DSB signal to synchronize the local oscillator used for demodulation. Alternatively, two extremely stable oscillators can be used to generate the carrier and demodulating waveforms. (Periodic synchronization of the oscillators is necessary to ensure adequate stability.) In most laboratory measurements using DSB, the method used is to transmit the carrier wave, or a waveform phase-locked to it, directly from the modulator to the demodulator via a separate connection. This is the approach used, for example, in the lock-in amplifier measurement system described in the next section.

## 8-7   Application: Lock-In Amplifiers

The **lock-in amplifier** is a complete signal measurement and processing system that is very efficient in discriminating against noise components in a signal. A complete lock-in amplifier measurement system consists of four main operations: modulation, selective amplification, synchronous demodulation, and low pass filtering. The lock-in amplifier itself normally carries out only the latter three operations.

### Lock-In Amplifier

The typical signal measured with a lock-in amplifier is a relatively slowly varying signal with a substantial dc frequency component. The signal is modulated to put its information on a carrier wave whose frequency is chosen to be well removed from $1/f$ noise, environmental noises such as 60 Hz, and other interferences. This enables the signal information to be amplified in a frequency region of minimal noise.

A block diagram of the basic components of a lock-in amplifier is shown in figure 8-27. The modulated carrier is selectively amplified (often with an amplifier tuned to the carrier frequency) before being demodulated. Any noise components outside the modulated carrier bandwidth are strongly attenuated at this step. Then the amplified carrier wave is synchronously demodulated. This is accomplished by multiplying the modulated carrier wave by a bipolar reference square-wave signal equal in frequency and phase to the carrier wave. Synchronous demodulation is very powerful in its ability to discriminate against random noise components, because on the average only the in-phase, or "phase-locked," carrier wave is demodulated by this multiplication operation. It is this step that has given the name lock-in amplifier to this signal processing technique. Finally, the output from the synchronous demodulator is low pass filtered to regenerate an amplified form of the original signal. Note that the basic steps in the lock-in amplifier are directly analogous to those of DSB modulation and demodulation, discussed in the previous section.

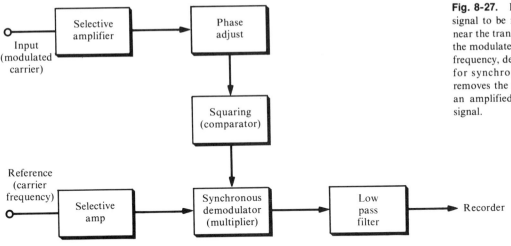

**Fig. 8-27.** Block diagram of a lock-in amplifier. The signal to be measured modulates a carrier wave at or near the transducer. The amplifier selectively amplifies the modulated carrier. A reference signal of the carrier frequency, derived from the modulating device, is used for synchronous demodulation. A low pass filter removes the carrier frequency component to produce an amplified output proportional to the modulated signal.

## Practical Lock-In Systems

In general, the basic signal sources to which lock-in amplifier techniques are applicable are those in which the signal frequencies are at or very near dc. Fundamentally, the modulation can take place anywhere before the lock-in amplifier, but in a practical sense, just where in the signal conditioning modulation takes place is important because often the carrier can be selectively modulated by the signal with respect to various noise sources in the system.

It is usually best to carry out the modulation step as soon as possible. Spectrophotometry provides a good example. In a spectrophotometric system light from a source illuminates a sample, and the light transmitted through, reflected by, or emitted by the sample is measured for various times, sample conditions, or wavelengths of light. In such a system the modulation step can be carried out by inserting a mechanical chopper between the light source and the sample cell. Modulation can also be achieved by electronic modulation of the light source power supply. If modulation takes place between the sample and the detector, any interfering signals originating at the sample modulate the carrier. Modulation between the detector and the lock-in amplifier results in modulation of the carrier by the detector noise and by all other noise sources and interferences in the optical part of the system. Thus, the location of the modulation step in the measurement system is far from trivial, and selective modulation can greatly aid the signal-to-noise ratio enhancement.

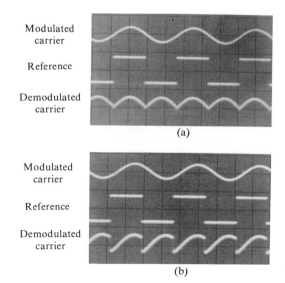

Modulated carrier

Reference

Demodulated carrier

(a)

Modulated carrier

Reference

Demodulated carrier

(b)

**Fig. 8-28.** Lock-in amplifier demodulator waveforms. When the reference signal is exactly in phase with the carrier waveform as in part (a), the demodulated carrier waveform is unipolar as in synchronous full-wave rectification. The out-of-phase condition, illustrated by the waveforms of (b), shows asymmetry and the partial self cancellation that occurs when the negative and positive parts of the waveform are averaged in the low pass filters.

**Note 8-11.   Phase-Locked Demodulation.**
When two sinusoidal waveforms are multiplied, the resulting waveform contains frequency components equal to the sum and difference frequencies. Consider $v_c = V_c \sin(\omega_c t + \theta_c)$ and $v_r = V_r \sin(\omega_r t + \theta_r)$. The product $v_p$ of $v_c$ and $v_r$ is given by

$$v_p = \frac{V_c V_r}{2} \cos(\omega_c t + \omega_r t + \theta_c + \theta_r) - \cos(\omega_c t - \omega_r t + \theta_c - \theta_r)$$

If the multiplier output is filtered to eliminate the sum term and $\omega_c = \omega_r = \omega$, the filter output is a dc signal given by

$$v_{dc} = \frac{V_c V_r}{2} \cos(\theta_c - \theta_r)$$

Thus the filter output voltage is a function of the phase difference between $v_c$ and $v_r$ and is a maximum when $\theta_c = \theta_r$.

The best regions for the choice of carrier frequency are in the ranges 10–35 Hz and $250–10^5$ Hz. The lower region is used mainly for detectors that have poor frequency response or for modulation with slow-response sources.

An important aspect of the modulation step is the generation of a reference signal that is the same frequency as the carrier wave (chopping frequency) and is phase-locked to it. (It need not be exactly in phase with the carrier but only phase-locked to it, as their relative phases can be adjusted later in the measurement system.) This reference signal can be generated, for example, at a rotating mechanical chopper by an auxiliary light source and detector combination. If the source is electronically modulated, the waveform used to modulate the source power supply can also be used as the reference signal. Many lock-in amplifiers have an internally generated reference signal that can be used to drive an external modulator. The resulting modulated carrier wave is demodulated using this internal reference.

The next three steps, amplification, demodulation, and filtering, are normally carried out in the lock-in amplifier. First, the modulated carrier wave is selectively amplified. Traditionally this has been done by using a tuned amplifier with a bandpass sufficient to pass the carrier wave and its signal sidebands. However, if the modulation frequency drifts within the bandpass of the tuned amplifier, its amplitude fluctuates and it is impossible to distinguish these amplitude fluctuations from those caused by the modulation step. In some measurement situations the amplifier can be rather broadly tuned to the carrier frequency, making it less sensitive to frequency drifts. Any random noise is effectively discriminated against at the synchronous demodulation or low pass filtering step.

The demodulation step of the lock-in amplifier provides the lock-in aspect of the measurement. Some form of selective amplification is often applied to the reference signal; then its phase must be adjusted relative to that of the modulated carrier. Finally, the reference is converted to a bipolar square wave before multiplication. Actual signal, reference, and demodulated carrier waveforms are shown in figure 8-28. For maximum output the phase of the reference wave should be adjusted to be exactly in phase with the carrier wave as illustrated in the figure (see note 8-11).

The final step in the recovery of the signal information is low pass filtering. This step simply converts the synchronously full-wave rectified carrier to a dc level, the magnitude of which is representative of the amplitude of the carrier wave. Further signal-to-noise ratio enhancement is obtained when the low pass filter time constant is very long compared to one period of the carrier. Then the output is actually the average of the demodulation of hundreds or thousands of cycles of the carrier. Time-constant selection allows a choice in the trade-off between output averaging and system response speed.

Lock-in amplifiers are now used routinely to make meaningful measurements of signal components that are literally buried in noise—unidentifiable in an oscilloscopic observation of the original signal. However, the lock-in measurement system does have some limitations. The signal must be capable of modulating a carrier wave. For many signals this cannot be done very effectively; transient signals, signals with a high repetition rate, low duty cycle signals, and fast pulse signals are all signals for which it is essentially impossible to use lock-in amplifiers.

## Chopper-Input Amplifier

The chopper-input amplifier is an example of the lock-in amplifier principle applied to eliminate the zero drift (a form of $1/f$ noise) in a difference amplifier. The basic form of the amplifier is shown in figure 8-29. The **chopper** is a continuously alternating switch connected to produce a waveform that has an amplitude equal to the voltage difference at the inputs and a frequency equal to the chopper drive rate. This modulated signal is connected to a basic lock-in amplifier. The dc drift due to the amplifier is eliminated because the amplifier is an ac amplifier and the ac input signal always has zero amplitude when $v_+ = v_-$. The common mode rejection ratio is likewise very high since input capacitor $C$ charges to the common mode voltage and the two inputs are symmetrical. The chopper-input amplifier would appear to have almost infinite resistance between the inputs since the $v_+$ and $v_-$ chopper contacts are never connected together. In practice, the resistance between inputs is limited only by the resistance of the open chopper switch. Carrier frequencies up to 1 kHz can be achieved with vibrating reed choppers, but field-effect transistor and photoconductor switches are also used. The quality and symmetry of the chopper is very important. Its $1/f$ noise appears in the modulated carrier. Amplifiers with submicrovolt stabilities can be achieved with this principle. They are especially useful as null detectors and as amplifiers for detecting and correcting drift in dc systems.

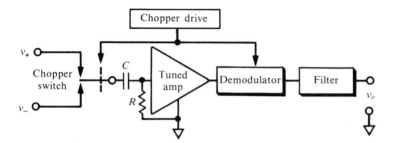

**Fig. 8-29.** Chopper-input amplifier. A modulated signal with an amplitude $v_+ - v_-$ is obtained by alternating the chopper switch at the carrier frequency. The difference-modulated carrier is coupled to the tuned amplifier with a high pass filter and demodulated as in the lock-in amplifier. The resulting amplifier has very low drift and excellent CMRR.

## Suggested Experiments

**1. Difference and instrumentation amplifier.**
Wire the difference amplifier of figure 8-2 for a gain of 10 (with 100 kΩ and 1 MΩ resistors). Measure the gain. The CMRR can be determined from the output change as a signal connected to both inputs is varied. Connect the instrumentation amplifier of figure 8-2 with $R_1 = 10$ kΩ. Measure the gain and CMRR for various values of $a$.

**2. Waveshaping.**
Connect the inverting limiter circuit shown in figure 8-4. Test its response for several positive and negative input voltages. Connect the function generator (FG) to $v_{in}$ and the VRS to $V_R$, and observe the waveform at $v_o$ as $V_R$ is varied. Add the noninverting limiter to this circuit to obtain the absolute value circuit of figure 8-6. Test the accuracy of its response function, and observe the result of the absolute value operation on the various waveforms of the FG.

**3. Logarithmic amplifier.**
Wire the log amplifier of figure 8-7 using a diode or a transistor, or a logarithmic function module. Determine the transfer function and the useful dynamic range for the log amplifier.

**4. Analog multiplier.**
Connect a four-quadrant multiplier IC as a multiplier. Test its response and useful range. Repeat for the square, square-root, and divider applications of the IC.

**5. Operational transconductance amplifier.**
Wire an input circuit for an OTA as shown in figure 8-10. Connect a current meter to the OTA output, and measure the transconductance as a function of the programming current input. Connect and operate the OTA as a variable gain voltage amplifier. Determine its useful range of gain and evaluate its usefulness as an analog multiplier in two quadrants.

**6. Active filters and tuned amplifiers.**
Determine the frequency response function for first order active high and low pass filters. Connect a state variable filter as shown in figure 8-21, and measure the frequency response function for the LP, HP, and BP outputs. Sum the LP and HP outputs with a summing amplifier, and verify the notch filter response.

**7. Oscillator.**
Connect the Wein bridge oscillator of figure 8-22, and determine the frequency of oscillation for several values of $R$ and $C$. Observe the quality of the sine-wave output by a Lissajous figure or by spectrum analysis, and adjust the feedback ratio for minimum distortion. Modify the state variable filter of experiment 6 to produce the oscillator of figure 8-23. The multipliers may be analog multipliers or OTAs, or they may be omitted. Observe the quadrature outputs on dual-trace and $x$-$y$ displays. Determine the range of frequencies available from voltage-controlled oscillator operation.

**8. AM modulation.**
Use analog multipliers to produce AM modulated and double-sideband modulated waveforms. If one is available, use a spectrum analyzer to observe the output signals. Connect a multiplier as a demodulator, and use it to regain the modulating signal.

**9. Lock-in amplifier.**
Use the tuned amplifier from experiment 6 and the demodulator from experiment 9 to build the amplifier-demodulator part of the lock-in amplifier of figure 8-27. Use a square-wave signal to light a LED light source, and observe a photodiode or phototransistor output with the lock-in amplifier. Use the square-wave modulation source for the demodulating signal. Compare the sensitivity of the lock-in amplifier detection system with that of (a) a constant source and measurement of dc detector output level and (b) a modulated source with ac detection using an oscilloscope.

## Questions and Problems

**1.** The gain of an IC instrumentation amplifier package can be set by a single external resistor. The CMRR for this amplifier is 74 dB, 94 dB, 104 dB, and 110 dB for gains of 1, 10, 100, and 1000, respectively. Explain why the CMRR increases as the gain increases.

**2.** For the instrumentation amplifier of figure 8-3, calculate the appropriate value for the gain-setting resistor $R_1/a$ to obtain an amplifier with a gain of 1000 if $R_1 = 50$ kΩ, $R_2 = 100$ kΩ, and $K = 10$.

**3.** Sketch the response curve for the precision inverting limiter of figure 8-4 if $R_f = 100$ kΩ, $R_{in} = 10$ kΩ, $R_R = 500$ kΩ, and $V_R = +15$ V. Show how the response curve is changed if diodes $D_1$ and $D_2$ are reversed but all other values remain the same.

**4.** (a) For the absolute value circuit of figure 8-6, suggest a method of adjusting the gain of one of the limiters so that the output slopes are both exactly equal in magnitude but opposite in sign. (b) Is it possible to change the resistance values to produce an absolute value circuit with a gain of ten for both polarities? (c) An input voltage polarity indicator can be made by connecting the inputs of a comparator to the output of each of the op amps. Explain why the state of the comparator output changes with the polarity of the input voltage.

**5.** For a silicon diode in the logarithmic amplifier of figure 8-7a, what temperature increase from $25°C$ causes the same change in $v_o$ as a 1% change in the current $i$?

**6.** Sketch the input/output transfer function for the multiplier applications shown in figure 8-8. For the divider the input quantity is $v_z/v_x$; for the multiplier, $v_x \times v_y$; for the square, $v_x$; and for the square root, $v_z$. Consider both positive and negative values of the input quantity.

**7.** If the transconductance of an OTA is $17 \times I_{ABC}$ and a current of 2.0 mA is applied to the $I_{ABC}$ input, what is the maximum value of the output current, and what voltage difference at the OTA input will produce the limiting output current?

**8.** (a) Design a noninverting active high pass filter such as that of figure 8-15 so that the cutoff frequency $f_1$ is 500 Hz and the input impedance at the cutoff frequency is 10 k$\Omega$. (b) Calculate the input impedance for 0.01 $f_1$ and 100 $f_1$.

**9.** (a) For the active second order filter of figure 8-19, what should $R'$ and $(K-1)$ $R'$ be for a damping factor of 0.2? Of 0.01? (b) What would the $Q$ be of a second order filter with a gain of 2?

**10.** A second order low pass filter has values of $R$ and $C$ of 10 k$\Omega$ and 0.01 $\mu$F. (a) What is $\omega_c$? (b) What is the response of the filter at $0.1\omega_c$ and $10\omega_c$?

**12.** (a) What is the resonant frequency of a parallel $LC$ circuit with $L = 35$ $\mu$H and $C = 100$ pF? (b) What is the $Q$ of the circuit if the resistance in the inductor is 5 $\Omega$? (c) What is the impedance of the circuit at $\omega_0$?

**13.** It is desired to make a tunable filter following the state variable design of figure 8-21. The op amps have a maximum output voltage of $\pm10$ V and an output current limit of 5 mA. (a) What is the lowest practical value for $R$? (b) Assuming the maximum practical value of $R$ is 2 M$\Omega$, what is the ratio of the highest to the lowest frequencies for a given value of $C$? (c) What value of

$C$ should be chosen so that 1000 Hz is the mid-frequency in the range?

**14.** Design a state variable filter that has a notch frequency at 60 Hz to remove power-line noise from a signal. What components in the circuit of figure 8-21 affect the sharpness of the notch filter?

**15.** Op amp experimenters often find the op amp they have patched together is acting as a high-frequency oscillator. (a) If the open-loop gain of an op amp is $5 \times 10^5$, what fraction of the output signal must be coupled to the noninverting input in order to sustain oscillation? (b) Why is this most likely to occur unintentionally at high frequencies?

**16.** (a) Discuss the relationship between the feedback ratio of an oscillator and the quality of the sine wave produced. (b) Why is it difficult to maintain the optimum feedback ratio in a practical oscillator?

**17.** (a) Discuss the choice of carrier frequency for the AM modulation of a signal in terms of avoidance of likely noise sources and the highest frequency components in the signal source. (b) AM radio stations are limited to a total bandwidth of 20 kHz. What is the approximate upper limit of the frequency content of a demodulated AM radio signal?

**18.** Identify the figure numbers of circuits from this text that could be used for the functions shown in the blocks of the lock-in amplifier of figure 8-27. (Hint: for the phase adjust consider a summing amplifier with variable weighting of quadrature input signals.)

**19.** (a) For the phase-locked demodulator, show that the filtered dc output is a maximum when $\theta_c = \theta_r$. (b) What fraction of the maximum output results when the difference in phase is $45°$, $90°$, $180°$?

**20.** A chopper-input amplifier like that of figure 8-29 is used as a floating null detector for the voltage comparison measurement of an unknown voltage of about 5 V. (a) Assume that the difference signal ($v_+ - v_-$) is 3 V. Describe the waveforms, and give the average dc value of the signals on both sides of the input capacitor. (b) Given a chopper frequency of 1 kHz, what would you estimate to be the upper frequency limit of input signal variation that could be followed with this amplifier?

# Chapter 9

# Frequency, Time, and the Integrating DVM

Signals in which the information is encoded as the frequency or as some other time relationship of the waveform are time-domain signals. The data in time-encoded signals are much less dependent on absolute amplitude than those in analog signals. They are, therefore, much less affected by electrical noise and transmission attenuation. The many examples of time encoding in modern electronics include the frequency encoding of FM radio, the pulse-width encoding in radio remote-control (RC) systems, and the audio-frequency encoding of push-button telephone dials. In many measurement systems the information sought is inherently in the time domain. Examples are the Geiger tube, which converts the level of radioactivity to a pulse repetition rate, and radar, in which the time between a transmitted pulse and its echo is related to the distance of the reflecting object.

Time-domain signals are often measured by comparing the time-encoded signal variations with counted increments of time obtained from a precision clock. Very precise clocks and very high speed counters allow the measurement (digital conversion) of time-domain signals to an accuracy of eight significant digits or more. It is customary to convert the time-encoding variations into logic-level transitions as early in the signal processing as possible. Advantage is then taken of inexpensive and high-speed digital integrated circuits for further signal processing. Time and digital signals are often confused in the literature because digital circuits are used for both.

Frequency, pulse width, period, and other time-encoded signals can be digitally measured with various interconnections of a counter and a clock as illustrated in the first section of this chapter. The techniques for converting signal variations to logic-level transitions are explored in the section on comparators. The next two sections describe two of the most popular schemes for analog-to-digital conversion. The chapter ends with a discussion of several very useful operations that can be performed on time-domain signals. These include the phase-locked loop (a frequency-domain servo system) and FM modulation and demodulation.

## 9-1  Digital Measurement of Frequency and Time

The digital measurement of time-domain signals is readily accomplished by counting techniques. The quantity to be measured is counted over a period specified by the appropriate boundary condition. The functional blocks needed for a variety of counting, timing, and frequency measurements are comparators, a crystal oscillator and scaler, a counting gate, and a counter with latch and display. As illustrated in this section, these basic building blocks are interconnected in different ways to perform measurements of frequency, frequency ratio, period, multiple period average, and time interval. This section begins by describing the versatile crystal-controlled time base.

### Precision Time Base

The precision time base used in digital time and frequency measurements is usually derived from a crystal oscillator. The crystal oscillator can produce a reference frequency that is accurate and stable to better than one part per million. A variety of convenient time periods are derived from the basic oscillator frequency by a multi-decade frequency divider known as a **scaler**. Modern integrated circuit time base chips are available with all the needed circuitry (Fig. 9-1). In this example, a 1-MHz crystal oscillator is divided by the scaler to provide periods from 1 $\mu$s to 1 h. The output frequency is selected by a programmable switch, which is set by the binary signals applied to the external clock control inputs. The time base output is then available to gate the counter or to provide the basic time increments to be counted. An external input allows scaling of an external signal rather than that from the oscillator.

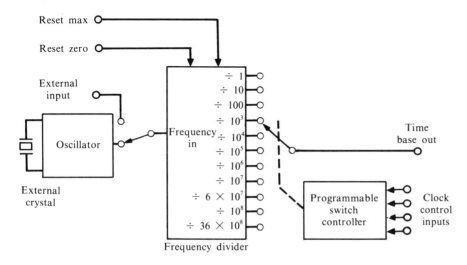

**Fig. 9-1.** Precision time base. This integrated circuit clock provides the necessary circuitry for a precision time base. An external crystal is needed. With a 1-MHz crystal the output period ranges from 1 $\mu$s to 10 s in decade steps with additional periods of 1 min, 100 s, and 1 h. Logic-level signals at the clock control inputs select the time base output.

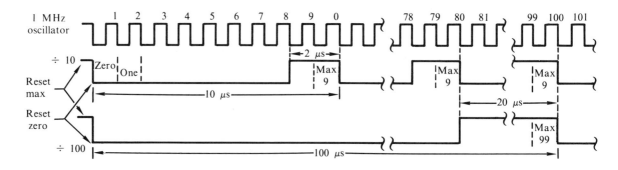

**Fig. 9-2.** Precision time base waveforms. The outputs are shown. The state of the scaler increments on each HI→LO transition at the oscillator input. The Reset zero function puts all scaler circuits in the zero state (LO output). The scaler advances to state one on the first HI→LO oscillator transition after the Reset zero signal is removed. The Reset max puts all scaler circuits in their maximum state (HI output). All outputs undergo a HI→LO transition to the zero state on the first oscillator HI→LO transition after the Reset max signal is removed.

Figure 9-2 shows the output waveforms for a few of the scaler outputs. The functions of the two reset modes, Reset zero and Reset max, are also illustrated. The Reset max function is used if a complete timing cycle, including the initial HI→LO edge, is desired within one oscillator period after the application of a Reset max signal. This eliminates waiting up to one complete cycle for the next initial edge to occur. The Reset max mode is often used when the time base output controls some external device such as a counting gate. The Reset zero mode provides a complete timing cycle within one oscillator period of the release of the reset signal. This mode is used when some external event is to trigger the time base.

## Frequency and Frequency Ratio Measurements

A frequency measurement determines the number of cycles of a signal that occur per unit time. Basic counting and timing functions are arranged so that a signal is counted for the time period determined by the precision time base. The scaler output is therefore connected to the Start and Stop inputs of the counting gate so that the gate is activated for one complete clock cycle as shown in figure 9-3a.

**Fig. 9-3.** Frequency meter (a) and frequency ratio meter (b). In the frequency meter the number of cycles of $A$ is counted per cycle of the precision time base (i.e., per unit time). In the frequency ratio measurement the number of cycles of $A$ is counted for $N$ cycles of $B$, where $N$ is the scaling factor selected on the scaler.

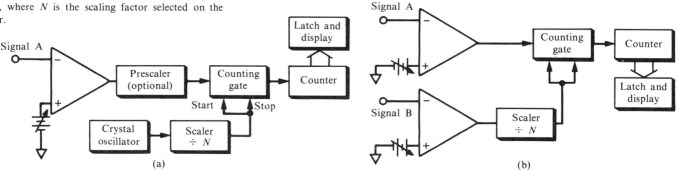

The frequency mode measures the average frequency over the time base period. Any variations in signal frequency are averaged during this time. Very high frequency signals may exceed the maximum counting rate of the counting circuitry. For these signals a high-speed prescaler (usually one or a few decades) is used to reduce the input frequency to a value low enough for accurate counting.

In the frequency ratio mode a second signal frequency connected to the $B$ input is substituted for the internal time base as shown in figure 9-3b. The ratio $f_A/f_B$ is obtained. For signals close together in frequency, a scaler is usually introduced in the $B$ channel to increase the number of counts in the ratio measurement; the number of counts obtained is then $Nf_A/f_B$.

## Period and Period Average Measurements

In a period meter the number of unit time increments is counted during one complete cycle (period) of the input signal. The inputs to the gate and counter are thus reversed from the frequency mode as shown in figure 9-4a. With a 1-MHz crystal oscillator, the unit time increment is 1 $\mu$s. For a 100-Hz input signal, the period is $1/100$ s $= 10^4$ $\mu$s. Therefore, the counter would accumulate $10^4$ counts during one period of the input signal. The resolution of this measurement mode in 0.01 s is thus 100 times greater than that of a frequency measurement of the same signal taking 1.0 s. The period mode is thus preferable for low-frequency signals, and the frequency mode is better for high-frequency signals. It should be noted that in a single period measurement no averaging of the input signal occurs, and the measurement is only as precise as the reproducibility of a single period.

In order to improve the measurement accuracy and precision, the time of 10, 100, 1000 or more successive periods can be measured by inserting a scaler to divide the input signal prior to the gate connection, as shown in figure 9-4b. This measurement of the average duration of multiple periods reduces the relative error from time errors in the comparator triggering by $N$ times, where $N$ is the scaling factor.

**Fig. 9-4.** (a) Period measurement and (b) multiple period average measurement. In the period measurement the number of time increments is counted per cycle of the input signal. In the multiple period average mode, the number of oscillator pulses per $N$ periods of the input signal is counted.

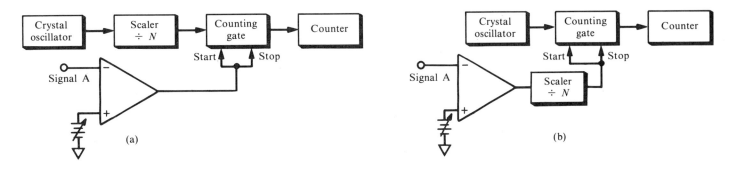

**Fig. 9-5.** Pulse-width measurement. The number of time increments is counted from the leading edge of an input signal *A* to its falling edge. The counting gate is opened on the rising edge of the signal and closed on the inverted falling edge.

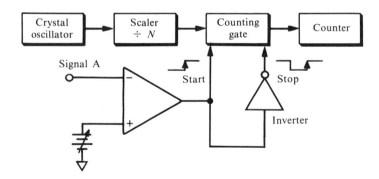

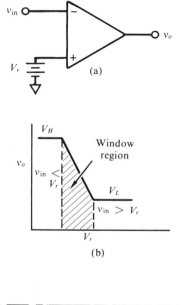

(a)

(b)

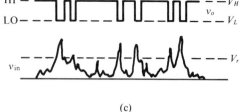

(c)

**Fig. 9-6.** A basic comparator. Its symbol (a), its transfer characteristics (b), and its response to a varying input signal (c). When $v_{in} < V_r$, the comparator is at its positive limit or HI output state, $V_H$. When $v_{in} > V_r$, the comparator is at its negative limit or LO output state $V_L$. In the narrow window region, where $v_{in} \simeq V_r$, the amplifier is in its linear region.

## Time Interval Measurements

In a time interval measurement the number of time increments is counted for the interval between Start and Stop signals applied to the counting gate. Separate *A* and *B* signals can be used to obtain the time interval between event *A* and event *B*. Another use of the time interval mode is to obtain the width of a pulse as shown in figure 9-5. Here the signal is applied directly to the Start input and through an inverter to the Stop input. If the Start and Stop inputs are activated by LO→HI logic-level transitions, the gate opens on a LO→HI transition and closes on a HI→LO transition. Thus time increments are counted for one pulse width (one-half cycle).

## 9-2   Comparators and Schmitt Triggers

The comparator is a device whose output state indicates whether an input voltage is larger or smaller than a reference voltage. As such, the comparator serves to produce logic-level outputs from an analog signal input. The comparator can also provide noise discrimination by producing output transitions only when the input signal exceeds or falls below an adjustable threshold level. A comparator whose HI→LO output transition occurs at a different threshold value than its LO→HI transition is known as a **Schmitt trigger**. It is frequently used as a snap-action comparator with fast rise and fall times and good immunity to spurious noise. A comparator that produces a logic-level transition whenever an input signal crosses a threshold voltage of zero is called a **zero-crossing detector.**

### Basic Comparator

The basic op amp comparator circuit is shown in figure 9-6. The amplifier is operated open loop, and only a small voltage difference $(V_H - V_L)/A$ is required to change the output state. For connection to circuits with logic-level inputs, the normal ±10-V comparator output can be converted to a logic-level output with a simple transistor switch. Alternatively, integrated circuit comparators are available with ±10-V analog inputs and standard logic-level outputs.

The ideal comparator would have infinite gain and would change output states instantaneously. In reality, comparators have limitations due to their finite open-loop gains, their response times, and their input characteristics (bias currents, offset voltages, and common mode rejection ratios). The finite open-loop gain means that for a small range of input voltages the comparator is in the linear region between its output limits. This can cause a small uncertainty in the output transition time, which for slowly changing signals depends on the rate at which the input signal traverses the linear region. For example, a comparator with logic-level outputs of +5 and 0 V and an open-loop gain of 2500 has a threshold window of $5\,\text{V}/2500 = 2\,\text{mV}$. If the input signal changes at a rate of 10 V/s through the window region, the transition time is uncertain by $2\,\text{mV}/10\,\text{Vs}^{-1} = 200\,\mu\text{s}$. For faster signals the transition time becomes limited by the basic response time and **slew rate** (maximum rate of change of output voltage) of the amplifier.

In addition to the uncertainty caused by response time and input signal rate, any noise present in the comparator, in the reference voltage or in the signal will cause an additional time uncertainty or "jitter" in the comparator output (recall fig. 1-4). If the signal noise is large enough to cause the threshold region to be crossed several times, the comparator output flips from one state to the other until the difference between the signal voltage and the reference voltage is larger than the noise (fig. 9-7). The Schmitt trigger described next can eliminate this effect and provide faster output transition times.

## The Schmitt Trigger

**Hysteresis** is used to make the threshold level for a LO→HI transition different than that for a HI→LO transition in the Schmitt trigger. The Schmitt trigger and its transfer characteristics are shown in figure 9-8. The positive feedback loop makes the threshold voltage dependent on the comparator output state. If the positive feedback loop were absent, the comparator threshold level would be the reference voltage $V_r$. In the presence of positive feedback, the threshold for the HI→LO output transition is increased to $V_r + \Delta V/2$, and that for the LO→HI output transition is decreased to $V_r - \Delta V/2$. The amount of hysteresis or **hysteresis lag** $\Delta V$ is given by

$$\Delta V = \left(\frac{\beta A - 1}{A}\right)(V_H - V_L)$$

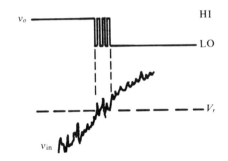

**Fig. 9-7.** Multiple triggering of comparator by noise. This effect can be minimized by reducing high-frequency noise, by increasing the rate of signal change through the threshold, or by using a Schmitt trigger.

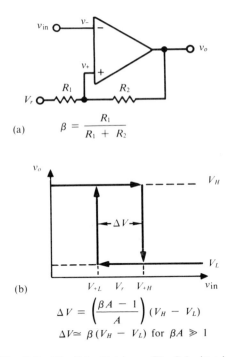

(a) $$\beta = \frac{R_1}{R_1 + R_2}$$

(b) $$\Delta V = \left(\frac{\beta A - 1}{A}\right)(V_H - V_L)$$

$$\Delta V \approx \beta(V_H - V_L) \text{ for } \beta A \gg 1$$

**Fig. 9-8.** The Schmitt trigger. The Schmitt trigger circuit (a) uses positive feedback to provide hysteresis to the threshold level. (b) The amount of hysteresis $\Delta V$ depends on the feedback fraction $\beta = R_1/(R_1 + R_2)$, the open-loop gain of the comparator $A$, and the difference between the HI and LO logic level voltages $V_H - V_L$ (see note 9-1).

**Note 9-1.   Derivation of Schmitt Trigger Hysteresis.**

When the output of the Schmitt trigger of figure 9-8 is in the HI state, the threshold level is given by

$$V_{+H} = V_r + \beta(V_H - V_r)$$

In the LO state, the threshold $V_{+L}$ is

$$V_{+L} = V_r + \beta(V_L - V_r)$$

The difference signal $v_- - v_+$ required to achieve the HI output state is

$$v_{+H} - v_{-H} = \frac{V_H}{A} = v_{+H} - v_{inH}$$

The signal required to achieve the LO output state is

$$v_{+L} - v_{-L} = \frac{V_L}{A} = v_{+L} - v_{inL}$$

The input signal voltage required for the output to be HI, $v_{inH}$, is thus

$$v_{inH} = \frac{-V_H}{A} + v_{+H} = \frac{-V_H}{A} + V_r + \beta(V_H - V_r)$$

or

$$v_{inH} = -V_R(1 - \beta) + \left(\frac{\beta A - 1}{A}\right)V_H$$

Similarly, the input signal required for a LO output, $v_{inL}$, is

$$v_{inL} = -V_r(1 - \beta) + \left(\frac{\beta A - 1}{A}\right)V_L$$

The comparator hysteresis $\Delta V$ is

$$\Delta V = v_{inH} - v_{inL} = \left(\frac{\beta A - 1}{A}\right)(V_H - V_L)$$

where $\beta$ is the feedback fraction and $A$ is the open-loop comparator gain (see note 9-1).

With a Schmitt trigger the speed of the transition is determined by the response time of the amplifier itself and not by the time required for the input signal to pass through the threshold region. For this reason the Schmitt trigger finds use where the input signal changes are slow and it is necessary to have very fast output transitions.

## Zero-Crossing Detector

The **zero-crossing detector** is a comparator circuit that changes state each time an ac input signal changes polarity. It is frequently used at the input of inexpensive frequency meters for ac signals because it amplifies and squares the input signal.

A circuit that accomplishes the zero-crossing function for the ac component of signals is shown in figure 9-9. The ac-coupled preamplifier removes any dc component from the signal, and its approximately logarithmic response produces an output signal amplitude that is relatively independent of the input signal amplitude. The zero-based comparator then produces a logic-level transition for each reversal of the ac component of the input signal.

## Comparator Application Requirements

The various comparator types find widespread use in frequency and time measurements. The requirements for the comparator are highly dependent on the application. In a frequency measurement, for example, the comparator is not critical. It should not allow multiple triggering or spurious trigger signals, but the measurement is relatively independent of the comparator response speed and time jitter. A frequency ratio measurement likewise does not depend on the comparator timing when the number of cycles of $A$ is measured for a large number of cycles of $B$. A single period measurement, on the other hand, is highly dependent on the comparator. It must start and stop the counter at the same point in the waveform on each edge. Jitter in the trigger point and nonreproducible time delays can cause errors. Multiple period averaging reduces the influence of comparator stability. In a pulse-width measurement the trigger point must occur at the same voltage on the leading edge as it does on the falling edge. Any time delay must be the same for both rising and falling signals.

## 9-3   Voltage-to-Frequency Converter

The conversion of an analog quantity into a digital domain signal requires the determination of the number of units or increments that comprise the

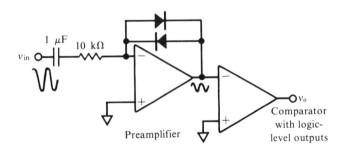

Preamplifier

Comparator with logic-level outputs

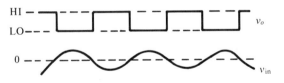

**Fig. 9-9.** Zero-crossing detector. The preamplifier and comparator form an ac zero-crossing detector with wide dynamic range and logic-level outputs. The ac log amplifier maintains a high gain for low-level signals but will not overload when a large signal is applied.

measured quantity. This is accomplished by determining how many increments of that quantity must be generated to match the analog quantity. The comparison could be made in any of the analog domains, but charge and voltage comparisons are most common. The voltage-to-frequency and dual-slope converters in this section and the next are representative of integrating or charge comparison converters. Voltage comparison analog-to-digital converters are discussed in chapter 13.

## Charge-to-Count Converter

The **charge-to-count converter** is an example of a null comparison measurement of charge. In the system shown in block form in figure 9-10, the unknown and reference charges are compared by the charge difference detector. The reference charge $q_r$ is generated in increments of $q_s$. Each generated increment is counted by the counter. When the difference detector output is zero, the counter displays the number of charge increments $q_s$ contained in the unknown charge $q_u$.

An op amp integrator can be used as a charge difference detector as shown in figure 9-11. The unknown and reference charges are integrated

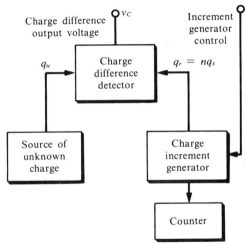

**Fig. 9-10.** Charge-to-count conversion. When the charge increment generator has generated $n$ charges of known value $q_s$ such that the charge difference detector's output is zero, the charge $nq_s$ is equal to the unknown charge $q_u$. The number $n$ is determined by counting the increments of charge as they occur.

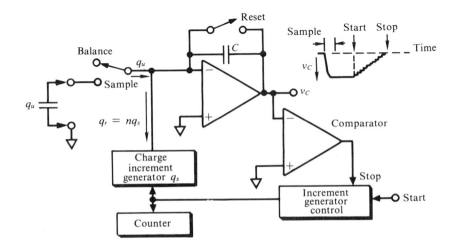

**Fig. 9-11.** A charge-to-count converter. The charge $q_u$ is transferred to the integrating capacitor when the input switch is in the Sample position. In response to a Start command the charge increment generator adds charge increments of the opposite sense until the comparator, indicating charge equivalence, stops the generator. The counter displays the number of reference charge increments $n$, where $n = q_u/q_s$.

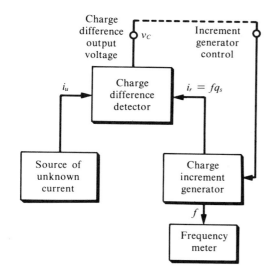

**Fig. 9-12.** Current-to-frequency conversion. The charge difference detector is at balance when $v_C$ is a constant value. In this state the unknown and reference currents are equal, and the reference current can be measured as the frequency of generating charge increments $q_s$.

sequentially, and a comparator is used to indicate when the integrator has been returned to its initial charge. The charge increment generator can be a generator of reproducible charge pulses where the number of pulses is being counted, or it can be a constant current generator turned on for successive increments of time where the time increments are being counted. These two versions of the charge-to-number converter concept are used in the current-to-frequency converter and the dual-slope converter described in this chapter.

## Current-to-Frequency Converter

The **current-to-frequency converter** is an application of the charge balance principle in which the input signal current is balanced by the rate of reference charge pulse generation so that the charge difference detector output remains near zero. As shown in the block diagram of figure 9-12, it is the continuous balance that distinguishes this converter from the charge-to-count converter and results in an output pulse rate proportional to the flow of charge at the input.

The reference current $i_r$ is generated by repeated triggering of the charge increment generator. The average current is the charge per pulse $q_s$ times the pulses per second $f$. Thus $i_r = fq_s$. The currents $i_u$ and $i_r$ are connected to the charge difference detector simultaneously. Any difference between $i_u$ and $i_r$ causes a change in the charge difference output voltage $v_C$. The null condition is achieved when $v_C$ is constant. At the null frequency, $i_u = i_r = fq_s$. A number proportional to the unknown current is read from the frequency meter display.

A practical current-to-frequency converter (IFC) is shown in figure 9-13. The charge $q_s$ in each reference charge pulse is $t_p i_s$. The average rate of reference charge addition is $f_o t_p i_s$ which at balance is equal to $i_{in}$. Therefore,

**Fig. 9-13.** Current-to-frequency converter. The integrator output is negative-going while $i_{in}$ is integrating. When the comparator threshold is crossed, the monostable multivbibrator (MS) is triggered, and the current switch connecting $i_s$ to the integrator is closed for a time, $t_p$. At balance the average rate of charge addition from the reference charge pulses is equal to $i_{in}$. Therefore, the frequency of charge pulse addition $f_o$ is proportional to $i_{in}$. For voltage-to-frequency conversion a resistor is used in series with the input.

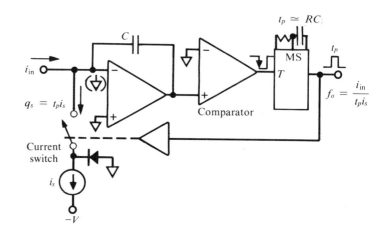

$f_o = i_{in}/t_p i_s$. The parts of the circuit that affect the current-to-frequency ratio, and that should then be stable, are the parts that determine $i_s$ and the pulse width $t_p$ of the monostable multivibrator. It is interesting that the integrating capacitor and the comparator threshold are not critical. With reasonable care in component selection, linearity to a few hundredths percent is possible using the simple circuit shown.

The upper frequency limit for the IFC is determined by the rate at which the reference charge pulses begin to deviate significantly from their low-frequency magnitude. Full-scale frequencies $f_{fs}$ of 10 kHz or 100 kHz are common, but converters that operate up to 10 MHz are available. The value of $f_{fs}$ sets some limits on $t_p$ and $i_s$. The pulse width $t_p$ must be shorter than one output cycle at $f_{fs}$; generally $t_p \leq (1/1.5f_{fs})$. During time $t_p$ the reference generator must add as much charge to the integrator as the full-scale input current $i_{fs}$ does in one output cycle. Therefore, $t_p i_s = i_{fs}/f_{fs}$. A typical value for $i_{fs}$ is 1 mA. If the above guideline for $t_p$ applies, then $i_s$ must be at least equal to $1.5i_{fs}$. If the input current exceeds the maximum feedback charge rate, the integrator is not discharged, and the comparator output does not return to HI to allow the next trigger. This would be a permanent hangup if the monostable multivibrator were not one for which the output pulse width is equal to $RC$ or to the trigger pulse width, whichever is longer. So as long as the comparator output stays LO, the current switch remains closed, and the system comes out of overload when the excessive current is removed.

Though it is inherently a current-to-frequency converter, this circuit is often provided with an input resistor and called a **voltage-to-frequency converter (VFC)**. In this case $i_{in} = v_{in}/R_{in}$. Therefore, $f_o = v_{in}/t_p i_s R_{in}$, and the output frequency is proportional to $v_{in}$. If the converter has a full-scale current of 1.0 mA, a 10-k$\Omega$ resistor at $R_{in}$ would give a full-scale input voltage of 10 V. Note that the input voltage or current cannot be bipolar and that at zero input current the output frequency is zero.

Voltage-to-frequency converters have many applications in modern electronic systems. They are used as voltage-controlled oscillators of exceptional linearity and dynamic range, as analog-to-frequency converters for reliable data transmission over a single connection, and as a basic function in analog-to-digital converters (ADCs). These applications have been made all the more attractive by the availability of the complete VFC in integrated circuit form.

## Digital Integration

The charge balance technique is a very convenient way to obtain the time integral of an input signal. Because the current-to-frequency converter produces a pulse for every increment of input charge, the total number of pulses

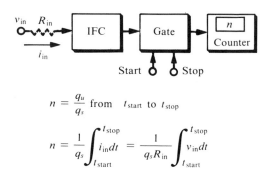

$$n = \frac{q_u}{q_s} \text{ from } t_{\text{start}} \text{ to } t_{\text{stop}}$$

$$n = \frac{1}{q_s} \int_{t_{\text{start}}}^{t_{\text{stop}}} i_{\text{in}} dt = \frac{1}{q_s R_{\text{in}}} \int_{t_{\text{start}}}^{t_{\text{stop}}} v_{\text{in}} dt$$

**Fig. 9-14.** Digital measurement of charge with an IFC. The count total is equal to the number of charge increments $q_s$ that were used to balance the charge input to the IFC over the time interval between the Start and Stop commands to the gate. The total charge is related to the integral of the input current or voltage over that same period as shown. A clock is used as a gate control for precision control of the integration period.

**Fig. 9-15.** Input current, $i_{\text{in}}$ output frequency, $f_o$ and digital integral (count $= n$) for two waveforms applied to the system of figure 9-14. Note that the frequency output tracks the input amplitude variations. Waveform (a) illustrates the use of integration to obtain the area under the signal peak. The count obtained in waveform (b) is proportional to the average value of the input signal over the count interval.

produced in a given period is equal to the total charge (the integral of the current) applied to the input over that period. As shown in figure 9-14, all that is required for digital integration is an IFC and a counter. The input current, output frequency, and count value vs. time are shown in figure 9-15 for two different input waveforms. This integrating property is maintained in the ADC application of the VFC.

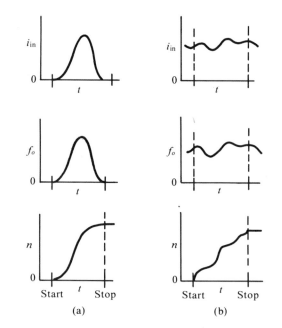

(a)                    (b)

## Analog-to-Digital Conversion

The conversion of data from the analog to the digital domain is completed for the VFC by connecting a digital frequency meter to the VFC output. The resulting ADC is shown in figure 9-16. The number of output oscillations that occur during one clock period $t_c$ is counted. The output count is thus $t_c f_o$, or from the $f_o / v_{\text{in}}$ relationship of the VFC, $n = v_{\text{in}} t_c / t_p i_s R_{\text{in}}$. This equation indicates that the overall conversion accuracy depends upon the stability of $t_c$, $t_p$, $i_s$ and $R_{\text{in}}$. The resolution of the conversion (the number of significant digits) can be increased at the expense of increased conversion time by increasing $t_c$. Any digits in excess of the stability of the least stable of the critical components, however, are not significant.

Consider now the application of this ADC in a situation where $v_{\text{in}}$ is not constant. The variations in $v_{\text{in}}$ may be of interest, or they may be due to noise components in the signal. When the digitization of varying analog- or

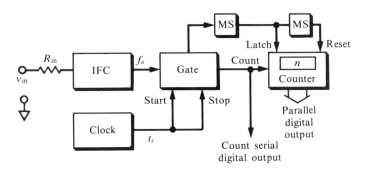

**Fig. 9-16.** Analog-to-digital conversion with a VFC and frequency meter. The output frequency of the IFC is determined by counting the output pulses that occur during one clock period $t_c$. The frequency $f_o$ is proportional to $v_{in}$ at any instant, and $n$ is $t_c$ times the average value of $f_o$ during the clock period. Increasing the clock period increases the resolution (number of pulses counted) and the time over which $v_{in}$ is averaged. The count $n = v_{in} t_c / t_{pis} R_{in}$.

time-domain signals is to be considered, a third dimension (real time) needs to be added to the data domain map of figure 1-12 as shown in figure 9-17. Here each interdomain conversion is shown as a slice through real time. The number that results from each digitization of the varying quantity is true only for the instant or period in real time during which it was sampled. It is not possible, therefore, to make a truly continuous digital record of a varying quantity. What is done is to sample and convert the quantity at regular intervals. The numerical result of each measurement is then recorded manually, by printer, on punched paper tape, or by a magnetic recording device. If the samples are taken frequently enough that the quantity changes only slightly between each sampling interval, the digital record can quite accurately represent the time variation of the measured quantity. As will be shown in chapter 14, the sampling rate (the reciprocal of the sampling interval) must be at least twice the highest frequency component of interest, and signal frequencies above twice the sampling rate should be filtered out.

For the converter in figure 9-16, the signal is sampled at successive periods each of which is $t_c$ in length. The digitization time is also $t_c$. The sampling interval is $2t_c$ as shown since the counter is gated on during alternate clock cycles. This could be shortened to approach $t_c$ by presetting the clock divider with the counter reset signal. Therefore, if $t_c = 0.1$ s, the maximum signal frequency variation that can be followed would be $1/(2t_c) = 5$ Hz. If the full-scale frequency for the IFC in this same converter is 100 kHz, the converter can fill a maximum of four digits in the counter. The converter resolution will then be 1 part in 9999 assuming other components are not limiting.

The effect of the integrating nature of the VFC type of ADC on the signal variations of higher frequency is also important. Recall that each reading of the counter is proportional to the average signal value over the gate period $t_c$. Thus signal variations of frequencies greater than $1/t_c$ tend to be smoothed out by the averaging process. The effect is quite similar to that of a low pass filter of time constant $t_c/2$. In addition, signal frequency components for which the periods are exact sub-multiples of $t_c$ are rejected

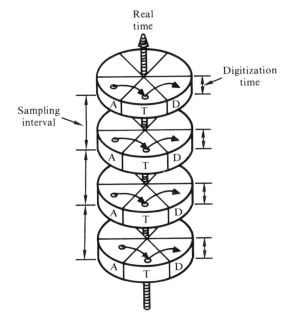

**Fig. 9-17.** Successive data domain conversions in real time. The digitization of varying analog- or time-domain data produces a sequence of digital readings, each of which represents the measured quantity at the specific instant or period in real time during which it was sampled. The time required for the digitization is one of the limits on the maximum rate at which the quantity can be sampled.

completely since the integral of one complete sine wave is zero. This characteristic can be very useful in noise rejection. For example, if a signal has a major noise component of 60 Hz and $t_c$ is 0.1 s, the 60-Hz component completes exactly six cycles during the integration time, and the noise from that source is rejected. This would also be true for any $t_c$ that is a multiple of $1/60$ s, but the minimum $t_c$ for rejection of 60-Hz noise is 16.67 ms.

An interesting converter that uses the same oscillator to determine the charging pulse width $t_p$ and the clock period $t_c$ is shown in figure 9-18. The charge balancing converter applies the reference charge to the integrator during single or successive clock cycles as needed to balance the charge from the input signal. The oscillator is connected to the clock of a $JK$ flip-flop and to the input of a frequency divider of $d$ decades. On each falling edge of the oscillator output, the $JK$ flip-flop (FF) responds to the information from the comparator. If the comparator output is HI (more charge from $i_{in}$ than $i_s$), the FF is set ($Q$ is HI) and the $\overline{Q}$ output (now LO) ceases to sink $i_s$ and causes $i_s$ to be integrated. This condition persists for as many clock cycles as necessary until the charge from $i_s$ exceeds $i_{in}$. The FF clears on the next falling edge. The gate is open for $10^d$ cycles of the oscillator but the $Q$ output of the FF allows only those clock cycles for which $i_s$ is on to be counted. Thus the counter displays the fraction of the integration time during which $i_s$ was applied to the integrator.

**Fig. 9-18.** Charge balancing ADC. The comparator senses when the reference charge to the integrator summing point is less than the charge from the input signal. The $JK$ flip-flop then turns on the $i_s$ reference current diode switch for an integral number of clock cycles until the cumulative charge from both sources is balanced. The counter counts the fraction of oscillator cycles during which the reference current switch is on. The input signal is integrated for $10^d/f_o$ seconds, but the conversion accuracy does not depend on the oscillator frequency.

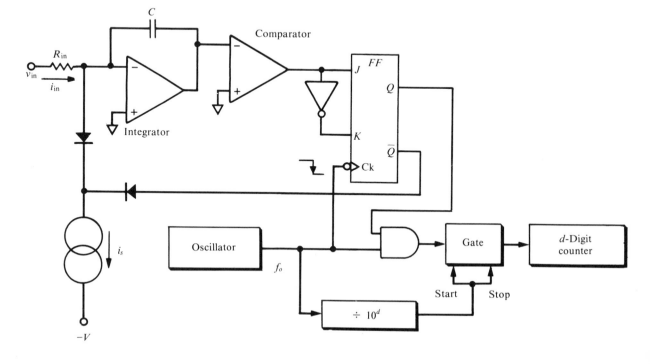

The accuracy of the charge balancing converter depends on remarkably few components. Because the gate period is $10^d/f_o$, the total charge from the signal is $v_{in}10^d/f_oR_{in}$. At balance this is equal to the total reference charge $ni_s/f_o$. Equating input and reference charges gives $n/10^d = v_{in}/i_sR_{in}$ which shows that $i_s$ and $R_{in}$ are the only critical quantities. The full-scale input current is exactly equal to $i_s$. Because the oscillator frequency $f_o$ cancels in the charge balance equation, it can be set for any convenient integration time without changing the converter resolution. It could even be tuned for maximum interference noise rejection through integration.

## 9-4 Dual-Slope Converter

In the dual-slope technique the input voltage is first converted to a charge by integration over a set period, and then the quantity of charge is determined by counting the number of charge units required to discharge the integration capacitor. In this section the basic operating principles of the dual-slope ADC are presented and the relationship between the readout count and the input voltage is developed.

### Basic Operation

A typical dual-slope ADC is shown in figure 9-19. The complete conversion takes place in three phases; auto zero, signal integration and reference integration. At the beginning of the conversion, in the auto zero cycle, switch $S_1$ connects the input of the converter to common, and $S_2$ closes to allow the

**Fig. 9-19.** Dual-slope ADC. During the auto zero phase, error information (buffer and integrator offset voltages, etc.) is stored on the auto zero capacitor $C_{AZ}$, and the integrator output $v_C$ is forced to zero. During the signal integration phase $v_{in}$ is integrated for a fixed number of clock pulses. In the reference phase reference voltage $V_r$ is connected to the integrator, and the integration capacitor discharges towards zero. The number of clock pulses required to discharge the integrating capacitor to zero is directly proportional to $v_{in}$.

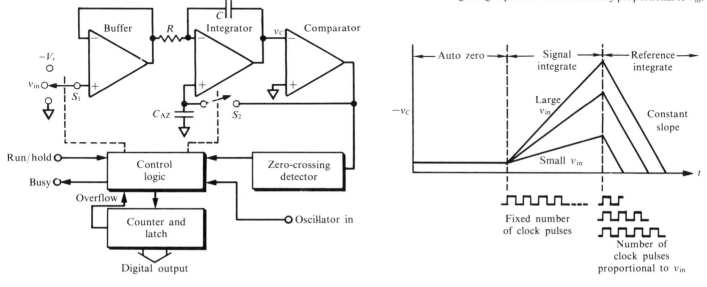

auto zero capacitor $C_{AZ}$ to charge. The auto zero capacitor then charges from the analog offsets in the system until the integrator output $v_C$ is zero, and the rate of change of $v_C$, $dv_C/dt$, is zero. The auto zero phase is usually one complete cycle of the counter (1000 clock pulses for a 3½ digit counter, 10 000 for 4½, etc.), but timing for this phase is not critical. At the end of the auto zero cycle (counter overflow), the control logic connects $v_{in}$ to the input and opens switch $S_2$. The input signal is then integrated for one complete counter cycle. Timing for this phase begins when the comparator indicates the integrator output has crossed zero. Counter rollover to all zeros (overflow) again indicates the end of the signal integration phase. As shown in figure 9-19, the integrator output $v_C$ is directly proportional to $v_{in}$ at the end of this phase. The control logic then changes switch $S_1$ to the reference voltage $V_r$ position, and the integration capacitor is discharged towards zero with a constant current of $V_r/R$. The number of clock pulses required for the integrating capacitor to discharge to zero is counted and is proportional to $v_{in}$. The end of the conversion occurs when the comparator indicates that $v_C$ has reached zero. The dual-slope converter is so named because of the shape of the integrator output voltage shown in figure 9-19.

The waveforms in figure 9-19 are shown for a positive $v_{in}$ and a negative $V_r$. Signals of either polarity can be accommodated if $V_r$ can be made bipolar. In many converters this is done with a single reference voltage by charging a capacitor to $V_r$ during the auto zero cycle. The polarity of $v_{in}$ is then sensed by the comparator during the signal integration cycle, and the capacitor charged to $V_r$ is connected by switches with the polarity required to discharge $v_C$ during the reference integration stage.

## Readout Relationship

The relationship between the readout count $n_r$ and the input voltage $v_{in}$ reveals many of the advantages of the dual-slope technique. At the end of the integration cycle, the charge on capacitor $C$ (if offsets are nulled during auto zero) is $q_C = Cv_C$, which is given by

$$q_C = \frac{v_{in}}{R} \Delta t = \frac{v_{in}}{R} \frac{n_m}{f}$$

where $\Delta t$ is the integration time equal to the maximum number of counts $n_m$ divided by the oscillator frequency $f$. Typical values for $n_m$ and $f$ are 10 000 and 120 kHz for an integration time of 83.33 ms. During the reference cycle the charge $q_C$ is discharged to zero in a time $n_r/f$, where $n_r$ is the readout count. Thus

$$q_C = \frac{V_r}{R} \frac{n_r}{f} = \frac{v_{in}}{R} \frac{n_m}{f} \qquad \text{or} \qquad n_r = \frac{v_{in}}{V_r} n_m \qquad (9\text{-}1)$$

For a 10 000 count cycle, $n_r = 10\,000\,v_{in}/V_r$. Note from equation 9-1 that the capacitance $C$, the resistance $R$, and the oscillator frequency $f$ do not influence the readout. The reason, of course, is that the reference and the input signals are integrated by the same integrators, timed by the same oscillator, and referenced to the same comparator threshold voltage. Voltage and current offsets in the buffer amplifier and integrator are balanced during the auto zero cycle.

The dual-slope technique inherently provides excellent noise rejection because of the signal integration. Interference noise from the 60-Hz power line can be almost eliminated by choosing the integration time to be an integral number of power-line cycles. The major disadvantage of the dual-slope technique is the rather long conversion time ($\sim 250$ ms for a three-cycle converter of 83.33 ms/cycle).

Further evolution of the dual-slope concept has included the "quad-slope" converter which provides four phases of integration and very high precision. Inexpensive *IC* forms of the dual-slope (and quad-slope) converter are the basis of many of the digital panel meters and digital multimeters in use today.

## 9-5 Time-Domain Operations

Analog signals are often converted to time domain signals for purposes other than analog-to-digital conversion. Once the signal is time encoded, it can be transmitted over long distances, recorded, or its frequency can be multiplied, divided, or shifted with little influence from electrical noise. The versatile phase-locked loop is shown in this section to be capable of a variety of operations on time-domain signals. In many cases it is desirable to convert a time-domain signal to an analog signal for display purposes or to perform various control functions. Various frequency-to-voltage converters and the time-to-amplitude converter are useful circuits for these functions. Many of the circuits described in this section are very useful in frequency modulation and demodulation schemes as described in section 9-6.

### Voltage-Controlled Oscillators

A circuit that produces an output frequency proportional to a dc control voltage is a **voltage-controlled oscillator** (**VCO**). The VCO performs an analog-to-time domain conversion. The voltage-to-frequency converter described in section 9-3 is, of course, one type of voltage-controlled oscillator. It produces a pulse output frequency proportional to the input voltage. Other types of VCOs are based on the development of sine-wave outputs,

**Note 9-2.    Transistor Constant Current Source.**
A popular constant current source used in integrated circuits such as the VCO of figure 9-20 is shown below. The current to be controlled is the collector current $I_{C2}$ of transistor $Q_2$. Transistor $Q_1$ is connected as a diode and acts as temperature compensation for transistor $Q_2$. The reference current is $I_R = (V^+ - V_{BE1})/R$. Since the base of $Q_1$ is connected to the base of $Q_2$, $V_{BE2} = V_{BE1}$, and if the transistors are identical, $I_{B1} = I_{B2}$. Then if the transistors have equal current gains, $I_{C1} = I_{C2}$. If both current gains are large, $I_{C1} \approx I_{C2} \approx I_R$, and the current through $Q_2$ is controlled. If the temperature rises, the nearly matched current gains of $Q_1$ and $Q_2$ both increase. Transistor $Q_1$ then draws more current. This reduces the base current of $Q_2$ to compensate for the increased current gain of $Q_2$. The temperature variation of $V_{BE1}$ is usually small compared to $V^+$, and the constant current is thus stabilized against temperature variations.

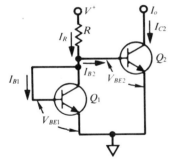

square-wave outputs or triangular-wave outputs. With most of these VCOs, the change in frequency from a base or center frequency depends on the sign and magnitude of the signal voltage.

The state variable oscillator described in section 8-5 is a VCO based on a sine-wave oscillator. Its output frequency can be controlled over a 100:1 range by the control voltage. The VCO of the function-generator type described in section 6-5, produces a triangular- and square-wave output with voltage-controlled frequency. With proper design it can also provide a 100:1 dynamic range.

A popular integrated circuit VCO is illustrated in figure 9-20. This VCO is also of the function-generator variety. The VCO consists of a precision constant current source (see note 9-2) and a Schmitt trigger. The current source alternately charges and discharges external capacitor $C$ between the two threshold levels of the Schmitt trigger. The Schmitt trigger output state controls the direction of the constant current with respect to the external capacitor. Because the capacitor is charged and discharged by the same current, the VCO produces a highly linear triangular-wave output as well as a square-wave output. This simple VCO has an output frequency adjustable over a 10:1 range with a triangular-wave output linear to better than 0.5%.

## Time-to-Analog Converters

Several input transducers used in scientific experiments produce a time-domain signal directly. Such devices as photomultiplier tubes and radioactive particle detectors are direct radiation intensity-to-frequency converters. Often it is desirable to display or record the frequency on a strip

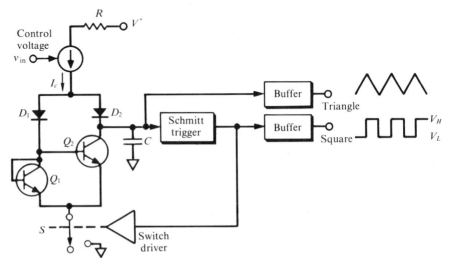

**Fig. 9-20.**    Voltage-controlled oscillator. Initially transistors $Q_1$ and $Q_2$ are off, and capacitor $C$ is charged by the control current $I_c$. When the HI threshold of the Schmitt trigger is reached, the Schmitt output closes switch $S$ and grounds the emitters of $Q_1$ and $Q_2$. Transistors $Q_1$ and $Q_2$ now conduct, and capacitor $C$ is discharged by the same control current $I_c$ until the LO threshold of the Schmitt trigger is reached. The cycle then repeats. The output frequency $f_o$ is determined by the control voltage and the values of $R$ and $C$ according to $f_o = 2[V^+ - v_{in}]/RCV^+$. This circuit is used in many phase-locked loop ICs and in the popular 566 VCO IC.

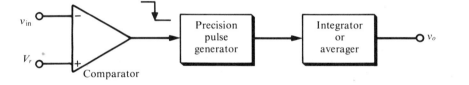

**Fig. 9-21.** Basic frequency-to-voltage converter. Each time the input signal crosses the comparator threshold, the comparator output triggers a precision pulse generator that sets up a reproducible current or voltage pulse. Then these pulses are either integrated for known periods to yield the average frequency for that period or they are low pass filtered to provide a dc output proportional to the input frequency.

chart recorder or other analog measurement device. Circuits that convert frequency-domain signals to analog signals are known as **frequency-to-voltage converters** or **count rate meters**. In some cases it is desirable to have an analog output proportional to the time interval between pulses or proportional to the pulse width. Devices that accomplish this latter function are often called **time-to-amplitude converters**. In other cases such time-to-analog conversion techniques are used to recover signals that have been transmitted, recorded, or otherwise manipulated in one of the time domains.

**Frequency-to-Voltage Converter.** Many frequency-to-voltage converters use the basic design shown in the block diagram of figure 9-21. Specific converters differ in the makeup of the precision pulse generator and in the use of integrating or low pass filtering techniques on the output. A popular frequency-to-voltage conversion technique is illustrated in figure 9-22. If the low pass filter has a long enough time constant, the steady dc output voltage obtained is directly proportional to the input frequency $f_{in}$. For reliable operation the input period $1/f_{in}$ must be less than the pulse width of the monostable $t_p \simeq RC$. This places an upper limit on the input frequency that is typically 100 kHz (although some converters may operate as high as several megahertz). The frequency-to-voltage ratio of the converter can be varied by changing the MS pulse width $t_p$, the reference current $i_s$, or the feedback resistor $R_f$. Note that the components of the frequency-to-voltage

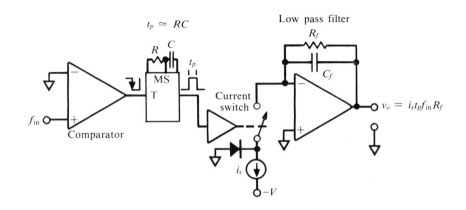

**Fig. 9-22.** Frequency-to-voltage converter. Each time the input signal goes negative the comparator triggers a monostable multivibrator (MS), which closes the current switch for a time $t_p \simeq RC$. The average current to the low pass filter is $i_s t_p f_{in}$. Thus $v_o$ is directly proportional to $f_{in}$.

converter are exactly the same as the components of the current-to-frequency converter of figure 9-13. Many integrated circuit VFCs can be operated in either mode just by changing a few connections.

**Time-to-Amplitude Converter.**   The time-to-amplitude converter is used to produce an analog output pulse with height proportional to the time between two input pulses. It finds many applications in nuclear measurements and in time-correlated spectroscopic measurements. A circuit diagram of a time-to-amplitude converter is shown in figure 9-23. The integrating capacitor $C$ charges during the time between Start and Stop pulses. The voltage on the capacitor is then switched to the output briefly to produce the output pulse. Flip-flops 1 and 2 insure that a valid pulse pair starts and stops the integrator. The delay allows the output pulse to appear a variable length of time after the stop pulse has been detected. Overrange detection circuitry is usually included to reset the system automatically if no stop pulse is detected within a certain time limit.

The pulse pair time resolution of commercial time-to-amplitude converters may be as low as a few picoseconds. Thus the logic components and switches usually are either high-speed devices or are made from discrete components. The output of the time-to-amplitude converter is often connected to a single or multichannel analyzer for further data treatment and storage.

**Fig. 9-23.**   Time-to-amplitude converter. A HI→LO transition at the Start input sets FFI and closes the current switch to allow the reference current $i_s$ to charge integrating capacitor $C$. Integration continues until a HI→LO stop pulse is received. The stop pulse causes $Q$ of FF2 to go LO, opening the current switch. The voltage $v_C$ on the integration capacitor after the stop pulse is $v_C = i_s \Delta t / C$, where $\Delta t$ is the time between start and stop pulses. The stop pulse triggers a delay that subsequently triggers the read timer. The read timer closes the output switch for a short time, allowing the amplified capacitor voltage to be transmitted as a pulse to the output. The output voltage during the pulse is $v_o = i_s \Delta t (R_1 + R_2)/CR_2$. After the read timer opens the output switch, it triggers the reset timer, which shorts integrating capacitor $C$ and clears the flip-flops.

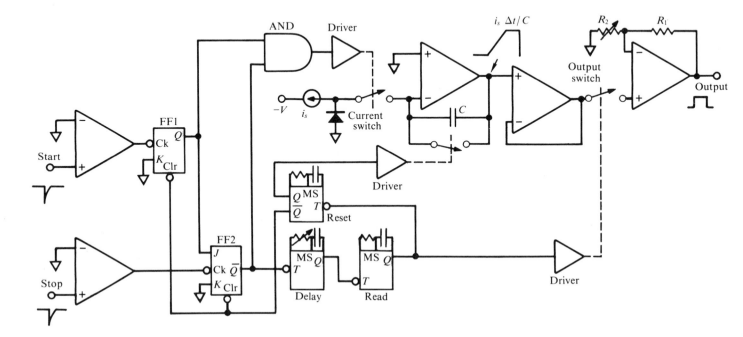

## Phase-Locked Loop

The **phase-locked loop (PLL)** is an extremely versatile circuit that is used for frequency comparison and synchronization, frequency multiplication and division, frequency-to-voltage conversion (FM demodulation), frequency shifting, and AM demodulation. The basic phase-locked loop consists of an analog multiplier, a low pass filter, and a voltage-controlled oscillator (VCO) as illustrated in figure 9-24. The ability to insert various components into the feedback loop before or after the VCO makes the PLL the basic building block for a variety of useful circuits.

The analog multiplier and low pass filter make up a phase angle-to-voltage converter (see note 9-3). The amplifier output $v_o$ is zero if the VCO output is of identical frequency to and exactly 90° out of phase with the input signal $f_{in}$. If the frequency of the two signals begins to differ, a change in phase angle results, and the amplifier output voltage changes. The amplifier output voltage is applied to the VCO with a sense such that the phase shift is decreased. Thus as the frequency of $f_{in}$ changes, the VCO frequency is controlled to track it precisely. When the loop is locked, the input signal and the VCO output are 90° out of phase as shown in the waveforms in figure 9-25. The frequency of an unstable power VCO can be made highly precise by incorporating it into a PLL and using a precision low-power oscillator as $f_{in}$. The VCO output is of exactly the same frequency and stability as the precision oscillator but capable of driving much heavier loads.

The PLL can be used as a narrow bandwidth tracking filter for recovering weak signals buried in noise. If the bandwidth of the loop is narrow enough, the signal-to-noise ratio at the VCO output can be much higher than that of the input signal, and the PLL will track any frequency drifts of the input signal.

Another application of the PLL is in frequency synthesis. The VCO output can be made to be a multiple of the input signal by locking to a harmonic of the input signal or by inserting a scaler in the loop as shown in figure 9-24. The harmonic locking technique is limited in the range of multiples that it can provide because harmonics are weaker in amplitude than the fundamental. With the scaler technique the VCO frequency is $f_o = Nf_{in}$ since $f_o/N$ is controlled to equal $f_{in}$. The scaler technique requires excellent

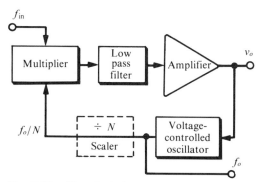

**Fig. 9-24.** Phase-locked loop. The input signal and the VCO output (or the scaled VCO output) are multiplied together by an analog multiplier. The multiplier output is low pass filtered and amplified. When locked, the frequency $f_o$ (or $f_o/N$) is equal to the input signal frequency $f_{in}$.

**Note 9-3.  Phase Detection.**
When two sinusoidal waveforms are multiplied, the resulting waveform contains frequency components equal to the sum and difference frequencies. Consider $v_1 = V_1 \sin(\omega_1 t + \theta_1)$ and $v_2 = V_2 \sin(\omega_2 t + \theta_2)$. The product $v_p$ of $v_1$ and $v_2$ is given by

$$v_p = \frac{V_1 V_2}{2}[\cos(\omega_1 t + \omega_2 t + \theta_1 + \theta_2) - \cos(\omega_1 t - \omega_2 t + \theta_1 - \theta_2)]$$

If the multiplier output is filtered to eliminate the sum term and $\omega_1 = \omega_2 = \omega$, the filter output $v_f$ is given by

$$v_f = \frac{V_1 V_2}{2} \cos(\theta_1 - \theta_2)$$

Thus the filter output voltage is proportional to the phase difference between $v_1$ and $v_2$.

(a)

(b)

(c)

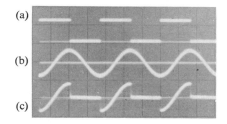

**Fig. 9-25.** Waveforms for a phase-locked loop. The VCO output in (a) is 90° out of phase with the input signal in (b). The multiplier output in (c) is a minimum when the phase shift is 90°. However, the phase shift is not exactly 90° because some error signal is required to drive the VCO to the required frequency. High gain in the loop keeps this error signal very low.

filtering because the VCO is running at $N$ times the frequency of the input signal. Any sum frequency component that appears at the input of the VCO causes a variation in its output frequency around the desired multiple. A wide range of multiplication factors can be obtained with a scaler in the loop, and automatic digital control over the output frequency can be achieved with a programmable scaler. Complete PLL synthesizers are available in integrated circuit form. They contain the PLL, the frequency divider, and the logic to program the divider. Only an external crystal need be added.

Frequency division can be accomplished by inserting a frequency multiplier instead of a frequency divider in the loop; frequency division can also be accomplished easily by conventional digital techniques.

A PLL is also capable of shifting an input signal frequency $f_{in}$ by a specific amount. A circuit to accomplish this is shown in figure 9-26. The frequency to be shifted $f_{in}$ is multiplied by the VCO output, which is initially set to the approximate shifted frequency ($f_{in} + \Delta f$). The filtered output of the first multiplier is fed to a second multiplier, where it is multiplied by the desired offset, or shift, frequency. The output of this second multiplier provides the error signal to close the loop. The output of the VCO is locked to $f_{in} + \Delta f$.

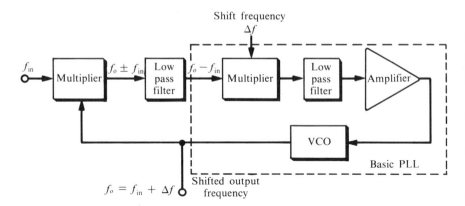

**Fig. 9-26.** Frequency shifting with a phase-locked loop. Adding an external multiplier and low pass filter to the basic PLL can translate the frequency of $f_{in}$ by a small amount $\Delta f$. When the system is locked, $f_o = f_{in} + \Delta f$.

The basic PLL is inherently a frequency-to-voltage converter or FM demodulator. The amplifier output voltage $v_o$, which is the control voltage to the VCO, is related to the input frequency by the transfer function of the VCO. The PLL can also be used as a synchronous demodulator for AM signals as shown in figure 9-27. Here the VCO output provides a phase-locked reference for synchronous demodulation. This complete circuit including the AM multiplier and low pass filter is also available in a single integrated circuit.

The versatile phase-locked loop is available in a variety of IC packages. One of its major uses, FM demodulation, is described in the next section.

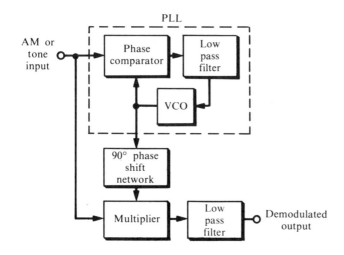

**Fig. 9-27.** Synchronous AM demodulation with a phase-locked loop. The PLL locks to the AM carrier so that the VCO output has the carrier frequency without amplitude modulation. The VCO output is phase-shifted and multiplied with the AM input. Since the PLL locks 90° out of phase with its input signal, the phase-shifted PLL output and the AM signal are in phase at the multiplier, and the filter output is proportional to the amplitude of the input signal.

## 9-6    Frequency Modulation and Demodulation

The alteration of the instantaneous frequency of a carrier wave by a signal is **frequency modulation**. The amplitude of the signal determines the extent of the carrier frequency change, and the rate of signal amplitude change determines the rate at which the carrier frequency changes. A simple method of carrying out frequency modulation is to connect the signal to a voltage-controlled oscillator. The oscilloscope traces of a signal and the FM carrier output from a VCO are shown in figure 9-28. Frequency modulation can also be carried out with a voltage-to-frequency converter.

### Frequency-to-Voltage FM Demodulators

One useful approach for FM demodulation is to convert the frequency to a voltage by an averaging frequency-to-voltage converter or by a phase-locked loop. The frequency-to-voltage converter discussed in section 9-5 and illustrated in figure 9-22 is often used when the carrier frequency is changing at a relatively slow rate or when the desired information is an average rate over several hundred cycles. Circuits of this type can provide a simple analog output for pulse counting measurements when the average frequency over a few seconds or tenths of seconds is desired.

   If faster changes must be measured, the phase-locked loop can be used. Inexpensive IC phase-locked loops can operate in the frequency range 0.001 Hz —500 kHz although general purpose PLLs that operate up to 50 MHz are readily available. The basic PLL of figure 9-24 can be used directly for FM demodulation. The FM carrier is connected to the multiplier, and the amplifier output signal is the demodulated output. The VCO base frequency

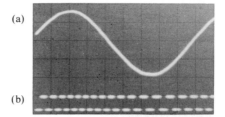

(a)

(b)

**Fig. 9-28.** Waveforms in frequency modulation. The signal in (a) is connected to a VCO to produce the modulated carrier waveform shown in (b). Note that the frequency of the FM carrier is greatest when the signal amplitude is at its maximum value and decreases to its minimum frequency at the signal minimum.

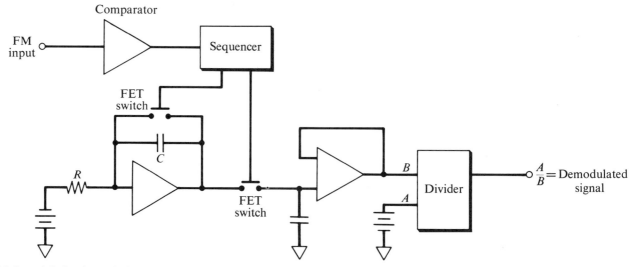

**Fig. 9-29.** FM demodulation by period measurement. The FM carrier, shaped by a comparator, controls a sequencer that gates a linear integrator for the duration of one period of the carrier wave. At the end of the period the integrator output is sampled and held, and the integrator is rapidly reset. The sample-and-hold output is then a series of analog levels proportional to the instantaneous period of the carrier. The reciprocal of the sample-and-hold output, obtained from an analog divider, provides an analog signal proportional to the frequency of the carrier.

**Fig. 9-30.** Waveforms for period demodulator. In (a) the modulating signal (upper trace) is shown with the modulated carrier wave (lower trace). The lower trace in (b) and (c) is the output of the sample-and-hold circuit. The stepped amplitudes represent the period of the carrier at each cycle. The upper trace in (b) is the output of the sample-and-hold circuit after low pass filtering. Note that it is not a sinusoidal waveform. The upper trace in (c) is the low pass filtered divider output and is thus the fully demodulated signal.

should be adjusted to be the center of the input signal frequency range. Specialized PLL circuits for tone decoding, FM stereo multiplex decoding and other purposes are commercially available in IC form.

## Period Demodulation

Another approach to demodulating a carrier wave is to measure the period of every cycle. This approach, like the phase-locked loop, is useful when relatively rapid changes in the carrier frequency are to be observed. A circuit for FM demodulation by period measurement is shown in figure 9-29. As can be seen from the waveforms in figure 9-30, the divider output is directly proportional to the frequency of the input signal measured over only one cycle. The primary limitation of a period demodulator is its small dynamic range. A demodulation frequency range of 10:1 is readily achieved; a range of 100:1 is possible with extremely careful design. These limitations result

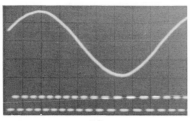

(a)

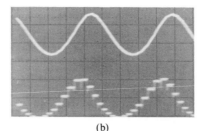

(b)

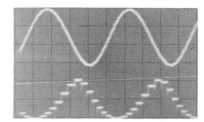

(c)

from the small dynamic range of the analog divider and the finite time required to stop the integrator, activate the sample-and-hold circuit, and reset the integrator. Despite these limitations, period demodulation is a valuable technique for measuring rapid frequency changes over a small range.

The period can also be measured by digital techniques as described in section 9-1. Digital methods can be used for FM demodulation provided the successive period measurements can be stored or read out rapidly.

## 9-7 Application: Touch-Tone Decoder

Touch-Tone telephone dialing is based on the generation of a frequency code for each numeral dialed. To minimize dialing error from noise frequencies, the code for each number is a pair of frequencies that must exist simultaneously. The frequency code for the dialing numerals is shown in figure 9-31. The decoding circuit must be able to detect all seven frequencies and decode each frequency pair into the corresponding dialed number. This is most often accomplished with the AM-detection phase-locked loop circuit of figure 9-27. Single-tone detectors based on the PLL are available in IC form. As shown in figure 9-32, only a few components are required to complete the circuit. The complete Touch-Tone decoder requires seven single-tone decoders, one for each of the seven encoding frequencies. When the numeral 1 is being dialed, the 697 Hz and 1209 Hz tone decoder outputs are both LO. This combination is detected by a logic gate connected to these two decoders. The other eleven tone combinations are similarly detected by eleven other logic gates.

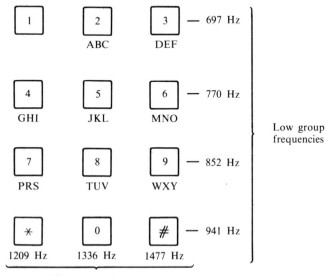

**Fig. 9-31.** The Touch-Tone keypad and corresponding frequencies. Depression of a key produces a dual-tone signal—one of the high frequencies from the column and one of the low frequencies from the row. If the 4 key is depressed, the tones 770 Hz and 1209 Hz are generated.

The Touch-Tone system achieves very high reliability through the use of simultaneous dual tones and the careful selection of frequencies that are not harmonically related. The decoder is interesting because all three domain classes are involved. The Touch-Tone encoder is an interesting application of digital techniques as we shall see in chapter 13.

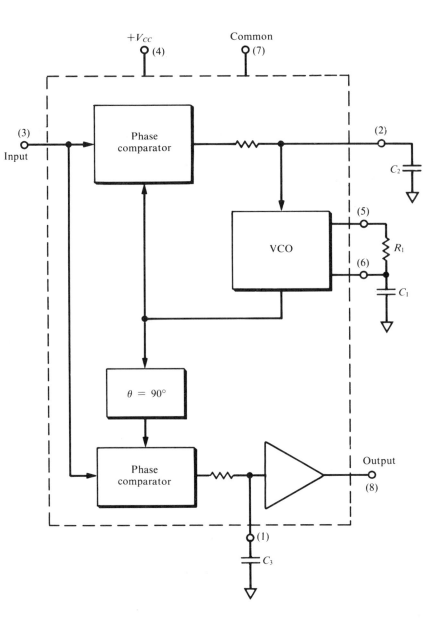

**Fig. 9-32.** A 567-type tone decoder IC. The detection frequency is equal to $1/(R_1C_1)$. When the input frequency is within the capture bandwidth (determined by $C_2$), the PLL is locked and the lower phase comparator AM demodulates the signal. The demodulated signal is filtered and connected to an analog comparator with a logic output. A LO logic level at the output indicates the presence of the selected tone at the input. $R_1$ should be between 2 kΩ and 20 kΩ, $C_1$ may select any frequency from 0.01 Hz to 500 kHz, $C_2$ is adjusted for the desired bandwidth for an input voltage greater than 250 mV rms, and $C_3$ is at least twice the value of $C_2$.

## Suggested Experiments

**1.  Precision time base.**
A crystal oscillator and a decade frequency divider are connected as a time base. Observe the time base output(s) for all available frequencies with the scope, and check for accuracy with the frequency meter. Observe the operation of reset and preset of the divider if available.

**2.  Scaling and ratioing in the frequency meter.**
Use a timer/frequency meter in conjunction with several decades of frequency scaling to measure the ratios of various frequency sources to the desired degree of accuracy. In a period measurement mode, use the prescaler to obtain measurements of multiple period average. Also use the prescaler to extend the range of the counter without overflow.

**3.  Schmitt trigger.**
Connect a comparator with positive feedback as in figure 9-8 to produce a Schmitt trigger. Observe the degree of hysteresis for several ratios of $R_1/(R_1 + R_2)$. Connect the comparator to a transducer output to detect discrete events. Connect the comparator output to a counter input to count the events.

**4.  Charge-to-count converter.**
Connect a charge-to-count converter as shown in figure 9-11. The charge increment generator is obtained from a monostable multivibrator, and the increment control generator is a counting gate between an oscillator and the MS trigger input. Apply a previously charged capacitor to the input as $q_u$, enable the start, and read the resulting count. Determine the linearity of response of this circuit. Modify this circuit to form a current-to-frequency converter by connecting the comparator to control both start and stop of the increment generator. Measure the relationship of $f_o$ to $i_{in}$.

**5.  Voltage-to-frequency converter.**
Connect and characterize an IC VFC circuit. Use both current and voltage inputs. Use the VFC with a frequency meter to make an integrating DVM. Observe the integrating nature of the DVM by measuring the standard deviation of a signal with random noise for several integration times, and observe the elimination of fixed-frequency noise for integration times that are integral multiples of the noise period.

**6.  Dual-slope converter.**
Connect and characterize an IC dual-slope converter. From observation of the control waveforms, determine the clock frequency, the integration time, and the count value at each change in function in the measurement sequence.

**7.  Phase-locked loop.**
Connect a phase-locked loop IC for operation. Test the response of the VCO part of the PLL. Connect and characterize the PLL for operation as a narrow bandwidth tracking filter and as a frequency multiplier. Test the range of control obtainable in each case.

**8.  Frequency modulation and demodulation.**
Use a VCO to obtain a frequency-modulated signal source. Connect and characterize both the PLL and frequency-to-voltage converter as FM demodulators for this signal. For the frequency-to-voltage converter, connect the VFC from experiment 5 to operate in this mode.

## Questions and Problems

**1.**  (a) For the period measurement of a 28-kHz signal, how many counts are displayed if the clock signal counted is 1.000 MHz? (b) In the period measured in (a) what percent uncertainty results from the $\pm 1$ count uncertainty? (c) How many periods should be averaged (decade numbers) to obtain a precision of 0.1% or better?

**2.**  For a frequency measurement of a 65-Hz signal, what gate period is required to give a precision of 0.1% or better?

**3.**  A voltage-to-frequency converter can be used in conjunction with a period measurement to give a digital readout proportional to the reciprocal of the input voltage, $1/V$. (a) With a 100-kHz/V VFC how many counts would be accumulated for a 650-mV signal if the period meter clock frequency is 1.000 MHz? (b) How many digits should the counter have to display $10^4$ periods of the signal in (a) without overflow?

**4.**  A digital frequency ratio measurement is to be made on two signals that are both about 100 Hz. It is desired to know the frequency ratio to within one part per thousand. What scaling factor should be used with the count gate control signal? Approximately how much time will be required for each frequency ratio measurement?

**5.**  Calculate the relative error due to the $\pm 1$ count uncertainty when the frequency ratio circuit of figure 9-3 is used to measure the frequency of 50-Hz, 3000-Hz and 250-kHz signals. The standard reference frequency that operates the counting gate is exactly 1.00 Hz. Similarly, calculate the relative error in the period measurement of the same signals with a standard reference frequency of exactly 1.00 MHz.

**6.** Over what range of differential input voltages is a comparator with a gain of 40 000 and output limits of $\pm 12$ V not at limit?

**7.** The basic comparator of figure 9-6a has logic-level outputs of $+5$ and 0 V. (a) If the open-loop gain is $1.6 \times 10^4$, what is the threshold window? (b) For a signal changing at 5.0 V/s through the window, how uncertain is the transition time?

**8.** In the Schmitt trigger circuit shown in figure 9-8, the op amp has an open-loop gain of $A = 1.0 \times 10^4$ and $R_1 = R_2 = 1$ k$\Omega$. The amplifier output limits are $V_H = +12$ V, $V_L = -12$ V. (a) Sketch the output voltage $v_o$ as a function of $v_{in}$. (b) What is the hysteresis lag $\Delta V$?

**9.** Design a Schmitt trigger circuit for a $\pm 12$-V op amp that has a 1-V hysteresis lag and a threshold of $+2.0$ V. Use reasonable resistance values (1 k$\Omega$ to 1 M$\Omega$).

**10.** In the charge-to-number converter of figure 9-11, the charge increment generator is made from a constant current source (5-V source through a 10-k$\Omega$ resistor) and an analog switch operated by a square-wave generator. The generator frequency is 100 kHz. (a) What is the charge content of each charge pulse from the charge increment generator? (b) If the unknown charge was 25.0 $\mu$C, how many reference charge increments would be required to balance the unknown charge? (c) After the Balance/Sample switch is thrown to the Balance position, how much time would be required before balance is reached?

**11.** In a charge-to-number converter of figure 9-11, it is desired to balance the converter in 25 ms for the maximum full-scale charge. It is also desired to have 12 000 counts for the full-scale reading. Capacitor $C = 0.1 \mu$F, and the op amp output limits are $\pm 12$ V. (a) What is the maximum full-scale charge that can be measured? (b) What should be the charge content of each pulse from the charge increment generator? (c) Describe how such a generator could be constructed from a 10-V source, a resistor, an analog switch, and a square-wave generator. Give the resistance required and the generator frequency.

**12.** A frequency of 1.5 kHz was obtained from the current-to-frequency converter of figure 9-13. If $C = 0.1 \mu$F and the monostable timing components were $R = 1$ k$\Omega$, $C = 10$ nF, and if $i_s$ was derived from a 10-V source and a 1-k$\Omega$ resistor, what was $i_{in}$? What is the transfer function in Hz/A? What determines the maximum full-scale current?

**13.** A voltage-to-frequency converter (figure 9-13) is to have a transfer function of 10 kHz/V and a maximum full-scale voltage of 10 V. If $R_{in} = 100$ k$\Omega$, find the value for the charge of each reference charge pulse. Choose appropriate values for the monostable multivibrator timing components and the reference current $i_s$. Remember that $t_p \leq (1/1.5)f_{fs}$.

14. The voltage-to-frequency converter of figure 9-13 was operated as an analog-to-digital converter as shown in figure 9-16. The reference charge of the VFC was 25 nC. The counter had a 0.1-s clock period. For $v_{in} = 250$ mV and $R_{in} = 10$ k$\Omega$, how many counts were accumulated?

**15.** (a) In the current-to-frequency converter of figure 9-19, what effect would a constant 100-mV offset in the comparator have on the output frequency? Explain. (b) In the current-to-frequency converter, the capacitance of the integrating capacitor does not enter into the final equation. Why? What effect does the capacitance have on the required sensitivity of the comparator? Explain.

**16.** In the charge-balancing ADC of figure 9-18, a four-digit divider and counter and a 1.00-mHz oscillator were used. The reference current was 1.00 mA and $R_{in} = 100$ k$\Omega$. (a) What is the full-scale voltage? (b) If $v_{in} = 1.00$ V, what count would accumulate? (c) What is the total integration time?

**17.** For a dual-slope converter, it is often desirable to make the integration period exactly one period of the 60-Hz line frequency. If a 3½ digit BCD counter (full scale 1999) is used and the signal is integrated until the two most significant bits of the counter are 1, what clock frequency must be used?

**18.** In the dual-slope converter of figure 9-19, $n_m = 1000$, $R = 10$ k$\Omega$, $V_r = 1.00$ V, $f = 100$ kHz and $C = 0.1 \mu$F. (a) If $v_{in} = 350$ mV, what is the readout count? (b) For $v_{in} = 350$ mV, what is the charge on capacitor $C$ at the end of the integration period? (c) What is the capacitor voltage $v_C$ at the end of the integration period?

**19.** In the dual-slope converter why must the reference voltage $V_r$ be of greater magnitude than the maximum input voltage $v_{in}$?

**20.** In the 566-type voltage-controlled oscillator of figure 9-20, timing resistor $R$ and timing capacitor $C$ are supplied externally by the user. If $V^+ = 10$ V, find appropriate values of $R$ and $C$ to give an output frequency of 10 kHz at $v_{in} = 8$ V given the restriction that 2 k$\Omega < R < 20$ k$\Omega$.

**21.** (a) In the VCO of figure 9-20, show that the slope of the output frequency vs. $v_{in}$ relation is $-4.26$ kHz/V for $R = 10$ k$\Omega$, $C = 0.047$ $\mu$F and $V^+ = 10$ V. (b) What values of $R$ and $C$

would give a slope of $-6.0$ kHz/V with $2$ k$\Omega < R < 20$ k$\Omega$? (c) For $R = 10$ k$\Omega$, $C = 0.047$ $\mu$F and $V^+ = 6.0$ V, what range in frequencies can be obtained with the restriction that $0.75 V^+ \leq v_{in} \leq 0.98 V^+$? (d) For $R$ and $C$ as in part (c), what supply voltage is necessary for the VCO sensitivity to be $10$ kHz/V?

**22.** In the frequency-to-voltage converter of figure 9-22, it is desired to obtain a transfer function slope of $1$ V/kHz. The reference current $i_s$ is made from a $-5$-V source and a $10$-k$\Omega$ resistor. Find appropriate and reasonable values for $R$ and $C$ (monostable), $R_f$, and $C_f$.

**23.** In the time-to-amplitude converter of figure 9-23, $i_s = 100$ $\mu$A, $C = 0.01$ $\mu$F, and the output follower has a gain of 50. (a) If the follower output is limited to $\pm12$ V and the circuitry after the follower can resolve $1.0$ mV, what range of $\Delta t$ values can be measured? (b) What are the voltages across capacitor $C$ for the highest and lowest $\Delta t$ value found in part (a)? (c) For the maximum $\Delta t$ found in (a), it is desired to obtain a higher voltage across capacitor $C$ at the end of $\Delta t$. How could this be accomplished without influencing the range of $\Delta t$ values measurable?

**24.** Compare the phase-locked loop AM demodulator of figure 9-27 with the lock-in amplifier of figure 8-27. How is the reference signal derived in each case? For the AM demodulator, draw waveforms for the AM input, the multiplier inputs, and the multiplier output.

**25.** Frequency multiplication with a phase-locked loop can be used to generate the local oscillator in CB radios. For channels 1–4 in a CB radio, the local oscillator must produce frequencies of 26.510 MHz (Ch. 1), 26.520 MHz (Ch. 2), 26.530 MHz (Ch. 3), and 26.550 MHz (Ch. 4). If a 10-kHz basic oscillator is available, describe how the PLL of figure 9-24 could be used to obtain the required frequencies. What does the channel selector dial on a CB radio change?

**26.** Frequency shifting with a phase-locked loop can also be applied to the generation of the local oscillator in a CB radio (see problem 25). Assume that the oscillator frequency for Channel 1 has been generated with a PLL frequency multiplier. Channels 2, 3, and 4 are to be generated by frequency shifting the Channel 1 local oscillator. A 10-kHz oscillator is available. Describe how the circuit of figure 9-26 could be used in this application. In this case what would the channel selector dial on the CB radio actually change? Would this circuit be any simpler than the frequency multiplier of problem 25?

# Chapter 10

# Logic Gates, Flip-Flops, and Counters

The amplitude of a digital signal must be in one of the two allowed states, HI or LO. Thus a digital signal level conveys the minimum amount of information that a signal could have and still be useful. The two states can be used to represent the two numerals (1 and 0) in the base 2 (binary) number system or the two states (TRUE and FALSE) in logic operations. Thus digital signals are also called **binary signals** or **logic-level signals**. The information represented by a single logic level is called a **bit**, which is a contraction of "binary digit." Larger amounts of information are represented or conveyed digitally by combinations of bits, which are called **words**. A combination of $n$ bits can have $2^n$ distinguishable states and thus can represent any of $2^n$ different numbers, characters, or other such specific pieces of information. The bits in a word can be sent in succession along a single channel of communication (a **serial digital signal**), or they can be sent along $n$ channels simultaneously (a **parallel digital signal**).

Because digital signals have just two states, all operations on them can be described as logic operations. The fundamental building block of digital circuits is thus the **logic gate**, a circuit that performs a simple AND (intersection) or OR (union) operation on two or more digital signals. In this chapter the basic logic functions and gates are introduced, and the algebraic expressions for these functions are summarized. Combinations of simple gates are shown to perform more complex functions such as equality comparison, addition, decoding, and multiplexing.

The basic digital memory circuit, the latch, is also made of simple logic gates. In turn, flip-flops are developed from latches, and combinations of flip-flops are used to make a variety of counters and shift registers. Almost all digital circuits from counters to computers are just combinations of logic gates—dozens, hundreds, or even millions of very simple functions.

## 10-1  Logic Gates

Two-level logic operations, however complex, are always combinations of three basic logic operations: AND, OR, and INVERT. The circuits that perform basic logic operations are logic gates. A gate produces an output voltage level (HI or LO) that is the result of a logic operation on the logic levels at the inputs. For example, an AND gate performs the AND logic function on its input signals.

The HI and LO voltage levels used in digital circuits are actually voltage ranges as shown in figure 10-1. In any given logic circuit family, the logic levels for all gate circuits are the same. The two voltage levels (HI and LO) represent the two logic levels, 1 or TRUE and 0 or FALSE. Usually the HI level is assigned to 1 or TRUE and the LO level to 0 or FALSE. The postulates of two-level logic operations, combinations, and equivalents are contained in Boolean algebra, in which logic functions can be expressed and manipulated as algebraic equations.

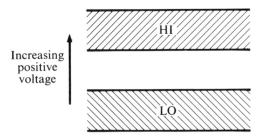

Increasing positive voltage

**Fig. 10-1.** Logic-level signal ranges. A voltage level of any value within the HI range or LO range is considered as HI or LO. The possibility of a level error due to noise decreases as the gap between the ranges increases. The voltage ranges for TTL logic are 0.0–0.4 V for LO and 2.4–4.0 V for HI.

## The AND Function

Consider the logic expression "if statement $A$ is true AND statement $B$ is true, then result $M$ is true." This is written in symbolic form as

$$A \cdot B = M \quad \text{or} \quad AB = M$$

All the possible combinations of conditions $A$ and $B$, and the result $M$ for each, are included in a **truth table** as shown in table 10-1 for the AND operation. There are four different combinations of conditions for the variables $A$ and $B$, and the $M$ column will have a 1 (TRUE output) only when $A = 1$ AND $B = 1$. A truth table is specific for the logic function performed. That is, the AND function always gives the truth table of table 10-1, and any function that has the same truth table as table 10-1 is by definition an AND function. The truth table can be expressed either in T and F or in 1 and 0. By convention, T and 1 are *always* equivalent.

The symbol for a circuit that performs the AND function is shown in figure 10-2. This symbol is used for any AND gate regardless of the actual gate circuit or the assignment of signal levels to the 1 and 0 states. An

**Table 10-1.** Truth table for the AND function.

| Variables | | Result |
|---|---|---|
| $A$ | $B$ | $M$ |
| **0** (F) | **0** (F) | **0** (F) |
| **0** (F) | **1** (T) | **0** (F) |
| **1** (T) | **0** (F) | **0** (F) |
| **1** (T) | **1** (T) | **1** (T) |

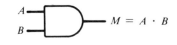

**Fig. 10-2.** The symbol for an AND gate. The output signal $M$ is the logical AND of input signals $A$ and $B$. That is, $M = A \cdot B$.

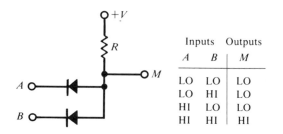

| Inputs | | Outputs |
|---|---|---|
| *A* | *B* | *M* |
| LO | LO | LO |
| LO | HI | LO |
| HI | LO | LO |
| HI | HI | HI |

**Fig. 10-3.** Diode AND gate and table of states. The application of a LO logic level ($\sim$ 0 V) to any input forward biases its diode and causes the output to be LO. Only when both inputs are HI ($\sim +V$ V) are both diodes reverse biased and the output HI. Additional diodes could be added to accommodate more input signals.

**Note 10-1.  Logic Conventions.**
The relationships among TRUE/FALSE, 1/0, and HI/LO and the use of these quantities in truth tables, tables of states, and logic circuits is often inconsistent in the literature. A rational convention that corresponds to the best current practice is followed in this text. This convention is as follows:

1. The relationship between TRUE/FALSE and 1/0 is always that TRUE = 1.

2. The truth table for a logic function is expressed in 1's and 0's (or T's and F's) and is independent of the implementation.

3. The table of states for a logic gate or circuit is always correct for that circuit (independent of the relationship between 1/0 and HI/LO) if expressed in HIs and LOs, not 1's and 0's.

4. The positive true convention (HI = 1) will be assumed unless otherwise indicated. When a signal is named for a particular condition or action, that condition will be true or that action produced when the signal is HI. When a gate circuit is given the name of a logic function (such as diode AND gate), it is assumed that the named function is performed on positive true signals.

5. The indication of LO-true signals in a logic diagram will follow the conventions described in note 10-2.

example of an AND gate circuit is the simple diode circuit shown in figure 10-3. The table of states for the diode AND gate indicates that a LO at any input produces a LO at the output. A comparison of this table with the AND function of table 10-1 shows that the diode AND gate performs the AND function if the HI logic level is assigned to the 1 or TRUE condition.

The assignment of the HI logic level to the TRUE condition is called **positive true logic** or **HI-true logic**. The current convention in logic circuits is that the positive true assignment is assumed unless the opposite, LO-true, is indicated. However, it is common and often unavoidable that both conventions are used in a single circuit. Therefore, we will use a convention that allows for both possibilities (see note 10-1).

The simple AND operation is basic to many of the operations in digital instruments and computers. The following example shows why it is often called a gate. Consider a two-input circuit. While one input is held at 1, the output follows the logic-level changes at the other input; that is, the gate is OPEN. On the other hand, if one of the inputs is held at 0, the output is 0 regardless of the other input signal level; that is, the gate is CLOSED.

## The OR Function

A circuit that produces a logic 1 output when $A = 1$ OR $B = 1$ OR both $A$ and $B = 1$ is said to perform the OR operation. Symbolically the OR function is written

$$A + B = M$$

where the $+$ sign is a logic symbol that is read as OR. The truth table and the circuit symbol for the OR function are shown in table 10-2. Note that $M = 1$ whenever $A$ is 1 OR $B$ is 1 OR both are 1. The output is 0 only when $A$ and $B$ are both 0.

The OR function can readily be implemented with diodes, as illustrated in figure 10-4. Note that the table of states for the gate corresponds to the truth table for the OR function for positive true signals. The OR gate can also function as a gate. If one input is held at 1, the output is 1 regardless of the levels at the other inputs.

**Table 10-2.**  Truth table for the OR function.

| Variables | | Result |
|---|---|---|
| *A* | *B* | $M = A + B$ |
| 0 | 0 | 0 |
| 0 | 1 | 1 |
| 1 | 0 | 1 |
| 1 | 1 | 1 |

| Inputs | | Outputs |
|---|---|---|
| $A$ | $B$ | $M$ |
| LO | LO | LO |
| LO | HI | HI |
| HI | LO | HI |
| HI | HI | HI |

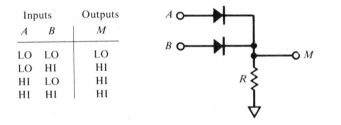

**Fig. 10-4.** Diode OR gate and table of states. When one input is HI, the corresponding diode is forward biased, and the HI input level (less the voltage drop across the diode) is transmitted to the output. Only when all inputs are LO are all diodes reverse biased and a LO output obtained. More diodes may be used for additional inputs.

## The NOT Function

The NOT function produces a variable that is exactly opposite in sense to the original. For example, if a variable $A$ is TRUE for the condition "the motor shaft is turning," the variable NOT $A$ will be TRUE for the condition "the motor shaft is not turning." The NOT of a variable is indicated by a bar over the symbol for the variable, e.g., NOT $A = \overline{A}$. Because $A$ and $\overline{A}$ are opposites of the same statement, when one is true the other must be false.

Operationally, the NOT function is achieved by reversing the logic level at the input. A HI at the input becomes a LO at the output, and vice versa. As we learned in chapter 4, the circuit that performs this function is called an **inverter**, and its symbol is as shown in figure 10-5. The circle at the output or input is a specific indication of the inversion of logic levels. A simple circuit that achieves the inversion operation is the transistor switch shown in figure 10-6.

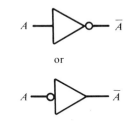

**Fig. 10-5.** The symbol for the NOT function. The input or output connection through a circle indicates a logic-level inversion. When inversion (NOT) is the only function performed, an amplifier symbol carries the circle inversion indication.

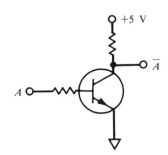

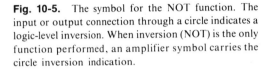

**Fig. 10-6.** A transistor inverter. A HI signal at the input turns the transistor on and produces a LO (~0.2-V) output signal. However, for a LO signal at the input (<0.6 V), the transistor is off, and the output level is HI.

The inverter operation may be represented either of two ways. One is the NOT function described above. The other is the conversion of the signal from HI-true to LO-true or vice versa. In other words, a HI-true signal representing variable $A$ at the inverter input is converted to *either* a HI-true signal representing $\overline{A}$ or a LO-true signal representing $A$ at the output (see note 10-2).

**Note 10-2.  Indication of LO-True Signals.**
The action of an inverter on a HI-true signal representing $A$ is to produce a signal that can be called either a HI-true signal representing $\overline{A}$ or a LO-true signal representing $A$. It is not desirable to adopt a convention that allows only one of these (for example, to require that all signals be HI-true) because both ways of considering inversion are convenient. In this text when a signal is LO-true, the symbol for that signal will be followed by an L such as $A$ L. All signals without an L suffix will be considered HI-true. Therefore, the inversion of signal $A$ will be either $\overline{A}$ or $A$ L (but not $\overline{A}$ L).

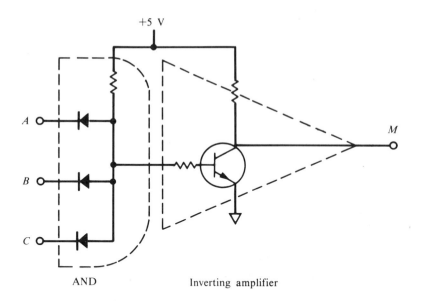

| Inputs | | | Output |
|:---:|:---:|:---:|:---:|
| A | B | C | |
| L | L | L | H |
| L | L | H | H |
| L | H | L | H |
| L | H | H | H |
| H | L | L | H |
| H | L | H | H |
| H | H | L | H |
| H | H | H | L |

AND                           Inverting amplifier

**Fig. 10-7.** Representation of an IC logic gate. The logic function and the output buffer amplifier are combined into a single circuit in an integrated circuit logic gate. There are several different types, or families, of IC logic gates. For each family the details of the logic and amplifier circuits are different. The buffer amplifier is designed so that its output signal is within the appropriate HI and LO ranges for its logic type and is capable of driving the inputs of a number of other gates of the same type. The HI and LO logic levels are often abbreviated H and L in complex tables of states.

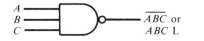

**Fig. 10-8.** The symbol for a NAND gate. The circle at the output indicates the inversion of the normal AND operation. The output inversion then either provides the NOT of $ABC$ or a LO-true-signal that represents the AND of the HI-true signals $A$, $B$, and $C$.

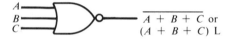

**Fig. 10-9.** NOR gate circuit symbol. The symbol is an OR gate with logic-level inversion at the output. As in the NAND gate, this inversion can be considered a logical NOT *or* a change in logic-level assignment.

## NAND and NOR Functions

Integrated circuit logic gates combine a logic function circuit and an output buffer amplifier as shown in figure 10-7. In this simplified representation, a diode AND gate is combined with a transistor inverting amplifier. The actual circuits of IC gates in the most popular logic families are discussed in chapter 11. The combination of the logic function and an inverting amplifier produces the inverse or complement of the logic function. The table of states in figure 10-7 shows that a LO logic level at any input produces a HI output and that a LO output is obtained only when all inputs are HI. This is the exact inverse of the AND gate as shown in figure 10-3 and is therefore called a NOT AND or NAND gate. The algebraic notation for the NAND function is the NOT of the AND function, or $M = \overline{ABC}$. The NAND gate symbol is shown in figure 10-8. Note that if only one of the NAND gate inputs is used, the output is the inverse of the function at that input. Thus the NAND gate is often used as an inverter for one signal.

The NOR function is similarly obtained by inverting the output of the OR gate. The NOR gate symbol is shown in figure 10-9. The NOR operation on two signals $A$ and $B$ is written symbolically as $\overline{A + B}$. The truth tables for the four basic logic functions are compared in table 10-3. From this table the inverted relationship between NAND and AND and between OR and NOR outputs is clear. From the symmetry of the table, a relationship between the AND and NOR functions and the NAND and OR functions might also be suspected. These pairs of functions are shown later to be equivalent if the input quantities are inverted.

**Table 10-3.** Comparison of four basic logic functions.

| Variables | | Results | | | |
|---|---|---|---|---|---|
| | | AND | NAND | OR | NOR |
| $A$ | $B$ | $A \cdot B$ | $\overline{A \cdot B}$ | $A + B$ | $\overline{A + B}$ |
| 0 | 0 | 0 | 1 | 0 | 1 |
| 0 | 1 | 0 | 1 | 1 | 0 |
| 1 | 0 | 0 | 1 | 1 | 0 |
| 1 | 1 | 1 | 0 | 1 | 0 |

## Boolean Algebra

Boolean algebra provides a simple means of designing, simplifying and analyzing logical networks. A list of the theorems of Boolean algebra is given in table 10-4. These theorems can be readily proven by means of truth tables. (For any two functions such as $\overline{A + B}$ and $\overline{A} \cdot \overline{B}$, if the truth tables are identical, the functions are equivalent.)

DeMorgan's theorems, which deal with the NOT function on entire terms, are especially interesting and useful. The theorems show the relationship between AND and OR functions and state that any logic expression can be inverted simply by inverting every term (change $A$ to $\overline{A}$, $B$ to $\overline{B}$, etc.), and changing every AND to an OR and every OR to an AND.

**Table 10-4.** Boolean algebra theorems

| | | |
|---|---|---|
| AND theorems | $0 \cdot 0 = 0$ | $A \cdot 1 = A$ |
| | $1 \cdot 1 = 1$ | $A \cdot A = A$ |
| | $1 \cdot 0 = 0$ | $A \cdot \overline{A} = 0$ |
| | $A \cdot 0 = 0$ | |
| OR theorems | $1 + 1 = 1$ | $A + 0 = A$ |
| | $0 + 0 = 0$ | $A + A = A$ |
| | $1 + 0 = 1$ | $A + \overline{A} = 1$ |
| | $A + 1 = 1$ | |
| NOT | $\overline{\overline{A}} = A$ | |
| Commutation | $A + B = B + A$ | $AB = BA$ |
| Absorption | $A + AB = A$ | $A(A + B) = A$ |
| Association | $A + (B + C) = (A + B) + C$ | $A(BC) = (AB)C$ |
| Distribution | $A + BC = (A + B)(A + C)$ | $A(B + C) = AB + AC$ |
| DeMorgan's theorems | $\overline{A + B} = \overline{A} \cdot \overline{B}$ | $\overline{AB} = \overline{A} + \overline{B}$ |

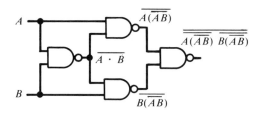

**Fig. 10-10.**   Analysis of a logic circuit using Boolean Algebra.

| | |
|---|---|
| $\overline{\overline{A(\overline{AB})}\ \overline{B(\overline{AB})}}$ | NAND operations indicated |
| $= A(\overline{AB}) + B(\overline{AB})$ | DeMorgan and $\overline{\overline{A}} = A$ |
| $= A(\overline{A} + \overline{B}) + B(\overline{A} + \overline{B})$ | DeMorgan |
| $= A\overline{A} + A\overline{B} + \overline{A}B + B\overline{B}$ | $A(B + C) =$ $AB + AC$ |
| $= 0 + A\overline{B} + \overline{A}B + 0$ | $A\overline{A} = 0$ |
| $= A\overline{B} + \overline{A}B$ | $0 + A = A$ |

As an example of the use of Boolean algebra in circuit analysis, consider the NAND gate circuit of figure 10-10. The application of the NAND operation on the signals yields an expression for the output quantity that is very complex. This expression can be reduced as shown in the figure caption by the application of the theorems of table 10-4. The result is a simple, and as we shall see, very useful function. The Boolean theorems can also be used to obtain an expression for the function of a logic circuit from its table of states. For example, the function of the diode AND gate of figure 10-3 for LO-true signals can be deduced from the table of states. First, we write a statement for $M$ that expresses all the combinations of $A$ and $B$ for which $M$ is TRUE. For the LO-true case, the output is TRUE when $A = 1$ AND $B = 1$ OR when $A = 1$ AND $B = 0$, OR when $A = 0$ AND $B = 1$. This is written $M = AB + A\overline{B} + \overline{A}B$. The distribution theorem is used to obtain $M = A(B + \overline{B}) + B(A + \overline{A})$. Since $A + \overline{A} = 1$ and $A \cdot 1 = A$, $M = A + B$. From this we see that an AND gate performs an OR operation on LO-true signals. Thus an AND gate could be drawn as an OR gate with inversion circles on both inputs and output. Algebraically, this equivalence is expressed as a variation of DeMorgan's theorem, $\overline{AB} = \overline{A} + \overline{B}$. Other gate equivalences are described below.

## Equivalent Gates

From DeMorgan's theorem, $\overline{ABC} = \overline{A} + \overline{B} + \overline{C}$, we see that the NAND operation on $A$, $B$, and $C$ is equivalent to the OR operation on $\overline{A}$, $\overline{B}$, and $\overline{C}$. Thus the NAND gate could be drawn as an OR gate with inversion circles on the inputs. This equivalence and the other basic gate equivalences obtained from DeMorgan's theorems are shown in figure 10-11. Through gate equivalences we see that every logic function could be achieved by using only NAND gates or only NOR gates. Figure 10-12 illustrates the achievement of

**Fig. 10-11.**   Gate equivalences. These four equivalent gate symbols are obtained from DeMorgan's theorems. The algebraic expressions are based on the NOT function being performed by all inversion circles. Of course, inversions could indicate a change to or from LO-true logic instead. Thus the NAND gate operation could be the LO-true AND of $A$, $B$, and $C$ ($ABC$ L) or the HI-true OR of LO-true inputs [$(A$ L$) + (B$ L$) + (C$ L$)$].

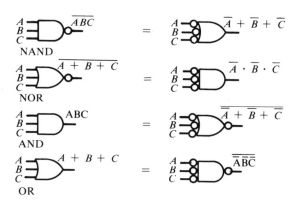

OR and AND functions with only NOR gates. Thus, very complex integrated circuit logic functions can be made from arrays of thousands of gates that are all the same basic subunit. In the TTL logic family, it is the NAND gate that is the basic subunit, and other functions are assembled by combining the basic logic part of NAND gates. In some other gate families, the NOR gate is the basic function. Logic gate families and their characteristics are discussed in chapter 11.

The use of equivalent gates can also be helpful in the analysis of gate circuits. Consider the gate circuit of figure 10-10 redrawn as shown in figure 10-13. NAND gates 1 and 4 have been drawn as inverting input OR gates, the inputs to gates 2 and 3 now are all from noninverting outputs, and the inverting outputs of gates 2 and 3 are connected to the inverting inputs of gate 4. The inverting circles at the input to gate 1 are shown to produce the NOT operation on $A$ and $B$, but the output inversions of gates 2 and 3 are shown as producing a signal that is temporarily LO-true. This gives gate 4 a simple OR function as shown.

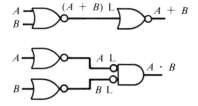

**Fig. 10-12.** OR and AND functions using only NOR gates. Necessary inversions are achieved by using only one input of the inverting gate. The use of the equivalent symbol with inverting inputs when the input signals are LO-true makes the gate function clearer.

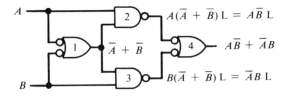

**Fig. 10-13.** Analysis of NAND gate circuit of figure 10-10 using equivalent gates. The analysis of the circuit function is simplified by substituting inverted input OR gates for some of the NAND gates. The choice of which gates to use equivalents for is based on minimizing the number of inverting outputs that are connected to noninverting inputs and vice versa.

## AND-OR-INVERT Gate

A very useful gate combination available in integrated circuit form is the AND-OR-INVERT (AOI) gate shown in figure 10-14a. A two-wide (number of AND gates), two-input gate is shown. Several different widths and numbers of inputs are also available in IC form. Many AOI gate packages have expansion inputs that allow the user to expand the width of the gate. In addition to AND-NOR gate packages, AND-OR gates as shown in figure 10-14b are also available. The AOI and AND-OR gates find extensive use in signal selection or multiplexing as described in section 10-2.

## Exclusive-OR and Equality Gates

The **exclusive-OR** function of the variables $A$ and $B$ can be stated as $A$ OR $B$ but NOT $A$ AND $B$. The truth table for this function is given in table 10-5. The exclusive-OR is not obtainable from the AND or OR functions by

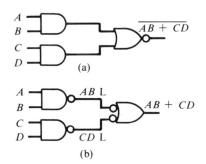

**Fig. 10-14.** AND-OR-INVERT gate (a) and AND-OR gate (b). The AND-OR gate is made from three NAND gates; the AOI gate has an inverted output.

**Table 10-5.** Exclusive-OR and equality functions.

| Variables | | Result | |
|---|---|---|---|
| $A$ | $B$ | $A \oplus B$ | $\overline{A \oplus B}$ |
| 0 | 0 | 0 | 1 |
| 0 | 1 | 1 | 0 |
| 1 | 0 | 1 | 0 |
| 1 | 1 | 0 | 1 |

Exclusive-OR

Equality

**Fig. 10-15.** Symbols for the exclusive-OR and exclusive-NOR, or equality, gate. The logic operation symbols are shown at the gate output.

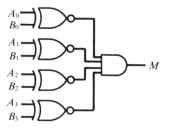

**Fig. 10-16.** Four-bit equality detector. Adding an AND gate to the outputs of four equality gates yields a four-bit equality detector.

simple inversion of inputs or outputs as its uniquely symmetrical truth table shows. Some people consider the exclusive-OR to be a fourth basic logic gate function (after AND, OR, and INVERT). The exclusive-OR function is used so frequently that the operation and the gate have been given special symbols (see fig. 10-15). From the truth table it can be seen that the output is true when $A = 0$ AND $B = 1$ OR when $A = 1$ AND $B = 0$. This statement is written in algebraic form as

$$M = A \oplus B = \overline{A}B + A\overline{B}$$

This function is the one produced by the gate circuit of figures 10-10 and 10-13.

Since the $A \oplus B$ operation is also $A \neq B$, the exclusive-OR is often used in comparing two signals and in anti-coincidence detection. The exclusive-OR can also be thought of as a controllable inverter. If one input is used as a control, the other input is inverted when the control is 1 and not inverted when the control is 0. As with all the other basic gates, the exclusive-OR is available in a convenient integrated circuit form, four gates to a package. The **equality gate**, also called the **comparator gate** or **coincidence gate** produces a 1 output whenever the two input signals have the same logic level. The truth table for the equality function is shown in table 10-5. From the truth table it is clear that the output function is the exact inverse of the exclusive-OR function. For this reason it is also called the **exclusive-NOR** function and is symbolized as shown in figure 10-15. In algebraic form the function is $AB + \overline{A}\overline{B}$.

## 10-2   Logic Gate Applications

Logic gates are used in digital instruments and digital information systems to control the flow of information, to encode and decode digital data, to detect and respond to particular combinations of conditions in control systems, and to perform mathematical and other logical operations upon digital data. The applications given in this section are only a few of the most basic and common of the endless variety of functions that can be performed by combinations of logic gates. Practical considerations in logic gate combinations are discussed in chapter 11.

### Digital Comparator

Comparison of the relative magnitudes of two binary numbers $A$ and $B$ is an important function in digital circuits. Digital comparators are important decision-making elements in computers and digital control systems. When two binary numbers are compared, there are three possible results: $A > B$, $A < B$ or $A = B$. To distinguish these three conditions, outputs are required for two of the conditions; the third condition can be assumed when neither of the others is TRUE. The equality condition is readily obtained by combining equality gates; the $A > B$ condition can be obtained by comparing the magnitudes of the most significant nonequal bits in the two numbers. The equality function can be expanded to additional pairs of bits so that the output is 1 when all pairs of inputs are equal. This is accomplished by combining the outputs of the equality gates for each pair into an AND gate as shown in figure 10-16. Only when the word represented by bits $A_3A_2A_1A_0$ is identical to the word represented by bits $B_3B_2B_1B_0$ will the output be 1.

The equality detector is combined with logic that decodes whether $A > B$ or $A < B$ in the **magnitude comparator** of figure 10-17. This circuit is an example of a **medium scale integrated circuit**. Note that the $A = B$ condition is just the four-bit equality detector of figure 10-16. The $A > B$ output is 1 whenever $A_3 > B_3$, OR $A_3 = B_3$ AND $A_2 > B_2$, OR $A_3 = B_3$ AND $A_2 = B_2$ AND $A_1 > B_1$, etc. Similarly the $A < B$ output is 1 when $A < B$. These latter conditions are decoded by 6-wide AOI gates. The magnitude comparator shown is fully expandable to any number of bits by cascading several ICs. The complete function table for a HI-true magnitude comparator is shown in table 10-6. The X entries in the table indicate that the indicated output is obtained whether that particular input is HI or LO. This is sometimes referred to as the "don't-care" condition.

**Fig. 10-17.** The 7485 magnitude comparator. The pin configuration is shown in (a) and the functional block diagram, in (b). The three outputs can be connected to the corresponding comparison inputs of another 7485 to expand the comparison in four-bit increments.

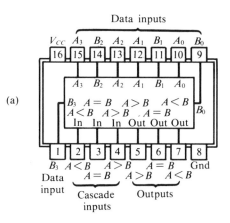

(a)

**Table 10-6.** Magnitude comparator function table.

| Comparing inputs | | | | Cascading inputs | | | Outputs | | |
|---|---|---|---|---|---|---|---|---|---|
| $A_3, B_3$ | $A_2, B_2$ | $A_1, B_1$ | $A_0, B_0$ | $A>B$ | $A<B$ | $A=B$ | $A>B$ | $A<B$ | $A=B$ |
| $A_3 > B_3$ | X | X | X | X | X | X | H | L | L |
| $A_3 < B_3$ | X | X | X | X | X | X | L | H | L |
| $A_3 = B_3$ | $A_2 > B_2$ | X | X | X | X | X | H | L | L |
| $A_3 = B_3$ | $A_2 < B_2$ | X | X | X | X | X | L | H | L |
| $A_3 = B_3$ | $A_2 = B_2$ | $A_1 > B_1$ | X | X | X | X | H | L | L |
| $A_3 = B_3$ | $A_2 = B_2$ | $A_1 < B_1$ | X | X | X | X | L | H | L |
| $A_3 = B_3$ | $A_2 = B_2$ | $A_1 = B_1$ | $A_0 > B_0$ | X | X | X | H | L | L |
| $A_3 = B_3$ | $A_2 = B_2$ | $A_1 = B_1$ | $A_0 < B_0$ | X | X | X | L | H | L |
| $A_3 = B_3$ | $A_2 = B_2$ | $A_1 = B_1$ | $A_0 = B_0$ | H | L | L | H | L | L |
| $A_3 = B_3$ | $A_2 = B_2$ | $A_1 = B_1$ | $A_0 = B_0$ | L | H | L | L | H | L |
| $A_3 = B_3$ | $A_2 = B_2$ | $A_1 = B_1$ | $A_0 = B_0$ | L | L | H | L | L | H |

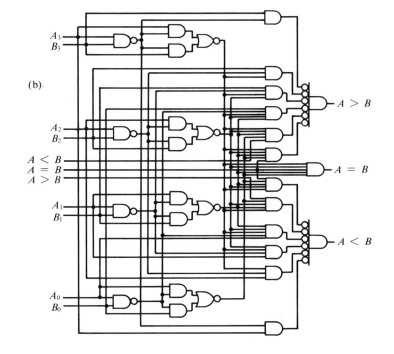

(b)

**Table 10-7.** Binary addition.

| Variables | | Results | |
|---|---|---|---|
| $A$ | $B$ | $S$ | $C$ |
| 0 | 0 | 0 | 0 |
| 0 | 1 | 1 | 0 |
| 1 | 0 | 1 | 0 |
| 1 | 1 | 0 | 1 |

**Table 10-8.** Truth table for full-adder.

| Variables | | | Results | |
|---|---|---|---|---|
| $C_i$ | $A$ | $B$ | $S$ | $C_o$ |
| 0 | 0 | 0 | 0 | 0 |
| 0 | 0 | 1 | 1 | 0 |
| 0 | 1 | 0 | 1 | 0 |
| 0 | 1 | 1 | 0 | 1 |
| 1 | 0 | 0 | 1 | 0 |
| 1 | 0 | 1 | 0 | 1 |
| 1 | 1 | 0 | 0 | 1 |
| 1 | 1 | 1 | 1 | 1 |

**Fig. 10-18.** Binary addition. The half-adder in (a) is an exclusive-OR gate and an AND gate. Half-adders are combined in (b) to form a full-adder.

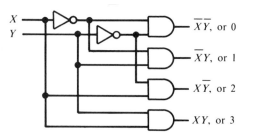

**Fig. 10-19.** Two-bit binary decoder. The circuit decodes the four possible states of the two-bit word $XY$. It is known as a two-line to four-line decoder.

## Binary Addition

The binary addition table can be summarized logically as follows: 0 plus 0 = 0; 0 plus 1 = 1; 1 plus 0 = 1; and 1 plus 1 = 0 and carry 1. The addition table is summarized in table 10-7, where $S$ is the sum and $C$ is the carry. It can be seen that $S = A \oplus B$ and $C = AB$. The addition function can be implemented with an exclusive-OR gate and an AND gate as shown in figure 10-18a. This circuit is called a **half-adder** because it will not accept a carry from the addition of the less significant bits in the two binary numbers being added.

The **full-adder** is made by combining half-adders as shown in figure 10-18b. Here the sum is generated from $A \oplus B$ and the carry in from the previous addition $C_i$ by $S = (A \oplus B) \oplus C_i$. The carry out $C_o$ is generated by $C_o = AB + C_i(A \oplus B)$. The complete truth table for the full-adder is shown in table 10-8. Complete four-bit full-adders are available in convenient IC packages.

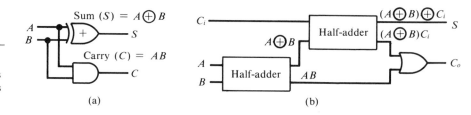

(a)   (b)

## Binary Decoders

A **binary decoder** produces a separate logic-level output for each relevant combination of input logic levels. A two-bit word can have four different states: 00, 01, 10, and 11. A complete two-bit binary decoder has four outputs, one to indicate the occurrence of each of the four possible states of the input word. Such a circuit is shown in figure 10-19. The outputs could be connected to other circuits or indicators that would then respond only to the appropriate combination of logic levels at the inputs. For instance, the outputs could be connected to lights labeled 0, 1, 2, and 3 to indicate the number represented by the two-bit word $XY$. If there were three inputs, there would be eight combinations. Thus eight three-input gates would be required for complete decoding, and eight outputs (0 through 7) would be produced. Such a circuit could decode a three-bit binary word into signals to turn on one of eight numerical indicators (0 through 7) corresponding to the decimal equivalent of the binary value of the three-bit binary word. The numerals 0 through 7 are all the numerals used in the **octal** (base 8) number system.

Thus a longer binary word can be decoded into its octal equivalent by converting each group of three bits of the binary number into a single numeral in the octal number. An octal number is written in about one-third the number of characters used by the equivalent binary number. Thus octal numbers are more efficient than binary ones for written copy and human recognition. Various 2-line to 4-line, 3-line to 8-line and 4-line to 16-line decoders are available in integrated circuit form.

In a **binary-coded decimal (BCD)** signal, four bits are encoded to represent each decimal digit. The 1 2 4 8 BCD decoding logic is given in table 10-9. The logic statements are those for the first ten terms in a hexadecimal decoder. An integrated circuit implementation with LO-true outputs is shown in figure 10-20. Note that the outputs conform to the logic of table 10-9 with the output logic being LO-true.

The BCD decoder is frequently used with a seven-segment LED display to provide a decimal readout. The seven-segment display is shown in figure 10-21. The BCD decoder and seven-segment driver must contain the logic to light the appropriate display elements in addition to the BCD-to-decimal decoding logic. For instance, element $a$ is to light for the numerals 0, 2, 3, 5, 7, 8, and 9. If the outputs of the BCD-to-decimal decoder for these numbers are combined in an OR gate and connected to a LED driver, segment $a$ will light when any of the above numerals are decoded. The other segments are driven in a similar manner. Complete BCD decoder-driver units are available in IC packages.

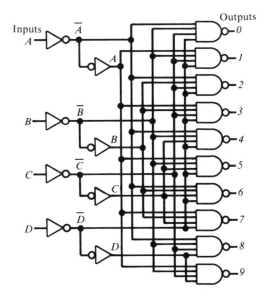

**Fig. 10-20.** BCD/Decimal decoder. This 7442 integrated circuit has LO-true outputs and full decoding to insure that all outputs remain OFF for invalid combinations (10 through 15).

**Table 10-9.** Binary-to-decimal decoding table, natural code.

| Decimal | BCD signals | | | | Logic statement | | | |
|---|---|---|---|---|---|---|---|---|
| | $D$(8) | $C$(4) | $B$(2) | $A$(1) | | | | |
| 0 | 0 | 0 | 0 | 0 | $\overline{A}$ | $\overline{B}$ | $\overline{C}$ | $\overline{D}$ |
| 1 | 0 | 0 | 0 | 1 | $A$ | $\overline{B}$ | $\overline{C}$ | $\overline{D}$ |
| 2 | 0 | 0 | 1 | 0 | $\overline{A}$ | $B$ | $\overline{C}$ | $\overline{D}$ |
| 3 | 0 | 0 | 1 | 1 | $A$ | $B$ | $\overline{C}$ | $\overline{D}$ |
| 4 | 0 | 1 | 0 | 0 | $\overline{A}$ | $\overline{B}$ | $C$ | $\overline{D}$ |
| 5 | 0 | 1 | 0 | 1 | $A$ | $\overline{B}$ | $C$ | $\overline{D}$ |
| 6 | 0 | 1 | 1 | 0 | $\overline{A}$ | $B$ | $C$ | $\overline{D}$ |
| 7 | 0 | 1 | 1 | 1 | $A$ | $B$ | $C$ | $\overline{D}$ |
| 8 | 1 | 0 | 0 | 0 | $\overline{A}$ | $\overline{B}$ | $\overline{C}$ | $D$ |
| 9 | 1 | 0 | 0 | 1 | $A$ | $\overline{B}$ | $\overline{C}$ | $D$ |

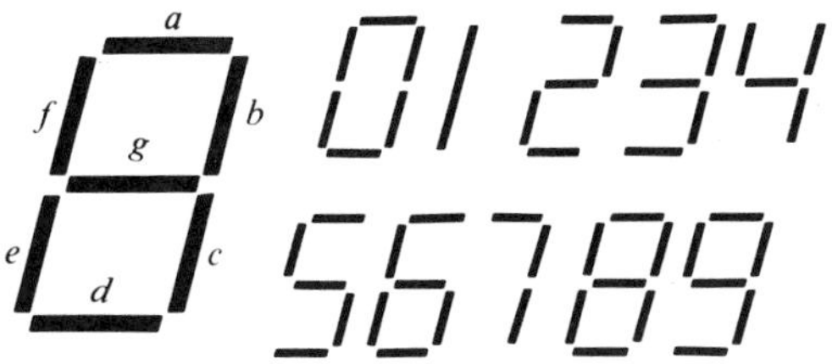

**Fig. 10-21.** Seven-segment decimal display. Combinations of the seven independent LEDs are lighted to form the numerals shown on the right.

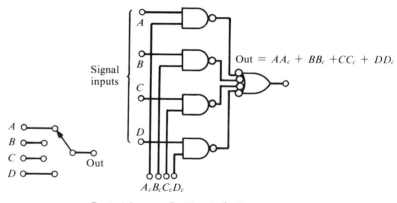

Signal inputs

$$\text{Out} = AA_c + BB_c + CC_c + DD_c$$

$A_c B_c C_c D_c$

Control inputs   **0**, Closed;  **1**, Open

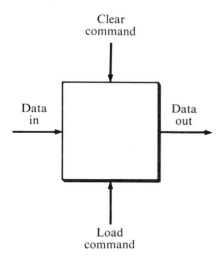

**Fig. 10-22.**   Four-line to one-line data selector and multiplexer. This four-wide AND-OR gate is equivalent to the multiple-throw switch on the left.

## Digital Multiplexers

The input of a logic circuit such as a counting register might have to be switched from one signal source to another to accommodate various desired applications. This requires the digital equivalent of a multiple-throw switch, as shown in figure 10-22. Such a gate circuit is called a **digital multiplexer**. The outputs of the four signal-switching gates are combined in an OR operation to provide a single output. This is drawn in figure 10-22 using a four-wide AND-OR gate. According to the output function, the state of the control inputs $A_c$, $B_c$, $C_c$, and $D_c$ determines which of the input signals $A$, $B$, $C$, and $D$ appears at the output. In practice only one control input is 1 at a time. This could be insured by using a two-line to four-line decoder at the control inputs. If $A_c = 1$ and $B_c$, $C_c$, and $D_c$ are all 0, the output function is $A$ regardless of the states of the other signal inputs. Thus the input signal that appears at the output depends on which of the control inputs is 1.

This circuit can be extended to any number of inputs and is readily modified for a LO-true output by the use of an AOI gate instead of the AND-OR gate. Integrated circuit multiplexers contain as many as 16 input lines switched to a single output line.

## 10-3   Latches and Registers

A **latch** is a kind of digital sample-and-hold. The digital signal source that is to be sampled is connected to the latch input and, at the time of the appearance of the data to be held, a load command signal is applied. The input logic level is now stored in the latch and appears at its output. The output remains at this level until the next load command signal. Connections to the latch are shown in figure 10-23. In most applications it is necessary to store

Clear command

Data in

Data out

Load command

**Fig. 10-23.**   A digital latch. The application of the load command signal (a logic-level pulse or transition) causes the logic level at the Data-in terminal to be stored and to appear at the Data-out terminal. The application of a clear command signal sets the stored data to 0.

more than the single bit of information that one latch circuit can acquire. One of the ways in which latches are combined to store more than one bit is shown in figure 10-24. The digital input and output data are in the parallel form. Most digital systems operate on the data in groups of bits called **bytes** (8 bits) or **words** (a specific number of bits that can differ from one system to another). A set of latches that can store a complete byte or word is called a **register**. In this section we shall see show the latch function is achieved with simple gates and how latches are combined in integrated circuit registers.

## Basic Latch

The basic bistable latch is the simple cross-coupled inverting gate circuit shown in figure 10-25. Recall that with the NAND gate the output will be 1 if any input is 0, and that the output can be 0 only when all inputs either are at logic 1 or are unused. Assume that the $R$ and $S$ inputs are 1. For either gate, then, a 1 at the cross-coupled input would produce a 0 at the output, or vice versa. Since the output of one gate is the input of the other, the two outputs must be in opposite states; that is, $Q = 1$ while $\overline{Q} = 0$, or $Q = 0$ while $\overline{Q} = 1$. Follow the logic levels through on the diagram to confirm that either of the above is a stable state, and that $Q$ and $\overline{Q}$ cannot both be 0 or 1 when $R$ and $S$ are 1.

If a 0 is applied to the $R$ (reset) input and a 1 to the $S$ input, $\overline{Q}$ must become 1, and $Q$ 0. If the $S$ (set) input is 0 and the $R$ input is 1, then $Q = 1$ and $\overline{Q} = 0$. If a momentary 0 level is applied to the $R$ or $S$ input, $Q$ becomes 1 or 0 as desired. Since the output remains in this state until a 0 is applied to the alternate input, the circuit remembers which input had the latest momentary 0 applied to it. If 0 levels are applied to both $S$ and $R$ inputs, both $Q$ and $\overline{Q}$ outputs must be 1. In a completely bistable circuit, this condition would be avoided.

Cross-coupled NOR gates can also be used to form the basic latch. The NOR gate has a 0 output level if any input is at logic 1. For this reason the truth table is very similar to the NAND gate latch except that the $S$ and $R$ inputs are reversed, the $Q$ and $\overline{Q}$ output levels are opposite, and the gate inputs are normally 0 between set and reset pulses.

## Gated Latch

By the addition of gates to the inputs, the basic latch can be made to respond to input levels only during a specified time interval, as shown in figure 10-26. So long as the load input is 0, gates 3 and 4 have 1 outputs regardless of the $S$ and $R$ input levels. To allow new information to reach the basic latch, the load input is changed to 1. A 1 at the $S$ input and a 0 at the $R$ input now gives a 0 input to gate 1 and a 1 input to gate 2, and $Q$ becomes 1. If $R$ were 1 and

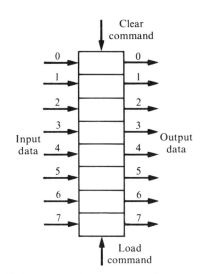

**Fig. 10-24.** Parallel input-output register. A number of latch circuits are combined with common load and clear connections. With such a register, an entire parallel digital word can be latched.

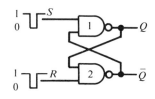

**Fig. 10-25.** Basic NAND gate latch. When both $S$ and $R$ inputs are 1, the outputs $Q$ and $\overline{Q}$ must be complementary. The level at $Q$ depends upon whether $S$ or $R$ had the most recent 0.

**Fig. 10-26.** Gated latch and table of states. A 0 at the load input keeps the outputs of gates 3 and 4 at 1. The basic latch of gates 1 and 2 can be changed by the $S$ and $R$ signals only when the load signal is 1. In the table, $t_n$ and $t_{n+1}$ refer to the times during and after the load pulse.

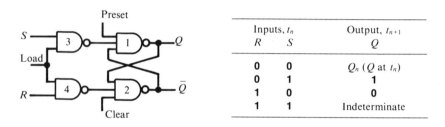

| Inputs, $t_n$ | | Output, $t_{n+1}$ |
| $R$ | $S$ | $Q$ |
| --- | --- | --- |
| 0 | 0 | $Q_n$ ($Q$ at $t_n$) |
| 0 | 1 | 1 |
| 1 | 0 | 0 |
| 1 | 1 | Indeterminate |

$S$ were 0, $Q$ would become 0. If both $S$ and $R$ are 0, the inputs to the basic memory remain 1's and the output does not change from its previous state. Load returns to 0 to end the input level sampling interval, and further changes in the $S$ and $R$ signals have no effect on the state of the latch.

This latch circuit is said to have a clocked input because the sampling interval can be timed to coincide with the appearance of the desired information at the $S$ and $R$ inputs. For this reason the load input is often called the **clock** input. It is also sometimes called the **enable** input. The table of states for a clocked circuit is usually written as shown in figure 10-26 when the output state after the load or clock pulse depends on its previous state. The $S$ and $R$ input signals to the gated latch do not have to be pulses or momentary level changes as required by the basic latch. In this case the input pulsing is accomplished by the load signal.

**Fig. 10-27.** Data latch. This variation of the gated latch of figure 10-25 ensures that the set and clear inputs are complementary. The levels at $D$ thus appears at $Q$ whenever clock is 1. If clock remains 1, $Q$ follows the changes of $D$.

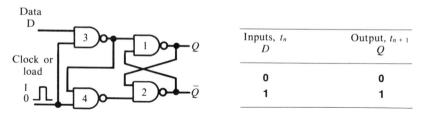

| Inputs, $t_n$ | Output, $t_{n+1}$ |
| $D$ | $Q$ |
| --- | --- |
| 0 | 0 |
| 1 | 1 |

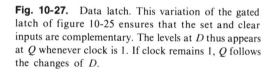

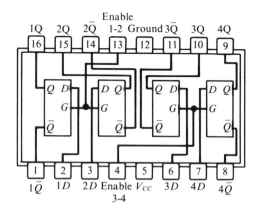

**Fig. 10-28.** Logic and pin connections for the 7475 quad latch. Note that each Enable or Load input operates two latches.

The normal form of the gated latch is the **data latch** circuit shown in figure 10-27. This circuit has a single data input. Gate 3 serves both to gate the input to gate 1 and to provide a $\overline{D}$ signal to gate 4. The $S$ and $R$ inputs of the gated latch are thus always $D$ and $\overline{D}$, respectively. The table of states is limited to those states in which the $S$ and $R$ inputs are complementary and the $Q$ output follows the $D$ input when the load input is 1. The availability of multiple latches in IC packages makes the construction of latches from gates generally unnecessary. The arrangement of latches in a popular quad latch IC is shown in figure 10-28. Other latch packages latch a full byte of information and provide some additional logic (input conditions) for the enable or load input.

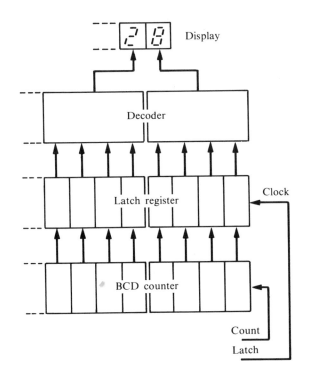

**Fig. 10-29.** Application of latch register in counting measurement. The quad latch packages are convenient for use with the binary-coded decimal counter because each package stores the binary information for one digit of display.

A common application of the latch register is in counting instruments. This application is shown in figure 10-29. At the end of the counting measurement, the latch register clock input is pulsed to load the count information into the latch register. The counter can now be cleared and the next measurement begun while the latch register holds the data for decoding and display, as shown, or for printing or transfer to a computer. If the clock input is held 1, the counter contents appear continuously at the latch outputs, and the counting operation can be observed in the display.

## 10-4   Flip-Flops

In many applications of the latch, the latch output is used in a way that can affect the level at the latch input. A common example is the toggle **flip-flop** first introduced in chapter 4 and shown in figure 10-30. The desired action is a reversal of state by the latch on each clock pulse. It is this successive alternation of state that gives this device the name, flip-flop. The reversing action is achieved by connecting the $Q$ and $\bar{Q}$ outputs to the $S$ and $R$ inputs in such a way as to reverse the current state of the latch on the next clock pulse. The simple gated latch described in the previous section cannot be used in the flip-flop application because it responds to its inputs as long as

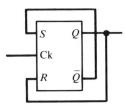

**Fig. 10-30.** A toggle flip-flop. The $Q$ and $\bar{Q}$ outputs apply a level at the $S$ and $R$ inputs that is opposite to the current state of the latch. The latch then reverses its state on every clock (Ck) pulse. The latch used for flip-flop operation must not change its output while its input is still enabled.

the enable signal is 1. During the enable pulse, the output is reversed; this, in turn, reverses the $S$ and $R$ input data, reversing the output again, and so on. Such a "race" condition can exist in logic circuits where a continuous data feedback path exists.

## Master-Slave Flip-Flops

One way to avoid the race condition is to use a latch that does not change its output until the input level has been stored and the input disabled. This is accomplished by using two gated latches in sequence, the inputs of which are never both activated at the same time. This technique is illustrated for the $JK$ master-slave flip-flop in figure 10-31. The master latch is used to hold the input data so that the data can be transferred to the slave latch after the master inputs are disabled. Setting the $1 \rightarrow 0$ threshold of the clock inverter lower than that of the latches ensures that the slave input is enabled only when the master inputs are disabled. The state of the master latch caused by the $S$ and $R$ inputs to the master while the clock was 1 is transferred to the slave at the $1 \rightarrow 0$ clock transition. The data appear at $Q$ and $\overline{Q}$ only nanoseconds later, but the master inputs are now completely disabled. The arrow in the $1 \rightarrow 0$ transition of the clock signal indicates the clock edge on which the transfer to the output occurs.

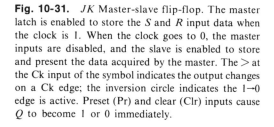
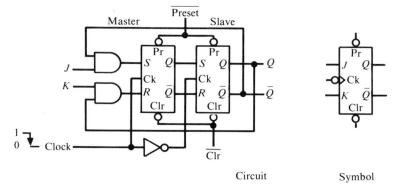

**Fig. 10-31.** $JK$ Master-slave flip-flop. The master latch is enabled to store the $S$ and $R$ input data when the clock is 1. When the clock goes to 0, the master inputs are disabled, and the slave is enabled to store and present the data acquired by the master. The $>$ at the Ck input of the symbol indicates the output changes on a Ck edge; the inversion circle indicates the $1 \rightarrow 0$ edge is active. Preset (Pr) and clear (Clr) inputs cause $Q$ to become 1 or 0 immediately.

Circuit        Symbol

**Table 10-10.**   Table of states for $JK$ flip-flop.

| Inputs, $t_n$ | | Output, $t_{n+1}$ |
|:---:|:---:|:---:|
| $J$ | $K$ | $Q$ |
| 0 | 0 | $Q$ at $t_n$ |
| 0 | 1 | 0 |
| 1 | 0 | 1 |
| 1 | 1 | $\overline{Q}$ at $t_n$ |

The AND gates at the inputs to the master latch of the $JK$ flip-flop provide additional control over the flip-flop action. If $J$ and $K$ are both 1, the flip-flop toggles, or alternates state, on each successive $1 \rightarrow 0$ edge of clock. A 0 at either $J$ or $K$ specifies the output state at the next clock edge as shown in table 10-10. In terms of transitions, a 0 at the $J$ input prevents a $0 \rightarrow 1$ transition at $Q$, and a 0 at the $K$ input prevents a $1 \rightarrow 0$ transition at $\overline{Q}$. The master latch is actually sensitive to 1 levels at its $S$ and $R$ inputs during the entire time of the 1 clock signal. This "1's catching" characteristic of the gated latch is often desirable or inconsequential, but if the data input levels are not fixed throughout the clock pulse, it can produce errors. In such cases, a true edge-triggered flip-flop is used.

## Edge-Triggered Flip-Flops

In an edge-triggered flip-flop, the input data are acquired on an edge of the clock signal. On that same edge, but slightly later, the output changes to its new state. This type of flip-flop is illustrated by the $D$ edge-triggered flip-flop shown in figure 10-32. As long as the Clock input (Ck) is 0, the outputs of gates 4 and 5 are 1, and the output latch (gates 1 and 2) is stable. When clock goes to 1 the state of the gates 5 and 6 latch depends upon the input $D$ level, the gate 5 output taking the level at $D$. Because of the connection between the gate 6 output and gate 3, the gate 3 output has the same value as $D$. Thus when Ck goes to 1, the gate 4 output will be $\overline{D}$. On the rising edge of the Clock input, then, gates 4 and 5 apply $\overline{D}$ and $D$ levels to gates 1 and 2 respectively, and the $D$ level appears at $Q$. Thus the flip-flop is truly triggered by the *edge* of Ck, and the value of $D$ at that instant is acquired. Later changes in $D$ have no effect. The Preset (Pr) and Clear (Clr) inputs act immediately and have precedence over all other inputs.

The $JK$ flip-flop can also be made in edge-triggered form as shown in figure 10-33. The integrated circuit has made the extra complexity of edge-triggering almost inconsequential in the cost, so almost all of the newer $JK$ flip-flop designs are edge-triggered. Note that only the polarity of the active edge is indicated by the symbol, not whether the flip-flop is edge-triggered or master-slave. The manufacturer's literature should always be consulted.

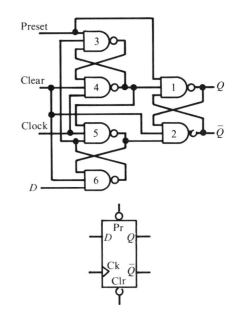

**Fig. 10-32.** *D* edge-triggered flip-flop. This circuit contains three basic latches. The gates 3 and 4 latch is also cross-coupled with the gates 5 and 6 latch. The output levels of gates 4 and 5 determine the state of the output latch when the Clock (Ck) signal has a 0→1 transition. The symbol shows the 0→1 edge activated clock input. The Preset (Pr) and Clear (Clr) are LO-true as shown by the inverting circle. They could have been labeled $\overline{Pr}$ and $\overline{Clr}$ with no inverting circle, but the representation shown is preferred. The $\overline{Q}$ output can also be shown as $Q$ with an inverting circle.

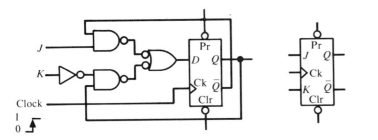

**Fig. 10-33.** *JK* edge-triggered flip-flop. The addition of input logic and connections to $Q$ and $\overline{Q}$ to an edge-triggered $D$ flip-flop creates an edge-triggered $JK$ flip-flop. The $J$ and $K$ inputs are active only on the 0→1 edge of Ck.

In flip-flop applications that require the output change to occur well after the input data are acquired, the edge-triggering and master-slave concepts are combined to form a flip-flop with **data lockout**. A $JK$ flip-flop with data lockout is shown in figure 10-34. The times of acquisition and output data change are separated by the width of the clock pulse in this type of flip-flop.

**Fig. 10-34.** A *JK* flip-flop with data lockout. The edge-triggered *JK* flip-flop, acting as a master, is triggered on the rising edge of Clock. The slave latch is disabled during Clock HI. The state of the master is transferred to the slave on the Clock falling edge. The inverter on Ck signifies that the output appears on the 1→0 transition. The double angle indicates that the acquisition and transfer occur on opposite edges.

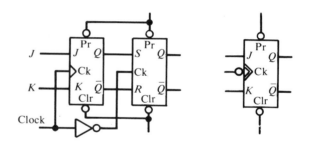

As a basic digital building block, the flip-flop in its various forms will be encountered frequently in subsequent sections and chapters. As the next section demonstrates, the counter is an important flip-flop application.

## 10-5 Counting Circuits

A counting circuit is a register that stores the count information in a coded digital form such as binary or BCD and that increments the stored number on command. After each increment command, the code that represents the number of counts received is revised to the next higher number. This concept is illustrated in figure 10-35a. The counter is initialized prior to the first count by a Clear signal after which the total accumulated number of input pulses can be obtained at any time from the parallel output. In digital signal terms, the counter converts count serial data to parallel data. The incrementing circuitry of a counter is specific for the parallel encoding used. In many counting applications, it is convenient to be able to begin counting from some value other than zero and to be able to count up or down. The data flow for a counter with all these capabilities is shown in figure 10-35b.

### Binary Counter

The register of a binary counter stores information about the accumulated number of input pulses as a binary coded number. It is a popular counter

**Fig. 10-35.** Counting registers. The number stored in the register of (a) can be increased by one (incremented) or changed instantly to 00000000 (cleared). The numerical value is continuously available at the parallel output. The more versatile counter of (b) can be preset to any value through the parallel data input and count up (increment) or down (decrement) from that value.

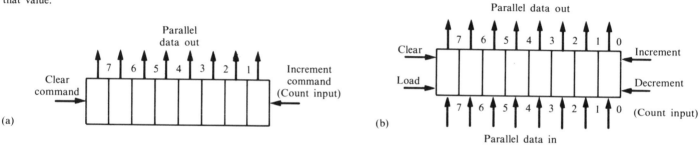

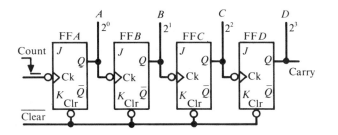

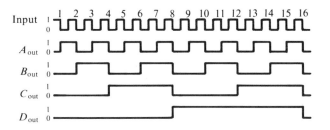

because it is very simple, stores the maximum possible amount of information per memory unit, and is related to the base 2 number system widely used in computers.

A basic binary counter is shown in figure 10-36. Toggling flip-flops that change state when the clock input signal goes from 1 to 0 are used. Assume that the four flip-flops have been cleared ($Q = 0$) by a 0 pulse on the $\overline{\text{Clear}}$ line. When the first pulse appears at the count input, flip-flop $A$ toggles and $A$ becomes 1 ($2^0 = 1$). On the next pulse, $A$ returns to logic 0, which toggles flip-flop $B$. Now $B$ is 1, and $A$ is 0 ($2^1 = 2$). After the next pulse, $A$ is 1 again and $B$ is unchanged ($2^1 + 2^0 = 3$), and so on. The table of waveforms in figure 10-36 shows how the $A$ flip-flop is set after every odd input pulse to represent the number 1. The $B$ flip-flop output represents the number 2, and the $C$ and $D$ outputs represent 4 and 8, respectively. The sum of the values of the set flip-flops represent the cumulative count at any instant. The outputs could be connected to indicators, and the instantaneous count read out in binary form. The counter of figure 10-36 can be extended to any desired number of flip-flops by adding flip-flops $E$, $F$, $G$, and so on, in like manner. Since each successive flip-flop output represents another binary digit (power of 2), $n$ flip-flops have $2^n$ states and can count from zero to $2^n - 1$. The waveforms of the four-flip-flop (four-bit) counter show the progression from 0 to 15. Then, on the sixteenth pulse the register returns to the zero state.

It is also possible to make a counter that counts down, or subtracts one from the count for each input pulse. A **down-counter** and its output waveforms are shown in figure 10-37. The circuit is identical to that of the up-counter except that $\overline{Q}$ instead of $Q$ is connected to the following clock input. Additional flip-flops can be added to produce a down-counter of any desired capacity.

## Synchronous Binary Counter

The counting circuits of figures 10-36 and 10-37 are not synchronous because each flip-flop is clocked at a different time. In each flip-flop there is a delay between the triggering edge of the clock pulse and the logic-level change at the output due to the propagation delay through the flip-flop gates. If the

**Fig. 10-36.** Binary counter and waveforms. The $1{\rightarrow}0$ transitions at the clock input of each flip-flop produce first a $0{\rightarrow}1$ and then a $1{\rightarrow}0$ transition at the output. Thus the output alternates level on successive $1{\rightarrow}0$ edges at the input. Using the first flip-flop output to clock the second flip-flop produces transitions that occur every fourth transition of the input, and for the third flip-flop, every eighth, and so on. The result is a counter for binary encoded numbers.

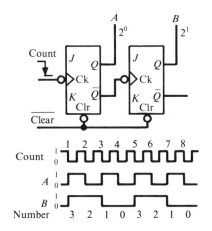

**Fig. 10-37.** Binary down-counter and waveforms. When the first $1{\rightarrow}0$ transition occurs at the Count input, $\overline{Q}$ of flip-flop $A$ changes from 1 to 0. This toggles flip-flop $B$ and fills the counter. The next input pulses reduce the count until the register is cleared.

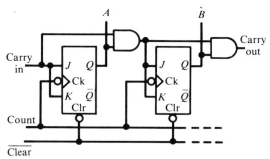

**Fig. 10-38.** Synchronous binary counter. Flip-flop *A* is gated to alternate states when clocked if the Carry in is 1. Flip-flop *B* should alternate only when Carry in AND *A* are 1. This is accomplished by the gate. The Carry out will be 1 only when the counter is full.

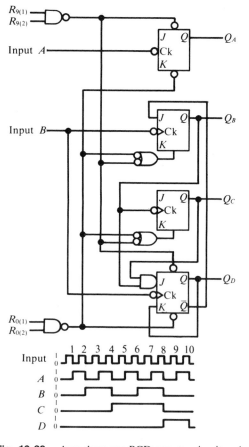

**Fig. 10-39.** Asynchronous BCD counter circuit and waveforms. The output waveforms show two deviations from the binary counting sequence. One is that *B* does not go to 1 on the tenth pulse, and the other is that *D* returns to 0 on the tenth pulse. Flip-flop *B* is inhibited from becoming 1 whenever *D* is 1 by the connection from $\overline{Q}$ of *D* to *J* of *B*. Flip-flops *D* and *B* are both clocked on every even-numbered pulse, but *D* cannot become 1 until *B* and *C* are 1 and then *D* returns to 0 on the next even-numbered pulse. A 1 at either reset zero ( $R_0$ ) input clears the counter while a 1 at either reset nine ( $R_9$ ) input sets the counter to 1001 (a count of 9).

output of either of these asynchronous counters is sampled while the new count information is "rippling through" the counter chain, serious errors in the apparent count could result. For instance, with the up-counter, as the eighth count pulse is applied, the waveforms of figure 10-36 show that the counter output will be 0111 (7), then 0110 (6), then 0100 (4), and finally 1000 (8). If the propagation delay is 50 ns through each flip-flop, a ten-bit counter will have error states for as long as 400 ns. This is only a problem if the parallel output must be able to be sampled at any time during the counting process. In such cases **synchronous counters** in which all the flip-flops are clocked simultaneously and all therefore change output state at essentially the same time are used.

A two-bit section of a synchronous binary counter is shown in figure 10-38. All flip-flops are clocked together, and the *J* and *K* inputs are used to inhibit all inappropriate transitions. This inhibit signal is obtained from the AND of the previous flip-flop output and its inhibit. In this way synchronous counting registers of any bit size can be made. The availability of four-bit synchronous counters in an integrated circuit package makes such counters very easy to build.

## BCD Counter

Counters that store numbers in the BCD code must automatically increment following that code. Each group of four flip-flops in a BCD counter is arranged to store one decimal digit. Up to 16 combinations of output states are available from four binary circuits, but for storing a decimal number, only ten of these states are used for the decimal numerals 0 to 9. A group of four flip-flops connected as a counting register with ten states is called a **decade counting unit (DCU)**. To store numbers up to 9999, four DCUs are required. Since the decoding for each DCU is identical, more DCUs and decoding circuits can readily be added to store and read out as large a decimal number as necessary.

The most common storage code for decimal numbers follows the first ten states of the binary counter. It is called the 1 2 4 8, or natural, code because the values of 1, 2, 4, and 8 can be assigned to flip-flops *A*, *B*, *C*, and

$D$, respectively, and the stored decimal number can be obtained by adding the values of the set (1 level) flip-flops. The waveforms and circuit for a 1 2 4 8 BCD counter are shown in figure 10-39. This simple circuit is the most popular integrated circuit BCD counter (7490 type). This counter can be obtained with maximum count rates of over 50 MHz. Generally the connection between the $Q$ output of $A$ and the Ck input of $B$ (and $D$) is completed by interconnecting two pins of the integrated circuit package. If unconnected, the $A$ flip-flop can then be used independently of the circuit composed of flip-flops $B$, $C$, and $D$, which has five stable states. Because this BCD counter is made up of a two-counter followed by a five-counter, it is sometimes referred to as a **biquinary counting circuit**.

## Presettable and Up-Down Counters

A counter that can be set to any desired value before the count begins is called a **presettable counter**. A common application for presettable counters is the determination of the time required for a particular number of counts to occur. The counter is preset, and the time required to achieve maximum

**Fig. 10-40.** Synchronous decade counter with synchronous preset and asynchronous clear (74160-type). The counter is synchronous because all flip-flops are connected to a common clock (count input). The BCD counting sequence is established by gating combinations of the $Q$ outputs to the $J$ and $K$ inputs of the flip-flops. When the Load input is 0, the data inputs are gated through to the corresponding $J$ and $K$ inputs so that the FFs take the data values on the next clock pulse. Thus Preset is synchronous with the count. The Clear, however, is asynchronous since its effect is immediate and independent of the clock signal. The Ripple carry output is 1 for a count of 9 and is used for cascading several DCUs. The $P$ and $T$ inputs enable the count operation and the carry output respectively.

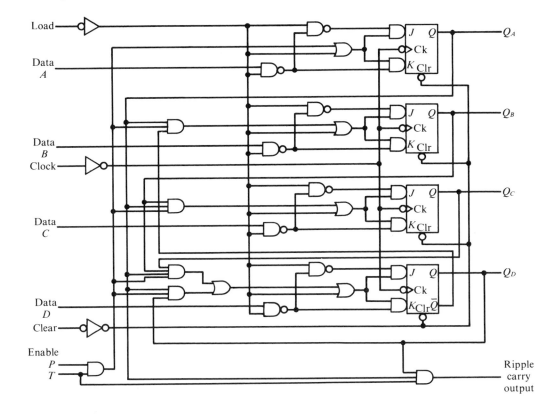

count (in an up-counter) or zero count (in a down-counter) is measured. A popular type of presettable decade counter is shown in figure 10-40. With this type of preset, the load signal level determines whether a 1→0 transition of the clock increments the counter or loads the counter flip-flops with the levels at the data inputs. This is called a **synchronous load** because the load operation is synchronized with the clock. By contrast, the reset input is asynchronous because it is connected to the direct clear inputs of the flip-flops and has an immediate effect.

In many applications it is desirable to have a counter that can either increment or decrement the count value. Such a counter is called an **up-down counter**. In practical integrated circuits, the up-down counting capability is combined with presettability. An example of such a versatile counting IC is shown in figure 10-41. One set of AND gates allows each flip-flop to change state if the previous flip-flops are all 1 *and DN/$\overline{UP}$* is 0 (the UP state). In the DOWN state, the other set of AND gates is enabled, and a flip-flop can change state only if the previous ones are all 0. The preset function in this counter is achieved by gating the data input levels to the direct set and clear inputs of the flip-flops. This results in an asynchronous load because the effect of the load input signal ($\overline{PL}$) is immediate and not synchronized with the clock. Up-down counters are available as binary or decade counters and with either synchronous or asynchronous load. Each combination of features has advantages for particular applications.

**Fig. 10-41.** Synchronous binary up-down counter with asynchronous preset (74191 type). The response of each FF to the clock is determined by the AND-OR gate at its *J* and *K* inputs. When the *DN/$\overline{UP}$* input is 0, the upper AND gates of the AND-OR gates are enabled, and an up-count progression is obtained. Conversely when *DN/$\overline{UP}$* is 1, the lower AND gates are enabled, and a down count results. A 0 at the $\overline{PL}$ input applies the data input levels to the FF Set and Clr inputs so that the data loading is immediate and asynchronous with the clock. Additional counters can be cascaded through the Max/min and Ripple clock outputs.

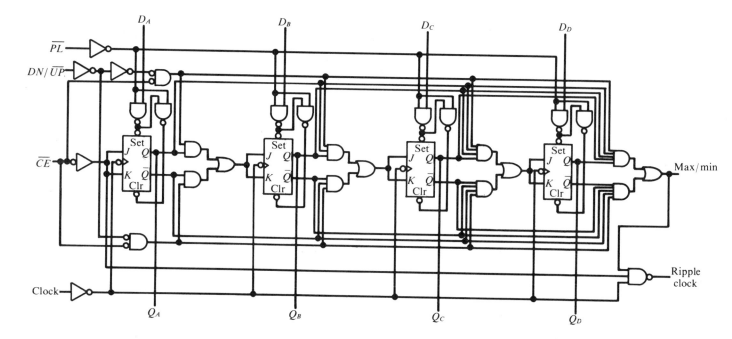

## Odd or Variable-Modulus Counters

The **modulus** of a counter is the number of count input pulses for a complete count cycle. For instance, the four-flip-flop binary counter of figure 10-36 has a modulus of 16. If it starts at 0000, in 16 pulses it is at 0000 again. Similarly, a three-flip-flop binary counter is a modulo-8 counter and the 1 2 4 8 BCD counter of figure 10-40 is a modulo-10 counter. Many counting and scaling applications require counters with a modulus other than $2^n$ or $10^n$. Furthermore, it is sometimes desirable to be able to select or vary the modulus of a counter as needed.

A counter can be made to have a particular modulus in several ways: the flip-flops can be wired to repeat the output cycle every $N$ counts (fixed-modulus counter); or a decoding circuit can be used to detect the $N$th count and reset the counter; or the counter can be preset to a value from which it will proceed until it is full or clear. An example of the fixed-modulus counter is the BCD counter described above. The necessary number of flip-flops is $n$ where $n$ is the smallest integer for which $2^n$ is larger than $N$. Then the clock and $JK$ inputs are connected so as to achieve the desired modulus and count sequence.

For some counting and scaling applications it is desirable to be able to change the modulus of the counter quickly and easily. This is generally done by allowing an ordinary binary or BCD counter to advance until the desired maximum count is detected and then stopping or clearing the counter. A general block diagram of such a variable modulus counter is shown in figure 10-42. A circuit is used to compare the outputs of the counting register with inputs that represent the desired maximum count $N$. When a count of $N$ is reached, the digital comparison circuit output resets the counter. If an immediate repeat of the counting cycle is not desired, the comparator output could instead be used to close the counting gate or to preset the counter to a different number. Another, rather similar approach, that is commonly used is to detect the zero state of a presettable down-counter and connect it to the preset control input. The counter then presets to the parallel data input automatically upon reaching zero and begins to count down again. The modulus of this counter is equal to the preset number.

## Counter Gating

The purpose of a counting gate is to provide control over the counting interval. The feature of the signal that causes the counter to advance is an edge, often the 1→0 transition. The counting gate should therefore transmit only those input signal edges that occur during the desired counting interval. The simplest counter input gate is the AND gate shown in figure 10-43. This simple gate is useful only when it can be assured that the Control input will

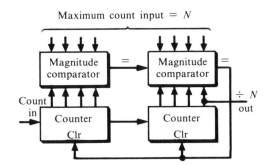

**Fig. 10-42.** Divide-by-$N$ or variable-modulus counter. When the counter reaches the value $N$ at the maximum count input, the comparator equals output applies a clear signal to the counter. Because the cleared counter is no longer equal to the maximum count input, the clear signal is removed, and the counter counts again from zero. The output frequency of the most significant bit flip-flop is the input frequency divided by $N$.

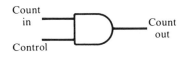

**Fig. 10-43.** AND gate. The output signal follows Count in as long as Control is 1. A 0 Control level produces a 0 output and closes the gate. From the symmetry, when Count in is 1, output edges are produced by level changes at Control.

not have a 0→1 or 1→0 transition while the Count input is 1. This is necessary in order to avoid an extra counted edge produced by the gate control transitions.

**Fig. 10-44.**    An asynchronous waveform gate. A 0→1 transition at Count sets the $D$ flip-flop only when Control is 1. The next 1→0 transition at Count clears the flip-flop to reproduce the count pulse at the output. Control changes cannot produce inappropriate or truncated pulses.

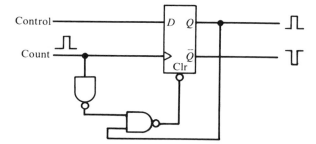

Far more common is the situation where the Count and Control signals are obtained from two unrelated or asynchronous sources such as a clock and an unknown signal. A flip-flop provides an excellent basis for a counting gate since it is sensitive to edges and readily controlled. Such a circuit is shown in figure 10-44. A 0→1 edge at the Count input causes the Control level to appear at $Q$. If Control is 0, the flip-flop remains cleared; if Control is 1, the flip-flop is set and will be cleared when Count returns to 0. If Control becomes 1 part way through a Count pulse, it is ignored because the 0→1 edge has already happened. If Control becomes 0 part way through a Count pulse, the flip-flop is still not cleared until Count returns to 0. This type of circuit, which only produces complete pulses, but is controllable by an asynchronous signal, is called an **asynchronous waveform gate**.

In counter-based measurement systems, additional circuits are required to generate the appropriate gate control signals and to sequence the data storage and measurement repetition. A general-purpose counter control system is shown in figure 10-45. A 0→1 transition at the Start input sets the Start flip-flop, enables the Stop flip-flop to be set and changes the gate control signal to 1. The gate remains open until a 0→1 transition occurs at the Stop input. The resulting 1→0 transition of the gate control signal closes the gate and triggers a chain of monostable pulse generators which store the data, reset the counter, and clear the Start and Stop flip-flops. The system is now prepared for another Start signal to begin the next measurement.

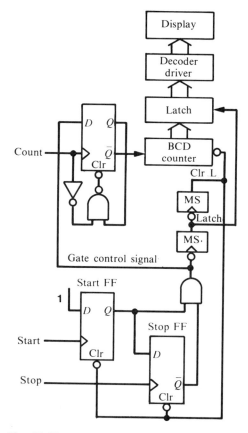

**Fig. 10-45.**    Input and control system for automatically recycling counting measurements. The counting gate is controlled by a signal obtained from the counter Start and Stop commands. At the end of each gate period, a monostable multivibrator (MS) generates a latch signal to store the count and triggers another monostable which clears the counter and resets the Start and Stop flip-flops.

## 10-6    Shift Registers

In a shift register, the flip-flops are connected so that the content of each one is transferred to the next one in line upon a shift command. This operation is shown in block form in figure 10-46a. Serial data appearing at the bit 0 input are shifted into the register by a shift command during each bit time of the signal. After eight shifts, the eight-bit register is completely filled with new

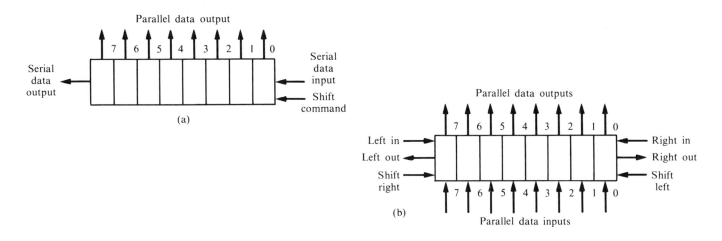

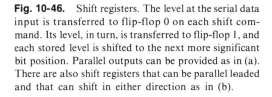

data, the first bit of which is now at the serial output. One function performed by the shift register is serial-to-parallel conversion since the most recent eight serial bits are available at the parallel output. Another function is as a serial delay element. The input serial data appear at the serial output delayed by exactly eight bit times. Different word lengths or delay times are obtained by using various length shift registers. A parallel-to-serial conversion can be performed by loading a parallel input shift register with the parallel word. As the shift command is repeated, the word appears in serial form at the serial output.

A basic four-bit shift register is shown in figure 10-47. The shift command is accomplished by the synchronous clock connected to all the flip-flops. The output of each flip-flop is simply connected to the input of the next. In this way, a shift register of any length can easily be constructed. A shift-right, shift-left register is made by connecting a two-wide AND-OR gate to each flip-flop input. The input of one AND gate is connected to the

**Fig. 10-46.** Shift registers. The level at the serial data input is transferred to flip-flop 0 on each shift command. Its level, in turn, is transferred to flip-flop 1, and each stored level is shifted to the next more significant bit position. Parallel outputs can be provided as in (a). There are also shift registers that can be parallel loaded and that can shift in either direction as in (b).

**Fig. 10-47.** Four-bit shift register and waveforms. The level present at the $J$ input of each flip-flop is transferred into that flip-flop at the $1 \to 0$ transition of the clock. After four clock pulses, the level originally at the serial input is at the $D$ output.

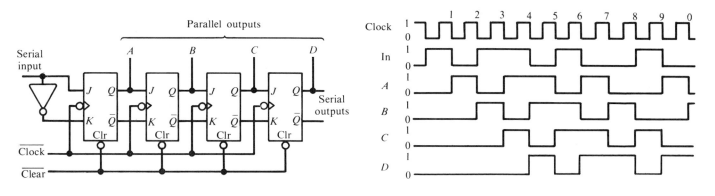

flip-flop output at the left, and the input of the other to the output at the right. A control signal to the AND gates determines whether the input data for each flip-flop come from the left or the right.

Parallel loading of the shift register is accomplished either of two ways. One is through the preset and clear connections to the flip-flops, exactly as in the counter of figure 10-41. This is called an asynchronous load, or direct set, since it occurs immediately and always takes precedence over a shift operation. The other technique is to use AND-OR gates to determine whether the data at each flip-flop are from the previous flip-flop or from the parallel input. If the AND-OR gates are set to accept the latter, the parallel word is loaded on the next clock edge. This form of parallel input is called a synchronous load.

Several types and lengths of shift registers are available in integrated circuit form. Many ingenious applications in counting, generating, and data manipulation have been devised for the versatile shift register. One early application was as the circulating memory register shown in figure 10-48. Parallel data outputs are often omitted from the shift register designed for circulating applications. This means that only five or six external connections to the register are required (input, output, clock, and power) no matter how many flip-flops are used. For this reason, serial input-output shift registers for circulating storage applications were one of the first circuits developed for large-scale integration (LSI). LSI shift registers of over 1000-bit lengths are made by several manufacturers and have been a popular form of data storage. Modern magnetic bubble memories are actually circulating serial memories that are accessed in much the same way.

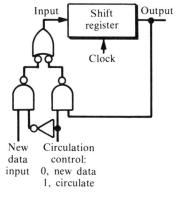

**Fig. 10-48.** A circulating memory register. A logic level in the shift register can be read serially at the output without being lost if it is connected back to the serial input. This input is gated so that the stored data or new data can be connected to the register input. If the clock were pulsed continuously, the entire register contents would appear at the output every $n$ clock cycles, where $n$ is the number of flip-flops in the register.

## 10-7   Application: Logic Test Probe

A very convenient test instrument for working with digital circuits is a **logic probe**. This device is shaped like a thick pencil with a metal probe at the tip. LED indicators on the probe indicate the logic level at the part of the circuit contacted by the probe. Additional circuits are sometimes available to detect pulses or pulse trains. The logic probe circuit shown in figure 10-49 has a convenient feature—the HI or LO logic level is indicated by an audible tone of high or low frequency.

## 10-8   Application: State Counter Sequencer

Automated instruments and control systems must often carry out a specific sequence of operations in order to perform their measurement or control function. Automatic sequencing is a necessity for total automation, but it may also provide increased convenience, speed, and reliability for many applications. It is assumed here that all required operations are ON-OFF operations (not continuous adjustments) and that each operation to be

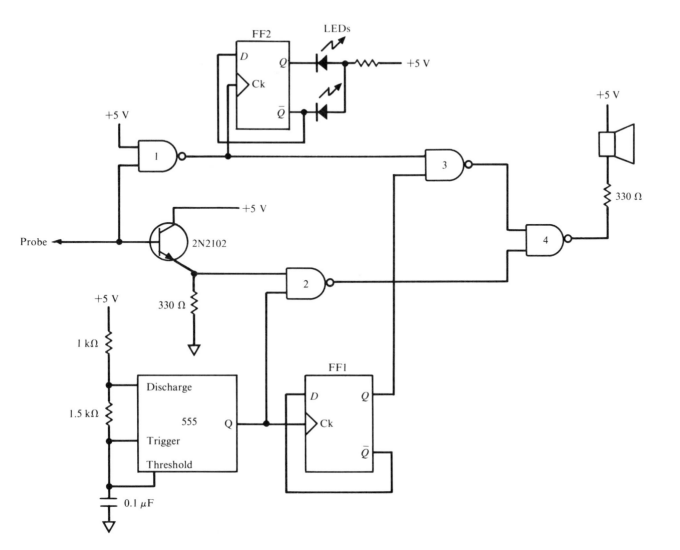

**Fig. 10-49.** Logic probe with audible output. The high frequency tone is generated by the 555 astable oscillator. A tone one octave lower (half the frequency) is obtained from the output of FF1. The high and low tones are multiplexed to the speaker by gates 2, 3, and 4. The multiplexer is controlled by gate 1 and the transistor. A LO at the probe input turns gate 3 off and gate 2 on to produce the high tone. When not connected, the control inputs to gates 2 and 3 are both LO and no tone is produced. The FF2 circuit alternates the illuminated LED on each LO→HI edge at the probe input.

performed can be controlled by a logic-level signal. Therefore a sequencer is a circuit or device that provides a logic-level signal output for each operation to be controlled and that determines the time relationship of the logic-level changes appearing at the outputs.

There are many approaches to the design of sequencers. Electro-mechanical devices with switch contacts operated by cams attached to a motor shaft were once quite popular and are still used. A series of mono-stable multivibrators as in the automatic recycling counter of figure 10-45 can be used for sequencing where the time relationships between operations and the reproducibility of the timing are not critical. Where timing is critical and complex operations must be sequenced a better approach is the state counter sequencer described here. The heart of the state counter sequencer is a counter, a multiplexer, and one or two decoders, a total of three or four MSI ICs.

To understand the state counter approach, consider a sequencer to turn on two lamps $A$ and $B$ for different time periods one after the other as might be encountered in a household security system. There are three distinct states to this sequence: State 0 = lamp $A$ OFF, lamp $B$ OFF; State 1 = lamp $A$ ON, lamp $B$ OFF; State 3 = lamp $A$ OFF, lamp $B$ ON. In State 0 the controller waits for a manual switch to be turned on to initiate the sequence. The ON condition of the switch is referred to as an input or transfer condition. When the input condition for State 0 is true, i.e., the switch is ON, the controller advances to State 1. The process of this advance causes an output or transfer function to occur: lamp $A$ is turned on, and a time delay is initiated. In State 1 the controller tests to see if the time delay for lamp $A$ has elapsed. If so, the sequencer advances to State 2 which causes lamp $A$ to turn off, lamp $B$ to turn on, and a second time delay to begin. The transfer condition for State 2 is the expiration of the second time delay. When this condition is met, no output function is required except to return to State 0 and repeat the cycle.

The application of this approach involves first converting the states, input conditions, and output functions into a standard flow chart format. The two basic flow chart units for each state are shown in figure 10-50a and b. The flow chart for the two-lamp sequencer is illustrated in figure 10-50c. The beauty of the state counter approach is that it can be implemented with the standard integrated circuits shown in figure 10-51. The heart of the sequencer is a $K$-bit counter, an $N$-bit multiplexer, and one or two $N$-bit decoders, where $N$ is the number of states in the sequence and $N \leq 2^K$. From the flow chart the number of states in the sequence and the various input conditions and output functions are known. If no secondary functions are needed, the secondary $N$-bit decoder can be eliminated. Both fall-through sequencers, which merely proceed from one state to the next, and branching sequencers, which can skip, repeat, or alter sections of the basic sequence, can be implemented with this simple approach. An eight-state

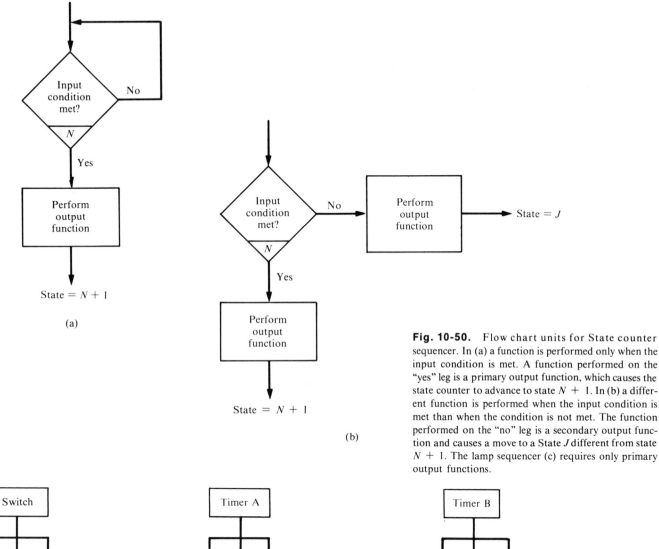

(a)

(b)

**Fig. 10-50.** Flow chart units for State counter sequencer. In (a) a function is performed only when the input condition is met. A function performed on the "yes" leg is a primary output function, which causes the state counter to advance to state $N + 1$. In (b) a different function is performed when the input condition is met than when the condition is not met. The function performed on the "no" leg is a secondary output function and causes a move to a State $J$ different from state $N + 1$. The lamp sequencer (c) requires only primary output functions.

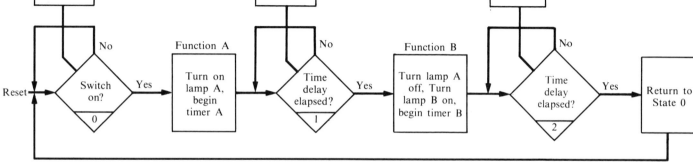

(c)

sequencer requires a 74163 counter, a 74151 multiplexer, and one or two 7442 decoders. Some additional circuitry is, of course, required for delay timing, for generation of the input condition signals, and for driving the various devices being sequenced.

**Fig. 10-51.**  General state counter sequencer. The $K$-bit counter keeps track of the state. The outputs of the state counter determine which input condition is supplied by the multiplexer to the decoders. The presence of a true input condition at the multiplexer output causes the appropriate primary output function to go true and the state counter to increment on the next clock pulse. A false input condition causes either a secondary output function line to go true or a wait period until the condition is true. When a jump or branch in the sequence is necessary, the load (LD) line of the state counter is activated, and a new state is entered through the data inputs.

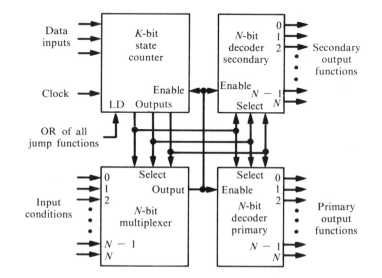

# Suggested Experiments

**1.  Basic logic gates.**
Determine the logic function performed for both HI-true and LO-true signals by the basic AND, OR, NAND, NOR, and INVERT gates.

**2.  Exclusive-OR and digital comparators.**
Verify the logic function for an exclusive-OR gate and the equality function for the inverted output. Combine equality gates to make a four-bit comparator. Observe the operation of a four-bit MSI magnitude comparator IC.

**3.  Logic operations.**
Combine various basic gate circuits to achieve a binary adder, a two-bit decoder, and a two-input multiplexer. Observe the operation of MSI decoder and multiplexer ICs.

**4.  Latches and registers.**
Connect NAND gates to obtain the basic latch, gated latch, and data latch of figures 10-25, 10-26, and 10-27. Confirm their operation. Observe the operation of a quad latch IC or other IC registers.

**5.  Flip-flops.**
Connect and verify the operation of various IC flip-flops including a *JK* flip-flop, a *D* edge-triggered flip-flop, a *JK* edge-triggered flip-flop, and a flip-flop with data lockout. (The latter FFs can be connected with gates and simple flip-flops as shown in figures 10-33 and 10-34.)

**6.  Counters.**
Connect four *JK* flip-flops in the binary counter of figure 10-36, and observe the output sequence. Modify the circuit to produce a down-counter. Connect and operate an IC binary up-down counter that can be preset to any particular modulus from 1 to 15.

**7.  Counter gating.**
Construct the counting gate that is part of figure 10-45, and verify its operation.

**8.  Shift registers.**
Connect a presettable eight-bit shift register. Demonstrate the operation of the shift register as a serial-to-parallel converter, a parallel-to-serial converter, and a shift left/shift right register. (Shift left can be obtained by connecting each output to the preset input of the previous flip-flop).

# Questions and Problems

**1.**  For a TTL gate an input that is unconnected acts as though it is connected to a HI signal. (a) If an input on a three-input AND gate is unconnected, what is the effect on the gate operation of the other two inputs? (b) What would be the effect if the same situation occurred with an OR gate?

**2.**  How many entries are there in the complete truth table for (a) a three-input gate, and (b) an eight-input gate?

**3.**  Make a truth table for the AND-OR-INVERT gate of figure 10-14.

**4.**  Explain why the truth table for a logic operation is always correct if expressed in 1's and 0's and a table of states for a particular logic circuit is always correct if expressed in HIs and LOs.

**5.**  Write the table of states (in HIs and LOs) for a two-input gate that performs the exclusive-OR operation for HI-true signals. Write the logic function that this gate performs on LO-true signals.

**6.**  Use the basic theorems of Boolean algebra to prove the following identities: (a) $A(\overline{A} + B) = AB$; (b) $AB + \overline{A}C = (A + C)(\overline{A} + B)$; (c) $(A + B)(\overline{A} + C) = AC + \overline{A}B$.

**7.**  Verify the following identities using Boolean algebra: (a) $\overline{A} + B + \overline{A} + \overline{B} = A$; (b) $AB + AC + B\overline{C} = AC + B\overline{C}$; (c) $AB + \overline{B}\overline{C} + A\overline{C} = AB + \overline{B}\overline{C}$.

**8.**  A lamp is to be controlled by logical combinations of three switches *A*, *B*, and *C*. Whenever switches *B* and *C* are in the same position, the lamp is to be ON. When *B* and *C* are in opposite states, the lamp is to be ON when switch *A* is CLOSED. (a) Draw the truth table for the above operation. (b) Write the Boolean expression for the lamp *M* in terms of *A*, *B*, and *C*. (c) Simplify the resulting expression, and suggest a logic circuit to accomplish the desired function.

**9.**  (a) What is the truth table for the circuit of figure 10-52a below? (b) Confirm that the circuit of figure 10-52b is a parity-detector; i.e., that the output depends on whether the number of inputs that are 1 is odd.

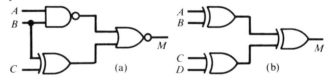

**Fig. 10-52.**

(c) Draw a circuit for an eight-bit parity detector.

**10.**   Draw a gate circuit that performs a binary subtraction for 2 one-bit numbers.

**11.**   Ready will pass this course if Able will help him study and if he does not have a date with Willing this weekend. However, Able and Willing's roommate are going skiing if it snows. Willing is planning to go out of town with her roommate this weekend if her car is working and if her roommate is not skiing. Under what conditions will Ready pass this course? Design a circuit with binary sources, gates, and indicators to solve this problem.

**12.**   It is important that no more than one of the control inputs of the multiplexer of figure 10-22 be 1 at a time. (a) Show that if the control inputs were obtained from the binary decoder of figure 10-19, this condition would be met. (b) Design a circuit including an oscillator and counter that will cause the multiplexer to sample each of the four inputs in sequence, repeating the cycle continuously.

**13.**   Develop a table of states like that of figure 10-26 for a gated latch made of NOR gates.

**14.**   (a) What is the table of states for the $D$ flip-flop of figure 10-32? (b) Combine flip-flops to produce a $D$ flip-flop with data lockout.

**15.**   Describe the difference in operation between a synchronous counter and an asynchronous counter.

**16.**   Draw a four-bit binary down-counter. Sketch the output waveforms for the complete count sequence.

**17.**   In the most popular asynchronous BCD counter IC, the $A$ output and the $B$ input are not connected internally. (a) Sketch the count sequence waveforms that result if the $A$ flip-flop is connected to follow, rather than precede, the $B$, $C$, and $D$ flip-flops in the counter. (b) Is the circuit still a decade counter? (c) What is the difference in the final output waveform between this circuit and the normal connection shown in figure 10-40?

**18.**   Some presettable counters are synchronous load and others are asynchronous load. Synchronous means that the loading occurs at the time of the clock signal. One type gates the preset data to the $J$ and $K$ inputs of the flip-flops and the other to the Pr and Clr inputs of the flip-flops. Identify the flip-flop inputs used for each type of presettable counter, and explain your reasons.

**19.**   Design a modulo-13 counter using $JK$ flip-flops, AND gates, and inverters. Show the complete count sequence.

**20.**   One of the many applications of the shift register is as a sequencer. As a 1 level is clocked along the register replacing 0's, circuits connected to the various outputs could be enabled in sequence. (a) Design a circuit that would ensure that when the 1 reached the last flip-flop in the register, the register would be completely reset to 0 and the 1 level would be ready to be clocked into the first flip-flop in the register. (b) Suppose it was desired to initiate four successive operations of which the second, third, and fourth occur 5 ms, 7.5 ms, and 15 ms after the first operation has begun. How many bits long should the shift register be, and what should be the clock frequency?

# Digital Devices and Signals

# Chapter 11

The logic gate is the basic circuit element in all modern digital circuits. There are several families of logic gates that differ in speed, logic levels, noise immunity, power consumption and other characteristics. The different gate families are described in this chapter, as well as their characteristics and limitations. Computer communication often occurs along a shared line or set of lines known as a bus. The open collector gate and the tristate gate allow gate outputs to be connected together for bus-oriented operations. Digital signal conditioning circuits are used to convert between different logic levels, to shape signals, and to discriminate against noise. Because of the rapid transitions that digital signals undergo, transmitting them over long distances poses special problems. The techniques and circuits for data transmission and reception are therefore discussed along with serial communication standards. The development of medium- and large-scale integrated circuits has made possible a tremendous increase in the complexity of circuits and functions available in IC form. This chapter concludes with a discussion of the arithmetic logic unit, the binary multiplier, and other selected examples of medium-scale integration (MSI) and large-scale integration (LSI) devices.

## 11-1 Logic Families

The characteristics of a computer or digital instrument are determined largely by the speed, noise immunity, versatility, and durability of the logic circuits that are used hundreds of times over in even the simplest systems. The basic logic functions (AND, OR, INVERT, NAND, NOR, etc.) can all be implemented by many different circuits. These logic circuits can be classified into families based upon how they perform the logic functions. This section describes the transistor-transistor logic family (TTL), the emitter-coupled logic family (ECL), the integrated-injection logic family ($I^2L$) and logic families based on MOSFET devices.

### TTL Gates

The first commercially available family of logic circuits was the **resistor-transistor logic (RTL)** family. Circuits based on RTL had relatively slow switching speeds (~50 ns) and poor noise immunity; they were also expensive to produce in IC form. A natural development from the diode logic gates mentioned in chapter 10 was the **diode-transistor logic (DTL)** gate made by adding a transistor inverter to the basic diode gate to reduce the output impedance and reestablish the logic levels. The DTL NAND gate shown in figure 11-1 was faster than RTL (~30 ns) and could be more economically manufactured in IC form.

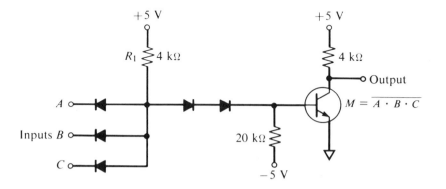

**Fig. 11-1.** DTL NAND gate. The input diodes and $R_1$ are the AND circuit. When the AND output is LO (~0 V), the transistor is cut off, and the output is HI (~+5 V). When the AND output is HI, the transistor saturates, and the output is LO. The gate function is thus $M = \overline{A \cdot B \cdot C}$ or the NAND function.

Along with the development of RTL and DTL circuits came the realization that some circuit improvements were possible with ICs that were not practical with discrete components. One such example is the multiple-emitter transistor of the TTL gate shown in figure 11-2. The multiple-emitter transistor replaces the input diodes of the DTL gate and provides superior speed. The output stage consists of transistors $Q_3$ and $Q_4$ and is called a **totem pole** or **push-pull** output because the output transistors are driven to be in opposite

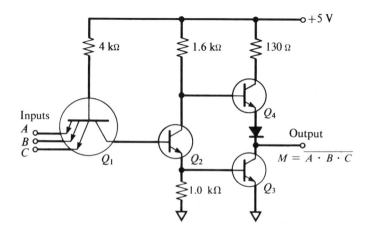

**Fig. 11-2.** TTL NAND gate. Transistors $Q_4$ and $Q_3$ are always in opposite states. When the output is LO, $Q_3$ is saturated, and $Q_4$ is cut off. When the output is HI, $Q_3$ is cut off, and $Q_4$ is saturated. This switching of $Q_3$ and $Q_4$ is accomplished by transistor $Q_2$, which provides the bases of $Q_4$ and $Q_3$ with voltages that swing in opposite directions. A LO signal at any one or more of the inputs forward biases $Q_1$, which turns $Q_2$ OFF and results in a HI output. When all inputs are HI or unused, $Q_2$ is ON, and the output is LO.

states by phase splitter $Q_2$. The development of TTL and its commercial introduction in 1964 as the 54/74 IC series soon made the older RTL and DTL families obsolete.

**TTL Gate Characteristics.**    When a LO signal source is connected to one of the TTL gate inputs, the base-emitter junction of $Q_1$ is forward biased, and a current must pass to ground through the signal source without raising the input voltage above the upper limit of the LO level. Therefore, the signal source for a TTL gate is required to absorb or "sink" current from the gate input in order to hold that input in the LO state. Gates of this type are called **current-sinking** gates.

Often the output of a logic gate is connected to several other gate inputs. There is a limit to the number of gate inputs that can be driven by a gate output without jeopardizing its logic level. This limit, called the **fan-out**, is set by the maximum current the gate output can sink before its output level exceeds the highest value for the LO state. Standard TTL gates have a fan-out of ten to other normal TTL gates.

The **noise margin**, or **noise immunity**, of a gate is the smallest HI level output voltage minus the minimum effective HI level input voltage, or the minimum effective HI level input voltage minus the maximum LO level output, whichever is smaller under the worst-case conditions. A noise pulse exceeding this value has a chance of affecting the state of the gate. The standard TTL gate has a noise margin of more than 1 V. An important switching speed characteristic is the gate **propagation delay time.** The HI→LO propagation delay is the time required for the gate output to fall to 50% of its HI value measured from the time the input waveform reaches 50% of its final value. The LO→HI propagation delay is similarly measured. Usually, the average of these two times is quoted in manufacturers' specifications. The standard TTL gate has a propagation delay of ~10 ns. Another important gate characteristic is the **power dissipation**, which for standard TTL logic is 10 mW per gate.

**TTL Gate Variations.**    Several versions of the basic TTL gate are available to meet special requirements. For lower power dissipation the resistances in the standard TTL gate are increased approximately tenfold, and a Darlington booster stage is added to the TTL "totem pole" output. The **low-power TTL** series (series 54L/74L) has a power dissipation of only 1 mW per gate at some sacrifice in propagation delay (typically 33 ns).

Conversely, for higher speed the resistances are reduced, and clamping diodes are added on the inputs to eliminate negative transients and transmission line reflections. The propagation delay for the resulting **high-speed TTL** gate is reduced to 6 ns, but the power dissipation in the series 54H/74H is increased to 20 mW per gate. Flip-flops in this series can be clocked at an average rate of 50 MHz.

**Note 11-1.   Schottky Diodes**
The Schottky diode has a rectifying metal-semiconductor junction instead of the usual p-n junction. This results in a small forward voltage drop (~0.3 V) and very short storage time.

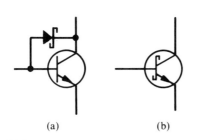

(a)                    (b)

**Fig. 11-3.**   Schottky-clamped transistor. The Schottky diode in (a) is connected from base to collector. Since the diode has a lower forward voltage drop than the base-collector junction, base current is diverted to the diode when the transistor turns on. The diversion of base current prevents the transistor from saturating by eliminating the excess charge stored in the base region. The circuit symbol for a Schottky-clamped transistor is shown in (b).

**Fig. 11-4.**   Low-power Schottky TTL NAND gate. Most low-power Schottky (LS) TTL circuits use the DTL input circuit with Schottky diodes to perform the AND function. This circuit is faster than the multiple-emitter transistor input structure and has a higher breakdown voltage. Other input arrangements in the LS series use diode clusters or pnp input transistors.

For still greater speed transistor saturation in the TTL gate is eliminated by using **Schottky diodes** (see note 11-1) as clamps on each transistor as illustrated for a single transistor in figure 11-3. As a result of the clamping, the Schottky TTL gate (series 54S/74S) has a propagation delay of only 3 ns with no increase in power dissipation over standard TTL.

A combination of low-power technology and Schottky clamping has resulted in the increasingly popular **low-power Schottky TTL** variation illustrated by the two-input NAND gate of figure 11-4. The 54LS/74LS series has the speed of standard TTL (~9.5 ns propagation delay) with a power dissipation one-fifth as great (2 mW per gate). For applications in which speed and low power are critical, low-power Schottky TTL gives the lowest product of propagation delay and power dissipation of any of the TTL variations. Table 11-1 compares all the TTL variations as to speed, power dissipation, and the product of speed and power. The **speed-power product**, a figure of merit for gate performance, is the product of the gate delay and the power dissipation. It is generally given in picojoules (pJ).

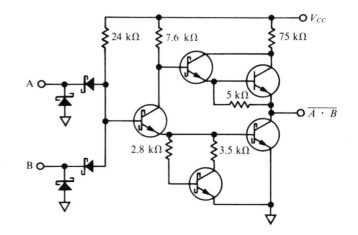

The fan-out of standard, high power, and Schottky TTL gates is 10 when they are driving other gates of the same type. Low-power gates and low-power Schottky gates have a fan-out of 20 in the same series. However, the output capacity and input load of the several TTL types are different when they are mixed in the same system. The fan-out capability for driving other types of inputs can be obtained from table 11-2. For example, a 74LS gate can drive one or two standard TTL inputs *and* twelve 74LS inputs. A 74S output can drive 12 standard TTL inputs or 50 74LS inputs.

In addition to the variations described above, TTL gates are available with open collector and tristate outputs, which are bus compatible outputs. These gate circuits are discussed in section 11-2.

**Table 11-1.**  Comparision of TTL families.

| Series | Gates | | | Flip-flops |
| | Propagation delay (ns) | Power dissipation (mW) | Speed-power product (pJ) | Frequency range |
|---|---|---|---|---|
| 54LS/74LS | 9.5 | 2 | 19 | dc to 45 MHz |
| 54L/74L | 33 | 1 | 33 | dc to 3 MHz |
| 54S/74S | 3 | 19 | 57 | dc to 125 MHz |
| 54/74 | 10 | 10 | 100 | dc to 35 MHz |
| 54H/74H | 6 | 22 | 132 | dc to 50 MHz |

**Table 11-2.**  Loading rules for mixing TTL families.

| Gate type | Fan-out to standard TTL load unit | Input load normalized to standard TTL gate |
|---|---|---|
| 54L/74L | 2.25 | 0.11 |
| 54LS/74LS | 5.0 | 0.25 |
| 54/74 | 10.0 | 1.0 |
| 54H/74H, 54S/74S | 12.5 | 1.25 |

## Emitter-Coupled Logic Gates

One of the limitations on the maximum switching speed of all the transistor gates described thus far (except the Schottky TTL gates) is the storage time of the saturated transistor. Several gates have been designed to eliminate this delay by not allowing the transistor to saturate. One of the earliest and most successful of these circuits is the **emitter-coupled logic (ECL)**. The principles of the ECL gate operation are illustrated in figure 11-5, where a difference amplifier is shown. In its logic application the fixed emitter current is switched from one transistor to another when the logic level of the input signal $V_{in}$ is changed. The transistor pair thus forms a nonsaturating **current mode switch**.

In the complete ECL gate shown in figure 11-6, parallel transistors $Q_1$, $Q_2$, and $Q_3$ provide for multiple gate inputs. The reference voltage is supplied

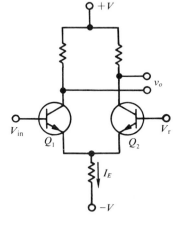

**Fig. 11-5.**  Nonsaturating current mode switch. This difference amplifier switches the emitter current $I_E$ from $Q_1$ to $Q_2$. When $V_{in}$ is LO ($V_{in} < V_r$), the emitter current is switched to $Q_2$. When $V_{in}$ is HI ($V_{in} > V_r$), the emitter current is switched to $Q_1$.

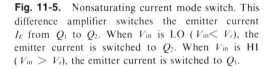

**Fig. 11-6.** ECL gate circuit. Transistors $Q_3$ and $Q_4$ form the nonsaturating current mode switch. When the inputs are all LO ($-1.5$ V), transistors $Q_1$, $Q_2$, and $Q_3$ are cut off, and the emitter current is switched to $Q_4$. This results in a LO at the OR output and a HI at the NOR output. If any input is HI ($-0.75$ V), $Q_4$ is cut off, and the OR output is HI.

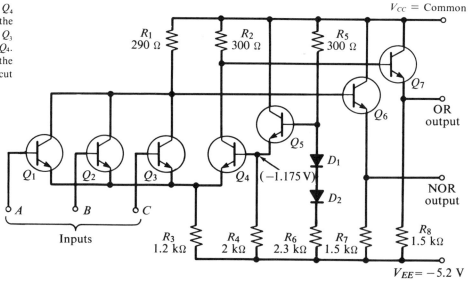

by the divider $R_5$ and $R_6$ and by emitter-follower $Q_5$. Diodes $D_1$ and $D_2$ compensate for the temperature dependence of the base-emitter voltages of $Q_5$ and $Q_4$. Outputs are taken at the collectors through emitter followers, which provide buffering and a low output impedance. Note that the collector resistors are connected to common and the emitter supply voltage $V_{EE}$ is negative. Thus all the voltages used are negative. If inputs $A$, $B$ and $C$ are all LO ($\sim -1.5$ V), $Q_1$, $Q_2$ and $Q_3$ are cut off, and the emitter current is switched to $Q_4$. The emitter current through $R_2$ causes a voltage drop across $R_2$ and a corresponding negative voltage at the base of emitter follower $Q_7$. The OR output is

$$\frac{-(5.2 - 1.9)\text{V}}{1.2 \text{ k}\Omega} \times 300 \text{ }\Omega - V_{BEQ_7} \simeq -1.5\text{V} \text{ (a logic LO)}$$

Since the parallel input transistors are cut off, there is almost no current through $R_1$, and the voltage at the base of emitter follower $Q_6$ is nearly 0 V. Thus the NOR output is $0 - V_{BEQ_7} \simeq -0.75$ V.

If one or more inputs is increased to the HI state ($\sim -0.75$ V), $Q_4$ is cut off, and the OR output is HI ($\sim -0.75$ V). The $IR$ drop across $R_1$ is (5.2 V $-$ 1.5 V) (290 $\Omega$ / 1.2 k$\Omega$) = 0.77 V, which results in $\sim -1.5$ V (logic LO) at the NOR output. Thus the outputs are the stated logic function of the input levels for HI-true logic. The great speed of the ECL gate is due to the elimination of the saturated state for the transistors and to the relatively small output voltage change between HI and LO ($\sim 0.75$ V).

The current-steering switch automatically provides differential, or "complementary," outputs. The gate designers have taken advantage of this to provide both OR and NOR outputs. Gates of this type are also called **current-steering logic (CSL)** and **current-mode logic (CML)**. The standard ECL propagation delay is about 2–4 ns, with flip-flop clocking rates of over 100 MHz and power dissipations of 25–50 mW per gate.

The noise margin is approximately 250 mV. Because the ECL gate has a very high input impedance and a very low output impedance, high fan-out is achieved. The relatively constant total current in the gate maintains a constant drain on the power supply and prevents internally generated noise. Its principal disadvantage is the small logic swing, which makes it susceptible to external noise.

## Integrated-Injection Logic

One of the newest logic families to be introduced commercially is **integrated-injection logic ($I^2L$)**. A typical $I^2L$ gate uses very little space and consumes very little power. The speed-power product of $I^2L$ is in the range of 0.1 – 0.7 pJ per gate compared to 100 pJ for normal TTL. Typical gate delays are ~15 ns. It is thus highly suitable for MSI and LSI applications. The logic levels of $I^2L$ gates are not compatible with other logic families. For this reason logic-level translators are provided on the IC for buffering and output drive capability. Most of these provide TTL compatible inputs and outputs. Because of the level translation requirements, $I^2L$ is only used in ICs that have more than ~75 gates.

The basic structure of $I^2L$ gates uses multiple-collector transistors as shown in figure 11-7. The $I^2L$ gate is a form of **direct-coupled transistor logic (DCTL)**. The NOR logic operation is performed on HI-true signals. If a collector of a gate whose input is $A$ is connected to a collector of a gate whose input is $B$ and the common collectors are connected to the base of

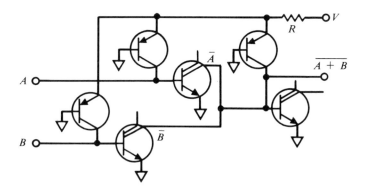

**Fig. 11-7.** Basic configuration of an $I^2L$ gate. The pnp transistors provide base current to the multiple-collector transistors. The logic is performed by the direct-coupled multiple-collector transistors.

another gate, the logic at the base is $\overline{A+B}$. The transistor whose input is $A$ has $\overline{A}$ at its collector, and the joining of various collectors yïelds the logical AND of the collector variables $(\overline{A} \cdot \overline{B}) = \overline{A + B}$. Thus, with the multiple-collector structure many different logic functions can be implemented by different collector connections in the IC.

Recent developments in $I^2L$ include the use of Schottky diodes to improve the speed-power product even further and the introduction of **isoplanar integrated-injection logic ($I^3L$)**, which increases the current gain of the gate transistors. With the latter technology, the speed-power product can be as low as 0.15 pJ per gate.

## C-MOS Logic

Gates made with field-effect transistors offer the advantages of high input impedance, low power consumption, greater circuit simplicity, and further reduced size. With these advantages goes a disadvantage—significantly slower speed due to high input capacitance. A number of FET gate types have been manufactured, including designs based on p-channel FETs (P-MOS) and n-channel FETs (N-MOS). Although they are highly successful for many LSI circuits such as microprocessors, large memories, calculators, long shift registers, etc., neither of these types has displaced TTL for basic gate and flip-flop applications. As a result, almost all P-MOS or N-MOS LSI circuits provide TTL compatible input and output logic levels for interfacing to other digital circuits.

A logic family made of complementary p-channel and n-channel pairs of MOSFET transistors (C-MOS) is, however, widely used for basic logic operations. The C-MOS family is based on the inverter shown in figure 11-8a. The simplicity of the inverter is apparent from the diagram. Only the

**Fig. 11-8.** C-MOS inverter (a), series A NOR gate (b), and series B NOR gate (c). In (a) a HI input voltage turns the n-channel FET ON and the p-channel FET OFF, giving a LO logic-level output. With a LO input voltage, the p-channel FET is ON, and the n-channel FET is OFF, giving a HI logic level. In the NOR circuit (b) the n-channel FETs are in parallel across the output so that if $A$ $or$ $B$ is HI, the output is LO. The p-channel FETs are in series with the supply voltage and ensure that $M$ is HI only when $A$ $and$ $B$ are LO. The series B NOR gate (c) has a pair of inverters on the NOR output for buffering purposes.

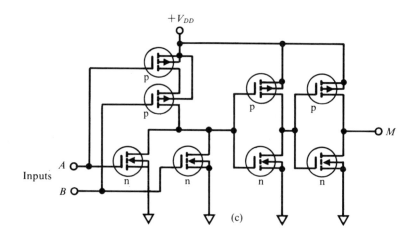

complementary transistors are required. C-MOS logic functions are designed around the basic inverter as illustrated by the two-input NOR gate circuit of figure 11-8b. The equation $\overline{M} = A + B$ applies for the n-channel FETs, and $M = \overline{A} \cdot \overline{B}$, applies for the p-channel FETs. Both equations represent (by DeMorgan's theorem) the NOR function. The B series C-MOS gate shown in figure 11-8c is a significant improvement over the older A series. The output buffers ensure that the output is driven to uniform values in both directions and they sharpen the rise and fall times. The B series devices are often slightly faster than the A series devices.

The C-MOS logic family has a variety of significant characteristics that make it attractive for many applications. One of the most significant factors is that C-MOS logic can operate over a range of power-supply voltages from 3 to 18 V. By contrast, TTL logic must operate in a narrow range of supply voltages (generally 4.75 – 5.25 V). The C-MOS family has the added advantage that the output logic-level transition occurs when the input voltage is halfway between $+V_{DD}$ and common. This gives an excellent noise immunity, typically 30–40% of the supply voltage. The logic-level swing of the output (unloaded) is from the supply voltage to common, the full range of the available voltage. Low power dissipation is also a significant advantage of C-MOS. Power dissipation is often two to three orders of magnitude lower than that of TTL logic. The exact power dissipation depends on the operating frequency, but is in the microwatt range for low-frequency operation. At frequencies above a few megahertz, low-power Schottky consumes less power than C-MOS.

Since the gate output is driven to both HI and LO levels and since the gate input current is very low, the fan-out of this series is very high ($>50$ to other C-MOS gate inputs). The propagation delay depends on the supply voltage, but it is about 60 ns for 5 V. The flip-flop clock rate for a 5-V supply is about 4 MHz.

Although the inputs to MOSFETs are essentially open circuits, the dielectric region of the gate capacitor is extremely thin. Thus, if static electricity were to reach the dielectric region, very high field strengths could result, and the gate could readily be destroyed. To alleviate these in-circuit problems with static electricity all C-MOS devices have built-in static protection and input gate protection. To avoid static outside the circuit, C-MOS ICs should be kept in conductive foam. Because the input protection networks are often diodes, it is necessary to limit the input current to 10 mA or less if the input voltage exceeds $V_{DD}$, the supply voltage. This latter condition may occur if the C-MOS device power is inadvertently turned off while the input voltage is on. For trouble-free use of C-MOS devices, the simple rules listed in table 11-3 are recommended.

When C-MOS circuits are operated from a $+5$ V power supply, their outputs are directly compatible with TTL input logic levels. The fan-out of a

**Table 11-3.** C-MOS usage rules.

1. Do not leave inputs floating.
2. Limit input currents to 10 mA if signals exceed $+V_{DD}$.
3. Avoid static when handling.
4. Debounce mechanical switch inputs.
5. Use sharp transitions on clocked circuits.

standard C-MOS gate is about 0.25 standard TTL load units, making it directly compatible with a single low-power TTL input. Driver gates having a fan-out of two standard TTL load units are available. The TTL gate outputs can drive C-MOS inputs if a 2-k$\Omega$ resistor is connected from the gate output to the +5-V supply. This is required to ensure a positive HI level signal under the low current drain of the C-MOS input. The dc current available would provide a very large fan-out. However, the C-MOS input capacitance causes a propagation delay increase of 3 ns per input connected to the TTL output.

The +5-V supply levels of TTL are not optimal for C-MOS; supplies at 9–12 V give faster operation, better drive capacity and higher noise immunity. Interfacing TTL circuits to C-MOS operating at these higher voltage levels requires that a high-voltage, open collector TTL driver be used (see section 11-2). To interface C-MOS operating at +10 V to TTL, a C-MOS to TTL driver (4049 or 4050) operated from the +5-V TTL supply should be used. These devices are internally protected to allow the +5-V supply to be exceeded without damage.

Series B C-MOS gates at lower power supply voltages can drive LEDs directly without the usual current-limiting resistors. With higher voltage operation or with an output that is not in the B series, the current-limiting resistor is required.

A relatively new form of C-MOS, called **C-MOS-on-sapphire** or **silicon-on-sapphire (SOS)** can achieve higher speeds than normal C-MOS because of a reduction in device capacitance. Power dissipation is also reduced to less than 0.1 $\mu$W per gate because of lower leakage currents. The SOS technology is still relatively immature, but it has great promise for future LSI developments.

## 11-2   Open Collector Gates and Tristate Logic

In digital systems it is often desirable to connect gate outputs together to allow them to share a common data line. This could, of course, be done with AOI gates or multiplexers, but it is often more convenient and economical to be able to interconnect directly all the gate outputs (and inputs) that are to share a particular line. A set of shared lines in a modern digital system is called a **bus**. In a bus-oriented system, the same data lines and control lines are common to each subsystem module. With normal TTL logic gates, the totem pole output (driven HI and LO levels) does not allow gate outputs to be tied together. The open collector TTL gate and the tristate logic gate, introduced in this section, have outputs that allow such wired connections. The logical consequence of connecting gate outputs together is discussed, and examples of bus-oriented operations are presented.

## Open Collector Gates

The TTL NAND gate of figure 11-2 has the standard totem pole arrangement of output transistors $Q_3$ and $Q_4$. In the LO output state $Q_3$ is ON, and $Q_4$ is OFF. In the HI output state $Q_4$ is ON, and $Q_3$ is OFF. This push-pull, or totem pole, arrangement can be visualized by simple switch equivalent circuits as shown in figure 11-9. It is this push-pull arrangement that gives TTL gates their high switching speeds and good fan-out characteristics. Because the output impedance is low in both states, capacitive loads can be driven without serious degradation of switching times. However, if two gate outputs are connected together and one gate output attempts to go HI and the other to go LO, the gate outputs would work against one another.

The **open collector** gate is shown in figure 11-10. Here transistor $Q_4$ of the normal TTL gate is missing so that only the LO output state is actively driven. In order to provide the HI output level when $Q_3$ is not conducting, the collector of $Q_3$ is connected to the supply voltage $V_{CC}$ through an external **pullup resistor**. When the gate output should be HI, $Q_3$ turns OFF, and the external resistive connection to $V_{CC}$ establishes the HI output level. In the LO output state $Q_3$ is conducting, and the pullup resistor serves to limit the current through $Q_3$.

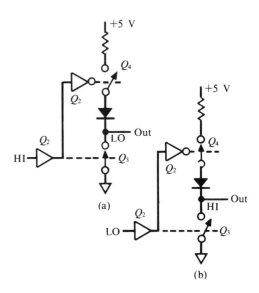

**Fig. 11-9.** Switch equivalent circuit for TTL totem pole output. In the LO output state (a), transistor $Q_2$ drives $Q_3$ to be closed and $Q_4$ to be open. In the HI output state (b), $Q_4$ is closed and $Q_3$ is open. In both states the output is derived from an ON transistor. Thus both states are low-impedance outputs.

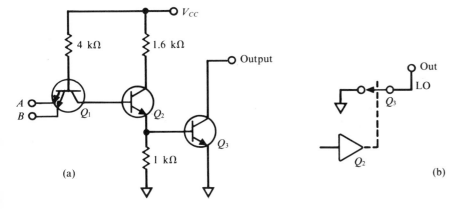
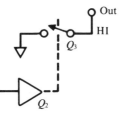

**Fig. 11-10.** Open collector TTL NAND gate. The actual gate circuit is shown in (a). Note that the output transistor acts simply as a switch to common as shown by the switch equivalents in (b). An external pullup to $V_{CC}$ is necessary to establish the HI output level when $Q_3$ is off.

Because the output of an open collector gate is simply a switch to common as shown in figure 11-10b, the outputs of several gates can be safely connected together with a single pullup resistor as shown in figure 11-11. If either of the gate outputs $X$ or $Y$ goes LO as a result of both inputs to that gate being HI, the output of the gate array is LO because the LO output gate has an ON output transistor that sinks the current through $R_L$ by connecting it to common. Only when both gate outputs are HI is the array output HI.

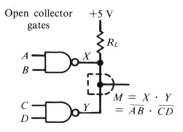

**Fig. 11-11.** Wired-AND connection of open collector NAND gates. If either gate output goes LO, the output at $M$ is LO. Only when both gate outputs $X$ and $Y$ are HI is the output at $M$ HI. The wire connection thus performs the AND logic function on gate outputs $X$ and $Y$. The overall operation is the NAND-AND function shown.

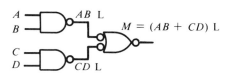

**Fig. 11-12.** Functional equivalent of wired-AND connected NAND gates for LO-true signals. The wired-AND connection performs the OR operation for LO-true signals.

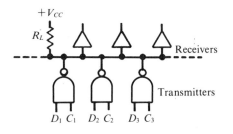

**Fig. 11-13.** Bus line with multiple transmitters and receivers. Multiple data sources can be connected to a single line as long as only one transmitter at a time is enabled to drive the line. The others are held in their nondriven state by the corresponding gate control inputs. The pull-up resistor $R_L$ holds the line HI when none of the transmitters is LO. The receivers present no data conflict, but each transmitter must be able to handle the total load.

Thus the HI-true logic condition of an AND gate is performed on the open collector gate outputs, as shown in the Boolean expression $M = X \cdot Y$. This implementation of the AND function is called wired-AND logic. The dotted AND gate in figure 11-11 is sometimes used to show the logic function performed by the connected outputs even though an AND gate is not physically present. From DeMorgan's theorems, an AND gate for HI-true signals performs the OR function on LO-true signals. This is shown by the equivalent gate circuit of figure 11-12. Thus the connected open collector gates are sometimes referred to as wired-OR logic when LO-true signals are used. Because gates are generally named for their HI-true function, the term wired-OR is somewhat misleading.

Open collector outputs permit several devices to share a bus line as shown in figure 11-13. Control inputs to the transmitting gates keep all but one of them in their nondriven state. This puts the bus line under the control of the enabled gate. With some open collector gates, it is possible to tie the outputs to a higher supply voltage than the +5 V used to power transistors $Q_1$ and $Q_2$. This allows the open collector gate to drive lamps, LEDs and C-MOS loads that require higher voltages than +5 V.

Because of the passive pullup and the size of the pullup resistor (usually $>1$ k$\Omega$), open collector gates can have substantially slower switching speeds than normal TTL gates particularly on the LO→HI transition. The open collector gate delay is typically about 30 ns with a nominal load capacitance of 15 pF and a pullup resistor of 4 k$\Omega$. As a rule the capacitance to be driven, particularly in a bused system, may be much higher than 15 pF. Thus typical bus delays due to $RC$ factors are often greater than 100 ns. (Note that in some logic families, such as ECL, the outputs are driven in only one state so that it is possible to connect the normal gate outputs in parallel).

## Tristate Gates

The tristate gate was developed specifically for bus-oriented operations. **Tristate logic** is fully TTL compatible and is actively driven in both HI and LO states. However, in contrast to normal TTL, the tristate gate can be turned completely OFF so that it exhibits a third high-impedance state that effectively disconnects the gate output from any subsequent circuits. It thus retains the speed and drive capabilities of normal TTL with the added flexibility of being bus compatible.

The tristate gate is similar to the totem pole TTL gate except that both output transistors can be turned OFF to give the high-impedance third state. Figure 11-14 shows a typical tristate gate structure and its output switch equivalent circuit. A HI logic level at the Disable input turns both output transistors OFF and forces the high-impedance third state no matter what

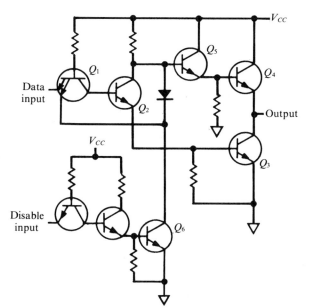

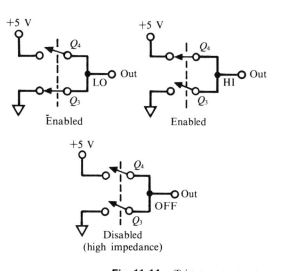

**Fig. 11-14.** Tristate gate structure and output switch equivalents. A HI logic level at the Disable input turns $Q_6$ ON. This turns both output transistors OFF and effectively disables the gate. When $Q_6$ is ON, a LO applied to input transistor $Q_1$ turns OFF $Q_2$ and $Q_3$. Transistors $Q_5$ and $Q_4$ are also turned OFF when the base of $Q_5$ is pulled LO by the $Q_6$ output. When the disable input is LO, $Q_6$ does not conduct, and normal TTL behavior is exhibited.

the logic level at the data input. When Disable is LO, the output is normal for a TTL gate.

The tristate gate eliminates the need for pullup resistors and significantly increases switching speeds over those of open collector gates. Typical bus delays with tristate gates are less than 10 ns since $RC$ time constants are smaller when both HI and LO states are actively driven.

A single bus line with tristate **bus drivers** is shown in figure 11-15. Here the Enable input is used to place the gate in either the active state or the high-impedance, disabled, state. Tristate gates are available with either HI or LO Enable inputs. Normally only one driver is enabled at a time by external control logic. One feature built into all tristate devices is a longer time delay in switching from the disabled to enabled state than in switching from enabled to disabled state. This prevents data interference by ensuring that a previously enabled device will be disabled a few nanoseconds before a newly selected device is enabled.

Tristate devices are frequently found on ICs that are likely to be used with data buses. For example, figure 11-16 shows the pin configuration of a quad tristate buffer and driver chip. Such ICs have adequate drive capability for the connection of up to 128 devices to a common bus line.

A four-bit parallel bidirectional bus driver is illustrated in figure 11-17. Each buffered line of the four-bit driver contains two separate tristate buffers in order to provide simple bus interfacing and bidirectional operation. The chip select signal ($\overline{CS}$) enables the chip when LO and forces the

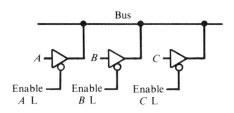

**Fig. 11-15.** Bus connections with tristate bus drivers. The signal at $A$ is driven on to the bus when a LO logic level is applied to Enable $A$. Only one bus driver should be enabled at any one time.

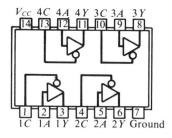

**Fig. 11-16.** Pin configuration of 74125 quad buffer-driver with tristate output. The inputs $1C$, $2C$, $3C$ and $4C$ are LO-true enables. When $1C$ is LO, the output at $1Y$ is $1A$. When $1C$ is HI, the output is in the high-impedance (disabled) state.

| $\overline{\text{DIEN}}$ | $\overline{\text{CS}}$ | FUNCTION |
|------|------|----------|
| 0 | 0 | $DI \rightarrow DB$ |
| 1 | 0 | $DB \rightarrow DO$ |
| 0 | 1 | High impedance |
| 1 | 1 | |

**Fig. 11-17.**   Logic diagram of four-bit parallel bidirectional bus driver. On the $DB$ side of the driver, the output of one buffer and the input of another are tied together. On the other side the inputs ($DI$) and outputs ($DO$) are separated to permit flexibility. The control signals $\overline{\text{CS}}$ and $\overline{\text{DIEN}}$ are device select and direction controls respectively, according to the truth table shown.

high-impedance state when HI. The *data in en*able signal $\overline{\text{DIEN}}$ controls the direction of data flow. The complete truth table is shown in figure 11-17.

In addition to the examples given here, tristate outputs appear on random-access memory chips, data selector-multiplexer chips, counter-latch chips and many microprocessor support chips. Additional examples of tristate logic outputs are given in later chapters of this book.

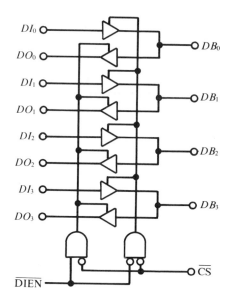

## 11-3   Digital Signal Conditioning

Signals that are to operate logic circuits reliably must meet two basic conditions: The logic levels must be well within the HI and LO logic-level ranges for the specific logic circuits used, and the transitions between logic levels must be clean and quick. The reason for the latter condition is that for several logic families a condition of instability can arise from slow movement of an input signal through the region between allowed logic levels. In this section, circuits for sharpening transitions, for discriminating against undesired signal components, and for converting from one set of logic levels to another are described. Several important pulse generation and shaping circuits are also presented.

### Level Conditioning

Ideally, the voltage level of a digital signal unambiguously identifies the signal as logic HI or LO. This clear distinction is achieved by a substantial

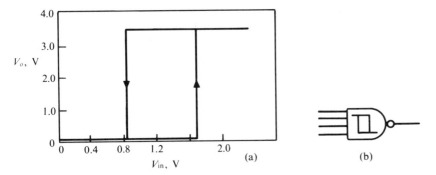

(a)

(b)

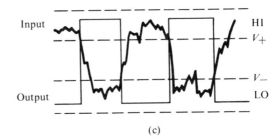

(c)

**Fig. 11-18.** Schmitt trigger input NAND gate. The transfer characteristic in (a) shows the ~800 mV of hysteresis present. The Schmitt input gate symbol in (b) reflects this hysteresis curve. A typical use for restoring degraded logic levels is shown in (c). A normal gate could be susceptible to multiple triggering.

separation of the HI and LO logic-level voltage ranges. In practice, many digital signal sources do not have outputs that conform completely to the standard for signals in the logic family. Also logic-level signals in a digital system may be degraded in amplitude or transition sharpness by loading or long transmission. In such cases the comparator and Schmitt trigger circuits can be used to establish sharp edges and appropriate levels.

**Schmitt Trigger Input Gates.** The operational amplifier comparator and the Schmitt trigger discussed in chapter 9 are useful for producing sharp logic-level transitions from a wide range of signal sources. Within the TTL family Schmitt trigger input gates are used for restoring the levels and edges of degraded logic signals. Figure 11-18 shows the transfer characteristics for the 7413 dual, four-input Schmitt trigger NAND gate along with the gate symbol and a typical application. Propagation delays for Schmitt trigger gates are typically less than 20 ns with normal TTL loads. Because of its speed and its typical TTL loading characteristics, a spare Schmitt trigger gate can readily be used as a standard gate. Schmitt inputs are also available on inverters and on line drivers and receivers.

**Level Translation.** The Schmitt trigger circuit is also an ideal circuit for converting a digital signal from the logic levels of one logic family to those of another. In pulse-counting applications, for example, ECL circuits can be used in the initial counting stages where high speed is necessary. Once the frequency is divided it is economical to use TTL circuits for the lower-speed counting states. The memory or computation circuits might be MOSFETs because of their low power requirements and higher component densities.

A typical Schmitt trigger level translator is illustrated in figure 11-19. Here the HI and LO threshold levels $V_{+H}$ and $V_{+L}$ of the Schmitt trigger are set for the input logic levels; the diode limiter is designed to provide the appropriate output logic levels. Interlevel converters for use between some of the most popular logic types are available in IC packages.

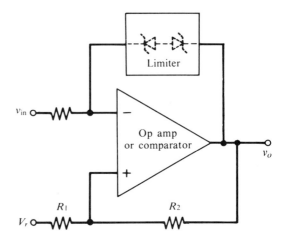

**Fig. 11-19.** Schmitt trigger circuit. For use as a level translator, the threshold levels $V_{+H} = V_r + \beta(V_H - V_r)$ and $V_{+L} = V_r + \beta(V_L - V_R)$ (see note 9-1 for the derivation) are set for the input logic levels. The limiter is designed to provide the appropriate output logic levels.

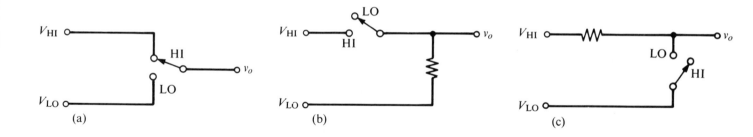

(a)          (b)          (c)

**Fig. 11-20.** Switch contact to logic level converter. The converter in (a) is general for any type of logic. The circuit in (b) is for current source logic, while that in (c) is for current-sinking logic. In (b) and (c), a resistor is connected directly to the level that supplies the lower current. The switch contact to the level that must supply the higher current overrides the resistive connection to the alternate level.

**Fig. 11-21.** Switch debouncer. Gates 1–4 are a NOR-gate gated latch, analogous to the NAND-gate gated latch of figure 10-26. Switch information is transferred to the latch on a HI Strobe signal. When the switch is moved to position $A_1$ and the contact first touches $A_1$, a LO is applied to NOR gate 1, forcing gate 3 to be LO and gate 4 to be HI even though the switch may bounce off the contact several times. This state remains until the switch is returned to $A_2$ producing the opposite state on first contact. The truth table shows the effect of the Strobe and Enable inputs.

**Contact Logic Interfaces.** A frequently encountered digital data source is the switch contact. The state of the switch contact (open or closed) is a form of digital data that must be converted to a logic-level signal in order to be used with gate logic circuits. The switch can be used as a logic-level voltage source as in figure 11-20. The circuits of figure 11-20b and c take advantage of the fact that the current required by most types of logic gate inputs is very different for HI and LO level signals.

A major difficulty with contact logic interfaces is contact bounce, which occurs when the switch state is changed. In each of the circuits of figure 11-20 contact bounce causes multiple transitions. When the switch has a positive contact at each state, a logic gate circuit can eliminate the multiple transitions caused by contact bounce. Such a contact bounce eliminator circuit is shown in figure 11-21. This circuit is a gated latch with a tristate output. The Strobe input allows sampling of the switch information on command of a Strobe signal. The circuit is available in IC form with four latches and resistive pullups (DM 8544).

The transition from logic-level signals to contact closures is also important in many applications. This may be accomplished by activating a relay drive circuit with a logic-level signal or by simulating a contact closure with a solid-state switch (see chapter 6).

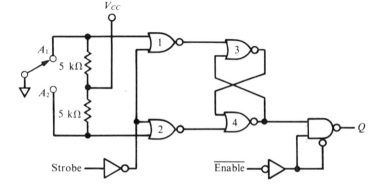

| $A_1$ | $A_2$ | Enable | Strobe | $Q$ |
|-------|-------|--------|--------|-----|
| X  | X  | HI | X  | HI-$Z$ |
| X  | X  | LO | LO | $Q(t-1)$ |
| LO | HI | LO | HI | LO |
| HI | LO | LO | HI | HI |
| HI | HI | LO | HI | $Q(t)$ |
| LO | LO | LO | LO | Indeterminant |

# Pulse Generation and Shaping

Digital signals may have the data encoded as the logic level (HI,LO), as the direction of the logic-level transitions (HI→LO, LO→HI), or as the presence of logic-level pulses (momentary changes in logic level). The principal kind of pulse generator is a monostable multivibrator introduced in chapter 6. As the name implies, the monostable multivibrator is stable in just one state. It can be triggered into its other "semistable" state, where it will remain for a predetermined time before returning to its initial stable state. The time spent in the semistable state is not determined by the triggering signal but by the choice of component values in the monostable circuit itself. This circuit can be thought of as a triggered, adjustable-width, pulse generator. It is used as a pulse shaper to provide uniform-width pulses from a variable-width input pulse train. Its other principal application is as a delay element since it provides a logic output transition at a fixed time after the trigger signal.

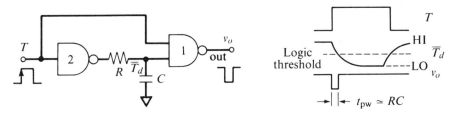

**Fig. 11-22.** $RC$ delay monostable. Before the trigger pulse, the $T$ input is LO, and the output of gate 2 is HI. The capacitor $C$ is charged through $R$ to this level. When $T$ becomes HI, the gate 2 output becomes LO, and the capacitor discharges toward $V_{LO}$ through $R$ and the gate 2 output impedance. The voltage at $\overline{T_d}$ decreases and crosses the HI→LO logic threshold for the gate 1 input. At this time the LO at $\overline{T_d}$ brings the gate 1 output HI. The pulse duration $t_{pw}$ thus depends on $R$, $C$, and the logic threshold level. It is generally of the same magnitude as the $RC$ time constant.

**Logic Gate Monostables.** Simple monostable pulse generators can be made from IC logic gates. They are based on the use of a low pass $RC$ circuit to delay the occurrence of a logic-level transition at one gate relative to another. The simplest such circuit is shown in figure 11-22. For most gate types $R$ in figure 11-22 is limited to small values. With current-sinking gates such as TTL, the LO level input current must pass through $R$ while at the same time the $IR$ drop must not bring the gate 1 input voltage too near the HI→LO threshold. Approximately 220 $\Omega$ is considered a safe upper limit input resistance for the standard TTL gate. This low $R$ limits the practical pulse duration from such a circuit, for even if a 1 $\mu$F capacitor is used the pulse is only about 220 $\mu$s long. In any case a Schmitt trigger gate should be used for gate 1 with time constants more than about 10 $\mu$s because of the slow threshold transition occurring at its input. For longer pulse widths the IC monostables described below or the IC timers discussed in chapter 6 should be used.

**IC Monostables.** Integrated circuit monostables are available in several different logic families. One popular TTL monostable (74121) is shown in figure 11-23. With this monostable the output pulse width can be varied

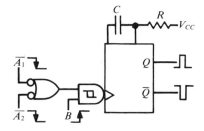

**Fig. 11-23.** Monostable multivibrator. This 74121 type monostable features dual HI→LO transition trigger inputs, $A_1$ and $A_2$, and a single LO→HI transition trigger input $B$, which can also be used as an inhibit. The Schmitt trigger input for the $B$ signal allows jitter-free triggering from signals with slow transition rates. The output pulse width, which is determined by the external timing components $R$ and $C$, is given by $t_{pw} \simeq 0.7\, RC$.

**Fig. 11-24.** Retriggerable monostable multivibrator. This type 74122 monostable features two HI→LO trigger inputs ($A_1$, $A_2$) and two LO→HI trigger inputs ($B_1$, $B_2$). The output pulse width is determined primarily by $R_{ext}$ and $C_{ext}$. The basic pulse width can be extended during the pulse by applying a retrigger pulse to the triggered input. An overriding clear can be used to terminate the output pulse when desired. The pulse width extension upon retriggering is the basic pulse width ($t_{pw}$) plus the propagation delay time ($t_{pLH}$) at the $B_1$ input.

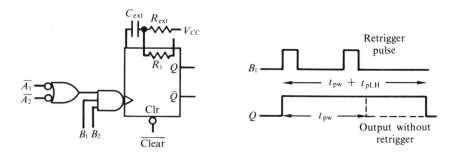

from ~40 ns to 28 s by appropriate choice of the external timing components. Once the monostable is triggered, the output pulse is independent of any further transitions at the trigger inputs. This monostable is also available in a dual version (74221) with an overriding clear that can be used to terminate the output pulse on command. The 74221 model is also available in low-power Schottky (LS) form.

Some monostables allow retriggering during the pulse to extend the pulse width. The 74122 type retriggerable monostable is illustrated in figure 11-24. This versatile monostable also has an internal timing resistor that allows the circuit to be used with only an external capacitor if desired. The retriggerable monostable is available in a dual package (74123) and in LS form.

The IC monostable is normally of moderate accuracy (5–10%) and is used for pulse widths as long as seconds. For higher accuracy and longer pulse durations the IC timer is often used (see chapter 6).

**Fig. 11-25.** Pulse-width discriminator. After being shaped by a Schmitt trigger and inverted, the input pulse triggers a monostable. The input and monostable pulse, thus inverted with respect to each other, are compared by NAND gates. If the LO-going pulse is the wider, it holds the gate output HI during the comparison; if not, a pulse results. Thus a pulse occurs at $p < t$ when the input pulse is narrower than the monostable pulse and at $p > t$ in the other case.

**Pulse-Width Discriminator.**    Sometimes the fact that a given pulse represents the occurrence of an event of interest is evidenced by its pulse width rather than its amplitude. In such cases it is useful to be able to sort input pulses according to their duration. Figure 11-25 shows a circuit that provides a pulse at one output or another, depending on whether the input pulse is

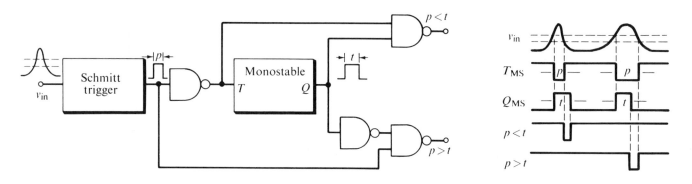

wider or narrower than a reference pulse from a monostable circuit. One application of such a discriminator is to exclude noise pulses of shorter duration than the monostable reference pulse. The pulse representing the true event would thus be taken from the $p > t$ output.

**Pulse Coincidence and Anticoincidence Gating.** Another criterion that is often useful in distinguishing pulses that represent phenomena of interest from those that do not is whether their occurrence is coincident (or not coincident) with a pulse from a related source. An elegant illustration of coincidence gating is the laser-beam reflection measurement of the distance to the moon (figure 11-26). A laser that emits a very short pulse of high-intensity light is aimed at a spot on the moon. A telescope observing the same area collects an almost infinitesimal fraction of the reflected light along with all the other light observed from that area. The intent is to start a precision timer with light from the laser and stop it with the pulse of reflected light. From the time delay and a knowledge of the speed of light, the distance to the moon can be calculated precisely. The monochromator selects only those photons of the wavelength emitted by the laser, and the detector PM tube then emits a pulse for each photon detected. Even so, the reflected laser photons are so few that the pulses that are not due to the laser reflection outnumber those that are by many orders of magnitude. Advantage is taken of the fact that the experimenter knows approximately when to expect the desired pulses. By means of monostables, a pulse is generated that is coincident with the expected reflection delay time and wide enough to encompass the time uncertainty. The generated pulse and the detected pulses are connected to a coincidence gate (AND gate) so that only those pulses occurring during the expected time stop the timer. The experiment was successful even though a reflected photon was detected, on the average, only every hundredth flash. The measurement was, of course, repeated thousands of times, and the data were analyzed by a computer for coincidence of measured times. For the round-trip time of about 5 s, a timer resolution of 1 ns results in a distance measurement resolution of about 16 cm!

An example of anticoincidence pulse gating occurs in the **pulse height amplitude window discriminator**, in which only pulses with amplitudes in a predetermined range are to be detected. To perform this function two Schmitt trigger circuits are used with their respective $V_+$ values set at the lower and upper boundaries of the amplitude range, or "window." Both Schmitt trigger circuits are connected to the pulse source as shown in figure 11-27. Only pulses with amplitudes between the two levels produce an output from the exclusive-OR gate. This type of discriminator is often found in nuclear pulse counting circuits.

**Crystal Oscillator.** By far the most stable oscillators are those that use a quartz crystal as the frequency-determining tuned circuit. A piece of quartz

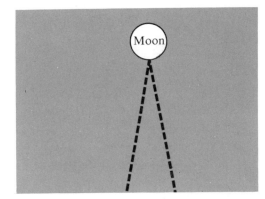

**Fig. 11-26.** Coincidence gating in measurement of the distance to the moon. A pulsed laser beam is reflected from the moon and collected by a telescope. The time interval between the flash of light and detection of the reflected light is measured.

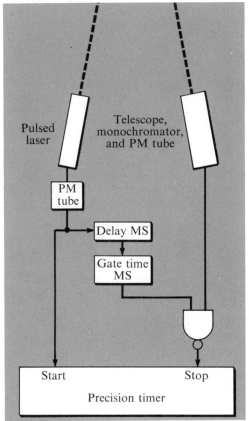

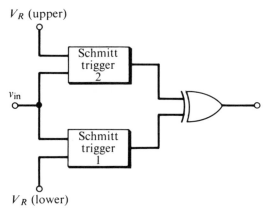

$V_R$ (upper)

$v_{in}$

$V_R$ (lower)

**Fig. 11-27.** Pulse height window discriminator. The window is the region between the upper threshold voltage and the lower threshold. Pulses below the window in amplitude trigger neither Schmitt, and pulses too large trigger both. Only when pulses are in the window is one Schmitt triggered and not the other. The exclusive-OR gate provides the anticoincidence logic.

**Fig. 11-28.** Quartz crystal oscillator. (a) equivalent circuit of quartz crystal; (b) phase-shift oscillator. In the oscillator circuit (b), the capacitors and the 10-k$\Omega$ resistor provide a 180° phase shift at the resonant frequency of the crystal. The 4001 C-MOS NOR gate (1) provides enough gain to overcome circuit losses. NOR gate 2 acts as an inverting buffer. The 20 pF capacitor provides a fine-tune of the output frequency. Appropriate values of $R$ range from 10 to 100 k$\Omega$, and values of $C$ should be 10 to 100 pF.

crystal sandwiched between metallic plates deforms mechanically when a voltage is applied between the plates. When the external voltage is removed, the relaxation of the crystal induces a voltage between the metal plates. Each crystal has a natural vibration frequency that depends on its cut and size. An ac voltage of the resonant frequency of the crystal applied across it maintains the resonant vibration in the same way that oscillation is sustained in an $LC$ resonant circuit. The crystal is superior to the $LC$ resonant circuit because the frequency is virtually independent of external circuit parameters and the resonance peak is much sharper ($Q$ much higher). An equivalent circuit of a quartz crystal is shown in figure 11-28a, and a simple C-MOS phase-shift oscillator circuit is shown in figure 11-28b.

The crystal oscillator has a fixed frequency. To change the frequency, another crystal of appropriate dimensions must be substituted. Most crystal oscillators are used to generate precision time bases in conjunction with digital frequency dividers as discussed in chapter 9. Extremely high precision pulse widths and delays can be obtained with a crystal oscillator time base, a presettable counter, and appropriate gating.

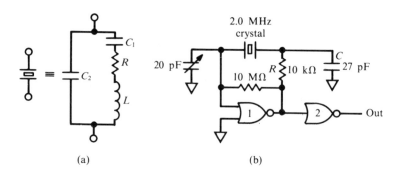

(a)                                   (b)

## 11-4 Digital Data Transmission

In modern digital systems it is often necessary to interconnect circuits that may be far apart. Even though digital signals are much less susceptible to noise and distortion than analog signals, precautions are necessary when signal leads are more than a few inches long. Because digital signals have transition times of only a few nanoseconds, it is important to understand the high-frequency characteristics of the lines used to transmit such signals. For that reason this section begins with a brief consideration of the impedance characteristics of transmission lines. Then line drivers and receivers that permit accurate transmission of digital signals over long distances are presented. The section concludes with a discussion of serial data transmission standards. Additional practical information concerning grounding and shielding techniques can be found in appendix A.

## Transmission Lines

The lines used to transmit electrical signals from one point to another in electronic systems are known as **transmission lines**. Some of the common types of lines are parallel-wire lines, single-wire lines above ground planes, and coaxial cables. These three types of transmission lines are illustrated in figure 11-29. The parallel-wire configuration usually consists of equal diameter wires held apart by an insulation material. The familiar antenna lead-in for a television set is a parallel-wire transmission line. An example of a single-wire line is the printed circuit board foil that interconnects various circuits on a board. The coaxial line is often used for input and output connections between laboratory instruments. It has the advantage that the signal on the cable is shielded from external radiation sources. Likewise, external sensors are shielded from the signal transmitted by the line.

Transmission lines are characterized by a series inductance $L$ per unit length and a parallel capacitance $C$ per unit length, both of which act to impede changes in the voltage applied to the transmission line. The load that the transmission line presents to a varying signal source is determined by the inductance and capacitance of a short section at the input end. As the equivalent circuit of figure 11-30 suggests, the effect on the input signal of the sections of series and parallel reactance farther from the input is increasingly less, so that a limiting **characteristic impedance** $Z_0$ is soon reached that is given by

$$Z_0 = \sqrt{L/C}$$

Note that $Z_0$ is independent of the length of the line and the frequency of the signal. For lines in which losses due to the resistance of the wire and leakages between the conductors can be neglected, $Z_0$ is a pure resistance, but it is clearly *not* the dc resistance of the line. Since $L$ and $C$ are functions of the line geometry type and dimensions, lines with various characteristic impedances are available. For parallel-wire lines $Z_0$ is typically about 300 $\Omega$; for coaxial and twisted-pair lines values from 50 to 100 $\Omega$ are most common. The popular RG58/U coaxial cable has a $Z_0$ of 50 $\Omega$, and $L$ and $C$ of 212 nH/m and 85 pF/m, respectively.

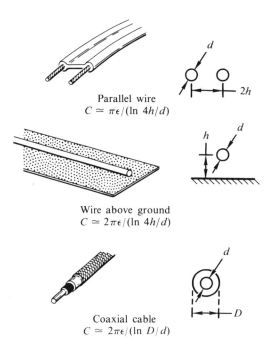

Parallel wire
$C \simeq \pi\epsilon/(\ln 4h/d)$

Wire above ground
$C \simeq 2\pi\epsilon/(\ln 4h/d)$

Coaxial cable
$C \simeq 2\pi\epsilon/(\ln D/d)$

**Fig. 11-29.** Several transmission lines and approximate equations of their capacitance per meter. Here $\epsilon$ is the dielectric constant of the material between the wires. The capacitance expressions are correct to about 5% if $h/d > 1$.

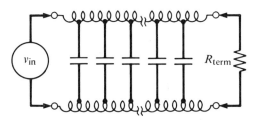

**Fig. 11-30.** Equivalent circuit of a transmission line. The line has a characteristic impedance given by $Z_0 = \sqrt{L/C}$, where $L$ is the inductance per unit length and $C$ is the capacitance per unit length.

**Note 11-2.  Transmission Line as a Delay Line**
Because transmission lines have characteristic delay times, $t_d = \sqrt{LC}$, they are sometimes used as intentional delay elements. For example, the RG 58/U cable has a delay time of $t_d = \sqrt{212 \text{ nH/m} \times 85 \text{ pF/m}} = 4.2$ ns/m.

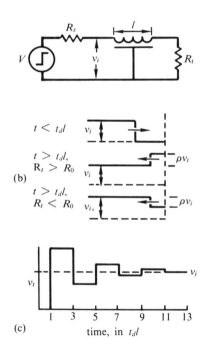

(b)

(c)     time, in $t_d l$

**Fig. 11-31.** Step function response of transmission line. (a) A source with internal resistance $R_s$ applies a step voltage $v_i$ to a transmission line of impedance $R_0$ terminated with $R_t$. (b) The voltage step moves along the line before $t_d l$, the delay time of the line. At the termination, a positive, negative, or zero reflection occurs depending upon whether $R_t$ is greater than, less than, or equal to $R_0$. (c) if $R_t > R_0$ and $R_s < R_0$, multiple reflections can occur before the output voltage reaches a steady state.

A change in $v_{in}$ applied to the transmission line travels down the line at a rate given by

$$t_d = \sqrt{LC}$$

where $t_d$ is the transmission delay time per unit length of line (see note 11-2). Consider the application of a step function to a transmission line as shown in figure 11-31a. The transmission line appears as a pure resistance of $R_0 = Z_0$ to the source $V$ with source resistance $R_s$. The initial voltage $v_i$ at the transmission line input is thus $V[R_0/(R_s + R_0)]$ as the step is applied. The current applied to the transmission line is a steady $v_i/R_0$ as the voltage step moves down the line as shown in figure 11-31b. When the edge reaches the end of the line, it simply terminates if the $IR$ drop across $R_t$ is exactly equal to the step voltage $v_i$, that is, if $(v_i/R_0)R_t = v_i$. For this to be true, $R_t$ must be equal to $R_0$. If $R_t$ is larger than $R_0$, the $IR$ drop across $R_t$ is larger than $v_i$ and a step edge is generated that is reflected back down the transmission line toward the source. The amplitude of the reflected step is a fraction $\rho$ of the original step where $\rho = (R_t - R_0)/(R_t + R_0)$. The quantity $\rho$ is called the **reflection coefficient** (see note 11-3). The voltage at the termination after the reflection is $v_i(1 + \rho)$. A value of $R_t$ less than $R_0$ causes the $IR$ drop across $R_t$, and thus the reflected voltage, to be less than $v_i$. The equation for $\rho$ indicates that it is negative for values of $R_t < R_0$. Figure 11-31b shows the reflected steps for values of $R_t \neq R_0$. In the extremes, if $R_t$ is very large (open circuit), $\rho = +1$, but if $R_t$ is very low (short circuit), $\rho = -1$. To avoid reflections altogether it is necessary for $R_t$ to be equal to $R_0$, for which $\rho = 0$.

A common case occurs when the source impedance is lower than $R_0$ ($0 > \rho' > -1$) and the transmission line is connected only to a high-impedance input ($R_t > R_0$ and $1 > \rho > 0$). A step voltage $v_i$ applied to this line produces a reflection of $+\rho v_i$ to give a terminal voltage of $v_i + \rho v_i$. When this reflection reaches the source, it sees a low terminating resistance, and a reflection edge of $\rho'$ times $\rho v_i$ is generated to give a net signal value of $v_i(1 + \rho + \rho'\rho)$ where $\rho'$ is the reflection coefficient for the source termination $R_s$. This multiple reflection process continues until the reflected edges become immeasurably small as shown in figure 11-31c. If $R_s$ were 0 and $R_t$ were $\infty$, $\rho$ and $\rho'$ would be 1 and $-1$, and a steady voltage would never be attained. In practice, resistive losses in the line eliminate the possibility of infinite oscillation, but the example shows that when neither end of the line is terminated properly, oscillations can be extensive. It should also be noted that when $R_s$ is not zero and $R_t$ is not equal to $R_0$, the final voltage approaches $VR_t/(R_s + R_t)$, which is different from $v_i$. Therefore, to transmit accurately signal voltage changes that are short compared to $t_d l$ (fast signals or long lines), proper termination is essential. For example, with ECL gates termination is advised when a connecting wire is longer than about 10 cm. For slow signals or short connections, termination is often unnecessary.

## Line Drivers and Receivers

In addition to termination precautions, special **line drivers** and **line receivers** should be used when signals are to be transmitted over distances of meters. For moderate distances the driving and receiving circuits are generally single-ended, and the line is a wire, printed circuit board foil, coaxial cable, or ribbon cable with possible commons at either end as shown in figure 11-32a. Driver type permitting, the cable is terminated at its receiving end in its characteristic impedance (for hookup wire or printed circuit board foil, about 200 $\Omega$) to avoid reflections. The terminating resistor on the driver end cannot always be used, but it is safest to do so. In the special case of open collector driver gates, the terminating resistor on the receiver end is connected to the supply voltage and serves the dual purpose of pullup resistor and line termination resistor.

For long distances balanced output drivers and differential input receivers are recommended. These are for use with balanced lines such as the shielded twisted pair shown in figure 11-32b. The balanced line transmission is less susceptible to induced noise (see appendix A) because both lines are driven identically, exposed to the same influences, and detected differentially. Note that both lines in a balanced pair must be terminated.

Integrated circuit line drivers and receivers as well as two-way communication devices called **transceivers** are available in many logic families. Many of these have tristate outputs and can therefore be used in shared line situations.

**Note 11-3. Derivation of Reflection Coefficient**
For a step voltage of $v_i$, the current step is $v_i/R_0$. The fraction of the voltage step reflected is defined as $\rho$. The voltage at termination is therefore $v_i(1+\rho)$. The current in $R_t$ is thus $v_i(1+\rho)/R_t$. The current in the reflected step is $\rho v_i/R_0$. The step current is equal to the sum of the terminating and reflected currents:

$$\frac{v_i}{R_0} = \frac{v_i(1+\rho)}{R_t} + \frac{v_i\rho}{R_0}$$

This equation can be solved for $\rho$ to give

$$\rho = \frac{R_t - R_0}{R_t + R_0}$$

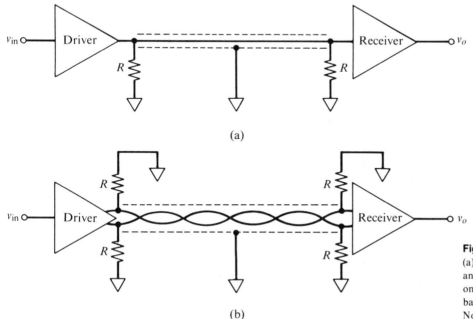

(a)

(b)

**Fig. 11-32.** Digital data transmission techniques. In (a) a single-ended line is shown with an input driver and output receiver. If possible, the line is terminated on both ends with its characteristic impedance. In (b) a balanced transmission line (twisted pair) is shown. Note that both lines are terminated.

For TTL systems the guidelines given in table 11-4 have been established to minimize transmission line effects (see also appendix A).

**Table 11-4.**   TTL transmission line considerations.*

---

1. Use direct wire interconnections that have no specific ground return for lengths up to about 10 in. only. A ground plane is always desirable.

2. Direct wire interconnections must be routed close to a ground plane if longer than 10 in. and should never be longer than 20 in.

3. When using coaxial or twisted-pair cables, design around approximately 100 $\Omega$ characteristic impedance. Coaxial cable of 93 $\Omega$ is recommended. For twisted pair, No. 26 or No. 28 wire with thin insulation twisted about 30 turns per foot works well. Higher impedances increase crosstalk, and lower impedances are difficult to drive.

4. Ensure that transmission line ground returns are carried through at both transmitting and receiving ends.

5. Connect reverse termination at driver output to prevent negative overshoot.

6. Decouple line driving and line receiving gates as close to the package $V_{CC}$ and ground pins as practical, with a 0.1 $\mu$F capacitor.

7. Gates used as line drivers should be used for that purpose only. Gate inputs connected directly to a line driving output could receive erroneous inputs due to line reflections, long delay times introduced, or excessive loading on the driving gate.

8. Gates used as line receivers should have all inputs tied together to the line. Other logic inputs to the receiving gate should be avoided, and a single gate should be used as the termination of a line.

9. Flip-flops are generally unsatisfactory line drivers because of the possibility of collector commutation from reflected signals.

---

*From R. L. Morris and J. R. Miller, *Designing with TTL Integrated Circuits* (McGraw-Hill Book Co., New York: 1971), p. *105*.

## Serial Communication Standards

Digital data can be transmitted between locations in several different ways. Where very high speed is required, parallel transmission lines are necessary with the substantial expense of multi-wire cables. Many digital instruments and computer peripherals do not require extremely high speed transmission and thus communicate via serial (single channel) transmission lines. Terminals, printers, plotters, teletypes, and data loggers are but a few of the many devices that rely on serial communications.

There are serial communication standards for the electrical characteristics of the transmitted signals, for the data formats, and for the transmission rates. These are summarized below.

**Electrical Standards.** Teletype communications occur via **current loops**, where a logic 1 is the presence of a current ($\sim$20 mA) in the loop and a logic 0 is the absence of current. Data rates are typically 110, 150 or 300 bits/s for teletype communication. Current loops are very low impedance lines that are quite insensitive to noise. Digital signals can be transmitted over distances approaching one mile without information loss.

The Electronics Industry Association (EIA) standard RS-232C covers electrical characteristics and physical specifications for serial transmission. The standard also defines control signals for standard telephone connection equipment and modulator-demodulators (modems). Electrically, nominal +12 V and −12 V signals are used for data and control signals. The bit rates may be any of the following standard rates: 19 200, 9 600, 4 800, 2 400, 1 200, 600, 300, 150, 110, 75, 50 bits/s. In table 11-5 the RS-232C standard is compared with two more recent long-distance standards RS-422 and RS-423. The advantage of these latter standards is that they are differential. This allows longer distances and higher data rates.

**Table 11-5.** Serial communication standards.

|  | RS-232C | RS-422 | RS-423 |
|---|---|---|---|
| Logic 1 | −1.5 to −36 V | $V_A > V_B$* | $V_A = +$ |
| Logic 0 | +1.5 to +36 V | $V_A < V_B$ | $V_B = +$ |
| Max. data rate | 20 kbits/s | $10^6$ bits/s | $10^5$ bits/s |
| Receiver input, minimum | 1.5V (single-ended) | 100 mV (differential) | 100 mV (differential) |
| Maximum line length | 100 ft. | 5000 ft. | 5000 ft. |

*$V_A$ is the voltage on wire $A$; $V_B$ is that on wire $B$.

Line drivers and receivers for conversion between TTL voltages and the EIA standard voltages are readily available. Figure 11-33 illustrates a typical RS-232C communication line.

**Fig. 11-33.** Typical TTL/RS-232C driver and receiver.

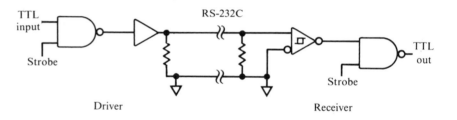

**Data Formats.** Serial data can be encoded in several different formats. One of these formats, the American Standard Code for Information Interchange (ASCII) uses seven binary bits to encode 128 possible characters. An eighth bit may be used for parity. The ASCII code is very popular for encoding characters for CRT display, teletypes, printers, etc. The ASCII code chart and various abbreviations used in the ASCII format are given in table 11-6.

**Synchronous and Asynchronous Serial Communication.** In order for the receiver to decode serial data, the beginning and end of each byte or word transmitted must be indicated. With asynchronous serial transmission this is done by sending a **start bit** to delineate the beginning of an eight-bit character and one or more **stop bits** to signal the end. Figure 11-34 illustrates the transmission of an eight-bit ASCII character by the asynchronous serial format. In this format a fixed time duration is used for each binary bit. The number of bits that can be transmitted per second is called the **baud rate**.

**Fig. 11-34.** Asynchronous transmission of an ASCII character (%). Each character is initiated by a start bit and separated by one or more stop bits. The duration of each bit is fixed.

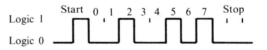

In synchronous serial communications the beginning of an entire message of data is indicated by a unique code that causes the receiver to lock in on the transmission and, by using a counter, to count the received bits and assemble them into characters. The synchronous technique is more efficient because it uses all the bit times after the starting code as data. However, any missing or faulty bits can cause the entire remaining part of the message to be in error rather than just a single character. Synchronous serial communication is more complex than asynchronous and is not useful when the data are to be transmitted in short units or at irregular intervals.

## 11-5 Applications: Selected MSI and LSI Devices

Previously we have considered many of the more common digital integrated circuit devices. This section examines some selected MSI and LSI devices

**Table 11-6.**   ASCII code.

| | 0000 | 0001 | 0010 | 0011 | 0100 | 0101 | 0110 | 0111 | 1000 | 1001 | 1010 | 1011 | 1100 | 1101 | 1110 | 1111 |
|---|---|---|---|---|---|---|---|---|---|---|---|---|---|---|---|---|
| | 0 | 1 | 2 | 3 | 4 | 5 | 6 | 7 | 8 | 9 | A | B | C | D | E | F |
| 0000  0 | NUL | SOH | STX | ETX | EOT | ENQ | ACK | BEL | BS | HT | LF | VT | FF | CR | SO | SI |
| 0001  1 | DLE | DC1 | DC2 | DC3 | DC4 | NAK | SYN | ETB | CAN | EM | SUB | ESC | FS | GS | RS | US |
| 0010  2 | SP | ! | " | # | $ | % | & | ' | ( | ) | * | + | , | - | . | / |
| 0011  3 | 0 | 1 | 2 | 3 | 4 | 5 | 6 | 7 | 8 | 9 | : | ; | < | = | > | ? |
| 0100  4 | @ | A | B | C | D | E | F | G | H | I | J | K | L | M | N | O |
| 0101  5 | P | Q | R | S | T | U | V | W | X | Y | Z | [ | \ | ] | ∧ | - |
| 0110  6 | ` | a | b | c | d | e | f | g | h | i | j | k | l | m | n | o |
| 0111  7 | p | q | r | s | t | u | v | w | x | y | z | { | ¦ | } | ~ | DEL |

| Col./Row | Symbol | Name | | Control Character | Function |
|---|---|---|---|---|---|
| 2/0 | SP | Space (normally non-printing) | | NUL | Null |
| 2/1 | ! | Exclamation point | | SOH | Start of heading |
| 2/2 | " | Quotation marks (diaeresis) | | STX | Start of text |
| 2/3 | # | Number sign | | ETX | End of text |
| 2/4 | $ | Dollar sign | | EOT | End of transmission |
| 2/5 | % | Percent | | ENQ | Enquiry |
| 2/6 | & | Ampersand | | ACK | Acknowledge |
| 2/7 | ' | Apostrophe (closing single quotation mark; acute accent) | | BEL | Bell (audible or attention signal) |
| | | | | BS | Backspace |
| 2/8 | ( | Opening parenthesis | | HT | Horizontal tabulation (punch card skip) |
| 2/9 | ) | Closing parenthesis | | | |
| 2/A | * | Asterisk | | LF | Line feed |
| 2/B | + | Plus | | VT | Vertical tabulation |
| 2/C | , | Comma (cedilla) | | FF | Form feed |
| 2/D | - | Hyphen (minus) | | CR | Carriage return |
| 2/E | . | Period (decimal point) | | SO | Shift out |
| 2/F | / | Slant | | SI | Shift in |
| 3/A | : | Colon | | DLE | Data link escape |
| 3/B | ; | Semicolon | | DC1 | Device control 1 |
| 3/C | < | Less than | | DC2 | Device control 2 |
| 3/D | = | Equals | | DC3 | Device control 3 |
| 3/E | > | Greater than | | DC4 | Device control 4 (Stop) |
| 3/F | ? | Question mark | | NAK | Negative acknowledge |
| 4/0 | @ | Commercial at | | SYN | Synchronous idle |
| 5/B | [ | Opening bracket | | ETB | End of transmission block |
| 5/C | \ | Reverse slant | | CAN | Cancel |
| 5/D | ] | Closing bracket | | EM | End of medium |
| 5/E | ∧ | Circumflex | | SUB | Substitute |
| 5/F | – | Underline | | ESC | Escape |
| 6/0 | ` | Grave accent (opening single quotation mark) | | FS | File separator |
| | | | | GS | Group separator |
| 7/B | { | Opening brace | | RS | Record separator |
| 7/C | ¦ | Vertical line | | US | Unit separator |
| 7/D | } | Closing brace | | DEL | Delete |
| 7/E | ~ | Overline (tilde, general accent) | | | |

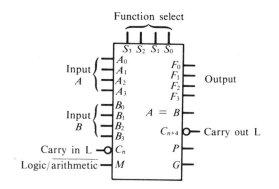

**Fig. 11-35.**  Arithmetic logic unit. The ALU performs logical and arithmetic operation on two four-bit binary words. When *M* is HI, logic functions are selected by inputs $S_0$ through $S_3$. When *M* is LO, 16 binary arithmetic operations can be selected by the function select inputs. The outputs are a four-bit word and a carry.

that were not previously discussed. The devices presented include the versatile arithmetic logic unit, the binary accumulator, and the binary and rate multipliers. These devices were selected because of their usefulness in many applications and because many similar functional blocks appear in the microprocessors discussed in chapter 12.

## The Arithmetic Logic Unit

The **arithmetic logic unit** (**ALU**) is available as an IC package in TTL, ECL, and C-MOS. It is a highly versatile unit capable of performing a variety of logical and arithmetic functions.

A functional diagram of a typical ALU chip (74181) is shown in figure 11-35. The inputs are two four-bit words $A$, $B$, and a carry input $C_n$. The mode control $M$ allows either logic functions ($M = $ HI or H) or arithmetic operations ($M = $ LO or L) to be selected according to the function-select inputs $S = S_3 S_2 S_1 S_0$. The outputs are available from $F = F_3 F_2 F_1 F_0$ and a ripple carry output $C_{n+4}$. The functions are given in table 11-7. For a typical case where $M = $ L and $S = $ HLLH, $C_n$ L $= $ H, the output at $F$ is $A$ plus $B$. If there had been a carry from the previous state, $C_n$ L $= $ L, the output at $F$ would be $A$ *plus* $B$ *plus* 1. Here *plus* is used to indicate addition while $+$ is reserved for the logical OR operation. When high speed addition is necessary, the chip can be used in conjunction with a 74182, **look-ahead carry generator.** The terminal marked $P$ on the ALU is for a carry that is to *propagate* through the ALU and appear at the $P$ output, while a carry generated in the ALU appears at the *generate* terminal $G$.

**Table 11-7.**  ALU function table.

| Selection $S_3$ $S_2$ $S_1$ $S_0$ | $M = $ HI Logic functions | $M = $ LO; Arithmetic operations | |
|---|---|---|---|
| | | $C_n$ L $= $ HI (no carry) | $C_n$ L $= $ LO (with carry) |
| L L L L | $F = \overline{A}$ | $F = A$ | $F = A$ plus 1 |
| L L L H | $F = \overline{A + B}$ | $F = A + B$ | $F = (A + B)$ plus 1 |
| L L H L | $F = \overline{A}B$ | $F = A + \overline{B}$ | $F = (A + \overline{B})$ plus 1 |
| L L H H | $F = 0$ | $F = $ minus 1 (2's COMPL) | $F = $ zero |
| L H L L | $F = \overline{AB}$ | $F = A$ plus $A\overline{B}$ | $F = A$ plus $A\overline{B}$ plus 1 |
| L H L H | $F = \overline{B}$ | $F = (A + B)$ plus $A\overline{B}$ | $F = (A + B)$ plus $A\overline{B}$ plus 1 |
| L H H L | $F = A \oplus B$ | $F = A$ minus $B$ minus 1 | $F = A$ minus $B$ |
| L H H H | $F = A\overline{B}$ | $F = A\overline{B}$ minus 1 | $F = A\overline{B}$ |
| H L L L | $F = \overline{A} + B$ | $F = A$ plus $AB$ | $F = A$ plus $AB$ plus 1 |
| H L L H | $F = \overline{A \oplus B}$ | $F = A$ plus $B$ | $F = A$ plus $B$ plus 1 |
| H L H L | $F = B$ | $F = (A + \overline{B})$ plus $AB$ | $F = (A + \overline{B})$ plus $AB$ plus 1 |
| H L H H | $F = AB$ | $F = AB$ minus 1 | $F = AB$ |
| H H L L | $F = 1$ | $F = A$ plus $A*$ | $F = A$ plus $A$ plus 1 |
| H H L H | $F = A + \overline{B}$ | $F = (A + B)$ plus $A$ | $F = (A + B)$ plus $A$ plus 1 |
| H H H L | $F = A + B$ | $F = (A + \overline{B})$ plus $A$ | $F = (A + \overline{B})$ plus $A$ plus 1 |
| H H H H | $F = A$ | $F = A$ minus 1 | $F = A$ |

*Each bit is shifted to the next more significant position.

The TTL ALU in conjunction with a look-ahead carry generator can perform a 4-bit addition in 24 ns and a 16-bit addition in 36 ns (with four ALUs). An ECL version can add two four-bit numbers in about 6.5 ns. The versatile ALU is the heart of every microprocessor unit. Combined with accumulators, shift registers and memory units it can perform complex sequences of programmed arithmetic and logic operations.

## Binary Accumulator

The Schottky TTL **binary accumulator** (74S281) combines an ALU with a shift and storage matrix in a single LSI integrated circuit. Figure 11-36 shows the pin configuration of the IC. The internal ALU can perform 16 logic or arithmetic operations on two four-bit words as shown in table 11-8. The ALU is controlled by three function-select inputs $AS2$, $AS1$ and $AS0$ and a mode control $M$. It has a carry input $C_n$, and propagate $P$ and generate $G$ outputs for use with a look-ahead carry generator.

The shift matrix is a bidirectional shift register with multiplexed input/output (I/O) lines. It can perform either logical or arithmetic shifts in either direction. Register control is accomplished by the register control input ($RC$) and two register select inputs ($RS0$, $RS1$). The I/O lines have tristate outputs multiplexed with an input. The least significant cascading bit is

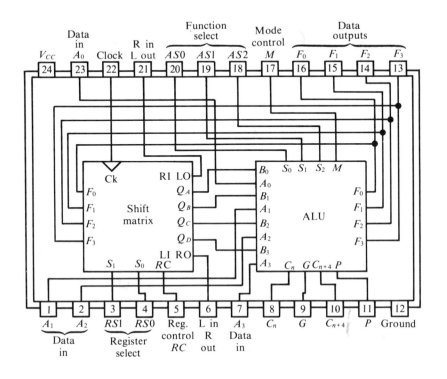

**Fig. 11-36.**  Pin configuration of four-bit binary accumulator. This 74S281 IC can perform 16 arithmetic and logic operations, has full bidirectional arithmetic and logical shift capabilities, and is capable of storage.

**Table 11-8.**   Four-bit binary accumulator function table.

Arithmetic functions
Mode control $(M)$ = LO

| ALU selection | | | Active high data | |
|---|---|---|---|---|
| | | | $C_n$ = HI (with carry) | $C_n$ = LO (no carry) |
| $AS2$ | $AS1$ | $AS0$ | | |
| L | L | L | $F_0$ = L, $F_1$ = $F_2$ = $F_3$ = H | $F_n$ = H |
| L | L | H | $F$ = $B$ minus $A$ | $F$ = $B$ minus $A$ minus 1 |
| L | H | L | $F$ = $A$ minus $B$ | $F$ = $A$ minus $B$ minus 1 |
| L | H | H | $F$ = $A$ plus $B$ plus 1 | $F$ = $A$ plus $B$ |
| H | L | L | $F$ = $B$ plus 1 | $\overline{F_n}$ = $B_n$ |
| H | L | H | $F$ = $\overline{B}$ plus 1 | $F_n$ = $\overline{B_n}$ |
| H | H | L | $F$ = $A$ plus 1 | $F_n$ = $A_n$ |
| H | H | H | $F$ = $\overline{A}$ plus 1 | $F_n$ = $\overline{A_n}$ |

Logic functions
Mode control $(M)$ = HI
Carry input $(C_n)$ = X (don't care)

| ALU selection | | | Active-high data function |
|---|---|---|---|
| $AS2$ | $AS1$ | $AS0$ | |
| L | L | L | $F_n$ = L |
| L | X | H | $F_n$ = $A_n \oplus B_n$ |
| L | H | L | $F_n$ = $\overline{A_n \oplus B_n}$ |
| H | L | L | $F_n$ = $A_n B_n$ |
| H | L | H | $F_n$ = $\overline{A_n + B_n}$ |
| H | H | L | $F_n$ = $\overline{A_n B_n}$ |
| H | H | H | $F_n$ = $A_n + B_n$ |

Shift mode functions
$C_n$ = $M$ = $S_0$ = $S_1$ = LO, and $S_2$ = HI

| Register selection | | Register control input | Shift-matrix inputs | | | | Clock input | Input/ output RI/LO | Shift-matrix outputs (internal) | | | | Input/ output LI/RO |
|---|---|---|---|---|---|---|---|---|---|---|---|---|---|
| $RS1$ | $RS0$ | | $F_0$ | $F_1$ | $F_2$ | $F_3$ | | | $Q_A$ | $Q_B$ | $Q_C$ | $Q_D$ | |
| L | L | X | $f_0$ | $f_1$ | $f_2$ | $f_3$ | ↑ | Z | $f_0$ | $f_1$ | $f_2$ | $f_3$ | Z |
| L | H | L | $Q_{Bn}$ | $Q_{Cn}$ | $Q_{Dn}$ | li | ↑ | $Q_{Bn}$ | $Q_{Bn}$ | $Q_{Cn}$ | $Q_{Dn}$ | li | li |
| L | H | H | $Q_{A0}$ | $Q_{Cn}$ | $Q_{Dn}$ | li | ↑ | $Q_{Bn}$ | $Q_{Bn}$ | $Q_{Cn}$ | li | $Q_{D0}$ | li |
| H | L | L | ri | $Q_{An}$ | $Q_{Bn}$ | $Q_{Cn}$ | ↑ | ri | ri | $Q_{An}$ | $Q_{Bn}$ | $Q_{Cn}$ | $Q_{Cn}$ |
| H | L | H | ri | $Q_{An}$ | $Q_{Bn}$ | $Q_{D0}$ | ↑ | ri | ri | $Q_{An}$ | $Q_{Bn}$ | $Q_{D0}$ | $Q_{Cn}$ |
| H | H | X | $Q_{A0}$ | $Q_{B0}$ | $Q_{C0}$ | $Q_{D0}$ | ↑ | Z | $Q_{A0}$ | $Q_{B0}$ | $Q_{C0}$ | $Q_{D0}$ | Z |
| X | X | X | $Q_{A0}$ | $Q_{B0}$ | $Q_{C0}$ | $Q_{D0}$ | L | X | $Q_{A0}$ | $Q_{B0}$ | $Q_{C0}$ | $Q_{D0}$ | X |

H = HI (steady state)

L = LO (steady state)

X = Don't care (any input, including transitions)

Z = High impedance (output off)

↑ = LO → HI transition

$f_0$, $f_1$, $f_2$, $f_3$, ri, li = The level of steady-state conditions at $F_0$, $F_1$, $F_2$, $F_3$, RI/LO or LI/RO, respectively.

$Q_{A0}$, $Q_{B0}$, $Q_{C0}$, $Q_{D0}$ = The level of $Q_A$, $Q_B$, $Q_C$, or $Q_D$, respectively, before the indicated steady-state input conditions were established.

$Q_{An}$, $Q_{Bn}$, $Q_{Cn}$, $Q_{Dn}$ = The level of $Q_A$, $Q_B$, $Q_C$, or $Q_D$, respectively, before the most recent transition of the clock.

combined with $A_0$ and $F_0$ to provide the shift-right input and shift-left output(RI/LO), and the most significant bit is combined with the $A_3$, $F_3$ circuitry to provide the shift-left input and the shift-right output (LI/RO).

Four of these units can be combined with one look-ahead carry generator to provide a 16-bit binary accumulator. Once loaded, the accumulator can perform a 16-bit addition in 29 ns.

## Binary Multipliers

Multiplication of two binary numbers can be accomplished with **binary multiplier** ICs. The two-IC circuit shown in figure 11-37 can produce an eight-bit product of two four-bit binary numbers in 40 ns. Expansion of these devices to provide multiplication of larger numbers, however, is not so easy. An 8-bit by 8-bit multiplier with a 16-bit product, for example, requires eight multiplier chips, four dual full-adder chips, three ALU chips and one look-ahead carry generator, a total of 16 ICs. The 16-bit product can be obtained in 70 ns. One manufacturer (TRW) has introduced a series of single-chip, high-speed TTL multipliers. These LSI chips are $N \times N$ bit parallel multipliers ($N = 8$, 12, or 16) with double precision ($2N$) outputs. Typical multiply times are in the 130 to 160 ns range, and the power dissipation is $\sim 3.0$ W for the 8-bit multiplier, 5.5 W for the 12-bit multiplier, and 8.0 W for the 16-bit multiplier.

## Rate Multipliers

**Rate multipliers** are MSI circuits that perform fixed or variable-rate frequency division. They contain counters with internal gating to allow only a controlled fraction of the applied clock pulses through to the output. Logic-level inputs determine the number of pulses applied. A decade rate multiplier (74167) is shown in figure 11-38. The counter is enabled when Clear, Strobe, Set-to-9, and Enable are LO. When enabled, the output frequency ($Z$ or $Y$ output) $f_{\text{out}}$ is equal to the input frequency times the decimal value of the multiplier $M$ divided by 10:

$$f_{\text{out}} = M f_{\text{in}} / 10$$

where $M = 2^3 D + 2^2 C + 2^1 B + 2^0 A$ for $M = 0$–9. Thus, when $M = 8$ ($1000_2$), $f_{\text{out}} = 0.8 f_{\text{in}}$.

The rate multiplier is also available in six-bit binary form (7497). In this case the output frequency is given by $f_{\text{out}} = M f_{\text{in}} / 64$, where $M$ is the decimal value of the six binary rate inputs ($F$–$A$). Figure 11-39 shows two binary rate multipliers cascaded to perform a scaling operation preceding a BCD counter. Scaling factors from 4095/4096 to 0 are selected by the rate inputs. These circuits also find use in digital-to-analog and analog-to-digital converters.

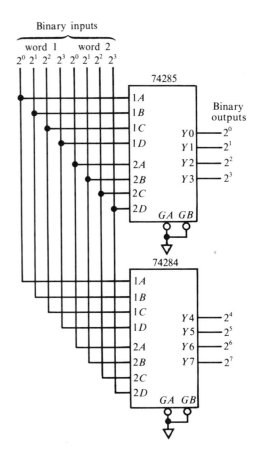

**Fig. 11-37.** A $4 \times 4$ multiplier. This two-chip circuit gives an eight-bit product of two four-bit numbers in 40 ns.

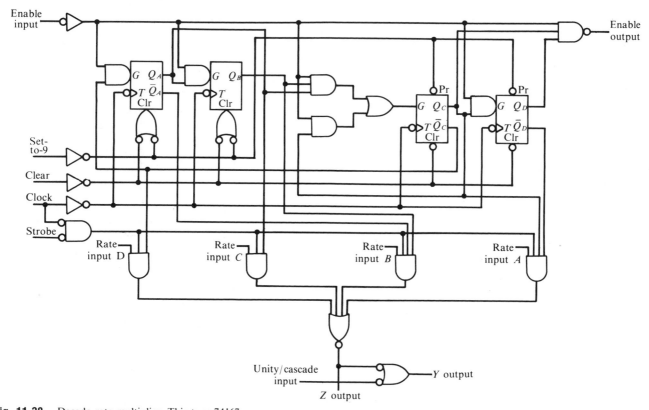

**Fig. 11-38.**   Decade rate multiplier. This type 74167 multiplier features buffered Clock, Clear, Enable and Set-to-9 inputs to the decade counter and a Strobe input to enable or inhibit the rate input AOI gates.

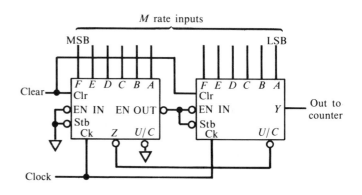

**Fig. 11-39.**   Scaler application of rate multipliers. Here two 7497 rate multipliers are used to provide a scaling factor of $M/2^{12}$. If $M = 3125$ (110000110101), for example, the scaling factor would be exactly 3125/4096 = 0.762 939 5.

# Suggested Experiments

**1. Logic families.**
Measure the output voltage vs. the input voltage for a 7400 NAND gate to determine the LO and HI level ranges. Measure the LO level input current, and note the direction. Repeat for a 74LS00 NAND gate. Investigate a C-MOS gate operated from several supply voltages (+5, +15V). Measure the gate input current.

**2. Open collector gates.**
Investigate the output states of the 7407 open collector driver by measuring the output resistance for both input states. Use the driver to control an LED. Combine three driver outputs with a pullup resistor to produce the "wired" logic effect. Wire 7403 open collector NAND gates together, and determine the resulting logic function. Measure the switching speed of the 7403 gate as a function of the size of the pullup resistor.

**3. Tristate gates.**
Investigate a 74125 tristate buffer and driver. Note the effect of the enable input. Connect several gate outputs together to form a bus. Measure the rise and fall times of the tristate gate output, and compare to the open collector gate.

**4. Level conditioning.**
Study a Schmitt trigger input inverting gate. Measure the output voltage vs. the input voltage, and note the hysteresis. Using a normal (not bounce-free) push button switch, note the contact bounce. Construct a bounce eliminator or use a switch debouncer IC, and note the sharp, clean transitions that result.

**5. IC monostables.**
Investigate several IC monostables. Determine the delay time for various $RC$ values. For retriggerable and nonretriggerable monostables, note the effect of a retrigger pulse applied during the delay.

**6. Window discriminator.**
Wire a pulse height window discriminator with two comparators, and an exclusive-OR gate. Measure the lower and upper threshold levels. Note the effect of changing the threshold voltages.

**7. The arithmetic logic unit.**
Investigate a 74181 arithmetic logic unit (ALU). Note several of the logic functions that can be performed as well as several of the arithmetic functions.

# Questions and Problems

**1.** The fan-out of a TTL gate is determined by the maximum current a gate output can sink in the LO state, $I_{oL}$. A standard 74 series TTL gate output can sink 16 mA in the LO state; a standard gate input requires that a LO signal source sink $-1.6$ mA ($I_{iL} = -1.6$ mA). (The minus sign indicates that the direction of current is out of the gate.) Thus a standard TTL gate has a fan-out of 10 in the LO state. A 74 LS gate has $I_{oL} = 8$ mA and $I_{iL} = -0.4$ mA. How many 74 LS gates can be driven by a standard 74 series gate in the LO state? How many standard 74 series gates can be driven by a 74 LS gate output in the LO state? How many 74 LS gates can be driven by another 74 LS gate?

**2.** In the HI state the maximum output current a standard TTL output can supply is $I_{oH} = -400$ $\mu$A; a standard TTL input requires a HI level input current of $I_{iH} = 40$ $\mu$A. Thus the fan-out of a standard gate in the HI state is also 10. For a 74 LS gate the HI state values are $I_{oH} = -400$ $\mu$A and $I_{iH} = 20$ $\mu$A. How many standard TTL gate inputs can be driven by one 74 LS gate output in the HI state? How many 74 LS gates can be driven by a standard 74 series gate output in the HI state? How many 74 LS gates can be driven by a 74 LS gate output in the HI state?

**3.** Schottky TTL gates, 74 S series, have maximum input currents of $I_{iL} = -2$ mA and $I_{iH} = 50$ $\mu$A. The maximum output currents are $I_{oL} = 20$ mA and $I_{oH} = -1000$ $\mu$A. How many 74 LS gate inputs can be driven by a 74 S gate output in each state? (See problems 1 and 2 for data on 74 LS gates.) How many 74 S gate inputs can be driven by a 74 LS gate output in each state?

**4.** Define the following terms: fan-out, power dissipation, noise margin, propagation delay time, and speed-power product. Compare the relative values of the above terms for standard TTL, low-power Schottky TTL, ECL, C-MOS, and $I^2L$.

**5.** What are the major advantages and disadvantages of the ECL logic family compared to the TTL family? Under what circumstances would ECL be used instead of TTL?

**6.** It is desired to obtain the exclusive-OR function on four input variables. (a) Show how the four-input exclusive-OR function could be implemented using only NAND gates. (b) Show how the four-input exclusive-OR could be implemented using open collector NAND gates in parallel and an inverter.

**7.** Compare the wired-AND connections of open collector NAND gates shown in figure 11-11 to the AND-OR-INVERT (AOI) function on signal pairs $AB$ and $CD$. Draw truth tables for both the wired-AND connection and the AOI gate. In what circumstances would one implementation be preferred over the other?

**8.** The outputs of two ECL gates can safely be tied together because the outputs are actively driven in only one state. (a) Verify that if the OR outputs of two 2-input ECL gates are tied together, the result is the same as a 4-input OR gate. (b) Verify that if the NOR outputs of two 2-input ECL gates are tied together the result is $\overline{AB} + \overline{CD}$. (c) How could the exclusive-OR function of two variables be implemented with ECL gates that provide only OR and NOR outputs?

**9.** Compare and contrast open collector gates and tristate gates as to speed, external component requirements, and versatility.

**10.** Verify the operation of the four-bit parallel bus driver shown in figure 11-17 by tracing the signal paths for the various conditions of $\overline{DIEN}$ and $\overline{CS}$.

**11.** A quad, two-input NAND gate IC with tristate outputs is available commercially. (a) How many pins would be required for such an IC if all four NAND gates were to have independent disables? (b) In a 16-pin IC, how many independent disables could be accommodated?

**12.** A quad switch debouncer similar to that shown in figure 11-21 is available in a 16-pin IC package. How many separate Strobe and Enable inputs are available?

**13.** Redraw the switch debouncer shown in figure 11-21 using NAND gates instead of NOR gates. Describe its operation by tracing the logic levels through the latch for the two conditions of the switch.

**14.** Challenge question: Consider a simple series $RC$ circuit. The capacitor initially has a voltage $v_o$ across it. A pulse signal is applied so that the capacitor charges to some final voltage $v_f$ (see figure 6-22). (a) Show that the voltage across the capacitor at any time $t$, $v_C(t)$ can be described by

$$v_C(t) = v_f - (v_f - v_o)e^{-t/RC}$$

(b) Derive an equation that describes the voltage across the resistor as a function of time $v_R(t)$. (c) If $v_{th}$ is a critical threshold voltage that is between $v_o$ and $v_f$, show that the time $t_{th}$ required for the capacitor voltage to charge from $v_o$ to the threshold $v_{th}$ is

$$t_{th} = RC \ln \frac{v_f - v_o}{v_f - v_{th}}$$

(d) In the 74121 monostable multivibrator the threshold voltage is approximately half the sum of $v_f$ and $v_o$. Find the time required to reach the threshold $t_{th}$ in units of $RC$.

**15.** Develop a truth table for the logic of the $A$ and $B$ inputs of the 74121 monostable multivibrator shown in figure 11-23. Describe how the $B$ input can be used as an inhibit.

**16.** Combinations of monostable multivibrators triggered in series or in parallel are sometimes used as moderate-accuracy sequence generators. Suppose that a pulsed measurement system consists of a gated integrator followed by a sample-and-hold circuit. The integrator is to be turned on for 1.0 ms by a HI signal applied to an analog switch 2.5 $\mu$s after the beginning of the experiment. (A sharp HI→LO transition is available at the start of the experiment.) When the integration period is complete, a sample-and-hold circuit is to be switched to the HOLD mode by a LO logic level, and the output held for 1.0 s for display and read-out. The integration capacitor should be shorted during the hold period in preparation for the next trigger pulse. Design a circuit with 74121 monostables to generate the needed pulses. State explicitly what inputs ($A$ or $B$) and what outputs ($Q$ or $\overline{Q}$) are to be used to activate the two analog switches. Draw the waveforms of all signals on a common time scale.

**17.** Discuss the similarities and differences between the IC monostable of the 74121 type and the IC timer of the 555 type (chapter 6). Is the 555 timer retriggerable? Can the 74121 be used in an astable mode? Explain.

**18.** It is desired to locate the time of the maximum in a Gaussian peak signal using an op amp differentiator whose output crosses zero at the time of the peak maximum. Differentiator circuits are known to be noisy, but it is suspected that most noise spikes on the differentiator output will be much faster than the derivative signal. Design a pulse-width discriminator circuit that can discriminate against rapid noise spikes but that allows the derivative to be measured for a "true" peak. "True" peaks are known to have half-widths on the order of 500 ms or larger.

**19.** Design a circuit that gives a HI logic-level output when two signals *both* have amplitudes within a predetermined window at the same time but not when either signal or both signals are outside the window. Describe how this circuit differs from the pulse-height window discriminator of figure 11-27.

**20.** (a) The dielectric constant (relative permittivity) of the coaxial cable shown in figure 11-29 is $\epsilon_r = 2.3$. The permittivity of vacuum is $8.85 \times 10^{-2}$ $C^2N^{-1}m^{-2}$. If the ratio of the outside diameter to the inside diameter is 4.0, find the capacitance of the cable per meter. (b) If the same coaxial cable has a characteristic imped-ance $Z_0$ of 50 Ω, what is the inductance per meter? (c) What length of cable gives a delay time of 7 ns?

**21.** The inductance of a given coaxial cable was found to be 200 nH/m. The dielectric constant is $\epsilon_r = 2.3$. What dimension ratio $D/d$ is necessary to give a 100-Ω characteristic impedance? What ratio would be necessary to give a 1-kΩ characteristic impedance?

**22.** The transmission line of figure 11-31 is a parallel-wire line in which the distance between the wires is ten times the radius of each wire. The material between the wire is air with a dielectric constant of 1.0. (a) Calculate the capacitance per meter of the line. (b) If the characteristic impedance is 300 Ω, calculate the induc-tance per meter of the line. (c) The line is terminated with a 1-kΩ resistance. Find the reflection coefficient.

**23.** In the 74181 arithmetic logic unit shown in figure 11-35, the $A$ input is 1011 and the $B$ input is 0110. There is no carry input. (a) If $M = L$ and $S = $ HLLH, what is the output? (b) If $M = H$ and $S = $ HLHH, what is the output? (c) If $M = H$ and $S = $ LLLH, what is the output?

# Chapter 12      Microcomputers

To many people the microcomputer is the most dramatic symbol of the semiconductor revolution. Because of the rapid and continuing decrease in size, cost, and power consumption that large-scale integrated circuits have brought to computers, we are becoming surrounded by computers in our offices, laboratories, and homes as they become incorporated in data and word processors, instruments, appliances, automobiles, and games. Indications are that this is just a foretaste of the "micro age."

Digital computers are well known for their ability to solve quickly problems whose length and complexity are well beyond the limits of manual computation. The computer achieves its great computational power by performing a predetermined list of very simple operations very quickly. The results achieved by even simple operations, if performed at the rate of several hundred thousand per second, can be very impressive. The computer performs these operations in response to commands called **instructions.** The list of instructions in order of execution is called a **program**. Instructions are encoded in binary words that must be decoded to be executed. When one instruction is completed, the control logic of the computer automatically fetches the next instruction in the program and begins its execution. Thus a complete program is executed automatically. The control logic that performs the sequencing of instructions is called the **central processing unit (CPU)**. The CPU also includes a circuit that performs the arithmetic and logic operations (the ALU) and a variety of registers used for temporary data storage. When the CPU is contained in a few ICs, it is called a **microprocessor.** A computer in which the CPU is a microprocessor is called a **microcomputer.**

The CPU is the principal manager of the main communication channel of the computer, a set of thirty or more leads called the **processor bus**. As shown in figure 12-1, all other parts of the computer are connected to the CPU through the data, address, and control lines of this bus. Interface circuits are often needed to adapt the bus signals to those used by standard input and output devices. Data are transferred in or out of the CPU as a parallel digital word along the data bus. The CPU also provides a code word called the **address** that specifies the source or destination of the data. Each device or memory location must monitor the address bus and supply or

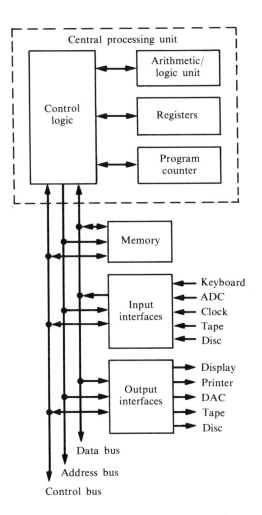

Data bus

Address bus

Control bus

**Fig. 12-1.** The structure of a basic digital computer. The central processing unit acquires and executes the instructions. The program is stored in a section of memory reserved for that purpose. The CPU supervises communication along the shared data bus by using the address bus to specify the source or destination of the data. The program counter supplies the address of the instruction to be executed. The control bus contains data direction, timing, and special control signals.

accept data as appropriate when its address code is present. The control bus includes signals that control the use of the address and data busses such as **read** for an input to the CPU and **write** for a CPU output. Other control signals are used to synchronize operations and to request or indicate a change in the CPU operation. A program being executed is located in a section of computer memory. The program counter in the CPU supplies the address of each instruction. Words in which instructions are encoded are acquired by the CPU one at a time through the bus. Memory is also used for the storage of data related to the purpose of the executing program.

The first sections of this chapter are devoted to **hardware**, i.e., the major elements and operations of microcomputers. The final section describes the aids that make the writing of complex programs (**software**) a practical task.

## 12-1    Microprocessors

The heart of the microcomputer is the microprocessor. It contains the control logic that fetches, decodes, and executes the instructions in the stored program. It manages the data, address, and control busses that connect it to the stored program and to all other sources and destinations of data. The elements of the microprocessor are shown in figures 12-2 and 12-3. In this example the data bus is eight lines ($D_0$ to $D_7$) and can thus transfer one byte of data at a time. More advanced microprocessors have a 16-bit data bus. The size of the address bus determines the number of unique address codes available. Microprocessor address busses are from 16 to 20 bits wide (63 536 to 1 048 576 individual addresses).

The clock is the "heartbeat" of the CPU; all operations are sequenced by its oscillations. The first step in any instruction is to read the instruction word from the address location specified by the program counter. The instruction word is put on the internal data bus and acquired by the instruction register. The instruction word is decoded to determine which operations are performed during the remaining clock cycles of that instruction. These

**Fig. 12-2.**   Structure of the microprocessor. The microprocessor is a combination of general and special-purpose registers with a system for controlling data flow and register operations. It is paced by a clock that also aids in timing communication with external devices.

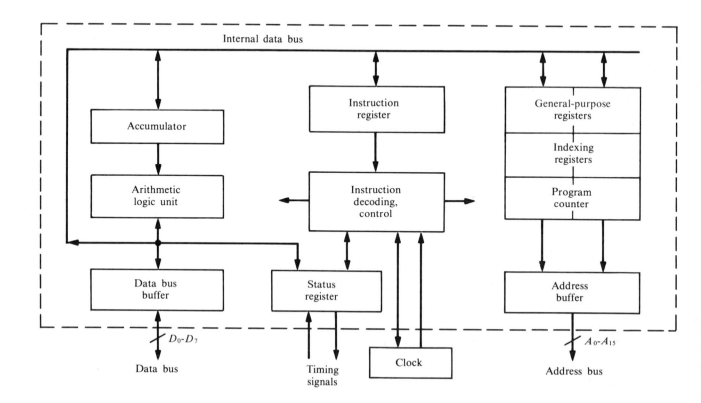

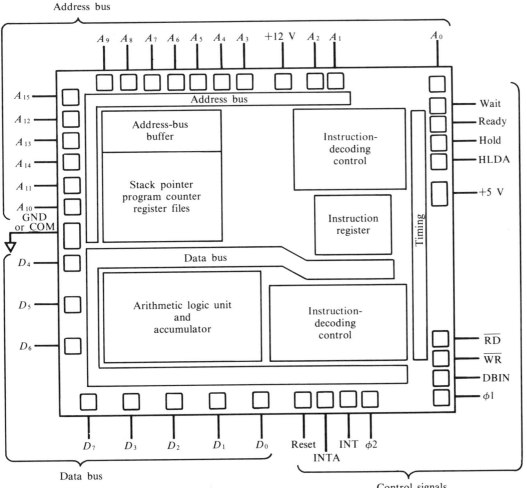

Address bus

$A_9$  $A_8$  $A_7$  $A_6$  $A_5$  $A_4$  $A_3$  +12 V  $A_2$  $A_1$   $A_0$

$A_{15}$

$A_{12}$

$A_{13}$

$A_{14}$

$A_{11}$

$A_{10}$
GND
or COM

$D_4$

$D_5$

$D_6$

Address bus

Address-bus
buffer

Stack pointer
program counter
register files

Data bus

Arithmetic logic unit
and
accumulator

Instruction-
decoding
control

Instruction
register

Instruction-
decoding
control

Timing

Wait

Ready

Hold

HLDA

+5 V

$\overline{RD}$

$\overline{WR}$

DBIN

$\phi1$

$D_7$  $D_3$  $D_2$  $D_1$  $D_0$  Reset  INT  $\phi2$
                                  INTA

Data bus

Control signals

operations may include transfers of data between external registers and internal registers, transfer of data between internal registers, operations on register data such as incrementing or clearing, and arithmetic or logic operations in the ALU. The ALU operations usually provide the principal computational power of the computer. The ALU has one or more primary registers called **accumulators**.

The general-purpose and indexing registers contain information that is repeatedly used by the program. The indexing registers are convenient for holding an address in a section of memory that is being accessed sequentially. The status register keeps track of various conditions in the CPU and in the

**Fig. 12-3.** Arrangement of microprocessor elements on the IC chip. Fine gold wires connect the square pads around the edge of the chip to the corresponding pins in the IC package. The IC chip is about 5 mm square.

rest of the computer. Generally, each bit in the status word has a particular significance. Individual bits may indicate an overflow or carry from the accumulator or the need of an external device for attention. Each microprocessor family handles the details of the interaction between the CPU and the rest of the computer a little differently. Therefore, there is no standard for the significance of the status register bits or for the particular timing and control signals on the control bus.

## Instructions and Operation Codes

Every CPU has a particular **instruction set**, that is, a list of specific instructions it is designed to execute. These instructions can be classified into the eight categories listed in table 12-1. It is these few types of relatively simple operations that give the microcomputer its great power and versatility.

**Table 12-1.** Classes of CPU instructions.

---

*Information transfer.*   Movement of a data word between registers, internal or external.

*Input/output.*   Movement of a data word between the accumulator and an input or output device.

*Arithmetic.*   Addition, subtraction, or comparison of a word with the word in the accumulator; a shift or complement operation on the accumulator; in advanced processors, multiplication and division.

*Logic.*   An AND, OR, or exclusive-OR of the bits in a data word with the corresponding bits in the accumulator word.

*Increment/Decrement.*   An increment or decrement of the word in an internal or external register.

*Jump.*   The setting of the program counter to a value other than that of the next address in the current sequence.

*Call subroutine.*   A jump to the beginning address of a utility program or subroutine in such a way that the main program can be easily reentered with an instruction to return from subroutine when the subroutine is finished. Allows common operations to be written as subroutines that can be accessed by any main program.

*Processor.*   Instructions that affect only the CPU operation, such as halt, reset, and enable interrupt.

---

The instruction is encoded in a binary format that includes an **operation code**, or op code, which specifies the operation or operations to be performed and which indicates the mode of locating the **operand** (the data to be operated on). One word in an eight-bit microprocessor can encode up to 256 different combinations of operations. In most microprocessors, all the instructions performed on operands in internal registers can be encoded using only the one-byte operation code. For example, in the 8080/8085 series by Intel, op code words with the form 00$xxx$lll are operations on the data in the accumulator. The eight different combinations for the bits $xxx$ allow eight different accumulator operations as shown in table 12-2. The accumulator register A has one more bit position than normal data registers in the CPU. This one-bit register is called the **carry,** or **overflow** register. In the rotate left instruction, the accumulator bits are shifted to the next more significant position, and instead of being lost, the most significant bit (MSB) is shifted around to the least significant bit (LSB) position. As shown in table 12-2, rotation can be in either direction, and the carry bit may be included or not.

An example of an op code that specifies both the operation and the source of the operand are the arithmetic and logic operations shown in table 12-3. The first two bits of the instruction code (10) denote an arithmetic/logic instruction. The bits $xxx$ specify one of eight operations that combine the contents of the accumulator with the contents of the register specified by bits $sss$. Seven internal registers can be used for these operations (including the accumulator itself). The eighth code is used to indicate that the desired operand is in computer memory at the 16-bit address given by the two 8-bit registers H and L.

A data transfer instruction must specify both the source and destination of the operand. In the 8085 one type of data transfer instruction has the form 01$dddsss$ where 01 indicates a transfer from the source specified by $sss$ to the destination specified by $ddd$. The code for $sss$ and $ddd$ are the same as that for $sss$ in table 12-3.

## Addressing Modes

Three types of information are handled in microcomputer systems: op codes, operands, and addresses. An address is an element of information that indicates where to find another element of information. **Addressing modes** are the various ways the location of an operand can be specified in an instruction. The number and types of addressing modes implemented in the computer's instruction set can significantly affect its programming and operating efficiency in many applications.

The common addressing modes can be classified as immediate, direct, and indirect. In the **immediate mode** the operand is found in the word(s) immediately following the op code in the program; it becomes, in effect, part of the instruction. This location is convenient for some types of operands (constant values), inconvenient for others (data from an input device), and

**Table 12-2.** Accumulator operations in the 8085.

| op code | 0 | 0 | $x$ | $x$ | $x$ | 1 | 1 | 1 |
|---|---|---|---|---|---|---|---|---|

| $xxx$ | Operation |
|---|---|
| 000 | Rotate A left |
| 001 | Rotate A right |
| 010 | Rotate A left through carry |
| 011 | Rotate A right through carry |
| 100 | Decimal adjust A |
| 101 | Complement A |
| 110 | Set carry |
| 111 | Complement carry |

**Table 12-3.** Arithmetic and logic operations in the 8085.

| op code | 1 | 0 | $x$ | $x$ | $x$ | $s$ | $s$ | $s$ |
|---|---|---|---|---|---|---|---|---|

| $xxx$ | Operation |
|---|---|
| 000 | Add (word in $sss$ to A) |
| 001 | Add with carry (word in $sss$ to A) |
| 010 | Subtract (word in $sss$ from A) |
| 011 | Subtract with borrow (word in $sss$ from A) |
| 100 | AND (word in $sss$ with A) |
| 101 | exclusive-OR (word in $sss$ with A) |
| 110 | OR (word in $sss$ with A) |
| 111 | Compare (word in $sss$ with A) |

| $sss$ | Register |
|---|---|
| 000 | B ⎤ |
| 001 | C ⎟ |
| 010 | D ⎟ incrementing registers |
| 011 | E ⎟ |
| 100 | H ⎟ |
| 101 | L ⎦ |
| 110 | M memory address indicated by H and L registers |
| 111 | A accumulator |

impossible for some (such as a variable operand when the program is to be stored in read-only memory). A **direct-mode** instruction specifies the location of the operand. A CPU register location can be specified in the op code itself; for a location in memory, the memory address is given in the next two bytes of the instruction. The direct mode may not be convenient if the operand address is variable. In the **indirect mode** the instruction provides the address of the register where the address of the operand is located. Again this indirect address may follow the op code in the instruction, or it may be in a CPU register indicated by the op code. The instruction format for each of these addressing modes is shown in figure 12-4.

**Fig. 12-4.** Addressing modes. The contents of the instruction registers and CPU registers are shown for the most common addressing modes. In this figure, L and H refer to the low and high bytes of a 16-bit address. The direct and indirect modes avoid the need for the instruction to include the operand itself. If the instruction specifies the location of the operand, it is called direct; if it indicates where the address can be found, it is indirect. Abbreviated, relative, and indexed modes are conveniences to save programming steps or computer time for various types of tasks.

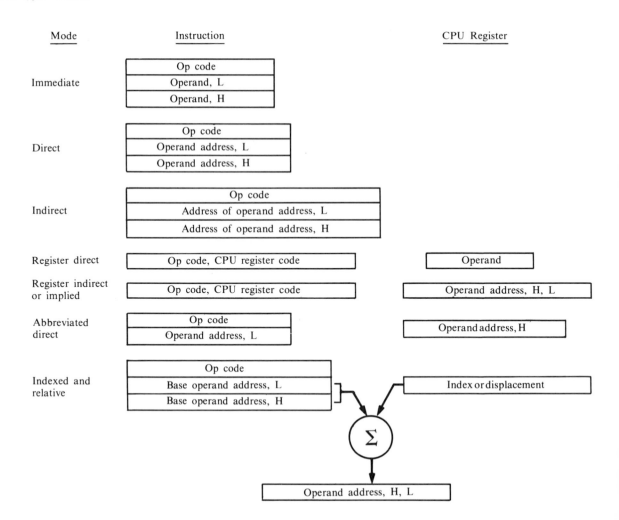

The op code must specify the addressing mode and identify the CPU register if one is involved. Examples of this are found in the 8085 arithmetic/logic op codes given in table 12-3. All these instructions (except where *sss* is 110) are register direct mode instructions; the operand is found in the CPU register specified. The case in which *sss* is 110 is an example of register indirect mode because the CPU registers H and L contain the address of the operand, not the operand itself. The 8085 also provides immediate addressing; the op code 11*xxx*110 performs all the *xxx* operations in table 12-3 on the word in the memory byte immediately following the op code.

The **abbreviated direct mode** shown in figure 12-4 is an example of a class of modes in which the instruction gives part of the address, the rest of which is either located in a CPU register or implied. In the 6500 series microprocessors from Rockwell and Mostek, there is an abbreviated mode called the "zero page mode" in which the high address byte is assumed to be 00000000. Although only the first 256 memory locations can be addressed in this mode, it effects considerable saving in program space and execution time. In the **indexed mode**, or **relative mode**, a base address and a displacement value are summed to obtain the operand address. The base address is sometimes part of the instruction, and the displacement is in a CPU register as shown in figure 12-4. In other cases (the 6800 series from Motorola, for example) the base is in the CPU and the displacement in the instruction. This mode is convenient where the operand address increments on successive passes through a program as, for example, in operations on a field of stored data. Differences in the addressing modes available in the instruction set are among the most often cited criteria for selecting one microprocessor over another.

## Control Instructions

In linear programming, the control of the processor passes from the present instruction to the one with the next higher address. If all programming had to be done this way, the program would have to repeat the routines for frequently used operations, and the program path could not depend on the results of operations or on changes in external signals. Therefore, all computers provide instructions that allow processor control to be shifted to programs at other specified locations in memory. These instructions all operate by simply changing the program counter contents to the address of the instruction that is to take control. The simple passing of control to a specified memory location is called a **jump**, or **branch**, instruction. The instruction to shift control to a new location may be either unconditional or conditional as shown in figure 12-5. The conditions available for jump or branch options are generally the value of the bits in the CPU status register such as "carry = 1 or 0," or "accumulator = positive, negative, or zero."

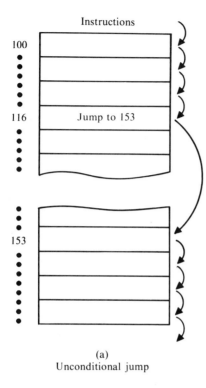

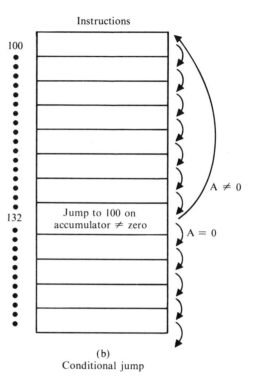

(a)
Unconditional jump

(b)
Conditional jump

**Fig. 12-5.** Illustrations of jump instructions. In example (a) the program encounters an unconditional jump to location 153 at location 116. The jump sets the program counter to 153. This allows program segments to be accessed in any order, not just in successive memory locations. In example (b) a conditional jump occurs at location 132. If the accumulator register is not zero, the jump (in this case backwards) to 100 will be executed, otherwise control continues to the following instruction. In this example the effect is to repeat the instructions in locations 100 to 132 until the accumulator register is zero and escape from the loop occurs.

The jump instruction includes the op code for the jump and its conditions (if any). It also indicates the jump destination by the instruction word(s) immediately following the op code. The 6500 and 6800 series CPUs distinguish between a jump (to a specified two-byte address) and a branch (to a one-byte displacement from the current address). Branch instructions are only two bytes long, and program segments are more easily moved to another section of memory.

Instructions to jump to a subroutine are similar to simple jump instructions, except that there is an intention to return to the main program at the end of the subroutine. A **subroutine** is a section of program that accomplishes a specifically defined task. *Call subroutine* or *jump to subroutine* instructions allow a subroutine task to be accessed from many points in the main program even though it exists only one place in memory. A library of commonly needed subroutine tasks can be written and tested independently. Individual subroutines are then invoked as needed in the main program, by a call to the subroutine. The operation of subroutine calls is illustrated in figure 12-6. The call instruction causes the current content of the program counter (PC) to be saved. It specifies the subroutine starting address in the same way as the jump or branch instructions do. This address is loaded into

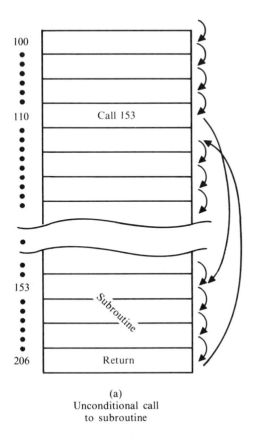

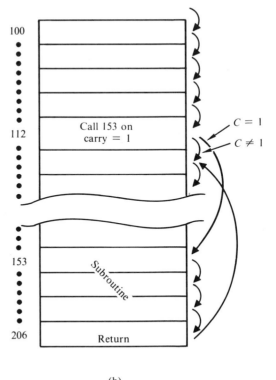

(a)
Unconditional call
to subroutine

(b)
Conditional call
to subroutine

the PC. Thus control passes to the subroutine. The last instruction in the subroutine is a *return* which reloads the PC with the saved value and returns control to the next instruction in the main program. Both the *call* and *return* instructions can be conditional.

## The Stack

In the *call to subroutine* described above, the PC contents are stored before the subroutine address is loaded into the PC. Most modern microprocessors use a section of memory called the **stack** for this purpose. As data are loaded into the stack, a stack pointer register (SP) in the CPU keeps track of the last address used. When data are retrieved from the stack, the SP contents serve as the address; that is, the most recently stored data are read first. Whenever the stack is written into or read from, SP is incremented or decremented so that it always points to the address at the top of the stack. A subroutine call always includes putting the PC contents on the stack (which advances SP by the number of bytes stored), and a subroutine return loads the PC register

**Fig. 12-6.** Illustrations of *subroutine call* instructions. In example (a) a *call 153* (that is, *jump to subroutine beginning at location 153*) is encountered at location 110. Before the program counter is set to 153 for the jump, the current program counter contents are saved. The last instruction in the subroutine (206) is *return from subroutine*, which restores the program counter to its previous state. The *call to subroutine* can also be conditional as in example (b). If the carry bit is 1 when instruction 112 is encountered, the subroutine at 153 will be entered; if not, the control passes directly to the instruction after 112. The return instruction can also be conditional.

**Note 12-1.    Stack Direction.**
The text describes a microprocessor in which the stack grows from the bottom up, i.e., the stack pointer register increments as the stack is added to. Many processors organize the stack from the top down. The first stack entry is at a high address and the stack pointer register decrements as the stack grows.

with the data stored at the memory address given by SP (and it reduces SP by the number of bytes returned). One of the features of the stack is that it allows a subroutine to be called from a program that is itself a subroutine. For example, a *call to subroutine* "LIST" puts the PC for return to the main program on the stack. Within subroutine "LIST," *call to subroutine* "PRINT" puts the PC for return to subroutine "LIST" on the stack. When subroutine "PRINT" is finished, control returns to subroutine "LIST," and when it is finished, control returns to the main program. If the stack space is moderately large, subroutines can be "nested" many layers deep. Control is always returned to the program that called the subroutine by the last-in, first-out operation of the stack. (See note 12-1.)

When a subroutine returns control to the calling program, it should leave the CPU registers as it found them (or with data needed from the subroutine) so that the calling program can continue. Therefore, the subroutine should store the contents of CPU registers that it will use so they can be reset to their original values. CPU register contents are conveniently stored on the stack by **push instructions** (*move register to top of stack*) at the beginning of the subroutine and then restored to the CPU registers by **pop instructions** (*move top of stack to register*) just before the return. As long as there are equal numbers of pushes and pops in every subroutine, the appropriate PC value is at the top of the stack at the end of the subroutine.

## Execution of an Instruction

The execution of instructions is paced by the CPU clock. All events are timed by the leading and/or trailing edges of the clock signal. Different operations in the CPU require different numbers of clock periods, and each instruction is a combination of one or more operations. Thus instructions vary widely in the number of clock periods they require for completion. This is illustrated in figure 12-7. The **fetch** operation is a reading of the op code from the address specified by the program counter. Every microprocessor provides an indication (through its timing signals) of when it is in the fetch state. From these signals the time of the beginning of each instruction can be determined. Following the fetch, the op code is decoded and executed. In the 8085 example shown, most instructions that do not require access to an external register, such as *sum the contents of the accumulator and another internal register*, will be completed in two clock periods following decoding or less. Each *read* or *write* to an external register requires three clock periods. For example, an instruction to *store the accumulator contents in the address specified by bytes 2 and 3 of the instruction* requires four periods to fetch and decode the op code, six periods to read the address bytes and put them in the address buffer, and three more periods to write the accumulator contents in that specified address—a total of thirteen periods.

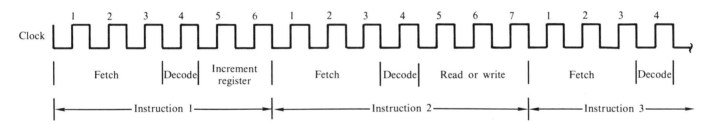

Clock

| Fetch | Decode | Increment register | Fetch | Decode | Read or write | Fetch | Decode |

←————— Instruction 1 —————→←————— Instruction 2 —————→←——— Instruction 3 ———→

From the above example, it can be seen that a substantial fraction of the instruction execution time is spent in fetching the instruction. Also, during much of the execution time the bus is not used. Instruction times can be reduced if unused bus time is used to fetch instruction words in advance. The 8086 CPU, a 16-bit processor from Intel, has separate controller circuits for instruction execution and bus control. The execution controller picks up instructions from an instruction register queue that is loaded by the bus control logic. Execution from internal instruction registers is faster, and the bus is used more efficiently than with single controller CPUs.

Most microprocessors provide a way of extending read and write cycles so that slower responding devices can take a few extra clock periods to recognize their address and supply or accept the data. If this were not possible, the CPU clock would have to be slowed to the rate required by the slowest device in the system. Some CPUs require two clock signals with the same period but different phase. The clock rates for common CPUs are such that each operation takes about 1 µs. Complete instructions, therefore, are executed in 1–5 µs. Higher speed versions are also available.

**Fig. 12-7.** Clock periods for various operations and instructions. In this example of 8085 operation, three clock periods at the beginning of each instruction cycle are used to fetch the op code, which is decoded in the fourth period. The number of remaining periods required for execution depends on the instruction. An internal operation, such as incrementing a register, requires two periods, and a *read* or *write* to a device on the bus requires three. Instructions can include several reads, writes, and other operations and can require up to 18 clock periods.

## 12-2  Numerical Data and Arithmetic Operations

A byte of digital data is a string of 1's and 0's eight digits long, such as 00110111. In this form the byte looks like a binary number, but actually it is just one of the 256 different states (combinations of 1's and 0's or T's and F's) for the byte. As shown in table 12-2, when the byte 00110111 is an op code in an 8085 microprocessor, it means *set the carry bit*. Even when a byte is known to represent numerical information, there are a huge number of ways its 256 different states can be interpreted. In the microcomputer the three numerical codes described in this section (unsigned binary, two's complement binary, and BCD) predominate. Clearly the significance of a byte of data cannot be understood unless one knows how the data were encoded.

### Binary Numbers

The most direct encoding and interpretation of a digital word is as a binary-coded digit in which the right-most position is the least significant bit (LSB) and the value of the bit increases two-fold with every space to the left until the most significant bit (MSB) is reached. Most people don't conceptualize binary numbers easily, and binary numbers are tedious to type when entering data into a computer. For this reason, **octal** and **hexadecimal numbers** are often used as a shorthand representation for binary numbers. In octal representation the bits are grouped in triplets starting at the decimal point, and each group is given its octal value. For example, the ten-bit binary number 1 110 011 101 is $1635_8$ where the subscript eight indicates base-8 or octal. Hexadecimal numbers are base-16, and thus each group of four binary digits translates directly into a hexadecimal numeral. The 16 numerals in base-16 numbers follow the decimal symbols 0–9 and then use A, B, C, D, E, F for the necessary six additional numerals. The same ten-bit binary number 11 1001 1101 is $39D_{16}$. Binary-encoded numbers are very useful for data that are positive integers, (count values, and memory addresses, for example). Even though op codes are not numbers, they are often written in the octal or hexadecimal shorthand. Thus the op code for a jump in the 6800 microprocessor is 01111110, but it may also be given as 176 (octal) or 7E (hexadecimal). Addresses are almost always given in hexadecimal notation. The hexadecimal representation of the address spaces available in a processor with a 16-bit address bus are 0000 to FFFF.

## Negative Numbers

To support even the most elementary arithmetic operations, there must be a method of encoding negative as well as positive numbers. By convention, the left-most bit is used to convey the sign of the number (0 for + and 1 for −) in **signed number codes**. Among the various binary signed number codes, two predominate. They are the sign-and-magnitude code and the two's complement code. In **sign-and-magnitude** encoding the left-most bit indicates the sign and the remaining bits, the magnitude in normal binary code. For example, 00100011 is $+0100011_2$ ($+35_{10}$ and $+23_{16}$), and 10101001 is $-0101001_2$ ($-41_{10}$ and $-29_{16}$). This code is conceptually simple and is used in some analog-to-digital converters, but it is not as convenient for counting or arithmetic operations as the two's complement code. To illustrate the latter code, consider first a bidirectional eight-bit binary counter. Zero will be represented by 00000000, all positive numbers by the equivalent number of counts up from zero, and all negative numbers by the equivalent number of counts down from zero. This is illustrated in figure 12-8. This notation is called **two's complement** because the negative numbers are written as a kind of complement of the positive numbers. The two's complement of a binary number is obtained by subtracting the number from the modulus of the hypothetical counter. The modulus of a counter is the full-scale value plus one, so the modulus of an eight-bit counter is 11111111 +1. The two's complement of 00000101 (+5) is thus 11111111 − 00000101 + 1 which is

| Two's complement | Decimal equivalent |
|---|---|
| 0 1 1 1 1 1 1 1 | +127 |
| ⋮ | ⋮ |
| 0 0 0 0 0 1 0 1 | +5 |
| 0 0 0 0 0 1 0 0 | +4 |
| 0 0 0 0 0 0 1 1 | +3 |
| 0 0 0 0 0 0 1 0 | +2 |
| 0 0 0 0 0 0 0 1 | +1 |
| 0 0 0 0 0 0 0 0 | 0 |
| 1 1 1 1 1 1 1 1 | −1 |
| 1 1 1 1 1 1 1 0 | −2 |
| 1 1 1 1 1 1 0 1 | −3 |
| 1 1 1 1 1 1 0 0 | −4 |
| 1 1 1 1 1 0 1 1 | −5 |
| ⋮ | ⋮ |
| 1 0 0 0 0 0 0 0 | −128 |

**Fig. 12-8.** Two's complement representation of eight-bit binary numbers with sign. The MSB is 0 for all positive numbers and 1 for negative values. Positive numbers follow the normal binary notation. Increasing negative numbers follow the pattern of a binary down-counter. To change the sign of a number, complement each bit, and add 1 to the result.

11111010 +1 or 11111011 (−5). This is confirmed in figure 12-8. In finding the two's complement of a number, it is easier to subtract first and then add 1 because the difference between a binary number and its modulus is simply the bit-by-bit complement of the original number. The same process is used to change the sign of a negative number. The two's complement of 11111101 (−3) is thus 00000010 +1 or 00000011 (+3). A number in a CPU register is readily changed to its negative by first complementing and then incrementing the register contents.

The great advantage of two's complement numbers is that the addition of positive and negative numbers follows the rules for ordinary binary addition, and subtraction is accomplished by changing the sign of the subtrahend and adding. Some examples are given in figure 12-9. Note that normal binary addition including the sign bit automatically produces the correct sign and magnitude except when the result exceeds the maximum value for the number of bits (−128 to +127 for an eight-bit number). If there is a carry into the sign bit but no carry out or if there is a carry out without a carry in, an overflow has occurred, giving a wrong answer. In some CPUs this condition sets a bit in the status register that the program can use to check for error.

## Decimal Arithmetic

Most microprocessors also provide addition and subtraction logic for numbers in BCD code. With four bits for each BCD digit, one byte encodes two digits exactly. The logic involved in the addition of a BCD digit is illustrated in figure 12-10. A binary addition is performed for each BCD digit. If a carry or an incorrect BCD code is produced, $0110_2$ is added to the digit to produce the correct answer. These instructions make no provision for sign representation or manipulation; if signs are involved, the program must keep track of them separately.

Of course, microcomputers perform far more complex arithmetic operations than the simple additions illustrated in this section. Multiplication, division, square roots, trigonometric functions, logarithms, and many other operations are performed on multiple precision numbers (16, 24, 32, or more bits per number), fractional numbers, or floating point numbers (scientific notation in magnitude and power-of-10 format). All these are achieved by programs that implement algorithms to perform the desired function by many simple operations on data taken one or two bytes at a time.

## 12-3 Memory

Memory, of course, is where information is stored for later retrieval. In a microcomputer, the **memory** is that part of the data storage and retrieval

| | | | | | | | | | |
|---|---|---|---|---|---|---|---|---|---|
| (a) | 0 | 1 | 0 | 1 | 0 | 1 | 1 | 0 | $+86_{10}$ |
| | 0 | 0 | 1 | 0 | 0 | 1 | 1 | 1 | $+39_{10}$ |
| | 0 | 1 | 1 | 1 | 1 | 1 | 0 | 1 | $+125_{10}$ |
| (b) | 1 | 0 | 1 | 0 | 1 | 0 | 1 | 0 | $-86_{10}$ |
| | 1 | 1 | 0 | 1 | 1 | 0 | 0 | 1 | $-39_{10}$ |
| | 1 | 0 | 0 | 0 | 0 | 0 | 1 | 1 | $-125_{10}$ |
| (c) | 0 | 1 | 0 | 1 | 0 | 1 | 1 | 0 | $+86_{10}$ |
| | 1 | 1 | 0 | 1 | 1 | 0 | 0 | 1 | $-39_{10}$ |
| | 0 | 0 | 1 | 0 | 1 | 1 | 1 | 1 | $+47_{10}$ |
| (d) | 1 | 0 | 1 | 0 | 1 | 0 | 1 | 0 | $-86_{10}$ |
| | 0 | 0 | 1 | 0 | 0 | 1 | 1 | 1 | $+39_{10}$ |
| | 1 | 1 | 0 | 1 | 0 | 0 | 0 | 1 | $-47_{10}$ |
| (e) | 0 | 1 | 0 | 1 | 0 | 1 | 1 | 0 | $+86_{10}$ |
| | 0 | 0 | 1 | 1 | 0 | 1 | 1 | 1 | $+55_{10}$ |
| | 1 | 0 | 0 | 0 | 1 | 1 | 0 | 1 | $+141_{10}$ |
| (f) | 1 | 0 | 1 | 0 | 1 | 0 | 1 | 0 | $-86_{10}$ |
| | 1 | 1 | 0 | 0 | 1 | 0 | 0 | 1 | $-55_{10}$ |
| | 0 | 1 | 1 | 1 | 0 | 0 | 1 | 1 | $-141_{10}$ |

**Fig. 12-9.** Two's complement addition. Examples (a) and (b) are like-sign additions. The arrows in (b) indicate a carry into the sign bit and a carry out of it. The sign bit thus remains correct. The carry out of the sign bit is ignored. The mixed sign additions of (c) and (d) show the correct sign bit is obtained by normal addition. In examples (e) and (f) the capacity of the seven-bit magnitude is exceeded by the addition and the result is wrong in both sign and magnitude. The overflow error condition is characterized by a carry going only into or only out of the sign bit.

**Fig. 12-10.** BCD addition. The initial result is wrong if the addition takes the result into or through the six missing codes (1010, 1011, 1100, 1101, 1110, and 1111) in the BCD notation. In such cases, the result is corrected by adding 6 as shown in (b) and (c).

(a)
$$
\begin{array}{cccc}
0 & 0 & 1 & 1 \\
0 & 1 & 0 & 1 \\
\hline
1 & 0 & 0 & 0 \\
\end{array}
$$

$3_{10}$
$5_{10}$
$8_{10}$

No carry; result is proper BCD digit; answer is correct.

(b)
$$
\begin{array}{cccc}
0 & 1 & 0 & 1 \\
0 & 1 & 1 & 1 \\
\hline
1 & 1 & 0 & 0 \\
0 & 1 & 1 & 0 \\
\hline
1\ 0 & 0 & 1 & 0 \\
\end{array}
$$

$5_{10}$
$7_{10}$
$12_{10}$

$2_{10}$

No carry; result is improper BCD digit; result is wrong.

Add 6 to obtain proper answer and a carry to the next digit.

(c)
$$
\begin{array}{cccc}
1 & 0 & 0 & 1 \\
1 & 0 & 0 & 0 \\
\hline
1\ 0 & 0 & 0 & 1 \\
0 & 1 & 1 & 0 \\
\hline
1\ 0 & 1 & 1 & 1 \\
\end{array}
$$

$9_{10}$
$8_{10}$
$17_{10}$

Carry generated; proper BCD digit; result is wrong.

Add 6 to obtain proper answer; take carry over to next digit.

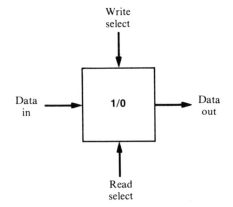

**Fig. 12-11.** One cell of memory. When the cell is selected for read, the stored value appears as a HI or LO at Data out. In memory types where the contents are alterable, the Write select allows a 1 or 0 to be "written into" the cell.

system to which the CPU has direct and immediate access. Since the memory contains both the program and the data needed for at least its current task, the CPU accesses memory at least once during each instruction. Thus the memory characteristics and the interactions between CPU and memory significantly affect the performance capabilities of the computer. One of the characteristics of a computer memory is its capacity. A microprocessor with an 8-bit word size and a 16-bit address bus can manage a memory of $2^{16} =$ 65 536 bytes, or over half a million bits. Most microprocessors operate with a memory that is much smaller than the maximum size. In addition to variation in capacity, microcomputer design allows great flexibility in the type of memory used and how the memory is structured in the system. The principal memory devices and their implementation in microcomputer systems are described in this section.

## Memory Types and Devices

All memory devices are made up of individual units or cells, each of which can store one bit of information. Such a cell is shown schematically in figure 12-11. A memory device contains a large number of single storage cells arranged for convenient access by computers.

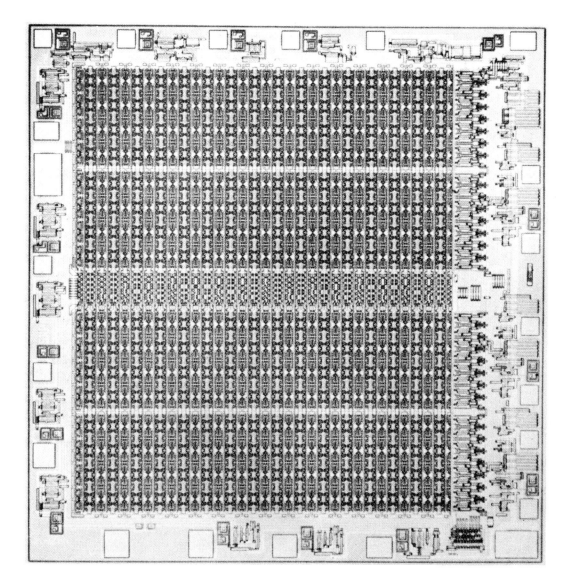

**Fig. 12-12.** Photomicrograph of a 5101 IC, a 1024-bit RAM. The cell array is 32 rows by 32 columns. Fine wires connect the square pads around the edge of the chip to the pins of the IC package. (Reproduced with permission of Intel Corp., Santa Clara, CA).

Generally, memory devices fall into one of two classes: RAM and ROM. **RAM** is an acronym for **random access memory**, which means that technically all storage cells are directly accessible in any order. Through common usage, however, RAM refers exclusively to **read-write memory**. In RAM memory the cells can be written into as easily as they are read. By contrast, the information stored in **ROM**, or **read-only memory** cannot be altered by the CPU at all (but it can be randomly accessed).

**RAM.** One class of semiconductor RAM devices is made of rows and columns of tiny transistor flip-flop circuits. The pattern of the cells and the criss-crossed select and data connections can be seen in figure 12-12. The flip-flop circuits hold their data as long as the circuit remains powered. Because the data remain unchanged, this kind of device is called **static RAM**. The maximum number of memory cells in a single IC has increased continually since their introduction. The 1024-bit device shown in figure 12-12 is now a relatively low-capacity device. A numerical abbreviation that expresses memory capacity in units of $2^{10} = 1024 = 1K$ is commonly used. Thus a 1024-bit device is a 1K chip. Static RAM ICs of 4K (4096 bits), 8K and larger capacity are available.

The 4K chips have a square array of 64 cells on each side, i.e., 64 rows and 64 columns. One way to organize the internal select and data connections is shown in figure 12-13. This device accesses only one cell at a time for read or write. Twelve address lines are needed to specify all 4096 cells ($4096 = 2^{12}$). When the computer addresses this device, only one bit is transferred. For a computer with eight-bit words, eight of these devices would be used to produce a 4K-byte memory as shown in figure 12-14. Another common arrangement for accessing the cells in a 4K chip is the $1024 \times 4$ organization shown in figure 12-15. Since this device reads or writes four bits at a time, only two chips are required for an eight-bit word computer. Memory capacity can then be increased in 1K increments with each additional pair of ICs.

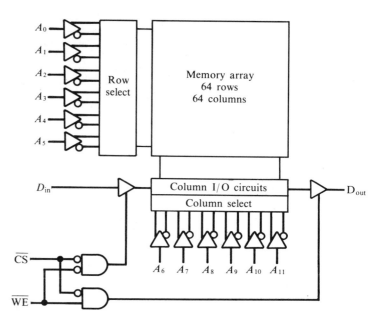

**Fig. 12-13.** Block diagram of 4096 $\times$ 1 bit static RAM (type 2141). Six address lines ($A_0$–$A_5$) select a specific row in the array and six more ($A_6$–$A_{11}$) select a specific column. Thus a single cell is selected for a read from $D_{out}$ or a write at $D_{in}$. When the chip select input $\overline{CS}$ is LO, the level at write enable $\overline{WE}$ determines whether a write or a read occurs.

As part of the effort to reduce the size of cells and thereby increase memory capacity per IC, memory cells that store information as a tiny charge rather than as a bistable state were developed. Because the charge eventually leaks off, data bits stored in these cells do not remain unchanged. This problem is overcome by reading each cell periodically and restoring its appropriate charge state. This type of memory is called **dynamic RAM**, and the circuit that performs the continual read-and-replace function is called the **refresh circuit**. Dynamic RAM ICs are available with capacities up to 64K bits. The refresh circuitry is not trivial, but it can be shared by whole sections of memory. Therefore, dynamic RAMs are most economical in memory sections of relatively large capacity (16K bytes and up). Because each cell must be refreshed within a particular time interval ($\sim$2 ms) the refresh circuit must intersperse refresh cycles with computer read and write operations. This can sometimes delay access or limit the duration of continuous access.

Earlier computers used a kind of read-write memory called **core memory**, in which each data bit is stored as the direction of residual magnetism in a tiny ferrite bead. This kind of memory had the advantage that when power was turned off, the stored data remained. By contrast, the semiconductor RAM devices described above are called **volatile memory** because

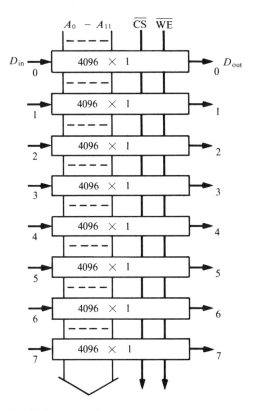

**Fig. 12-14.** Organization of 4K $\times$ 1 bit memory ICs into a 4K $\times$ 8 bit memory. The parallel connection of the address code $A_0$–$A_{11}$, chip select ($\overline{CS}$), and read/write command ($\overline{WE}$) to all eight chips allows eight bits to be written or read at the same time. A computer with a longer word size would require more chips to achieve a 4K-word memory.

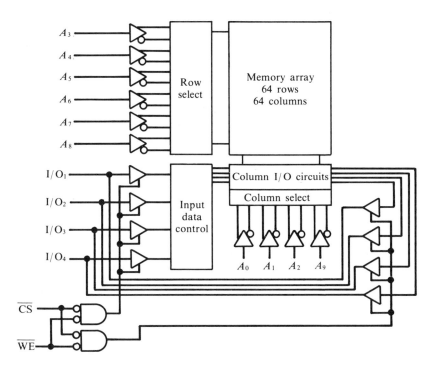

**Fig. 12-15.** Block diagram of 1024 $\times$ 4 bit static RAM (type 2114). Six address lines ($A_3$–$A_8$) select a specific row in this array. Four of the 64 bits in this row will be addressed, so only four more address lines are required ($A_0$–$A_2$, $A_9$). The internal connection of the input and output for each of the four data lines saves IC pins and allows easy connection to bidirectional data lines in the computer bus.

they lose all stored information when power is interrupted. If semiconductor RAM is the only memory, all programs and data must be reentered into memory when power is again turned on. Even when some kind of external permanent data storage is available, the program required to transfer the data into memory would not survive the no-power state. This problem is avoided in two ways: by using ROM, which is nonvolatile, or by protecting the RAM against power loss with a battery. In battery-backed systems, a battery is switched in to maintain power to the memory chips when the main power supply is off. The battery power required is greatly reduced by the use of memory chips that have a special low-power standby mode. This is the technique used in calculators with the "constant memory" feature.

**ROM.**   Read-only-memory was developed primarily as the solution to the problem of volatility in semiconductor memory devices, but its very successful evolution has significantly influenced the direction of microcomputer applications. ROM devices are randomly accessible arrays of storage cells in which a 1 or 0 has been permanently (or at least semipermanently) stored. These stored bits cannot be altered by a CPU write operation, and they survive the no-power state.

The ROM cell is considerably simpler and smaller than an equivalent RAM cell. ROM devices can therefore be made with large capacity, typically 8K–64K bits. The power required is also typically less for ROM than RAM. Because of their large capacity, ROM devices generally have multiple-bit outputs. For example, 32K and 64K devices are arranged 4K $\times$ 8 and 8K $\times$ 8, respectively, to produce a full byte of data from a single chip. In microcomputer systems ROM is used to store the programs needed for the normal functions of the computer system. In many specialized computer applications, ROM completely eliminates the need for external devices to serve as program sources. In such systems the ROM section of memory determines the specific capabilities of the computer. These capabilities can then be extended or altered simply by plugging more or different ROM ICs into the system. We see this idea in the plug-in preprogrammed memory modules that provide special complex functions in some hand calculators. ROM is also useful for the storage of data tables such as sines, logs, or conversion factors. (See note 12-2.)

The name read-only memory is somewhat misleading because to be useful the memory must be written in at least once. Depending on the type of ROM, the memory may be programmed in a variety of ways. In a true ROM, the memory contents are determined by the cell interconnection patterns at the time of manufacture. This form is very economical when relatively large quantities of the same ROM are required. The next type to be developed was a ROM that could be programmed by the user. The **programmable ROM**, or **PROM**, enabled designers to test their ROM programs

**Note 12-2.   ROM Look-Up Tables.**
In a ROM look-up table, the address is the input information, and the data word stored at that address is the output. For example, a ROM table for obtaining logarithms has at each address, the logarithm of the numerical value of the address. The processor then uses the number to be looked up as an address and obtains the log of that number from the data bus.

before committing to manufacture, but it also opened the door for individual users and low-volume applications of read-only memory. This inexpensive storage device for permanent lists and tables was soon recognized as an extremely versatile component. One type of PROM cell array is shown in figure 12-16. The blowing of the fuse links is, of course, an irreversible process. Bipolar PROMs are limited in capacity by their relatively large fuses. They also use more power than MOS memory, but they have the highest operating speed of any ROM.

MOSFET transistors are also used for PROM memory cells (fig. 12-17). The FET is made conducting by a programming step that stores negative charge in the gate. A quartz window through which ultraviolet (uv) light can pass is placed over the array. Exposing the array to ultraviolet light causes information stored in the PROM to be erased. This kind of device is called an **erasable PROM** or **EPROM**. This capability gives programmers that often-needed second (or third) chance. EPROMs have large capacity, low power requirements, and can be programmed quickly under computer control. They are currently displacing ROMs for all but very high quantity applications.

New PROM devices will continue to be developed for many years. Practical PROMs that can be selectively erased electrically have been introduced and dubbed EAROMs for **electrically-alterable ROM**. Still higher capacity, lower power and easier programming and reprogramming are to be expected in the future.

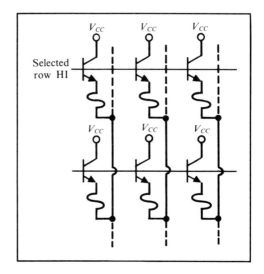

**Fig. 12-16.** Bipolar transistor PROM array. All the transistor collectors are tied together. The bases are connected together in rows. A fusible link connects the emitters to the common column connection. When a row is selected, the bases go HI, and a current occurs in each column lead for which the emitter fuse is intact. The ROM is programmed by "blowing" the desired fuses one-by-one using a special programming circuit.

**Fig. 12-17.** Cross-section of a MOS EPROM cell. In the programming step a high voltage injects electrons from the source or drain into the normally isolated gate. The trapped negative charge produces a p-type conducting channel at the substrate surface between the source and drain. The bit is read by testing for the presence or absence of conductance. Exposure to ultraviolet light allows the trapped charge to escape. The entire PROM is thus erased and can be programmed again.

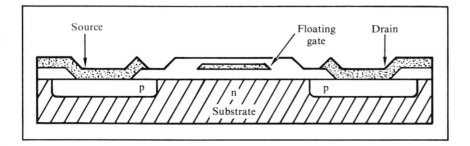

## Memory-Bus Interface

The connection of memory to the CPU bus involves the assignment of the memory to a particular subset of available memory codes and the use of control signals to direct that data to be supplied to or accepted from the data lines at the appropriate time. The range of possible CPU addresses can be thought of as a space into which memory and other devices can be placed according to various criteria. The addressable space for a computer with a

**Fig. 12-18.** Address space for the most significant bits of a 16-bit address. The most significant address bit ($A_{15}$) specifies whether the address is in the top or bottom half of address space. Bits $A_{15}$ and $A_{14}$ together divide the space into quarters; with $A_{13}$ the space is divided into eighths, and so on. The four most significant bits are required to define a particular 4K section of address space.

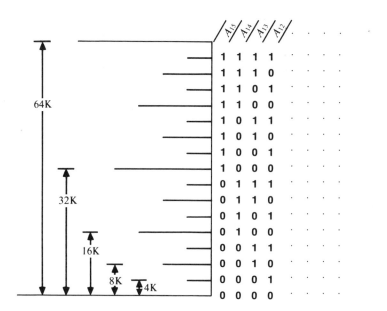

16-bit address bus is shown in figure 12-18. The most significant address bits define sections of address space. For example, if $A_{15}$ is set at 0, there are $2^{15} = 32\,768$ address codes available from all combinations of the remaining 15 bits. The same 32K codes can be used for $A_{15} = 1$. Thus specifying $A_{15}$ defines two spaces of 32K addresses each. Smaller spaces are defined by specifying more of the high order (most significant) address bits. Suppose an 8K × 8 ROM is to be assigned to a space. There are eight 8K spaces so three address bits ($A_{15}$, $A_{14}$, and $A_{13}$) must be defined and decoded to specify the desired space. The remaining address lines ($A_{12}-A_0$) are connected to the ROM address inputs. Thus all the address lines are used. A RAM and EPROM memory system with 4K-byte sections is shown in figure 12-19. The appropriate number of low order address lines are connected to the memory sections (12 lines in the case of 4K memory). The next three lines are connected to a 3-to-8 decoder. In this case, only $A_{15}$ remains, and it is connected to the decoder enable. Smaller sections of memory (or address busses with more bits) require the decoding of more high address bits to obtain the chip enable signals. For this reason eight 4K × 1 RAM chips are easier to interface than eight 1K × 4 RAM chips.

The timing of data transfer is very straightforward in microcomputers. Control signals from the CPU indicate when a memory transfer is taking place and whether the transfer is a read or write. The address and data direction are placed on the address and control busses first to prepare the addressed device for its response. In the case of read, the CPU allows the memory a short time to put the addressed data on the data lines, and then the data are transferred into the CPU. For a write operation, the CPU puts

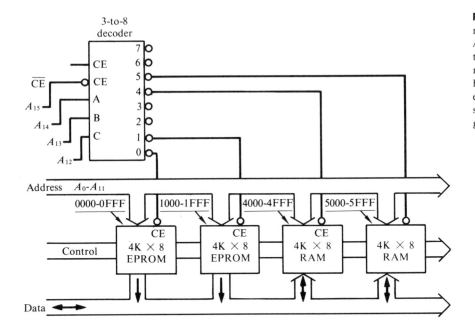

**Fig. 12-19.** Address space assignment for 4K-byte memory sections. A 3-to-8 decoder is used to decode $A_{12}$–$A_{14}$ into eight 4K spaces. The connection of $A_{15}$ to one of the decoder chips enables (CE or $\overline{CE}$) determines whether these spaces are in the top or bottom half of address space. The decoded high address bits enable the memory sections while $A_0$–$A_{11}$ address the space within them. The resulting address spaces are given in hexadecimal.

the data on the bus at the same time as the address and direction. It then issues a write command and later removes the data. The memory must respond to both *read* and *write* commands in the time allowed by the CPU (generally 500 ns or less). Some CPUs have their clocks slowed to suit the response time of the slowest memory. Others have a "ready" control line that a slow memory can use to extend the cycle and obtain more time for its response. Individual CPUs differ in the specific signals and timing used for control, but all follow this general pattern.

## 12-4   Input/Output Operations

To be generally useful a microprocessor must have ways of exchanging information with the rest of the world. This exchange occurs through external devices, or peripherals, such as keyboards, printers, tape recorders and CRT displays. (Common peripherals are described in the next section.) From the microprocessor's point of view (computer jargon is always ego-centric with respect to the processor), data are input from or output to such peripheral devices. The communication with peripheral devices uses the address, data, and control lines that comprise the CPU bus so that input/output (I/O) operations have much in common with the memory read and write operations described in the previous section. An input device places data expressed in parallel digital form on the data bus in response to its address and an input *read* command. Similarly, an output command causes the addressed output device to accept the word on the data bus. This section discusses some of the distinctions between memory and I/O operations and introduces the techniques used to coordinate the I/O operations with the communication needs of the peripherals.

## I/O Registers and Instructions

I/O devices are connected to the data bus of the CPU by input registers and output registers. An **input register** is a latch that is loaded with the input data and that can put that data on the data bus when a CPU I/O read instruction is executed. An **output register** is loaded by a CPU I/O write instruction; it supplies the loaded data to the output as needed. The interaction of the I/O registers and the CPU bus is illustrated in figure 12-20. The I/O instruction specifies the particular I/O register through the address lines and controls the data direction (read or write) and timing through the control lines. All microprocessors can accomplish the transfer of data to or from an I/O register on a *memory read* or *memory write* instruction. In such cases, each I/O register occupies an assigned address in memory address space. The address bus is decoded to provide a **device select signal** for each of the I/O registers.

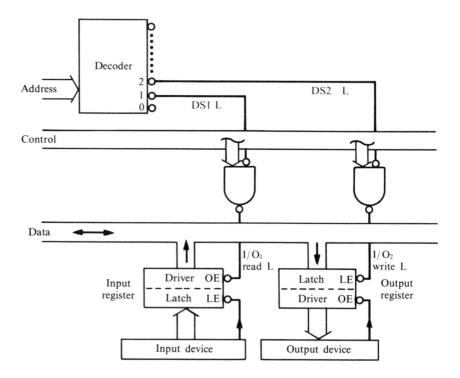

**Fig. 12-20.** Input and output registers. Data from an input device are latched into the input register by that device. The CPU transfers these data to a CPU register by an I/O read instruction directed to the address of that particular register. The decoded address and the read command are combined to enable the drivers to put the data on the bus. Data are supplied to an output device by enabling the latch of an output register with an I/O write instruction to the addressed register. The latched data remain available to the output device.

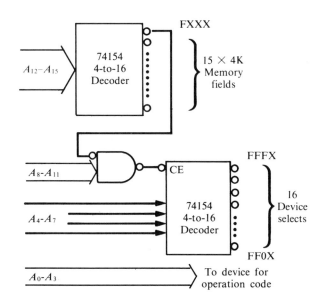

**Fig. 12-21.** Fully decoded address system for memory and I/O. The top 4-to-16 line decoder assigns I/O addresses to the top 4K section of memory space and provides select lines for 15 memory fields of 4K each. The I/O decoder is enabled for all addresses FF00 through FFFF (the top 1/4K of memory space). Address bits $A_4$ to $A_7$ provide 16 device select lines. These can be combined with $A_0$–$A_3$ at the device to enable up to 16 different operations for each device.

One of the many possible decoding schemes is the **fully decoded I/O address** system shown in figure 12-21. Three or four of the least significant address bits are often left undecoded because many specialized I/O interface chips require several of the low address bits to select the register or operation being enabled. In systems where memory and I/O requirements are much less than the address space capacity, simpler I/O decoding is possible. For example, in a 16-bit address bus system, $A_{15}$ can be used to enable one I/O device directly, $A_{14}$ can enable another, and still leave a 16K space for memory (when $A_{15}$ and $A_{14}$ are both 0). Direct selection from single address lines is called **linear addressing**. All techniques that put the I/O registers in memory space are called **memory-mapped I/O**. Some microprocessors have special I/O instructions for data transfers between I/O registers and the CPU accumulator. The ability to distinguish between a memory read or write and an I/O transfer allows I/O addressing that is independent of the memory address space. The 8085 CPU uses only $A_0$–$A_7$ for I/O addressing. This provides 256 I/O addresses and requires decoding only 8 bits for device select.

## Synchronization of the Computer and the World

A major difference between an I/O register and a memory register is that the computer does not have sole control over both read and write to the I/O register. An input register is loaded by the input device on its schedule and is read by the computer at some point in the input register tending program. No other coordination between times of read and write is necessary if the register always contains current data that the computer can access whenever they are needed. This is called **unconditional data transfer**, or **basic I/O**. On the other hand, if the input register contains an important word of data that will be lost if it is not read before the next word is loaded, some synchronization of loading and reading is required. Similarly, an output device that simply follows the output register, such as a digital-to-analog converter, may be able to be updated at any time, but a device such as a printer needs to know when a new character has been output by the computer. When it is necessary, synchronization of the I/O device and the I/O register tending program is achieved by the use of flags that indicate the status of the I/O register. The **service request flag** is a flip-flop that is set when the I/O device needs the computer to enter the service routine to tend that device. The service routine includes an instruction that clears the flag to indicate to the I/O device that its request has been acknowledged. This process, called **handshaking**, is illustrated in figure 12-22.

The computer's entry into the appropriate I/O service routine in response to the flag is accomplished by either polling or by an interrupt. A **polling** program periodically reads the status bits of all relevant I/O devices. The reading of a set service request flag causes a subroutine jump to the service routine for that device. The polling of each device must be frequent enough to ensure that no request is missed. An **interrupt** is a way for an external device to produce a subroutine jump by hardware rather than software. That is, the executing program does not have to include the polling routine in order to be responsive to service requests. The service request flag can be connected to the interrupt contact of the CPU. When an interrupt is requested, the CPU completes its current instruction and then jumps to a subroutine at a specified (often fixed) address. For subroutines initiated by an interrupt, it is particularly important to save CPU registers because there is generally no control over the CPU state at the time of the interrupt. For some microprocessors, all CPU registers are automatically pushed on the stack by the interrupt operation, but for most only the PC and status registers are saved. At the end of the service subroutine, a *return from interrupt* instruction restores all automatically saved registers and returns control to the interrupted program. Programs that should not be interrupted can include an instruction that disables the CPUs response to an interrupt request. The outputs from several flags can be connected to the CPU interrupt input to allow more than one device to request an interrupt. In this case the service subroutine must first poll the flags to identify the interrupting device. More sophisticated interrupt schemes may include multiple-level **priority interrupts**, which allow a service request from a higher priority device to interrupt a less urgent service routine, and **vectored interrupts**, in

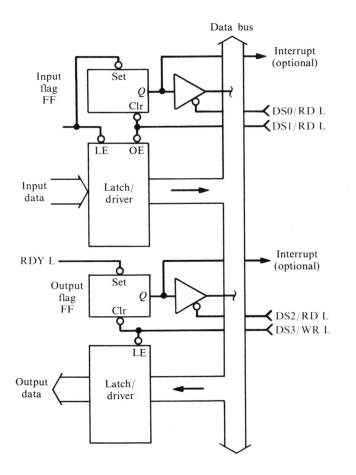

Data bus

Interrupt (optional)

Input flag FF

Set

Q

Clr

DS0/RD L
DS1/RD L

LE   OE

Input data

Latch/ driver

RDY L

Interrupt (optional)

Output flag FF

Set

Q

Clr

DS2/RD L
DS3/WR L

LE

Output data

Latch/ driver

**Fig. 12-22.**   A handshaking system for I/O synchronization. When the data are loaded in the input register, a service request flag FF is set. The CPU can read the status of the flag with a *read* DS0 and follow with a *read* DS1 if the flag is set. Reading the input register clears the flag to await new data. An output device can indicate its readiness for new data by setting a flag. The program can recognize this from a *read* DS2 and follow with a *write* DS3, which supplies new output data and clears the flag. In some cases an interrupt signal is used to direct the CPU to the I/O service routines.

which the interrupting device supplies the address of its own service subroutine to avoid the need for polling. Priority interrupt ICs allow easy implementation of these features in microcomputer systems.

## Direct Memory Access

The execution time for program-controlled input or output of a single data word (including storage in memory, address incrementing, flag checking, etc.) is about 50 $\mu$s for most CPUs. This maximum continuous transfer rate of 20 000 words per second is adequate for most applications but not for all. As we shall see later, it is not adequate to encode a full-range audio signal or to refresh the image on a high-resolution video screen. For such applications specialized hardware must be substituted for the slower but more versatile software. The hardware for fast memory transfers is called a **direct memory** access (DMA) controller. During a DMA transfer the controller takes over the CPU bus and generates the required address and I/O read or write signals. An address counter in the DMA controller is loaded with the desired starting memory address for the data transfer. The I/O device requests attention from the controller, which in turn requests the bus from the CPU. At the end of the current machine cycle, control of the bus is granted, and the DMA controller applies the address and an I/O read or write signal to the bus. Data are transferred through the data lines. The DMA address register is then incremented for the next transfer cycle. Because the DMA controller is cycled by the CPU clock, the transfer takes the same time as a CPU fetch. If the I/O device is ready for the next transfer, the DMA controller retains the bus; if not, bus control is returned to the CPU. Data transfer rates up to several million words per second are possible. This is an extremely efficient way to transfer large blocks of data. It has the advantages of an interrupt, but since the CPU is not involved, no subroutine housekeeping is required. Hardware complexity used to be a deterrent to DMA use, but DMA controller ICs are now part of standard microprocessor hardware lines.

## 12-5  Peripheral Devices

In microcomputer systems, peripheral devices perform two critical functions: they provide the mechanisms for data exchange with people, devices, and other computers, and they provide mechanisms and media for the storage and retrieval of large amounts of digital data. Most of today's peripheral devices are of types developed years ago for large central computers. Some of these have been scaled down and simplified first for minicomputers and then for microcomputers. The cost of microprocessor and memory chips is now so low that the major cost of most microcomputer systems is in the peripheral devices. The trend in peripheral development is to substitute electronic functions for complex or precise mechanical functions and then to implement the electronic functions with ICs or with limited function ROM-based microprocessor controllers. This section surveys briefly some of the most popular peripheral devices.

### I/O Devices

The most common I/O device for a general-purpose microcomputer is a **terminal**, a device with a keyboard and display for interaction between the operator and the computer. The keyboard is similar to a standard typewriter keyboard in layout. Some nonstandard keys such as CONTROL, ESCAPE and DELETE are included to provide additional characters and controls (fig. 12-23). In one form of keyboard each key is a SPST contact arranged to connect one of eight row lines to one of the eight column lines. This defines

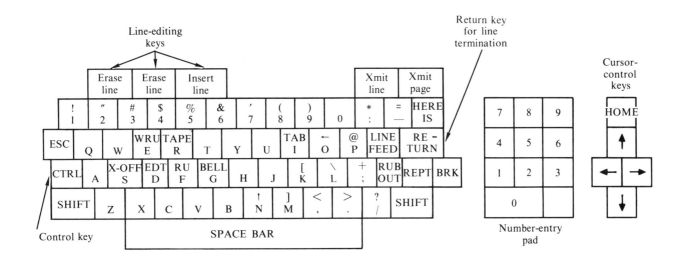

**Fig. 12-23.** A terminal keyboard. A number-entry pad duplicates the regular keyboard numbers for convenience. The cursor-control, line, and page function keys control operations in a CRT display terminal.

64 unique keys. A keyboard decoder IC identifies a depressed key by applying a signal to each row line in succession and watching for that signal on all of the column lines. Characters and controls beyond the basic 64 (up to 128) can be obtained by the depression of SHIFT and/or CONTROL in conjunction with the other keys. The keyboard decoder IC can be interfaced to the CPU bus directly, or it can be connected to a parallel-to-serial converter for serial data transmission to a more remote computer.

The terminal display may be either a cathode ray tube (CRT) or a printer. In the CRT display the image is formed by turning the electron beam on and off as it moves across the screen in successively lower horizontal traces. The complete "raster scan" image is composed of 144 to 275 horizontal traces, and, like a TV picture, the display is retraced 60 times per second. The number of characters that can be displayed varies from a minimum of 16 lines of 25 characters each (400 characters) to 25 lines of 80 characters (2000 characters). Each character is stored as a byte in memory at an address that corresponds to a particular location on the screen. Each character is displayed as a pattern of dots in a $5 \times 7$ or $7 \times 9$ matrix as shown in figure 12-24. As the CRT screen is scanned horizontally across the top of the first line, the first-line characters are read out of memory in order. The character is used as an address to a ROM that provides the CRT with the dot pattern for the top row of dots for the character. On the next scan, the first row characters are again read out in order, and the ROM address is changed to give the pattern for the second row of dots. The ROM that contains the dot patterns for the character is called a **character generator ROM**. Normal TV receivers can display a moderate number of $5 \times 7$ characters. Higher-density displays require a video monitor with a wide bandwidth. The CRT

**Fig. 12-24.** Dot patterns for selected characters (a) in a 5 × 7 pattern and (b) in a 7 × 9 pattern. The higher-density pattern requires a higher-resolution display, but it is easier to read and provides descenders and lower case characters.

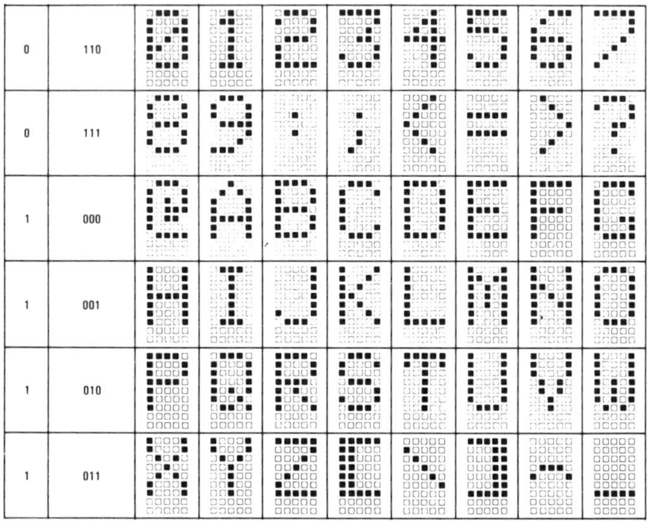

(a)

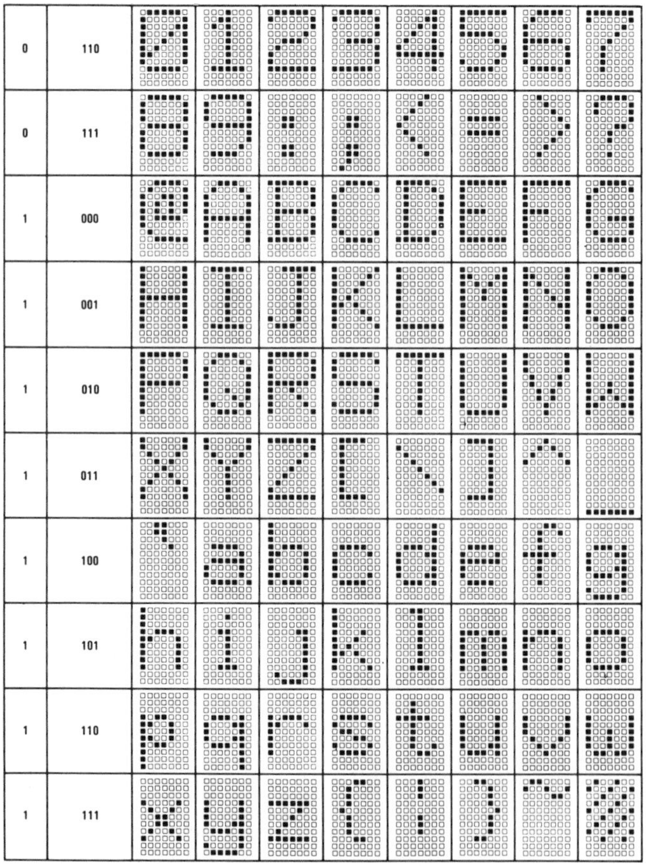

(b)

terminal has the advantages of responding rapidly and of not using paper to record nonmemorable interactions.

For many applications a permanent record of the computer output is essential. This function is performed by a printer, which may be a part of the terminal or separate. Printers are either **character printers**, which print each character as it is received from the computer, or **line printers**, which receive and store an entire line of characters before printing them in one operation. One type of printer makes characters from a matrix of dots as in figure 12-24. A print head with hammers for seven or nine vertical dots moves across the page horizontally with appropriate hammers activated for each column position of each character. Character printers stop after each character; line printers print the entire line in one horizontal motion (and some print the next line backwards on return). Other dot-matrix line printers have single-dot hammers at intervals across the page. The hammer assembly moves back and forth so that each hammer covers a horizontal section of the paper. Another class of character printers operates like a typewriter; a multi-character printing element in the shape of a sphere or wheel prints formed charcters as it moves across the page. These printers are slower, but give more neatly formed characters. Character printers typically print 30 to 150 characters per second. Line printers print two to ten 132-character lines per second. Line printers are thus much faster and, of course, more expensive. Character printers are useful as terminals when no CRT is available.

In many microcomputer environments, the inclusion of graphics capability in the system provides a very useful extension of the system functions. Data displays that are easier to interpret and the ability to manipulate data graphically are but two advantages. A CRT graphics display is defined by a matrix of resolution elements or **pixels**. For example, a graphics display might have an array of $512 \times 512$ pixels. This represents $512^2 = 262\,144$ dot positions on the screen that are either lighted or dark in the display. One bit of memory is required for each pixel; the example display thus requires 32K bytes. The $512 \times 512$ pixel graphics display can contain an $85 \times 56$ array of $5 \times 7$-pixel characters (including spaces). This same array could be stored as characters in less than 5K bytes of memory. Thus more memory and a different scanning system are required for graphics display than are used for character display. The same principle applies to dot-matrix printers in their application as graphics plotters. The programs required to convert numerical or vector graphics data to a raster pattern in memory are not trivial, and they require considerable time and memory capacity to operate if the graphics resolution is even a few hundred elements square. Some character CRTs provide limited resolution **character graphics** or **graphics characters** (see note 12-3). To provide hard copy graphics output $x$-$y$ plotters are very useful. Some plotters include multicolored writing capability and automatic line generation between successive coordinates.

## Mass Storage Devices

For general-purpose microcomputers mass storage devices serve several important functions: they provide a large data base capacity that can be read into and out of CPU memory relatively quickly under computer control; they provide nonvolatile storage of programs and data; if the storage medium is removable, they can provide off-line storage for permanent records or occasional applications; and they serve as a medium of data exchange with other computers. Most mass storage devices are some form of magnetic recording on discs or tapes of plastic coated with iron oxide.

The least expensive mass storage device in use is the audio cassette tape recorder. In order to use standard audio recorders, the data are recorded in

**Note 12-3.  Low-Resolution Graphics.**
Character-based CRT displays can include, in their character set, shapes which are useful in constructing graphics displays. In graphics mode, the byte for each character space determines the shape that is displayed in that space. For character graphics, the space is divided into six blocks ($2 \times 3$) that are independently selected by bits in the character byte. Other systems provide dozens of special graphics characters (line segments of various shapes, angles, and positions) that are selected by the character byte.

asynchronous serial format (see fig. 11-34). Each bit is either eight cycles of a 2400-Hz tone (a 1) or four cycles of a 1200-Hz tone (a 0). This type of encoding is called **frequency-shift keying (FSK)**. To allow for the tape speed variations encountered in inexpensive cassette recorders, the receiving circuit uses the recorded tones with a phase-locked loop oscillator to generate a decoded clock signal that is synchronized to the bit rate. The main advantages of this device are that it is simple, inexpensive, and universally available. Its disadvantages are that it is relatively slow (30 bytes per second, maximum) and that despite the above-mentioned precautions it is less error-free than desired for most critical applications.

Attempts have been made to use the Phillips cassette for digital recording at higher than audio bit density and tape speed, but they have not been completely successful. The difficulties were traced to the tape and cassette dimensions themselves, and higher reliability tape systems have now been developed in other formats. The 3M tape cartridge, designed specifically for digital recording, provides high capacity, much quicker access time, and high reliability. Large tape systems of the type one associates with central computers have not been popular with microcomputers because of their cost and complexity. Punched paper tape, popular as a low-cost off-line storage medium in the earlier days of minicomputers, has been replaced by the more reliable and much more convenient cassette tape.

The most widely used disc format in microcomputer systems is the **floppy disc**. The 8-inch square version of the floppy disc is shown in figure 12-25. In the normal drive data are recorded as serial bits on 77 concentric circular tracks of 41.7 kilobits each, to provide a total capacity of 3.2 megabits. Data are normally arranged on the disc in an IBM format that has 26 sectors per track with 128 bytes of data per sector. The remaining 36% of the track capacity is used for track and sector identification, interrecord gaps, check-sum data for error detection, etc. One of the 77 tracks is used for an index. In all, 265 kilobytes of "formatted" data can be stored per disc. Improvements in the disc and drive technology have led to dual density and two-sided discs. Some drives have read and write access to both sides of the disc at once. With dual density, such drives provide over one million bytes of available storage per disc. A 5-inch version of the floppy disc has been developed for smaller systems. It is slower and has only 23 tracks, but it is cheaper and adequate for some applications.

The cost breakthrough of the Winchester-type drive has made fixed discs attractive for some microcomputer systems. In a **fixed-disc** drive, the disc is not removable. In such systems the disc rotates faster (3600 rpm), and the recording density is greater than that of the floppy disc. This results in a relatively high data transfer rate of 0.6–1 million bytes per second and a capacity up to 34 million bytes for an 8-inch disc that fits in the same space as a standard floppy drive. The fast access time and transfer rates allow large

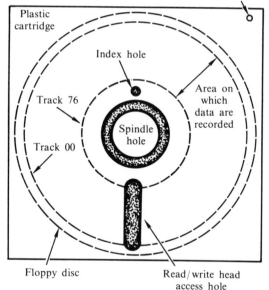

**Fig. 12-25.**    Floppy disc. A disc coated with magnetic iron oxide is permanently housed in a square plastic jacket with access openings as shown. When placed in the disc drive, the disc spins in the jacket at 360 rpm. During a data transfer, the read/write head is positioned radially at the desired track and contacts the disc surface through the access slot.

sections of memory to be loaded or dumped in fractions of a second. The read and write head of a fixed disc floats a few micrometers above the disc surface so there is no wear of the recording surface. In contrast, floppy discs and tape have a limited lifetime. A summary of mass storage device characteristics and economics is given in table 12-4.

**Table 12-4.** Characteristics of various mass storage devices.

| Device | Read/record rate | Capacity | Avg. access time | Price of system* | Price of medium† |
|---|---|---|---|---|---|
| Phillips cassette (audio quality) | 30 bytes/s | 54 kbytes per side C-60 cassette | 15 min. (C-60 cassette) | $50　　$1 | $4–6　　5¢ |
| 3M cartridge | 5000 bytes/s | 2.8 Mbytes | 100 s | $1500　　$.50 | $20–25　　1¢ |
| 8-inch floppy disc | 30 k or 60 k bytes/s | 256 k or 512 k bytes/side formatted | <0.5 s | $1000　　$2–4 | $5–10　　3¢ |
| 5-inch floppy disc | 30 kbytes/s | 89 kbytes or 178 kbytes/side | <0.5 s | $500　　$3–5 | $4–6　　5¢ |
| Winchester disc, 8-inch | 1 Mbyte/s | 4–34 Mbytes | 50–100 ms | $5000　　$.50 | N.A.　　N.A. |
| Winchester disc, 5.25-inch | 625 kbytes/s | 6.4 Mbytes | 170 ms | $3000　　$.50 | N.A.　　N.A. |

*Price of system is in total dollars at the upper left and in dollars per on-line kbyte at the lower right.

†Price of medium is in total dollars at the upper left and in cents per kbyte at the lower right.

The data on a fixed disc are not volatile, but neither are they completely safe. A speck of dust smaller than a smoke particle on a disc can cause a disc "crash" in which the read/write head grinds against the disc surface in a catastrophic failure. Data may also be lost if a record is unintentionally erased or written over. When this is an unacceptable risk, some off-line data storage is required as back-up. A cartridge tape system is a suitable device for storing valuable disc data. Its capacity and transfer rates are compatible with efficient movement of data from one to the other.

Many new technologies for mass memory devices are being explored and developed. We can look for continued dramatic improvements in performance and data storage costs. Electronic techniques such as bubble memory and plated wire memory are candidates for nonmechanical rapid-access multi-megabyte memory systems.

## 12-6   Programming, Languages, and Operating Systems

A computer is useless unless it has been programmed, that is, until it has a suitable instruction list in memory. An executable program must be in **machine language**, the binary op codes, addresses, and data that conform to the CPU's own instruction code. A computer cannot even load a program from one of its input devices unless the machine language program for that operation is already in memory. A general-purpose computer must therefore provide a way to place a program in memory and begin execution at the starting address of that program. In the earlier days of minicomputers, this was done by keying binary instructions into specified binary addresses through bit switches on the computer front panel. For humane and economic reasons this practice has been abandoned. Computers now have a ROM-based program, automatically entered on start-up, that allows the computer to receive commands regarding program loading and execution from the operator through the terminal. This is an example of a general trend; the decrease in cost of computer hardware (especially memory) has led to the development of many types of computer programs whose sole purpose is to make using the computer more convenient. This section summarizes the types of programs that have been developed to aid users in writing and executing programs.

### The Monitor

The **monitor** is a program in ROM that monitors the terminal keyboard and branches to subroutines indicated by keyboard entry. It also operates the display so that the results of the monitor or other programs can be seen. The subroutines available in the monitor allow the operator to: deposit data or instructions in RAM or CPU register locations, examine the contents of any memory or CPU register, begin program execution at any memory address, and interrupt program execution at any time. The monitor program is automatically entered on power-up, or reset, by a hardware-initiated jump to the monitor program's starting address. The operator can then run any other resident program by typing its starting address (in hexadecimal) and an *execute* command. The ability to load and to examine memory and CPU registers allows the operator to introduce or alter data or machine language programs directly from the keyboard. Because the functions provided are the same as the old front-panel switches, monitor ROMs are sometimes called "software front panels." Monitors may include additional programs such as subroutines that interrupt a program after each instruction to analyze its result, that load sections of memory from input devices (or "dump" them to output ones), and that stop execution when a specified location is addressed.

## The Assembler

One of the inconveniences that a computer can alleviate for programmers is the problem of remembering the binary op codes for the CPU instruction set. An **assembler** program accepts an instruction list written in **mnemonic code** or **assembly language** and converts it into a machine language program. The assembly language program is written using a three- or four-character mnemonic for each instruction (such as "ADD R" for "add the contents of CPU register R to the accumulator"). These instructions are written in the desired order of execution. Symbols or names can be used for data quantities and for addresses (such as LDA NUM for *load the quantity at the address NUM into the accumulator* and JMP START for *jump to the instruction labeled START*). The programmer need only specify the address of the first instruction and the location and initial values of all data quantities referred to. The resulting program is a **source code** for the assembler that produces an **object code** in machine language. The assembler program can produce a **listing** of the original mnemonic program, the corresponding machine language instructions and addresses, and a table of symbols used. Figure 12-26 shows an assembly language source program and the listing produced by the assembler. To run the program, the object code is loaded into the appropriate memory locations, and the monitor is used to begin execution at the starting address. Assembler programs may have additional convenience features such as editing capability for changing, inserting, or deleting lines in the assembly program and error detection when illegal or incomplete instructions are given. Although it is common to perform the assembly with the same computer that will run the object program, they need not be the same. It is often more convenient to use a **cross-assembler** program, which runs in a well-developed computer but produces object code for a completely different (and perhaps more limited) computer. This possibility emphasizes the difference between running programs that help generate the object program and executing the object program itself.

## High-Level Languages

Assembly level programming has significant disadvantages for general-purpose computer applications: Assembly codes are very dependent on the

```
LDA 2000        ;Load contents of loc. 2000 into the accumulator
MOV B,A         ;Move contents of register A into register B.
LDA 2001        ;Load contents of loc. 2001 into the accumulator.
ADD B           ;Add contents of register B to the accumulator.
STA 2002        ;Store contents of accumulator into loc. 2002.
RST 1           ;Return to monitor.
```

(a)

```
Address    Data    Mnemonic
  2003      3A     LDA 2000
  2004      00
  2005      20
  2006      47     MOV B,A
  2007      3A     LDA 2001
  2008      01
  2009      20
  200A      80     ADD B
  200B      32     STA 2002
  200C      02
  200D      20
  200E      CF     RST 1
```

(b)

**Fig. 12-26.** An assembly language program (a) and the listing produced by the assembler (b). This program adds the contents of address 2000 to the contents of address 2001 and stores the sum in 2002. The assembler ignores the comments that follow a semicolon symbol, but the comments are helpful for reviewing the program design. The assembler was given the starting address of 2003, but it assigned all the other locations. The listing shows the contents, in hexadecimal, at every address used.

computer because they correspond to the particular CPU's instruction set, and the exchange of useful programs is therefore limited. Many lines of assembly code are required to accomplish even a simple task such as converting a decimal number to binary form. It would be much more convenient if programs could be written in a universally understood, human-oriented language. The assembler program concept could then be extended into a program for assembling the machine language subroutines that correspond to the human language instructions into an object program. This goal has not yet been fully achieved because no human language is either universally understood or precise enough to be interpreted unambiguously by the translator program. Thus, computer scientists have invented languages capable of expressing complex data handling and mathematical operations in terms that humans can learn to use and that computer programs can translate. Several of these languages have been useful enough to become widely adopted and standardized. The most common ones are BASIC, FORTRAN, COBOL, APL, PL-1, and PASCAL. These are called **high-level**

**languages** because a single instruction in any of these languages is the equivalent of dozens or hundreds of machine language instructions. Thus it encodes a much higher level operation than does a machine instruction. Examples of high-level language programs are shown in figure 12-27. Since the time required to write a line of program is the same for any level language, programming efficiency is much greater with high-level languages.

The programs that interpret the high-level program to yield an object program are necessarily specific for the instruction set of the executing computer. Therefore, a particular high-level language cannot be used on a given microcomputer until the translation program for that combination of computer and language is available. There are two types of high-level language translation programs: interpreters and compilers. An **interpreter** program is a library of subroutines, one for each kind of command available in the language vocabulary. An interpreter program executes a line of the high-level program by finding the appropriate subroutine, executing it, and then returning to the next line of high-level code. The BASIC language is generally implemented by an interpreter. The BASIC program and the interpreter program must both be present in memory during execution. The subroutine look-up time is included in the time required to execute the program. The advantages of the BASIC language are that it is quite powerful, relatively easy to learn, and popular among microcomputer users. The availability of ROM-based BASIC interpreters for the popular CPUs has greatly supported this trend. However, when implemented as an interpreter it may be too slow for many applications.

A **compiler** program, like an assembler, translates the entire high-level program and produces an object program that will perform the programmed operations when loaded and executed. The compiler program does not need to be in memory during the execution of the compiled program. The compiled program execution time does not include any of the translation time, but it is not as short as that of the equivalent program written optimally in assembly language. Thus the compiler achieves efficiency in programming at the expense of some execution speed and program memory space. The language FORTRAN was designed to be compiled. An efficient FORTRAN compiler makes several passes through the program as it performs the translation and assembly in increasingly refined stages. A compiler can do this; an interpreter obviously cannot. Some compilers are available for programs written in BASIC. These provide the advantages of a compiler to programmers who prefer the BASIC language. Like assembly, compilation need not be done in the computer for which the object program is intended. **Cross-compilers** allow the power of a large computer to be applied to the creation of efficient object programs to be run on a small system. The choice among the high-level languages is a matter of style preference and of the nature of the tasks to be programmed.

FORTRAN

```
C
C      PROGRAM TO CALCULATE RESISTANCE OF COPPER WIRE
C
       REAL L,A,D,R
       WRITE (5,1)
1      FORMAT ('$ENTER THE LENGTH, DIAMETER: ')
       READ (5,2) L,D
2      FORMAT (2F15.5)
       A = 0.7854 * D**2
       R = 1.7241E-4 * L/A
       WRITE (5,3) R
3      FORMAT (' THE RESISTANCE IS',F15.5)
       STOP
       END
```

BASIC

```
10 REM PROGRAM TO CALCULATE RESISTANCE OF COPPER WIRE
20 PRINT "ENTER LENGTH: ";
30 INPUT L
40 PRINT "ENTER DIAMETER: ";
50 INPUT D
60 LET A = 0.7854 * D**2
70 LET R = 1.7241E-4 * L/A
80 PRINT "THE RESISTANCE IS ";R
90 STOP
100 END
```

**Fig. 12-27.** BASIC and FORTRAN programs to calculate the resistance of a copper wire of $A$ cm$^2$ area and $L$ meters in length. The FORTRAN program first defines the real variables $L$ and $D$ and their numeral format. Then follows the calculation, the output device specification, and the numerical format for expressing the value of $R$. The BASIC program asks the operator to enter the value for $L$, then $D$. After the calculation, the value for $R$ is printed.

## Operating Systems

An **operating system** is a collection of programs that provides for the efficient loading and execution of programs by the computer and that assists the user in the design, implementation, and testing of new programs. Programs that assist the user include:

1. *Language translators:* assemblers, interpreters, and compilers for various languages.

2. *Program editing and loading:* programs to list programs, change the list without rewriting it, and identify programs in such a way they can be located and loaded.

3. *I/O operations:* programs called **device handlers** that read files of data into or out of peripheral devices.

4. *File system:* a program to create and retrieve files of data in the mass storage device(s) and keep an index of the files. Additional functions may include file clearing, record compacting, and controlling user access.

5. *Debugging:* programs that can analyze program execution to help locate sources of error.

6. *System generation:* a program that adjusts the operating system options to conform to the specific characteristics (memory size, I/O devices, etc.) of the computer system of which the program is a part. This capability allows one set of operating system programs to serve an entire family of related computers and peripherals, and permits the operating system to be updated as peripherals are added.

7. *Diagnostics:* a set of programs that exercise and check for errors in various devices (such as the disc) or sections (such as memory) of the computer. These are very useful when hardware problems are suspected.

It is rare for all these programs to be needed in memory at one time. Thus to conserve memory space for executing programs, programs not in use are stored in the mass memory (disc or tape). The part that remains in memory is the **executive program**, that part of the operating system that communicates with the user and coordinates the loading and execution of programs. The executive program itself, is often stored on a particular disc or tape called the **system device**. The computer then has a ROM-based program that loads the executive program from the system device at start-up and begins its execution. This ROM program is called a **bootstrap** because it is used to "pull up" the system.

Operating systems vary considerably in their capabilities. The monitor described earlier is a minimum operating system. Most ROM-based BASIC interpreters include some editing, debugging, and I/O handler routines and thus qualify as operating systems. But fully capable operating systems rely on relatively fast mass storage devices such as discs for the storage of system and user programs that are ready to be called up. There are several types of operating systems that differ in the way the executive manages the execution of programs. The simplest is a **batch system** in which programs are executed in succession, each one continuing until it is completed. A **real-time system** (or **foreground-background system**) provides two levels of program execution. A noncritical task that is being executed in the background can be interrupted for the entry and execution of a time-critical task. This type of system uses, in the background, time that might be wasted while the real-time program waits for some peripheral to respond. In fact, most I/O devices are so slow, relatively speaking, that thousands of instructions can be executed between their needs for service. A **multiprogramming system** has an executive that coordinates the execution of several programs at once so that dead time for one program can be filled with activity on another. **Time-share multiuser** systems extend this concept to multiple simultaneous operators. The more complex the operating system, the more memory it occupies and the more time the computer spends in the executive (i.e., not the user) program. The trend for the future may be toward multiple inexpensive CPUs sharing more costly peripherals for true simultaneous program execution.

## Development Systems

A **development system** is a computer with an operating system designed to aid in developing programs for microprocessor systems that do not have a full operating system. In the applications of microprocessors in appliances, automobiles, games, intelligent computer peripherals, and most scientific instruments, the functions of the system are limited, predetermined, and accomplished by programs stored in ROM. When powered, the computer automatically begins the execution of the predetermined tasks (which are often affected by data it collects). Some of the data collected may be instructions from an operator who need not be aware that he or she is "running" a computer. The program development for such "intelligent" devices is very time-consuming (and expensive) when performed on the machine language level. Therefore the powerful programming aids of efficient operating systems have been expanded to help produce minimized object code for less complete systems. To provide emulation of the target processor's task and to avoid translation of the op codes, the development system often uses a CPU in the same family it supports. Development systems also provide diagnostic aids for application problems and an automatic ROM-programming facility for successful programs. These systems greatly facilitate the application of the versatile microprocessor to an ever-increasing variety of tasks.

## Suggested Experiments

**1.  Microcomputer overview.**
Become familiar with the microcomputer system, including the keyboard, display, printer (if available), terminal or teletype (if available) and any other peripherals. Learn about the internal CPU registers and the instruction mnemonics.

**2.  The monitor.**
Become familiar with the keyboard monitor. Learn how to input data and instructions. Use the monitor to examine registers and memory. Load a previously written program and execute it.

**3.  Instruction set.**
Execute a variety of instructions, and investigate the effect of each on the internal registers. Load a simple addition program, and use the single step facility to step through the program. Examine and record the contents of the registers after each instruction. Run a program, and after execution, examine all registers.

**4.  Assembly language programming.**
Write a program to subtract two numbers in assembly language. Convert the program to machine language by assembling it manually. Execute the program, and examine the registers after execution. Become familiar with inserting breakpoints into a program to assist in program debugging. If one is available, use the comput-er's assembler to assemble the subtraction program. Write a program to convert eight-bit numbers into ASCII characters and print the results.

**5.  Introduction to BASIC programming.**
Become familiar with BASIC language statements. Enter a program to print your name and social security number. List the program, and execute it. Edit the program, and insert your age and several other characteristics. Delete the program. Learn how numbers can be formatted in BASIC and how variables are assigned. Learn how arithmetic and logical operations are performed in BASIC and how arithmetic functions can be implemented. Write a simple program to add two numbers, to multiply two numbers, and to carry out AND, OR and INVERT operations on two numbers. Write a program to calculate the square root of all integers from one to nine.

**6.  Advanced BASIC programming.**
Learn about relational tests ("if-then" statements), looping ("for-next statements"), and subroutine calls. Learn how to use string variables and how to call assembly language subroutines. Write a program to increment a counter until a preset value is exceeded. Learn how to optimize the program.

## Questions and Problems

**1.**  Assume that the accumulator and carry registers in an 8085 CPU contain the data $C = 0$, $A = 1\ 0\ 1\ 1\ 0\ 1\ 0\ 1$. Give the contents of the accumulator and carry after the CPU has performed each of the following successive operations (use table 12-2):

```
op code: 0 0 0 1 0 1 1 1
         0 0 0 0 1 1 1 1
         0 0 1 0 1 1 1 1
```

**2.**  Write the op codes for the 8085 CPU that subtract the word in register D from the accumulator and exclusive-OR the result with the word in register B. What will the contents of the accumulator be if the content of register B is the same as the difference obtained in the first step? (use table 12-3).

**3.**  (a) What is the maximum number of unique addresses in an eight-bit address code? (b) How many are available if the address code is 16 bits? How many bytes of address code and how many address lines are required to provide at least one million unique addresses?

**4.**  What addressing mode would be used for an operand that is in the following locations?

a) CPU register A

b) at an address given by bytes 2 and 3 of the instruction.

c) at an address given by CPU registers H and L.

d) at an address given by summing bytes 2 and 3 of the instruction with register B in the CPU.

**5.**  (a) Describe the role of the program counter in executing an instruction list and in transferring control from one set of instructions to another. (b) Describe the consequences of the program counter being set to an address in a field of data just acquired from an ADC.

**6.** How many read operations are required in an instruction that reads a word from an address identified through an indirect addressing mode? How does the execution time of this instruction compare with a read instruction using an abbreviated direct addressing mode?

**7.** Give the two's complement number (eight-bit) for each of the following decimal numbers:

a) 17, b) −100, c) −127, d) +64

**8.** What is the range (in decimal) of the allowable sum of a 16-bit two's complement addition?

**9.** Give the number of address lines used by the memory chips and the number that must be decoded for the following arrangements of 8K of RAM. (Assume a 16-bit address bus.)

a) 8 chips each 8K × 1

b) 8 chips each 1K × 8

c) 16 chips each 1K × 4

d) 2 chips, 4K × 8

**10.** (a) If address lines $A_{15}$ and $A_{14}$ are used for individual device selects (linear addressing) how much space is left for addresses? (Assume a 16-bit address bus.) (b) What if $A_0$ and $A_1$ were used instead? (c) In another decoding scheme, $A_{15}$ is used to identify memory (0) and I/O (1). All memory chip device selects are ANDed with $\overline{A_{15}}$, and each other address line ANDed with $A_{15}$ provides an individual I/O device select. In this scheme, how many I/O device selects and how much address space is available? (Assume a 16-bit address bus.)

**11.** (a) Calculate the number of dot positions in a CRT display that is 16 lines of 5 × 7 characters, each line 32 characters long. What is the number of dot positions for one that is 25 lines of 6 × 9 characters, each line 80 characters long? Assume one dot space between each character and two dot spaces between each line. (b) If the display is refreshed 60 times per second, what is the required data rate (dots or bits per second) for the video signal for each of the displays in part (a)?

**12.** Which of the mass storage devices in table 12-4 must be interfaced through a DMA interface in order to transfer at maximum rate? Why?

**13.** A mass storage device is to be chosen for storing recorded spectra from an instrument. The instrument has a 12-bit converter and formats the data as two bytes per data point. Each spectrum has about one thousand points and also includes a 500-character identification text. Calculate the number of spectra that can be stored on each cassette, cartridge, or disc for each of the removable medium mass-storage devices in table 12-4.

**14.** Explain the role of the monitor, bootstrap, and hardware front panel in restarting a computer after power down. Compare the purpose and operation of each one.

**15.** Some computer systems for laboratory applications can be programmed in BASIC. Data can be acquired through an ADC by a *read in* instruction. In some cases the maximum input data rate is only a few hundred points per second while the same hardware, operating a FORTRAN program can acquire data at 10 000 points per second. Explain.

# Chapter 13

# DACs, ADCs and Digital I/O

Within the microcomputer data are transferred as parallel digital signals when they are needed for computer operation. In order to acquire and process information from the outside world and to control external devices, the microcomputer must accept and supply data at the times and in the data domains appropriate for the processes being monitored or controlled. Various domain converters are used to convert between the parallel digital domain and analog, time, and serial digital domains. Temporary storage and timing synchronization are conveniently provided by **parallel input/output (I/O) ports**. A complete interface is a combination of data converters and parallel I/O ports as shown in figure 13-1. The parallel input or output port provides a buffer register (latch and bus driver), handshaking control lines (Data ready and Data request), and sometimes computer-selectable functions. The parallel port greatly simplifies the interfacing of digital-to-analog converters (DACs), analog-to-digital converters (ADCs), displays, printers, and other devices that communicate via parallel digital signals. To interface devices that communicate by serial digital signals, a serial-to-parallel or parallel-to-serial conversion within the digital domain is required. This is usually implemented with a pair of I/O serial converters on an LSI chip. The

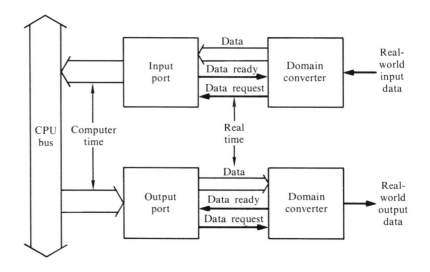

**Fig. 13-1.** Parallel I/O interface between the CPU bus and domain converters. The input and output ports connect domain converters appropriate for the real-world devices to the CPU bus. Data transfers are synchronized by the handshaking signals, Data ready and Data request.

combination of a serial converter and an I/O port on an LSI chip is a complete serial interface, or **serial port**. Serial ports provide standardized communication channels between the CPU and terminals, teletypes, modems, and other serial communication devices. Both parallel and serial ports are described in the first section.

Although the CPU is paced by its own clock, time information about real world signals is most accurately and efficiently obtained from an external time base called a real-time clock. Real-time clocks can differ greatly in complexity and versatility. Many of the most common types are discussed.

The DAC provides an analog output proportional to the numerical value of a coded digital signal. DACs find many applications in display and digital control. The DAC is also an integral part of the digital servo ADC, which operates on the null comparison principle. The principles and characteristics of various types of digital-servo ADCs including the popular successive-approximation converter are discussed. Sample-and-hold circuits, which provide the ADC with a steady input signal during the conversion process, are also described. An ADC with a sample-and-hold, an input multiplexer, an input amplifier, and a programmable real-time clock constitutes a complete analog I/O system. Several applications of converter systems in measurement and control are presented in the final section of this chapter.

## 13-1 Microcomputer Input and Output Ports

The availability of parallel and serial I/O interface circuits in LSI packages has greatly aided the development of microcomputer systems for measurement and control applications. The interface circuits are simply connected between the CPU bus and the external devices. They take care of most of the tasks necessary for timing synchronization, transfer of data, and passing of status information, all in response to simple program instructions. In this section simple parallel ports are considered first and then the versatile programmable I/O interface is discussed. The section concludes with a description of the serial I/O interface.

## Parallel Input and Output

The simplest parallel port is the input or output register introduced in section 12-4. The Intel 8212 IC shown in figure 13-2 is such a port. Although simple, the 8212 can be used for a variety of I/O operations, including basic I/O and strobed (handshaking) I/O. Figure 13-3a shows the 8212 as an input port with handshaking; figure 13-3b shows its use as an output port. As an input port, a Data ready signal (strobe) from the external device clocks data into the latch and generates $\overline{INT}$, which either interrupts the processor or

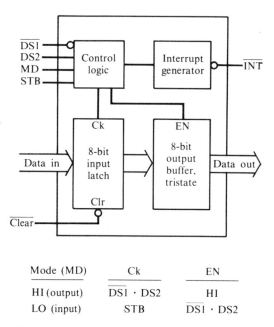

| Mode (MD) | Ck | EN |
|---|---|---|
| HI (output) | $\overline{DS1} \cdot DS2$ | HI |
| LO (input) | STB | $\overline{DS1} \cdot DS2$ |

**Fig. 13-2.** Simple I/O port. The data flow in the 8212 port is controlled by the mode select (MD), strobe (STB), and device select ($\overline{DS1}$, DS2) logic. To select the device, $\overline{DS1}$ must be LO and DS2 HI. The signals that clock the latch and enable the buffer in the two modes are given in the function table. The interrupt signal $\overline{INT}$ is generated when the data are strobed into the latch.

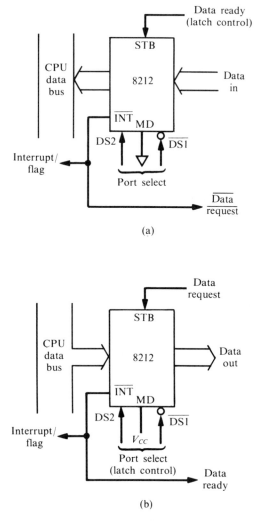

**Fig. 13-3.** Simple I/O port in the forms of (a) an input port and (b) an output port with handshaking. As an input port (a), the MD line is LO; as an output port (b) MD is HI. In either case, $\overline{\text{DS1}}$ is obtained from the control bus memory or I/O read or write line, and DS2 is derived from the address bus.

sets a flag to let the processor know that new data have been latched. The processor then enters a service routine in which the port is selected by $\overline{\text{DS1}} \cdot \text{DS2}$. This enables the buffer, drives the data onto the CPU bus, and drives $\overline{\text{INT}}$ HI, removing the interrupt condition or clearing the flag. A HI $\overline{\text{Data request}}$ line informs the input device that the processor has not yet collected the previous data. Thus new data should not be latched until $\overline{\text{Data request}}$ goes LO.

The same kind of handshaking can occur when the port is an output port as shown in figure 13-3b. Here the CPU data are latched when the port is selected by $\overline{\text{DS1}} \cdot \text{DS2}$. In the output mode the buffer is permanently enabled so that CPU data appear immediately on the Data out lines. The output device acknowledges receipt of the data by sending a Data request signal to the port, indicating that it is requesting new data. This sends $\overline{\text{INT}}$ LO, informing the processor that the data have been received. A LO Data ready line informs the output device that the new data have not been latched. When the port is again selected to drive $\overline{\text{INT}}$ HI, the Data ready line indicates the presence of new data.

The **programmable I/O (PIO)** interface chip provides much greater flexibility than the simple nonprogrammable port. It combines in a single LSI chip two or three parallel eight-bit I/O ports. Each port has its own buffer register, which can be directly addressed by the CPU as can registers with status information. The PIO is programmable in that the functions of the controlling logic can be specified by software. The programmer can specify which port connections are to be used for handshaking. The data lines in a PIO are also programmable. Each data line individually, or a group of data lines, can be specified by software as inputs or outputs. This versatility makes the PIO a truly general-purpose interface device.

An example of a PIO in the 6800 microprocessor series is the 6820 peripheral interface adapter shown in figure 13-4. The data input register latches data from the CPU data bus. The control registers allow the programmer to specify whether the control lines (CA1, CA2 for port A and CB1, CB2 for port B) are to generate interrupts, are to serve as handshaking lines, or are to serve as input or output lines. The control register can also be read to determine the status of the interrupt lines. The data direction registers define each line as an input or an output line. The output registers serve as storage registers for the data that appear on the port output lines. Three chip select inputs and two register select lines are used for addressing; a read/write input, an enable input, and a reset input provide chip control. The 6820 is a powerful and versatile LSI device for the 6800 family of microprocessors.

The 8255 programmable peripheral interface (PPI) shown in figure 13-5 is also a PIO. The PPI has 24 I/O lines, which can be grouped into either

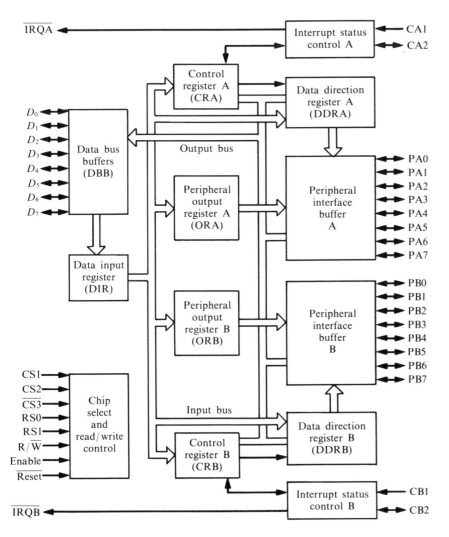

**Fig. 13-4.** Example of a PIO. The block diagram is for the 6820 peripheral interface adapter. This LSI device has two programmable eight-bit I/O ports. Each port has a data direction register, a control register, an output register, and a buffer.

three 8-bit ports or two 8-bit ports and two 4-bit ports. The data lines cannot be individually programmed as input or output lines, but each port can be programmed as either an input, output, or special function port.

In the 6500 microprocessor family, the 6522 versatile interface adapter (VIA) provides even greater flexibility and more complexity than the PIOs described above. The VIA has two 16-bit timers and a shift register in addition to two 8-bit I/O ports. The timers add great flexibility in events timing, pulse generation and other timing operations. The shift register allows serial-to-parallel and parallel-to-serial conversions to be made.

Each major microprocessor manufacturer markets a PIO-type device. Although these differ significantly in characteristics and features, all provide great convenience in interfacing external devices to the microprocessor.

**Fig. 13-5.** Intel 8255 programmable peripheral interface. This LSI device can be used in three software selectable modes: basic I/O (no handshaking), handshaking I/O, and bidirectional bus I/O with handshaking. In handshaking I/O operations, port C is used for control, status, and interrupt generation.

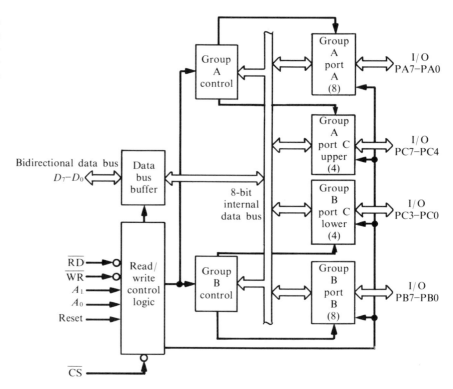

## Serial Input and Output

Serial communication devices such as teletypes, terminals, paper tape readers, cassettes, and printers require a serial I/O port in order to be interfaced to the parallel data bus of the processor. The signals and codes used in standard serial communications described in section 11-4 are referred to extensively in this section.

There are two basic ways of accomplishing parallel-to-serial and serial-to-parallel conversions: with appropriate software or with dedicated hardware. In software the program must accomplish the serial-to-parallel and parallel-to-serial conversions. The interface to a software serial converter is simply a pair of input and output level converters connected to single input and output data lines of a parallel I/O port. The I/O level conversions are between TTL and 20-mA current loop for a teletype or between TTL and RS232 for an RS232 device, etc. The program for an asynchronous serial transmission must transfer the following information to the output port: a start bit, the data bits, the stop bit(s). To accomplish the appropriate timing, the program must enter a delay loop so that each bit period is of the necessary length. For a 110 baud teletype, for example, each bit period is 9.1 ms. If the transmitted code is eight-bit ASCII delineated by one start bit and two stop bits, each character requires that eleven bits be transmitted. The software must then generate the start bits, load the character to be sent into the accumulator, shift the eight data bits out one at a time, and finally generate the stop bits. The flow chart for the simple subroutine necessary is shown in figure 13-6. Another subroutine tends the serial receiving port by sensing the start bit and sampling the port level at appropriate bit times thereafter. More sophisticated reception software can analyze the input signal to determine and adjust to the baud rate of the sender. The software serial transmission method is very simple to implement. However, it is a very inefficient use of the microprocessor since the CPU is tied up entirely during transmission and reception.

One of the first LSI chips to be generally available was the **universal asynchronous receiver/transmitter (UART)**. It performs the parallel-serial conversions and start and stop bit manipulations for both sending and receiving. A block diagram of a standard UART is shown in figure 13-7. An external clock is needed to establish the transmit bit periods and to synchronize the sampling of the receiver bits. The UART also takes care of start and stop bits, parity bits (odd or even), and various status outputs under program control. Interfacing the UART to a microprocessor bus requires parallel input and output ports. A further development of the UART for microprocessor applications is the UART with built-in I/O ports appropriate for the CPU bus. In the 8080/8085 series, the 8251 programmable communication interface is a serial port capable of both asynchronous and synchronous operation. It is often called a **USART** for **universal synchronous/asynchronous receiver transmitter**. A block diagram of the 8251 is shown in figure 13-8 along with a typical application for communicating with a CRT

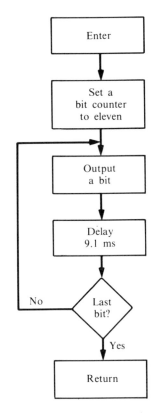

**Fig. 13-6.** Flow chart for software serial data transmission subroutine. A bit counter is set for the number of bits in the transmitted character (11 in this case) and new bits are generated at a constant rate (every 9.1 ms for 110 baud). When the last stop bit has been generated, control is returned to the main program.

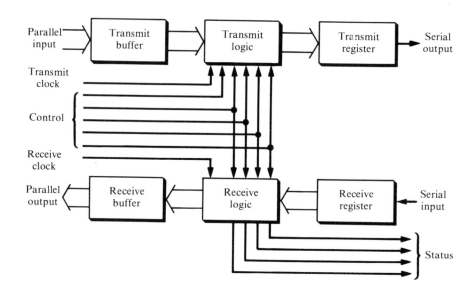

**Fig. 13-7.** A standard UART. This UART of the Western Digital 1602 type consists of a transmitter, a receiver, and control logic. An external clock is synchronized with the serial signal to establish the bit period. The control inputs are parity inhibit, odd/even parity, stop bit select, and word length select. The status outputs are data received, parity error, overrun error, and framing error.

terminal. The 8251 is programmed by the CPU to operate with virtually any of the serial data transmission techniques in use. In the synchronous mode it can be employed with transmission rates as high as 64 kbaud.

The parallel and serial input and output ports described above are basic building blocks in the interfaces to ADCs, DACs, data acquisition systems, motor controllers, etc., and they are thus critical in the achievement of automated measurement and control.

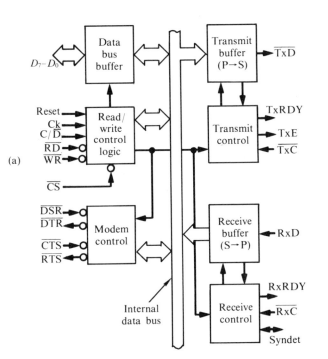

(a)

| Name | Function |
|------|----------|
| $D_7\text{-}D_0$ | Data bus (8 bits) |
| $C/\overline{D}$ | Control or data is to be written or read |
| $\overline{RD}$ | Read data command |
| $\overline{WR}$ | Write data or control command |
| $\overline{CS}$ | Chip enable |
| Ck | Clock pulse (TTL) |
| Reset | Reset |
| $\overline{TxC}$ | Transmitter clock |
| $\overline{TxD}$ | Transmitter data |
| $\overline{RxC}$ | Receiver clock |
| RxD | Receiver data |
| RxRDY | Receiver ready (has character for CPU) |
| TxRDY | Transmitter ready (ready for char. from CPU) |
| $\overline{DSR}$ | Data set ready |
| $\overline{DTR}$ | Data terminal ready |
| Syndet | Sync detect |
| $\overline{RTS}$ | Request to send data |
| $\overline{CTS}$ | Clear to send data |
| TxE | Transmitter empty |

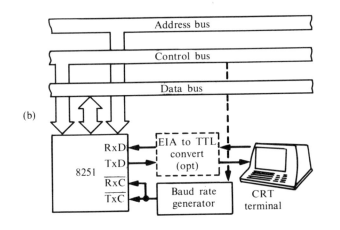

(b)

**Fig. 13-8.** Serial I/O port block diagram (a) and a typical application in a CRT terminal communication scheme (b). The 8251 programmable communication interface combines a USART with buffered I/O ports and supports virtually all the serial communications techniques in present use.

## 13-2    Real-Time Clocks

If a computer is to measure or control the time of events in the real world, it needs some way to keep track of time; that is, it needs a clock. A **real-time clock** is basically a counter that is incremented or decremented at regular intervals by a time-unit generator. A **world-time clock** is a real-time clock in which the number in the counter represents the actual time of day in hours, minutes, and seconds (and perhaps days, months, and years as well). A computer can use a world-time clock to record the actual time of events. It does so by reading the counter contents as part of its programmed response to the event. The computer could also be programmed to turn devices on or off at particular times by jumping to the control subroutine when the clock counter reaches the appropriate state. In many instances of time measurement or timed control, only relative time, the time difference between events, is important. In such cases a real-time clock that measures elapsed time, like a stop watch, is appropriate. In an **elapsed-time clock**, the counter contains the number of time units since the previous event (count-up) or until the next event is to occur (count-down). This kind of clock can be very simple and very precise, and it can be left inactive except when needed. Several common approaches to the implementation of the clock function are described in this section along with a discussion of the modes of interaction between the clock and the timing of world events. The section ends with examples of clock ICs available from several microprocessor manufacturers.

### Software and Hardware Clocks

A **software clock** uses an instruction loop as the time unit generator. The execution time of a single cycle through the loop is equal to the time unit to be counted. One of the steps in the instruction loop is to increment (or decrement) a register that serves as the clock counter. Examples of two software real-time clock programs are given in figure 13-9. The software clock is extremely simple and adequate for many purposes. However, it has serious limitations. The resolution of the time measurement is the execution time of one loop through the program. The maximum resolution thus equals the minimum loop execution time. To ensure reproducibility, no variable length instructions can be included in the loop part of the subroutine. This precludes any interrupts or any reads or writes to variable response-time devices during the timing operation. Finally, the timing routine monopolizes the processor for the duration of its execution. Only elapsed time clocks were illustrated for the software clock. If a world-time clock were implemented in software, the computer could do nothing else!

A clock design that relieves the computer of much of the time-keeping chore without much increase in complexity uses an electronic time-unit generator but keeps the counting operation in software. The time-unit generator

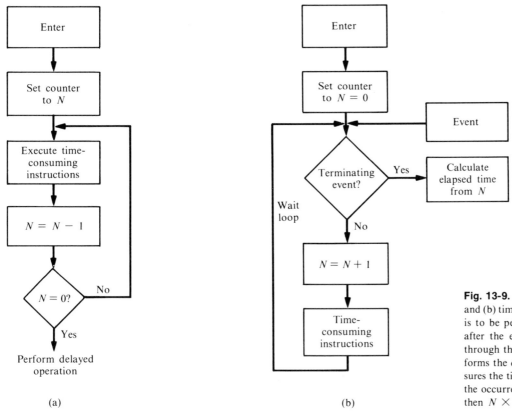

**Fig. 13-9.** Software clocks for (a) time delay control and (b) time interval measurement. In (a), an operation is to be performed $N$ loop-execution time intervals after the entry to the subroutine. On the $N$th pass through the loop, the program exits the loop and performs the delayed operation. The program in (b) measures the time between the entry to the timing loop and the occurrence of the timed event. The elapsed time is then $N \times t_L$ where $t_L$ is the loop execution time.

can be derived from the 60-Hz frequency of the power line, from a 555-type *RC* oscillator, or from a quartz crystal oscillator and scaler. Pulses from the time-unit generator set a clock status flag that is cleared between pulses by the processor. When set, the status register can either cause a program interrupt or, when its status is checked by an instruction to *check the clock status*, it can cause a software branch in the current program.

In the program the timing loop of the software interval generator is replaced by a *wait-for-clock* loop, or a background program may be executed between program interrupts from the clock. The major advantage of this clock is that the clock's time-unit period is independent of the computer's instruction execution time. Thus once the time-unit generator is enabled, the computer can ignore it and perform other functions until the generator completes its period and sets its status flag. This allows much more efficient use of the computer's time. Most computer world-time clocks use a hardware generator and a software counter. A crystal or line-frequency generator interrupts the CPU every 16.7 ms to update the clock counter. If a multi-programming or multiuser operating system is running, it will probably include a task queue check as part of its interrupt routine.

Having a time increment of only one size is often too restrictive, especially for elapsed-time operations. In such cases a quartz crystal oscillator with a programmable scaler can be used as a generator of presettable units of time. An example of a multiple-interval time-unit generator is shown in figure 13-10a. Software for the variable time-unit clock is the same as for the simple oscillator clock except that the clock period must be set before the clock is enabled. A typical flow chart for an elapsed-time measurement is shown in figure 13-10b. The variable time-unit clock can be set for a very short clock period when high-resolution timing is required and then set for a longer clock period to reduce the tending required when less resolution is needed.

**Fig. 13-10.** Variable time-unit clock (a) and program for elapsed time (b). In the variable time clock the computer can control the period by selecting the scaler output through the multiplexer (MUX). The program in (b) causes a counter to be incremented each time the clock flag is found to be set. A terminating event interrupts the program and causes a jump to a routine to calculate the time from the value of $N$.

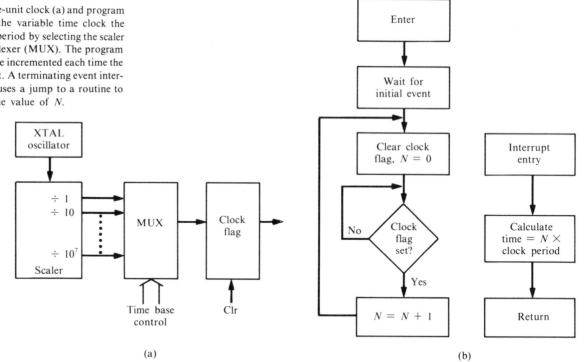

(a)

(b)

A third basic clock design uses an interfaced hardware counting register for the time-unit counter. This relieves the program of the task of advancing the time counter on every clock flag. A complete hardware clock is shown in

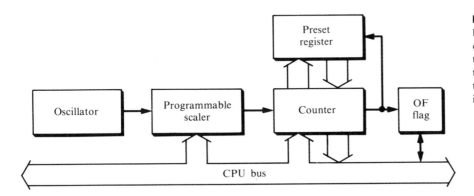

**Fig. 13-11.**  Hardware clock. The time unit is selected by programming the scaler. The counter can be cleared or preset at the beginning of the timed period and set to count either up or down. The CPU then either reads the relative time from the counter register or waits for the overflow (OF) flag to signal the end of a preset time interval.

figure 13-11. Both the time-unit selection scaler and the counter are interfaced to the CPU bus. The counter can be preset to any value by the CPU and instructed to count up or down. The CPU can read the clock counter register contents at any time. The programming for elapsed-time measurements becomes simple because following the initial event, all that is necessary is to clear and enable the clock, wait for the terminating event, read the clock, and calculate the time. Since program steps are at a minimum, the accuracy of the interval timer is good, even for short time intervals. In the timed control mode the program is also simple. The counter is preset to the desired interval and is decremented by the scaled oscillator signal until overflow occurs. The counter then automatically clears and presets itself, and the cycle is repeated.

## Clock and World Interactions

The characteristics of timed operations are affected not only by the design of the clock, but also by how the interactions between the clock and world events are implemented. Two types of interactions can be distinguished: computer *control* of the time of an event and computer *measurement* of the time of an event. In the event control case, the computer watches for the desired event time to be indicated by the clock counter and then enters a subroutine to command the event. The CPU can recognize the desired time by one of three methods: (1) comparing repetitive clock counter readings with the desired value, (2) presetting the clock counter to set its flag at the desired time, and waiting in a flag-check loop; or (3) presetting the clock counter to generate an interrupt at the desired time. The interrupt approach may use CPU time more efficiently, but the interrupt housekeeping makes the control subroutine response much slower than for the flag-check or comparison approaches. For timing on the microsecond time scale, all software response and control approaches are too slow. In these cases, the interaction between clock and event must be direct; the clock flag output can be connected to trigger the event directly. The computer is used to preset the appropriate delay between the clock counter reset and overflow, but the timing is in hardware, where it can be performed with nanosecond precision.

An analogous set of approaches is possible for interactions between clock and world in event-time measurement. In a programmed time measurement, the computer branches to a subroutine to read the clock counter in response to a flag or interrupt from the event to be measured. Subroutine execution time can be a factor in measurement accuracy for very short times or very accurate measurements. Direct control of a gate to the clock counter by the event detector allows timing in hardware alone. Real-time clocks vary considerably in the programmed and direct connection timing modes available.

## LSI Timers

In many microprocessor families, LSI devices known as programmable counter-timers are available. The majority of these are in the interval timer category. Typical of these is the 8253 programmable interval timer in the 8080/85 family illustrated in the block diagram of figure 13-12. There are

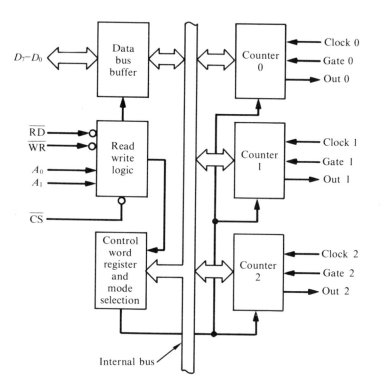

**Fig. 13-12.** Programmable interval timer (8253). This interval timer-counter chip contains three independent 16-bit, presettable counters. It can be used as a simple time-delay element, as an event counter, as an interval timer, and as a periodic signal generator.

three independent, presettable, 16-bit down counters for various counting and timing applications. Each counter can operate in either binary or BCD, and all modes of operation are controllable by software. Each counter has a clock input, a gate input, and output lines available for direct connection to external circuits. Among the five modes of timing that can be programmed are: the generation of an interrupt signal on terminal count with gate signal control over the count enable, the generation of a retriggerable delay of pulse width equal to the time required to reach a preset count, and the programmable scaling of signals with a continuous period equal to the time required to cycle through the preset value. The contents of the counters can be read by the CPU at any time. The counter-timer is treated by the system software as an array of I/O ports. External oscillators are needed for the clock inputs.

An even more versatile LSI timing chip is the AM 9513 system timing controller from Advanced Micro Devices, which contains an internal oscillator and five general-purpose, independent 16-bit counters. The internal oscillator frequency is scaled so that five different internal frequencies are available for input to any of the counters or a selected frequency may be brought out for external use. Counter modulus control can be readily achieved, and any of the counters can be ganged internally with an adjacent counter. Two of the counters can be arranged to operate as a 24-hour world-time clock. This general-purpose timing chip is thus capable of a wide variety of counting and timing operations.

## 13-3  Digital-to-Analog Converters

A **digital-to-analog converter (DAC)**, as the name suggests, is a circuit that converts a digital signal into an electrical analog quantity directly related to the digitally encoded number. The digital domain input signal may be in either serial or parallel form, but the parallel form is by far the more common. The analog output signal is generally in the current or voltage domain. Since the input quantity is a number, the basis of all the conversion techniques is to convert the number to a corresponding number of units of current, voltage, or charge, and then to sum these units in an analog summing circuit.

The standard approach for a digital-to-analog conversion is to generate a current of magnitude proportional to the value of each bit in the parallel digital word, and then to sum the currents of all the bits that are logic 1 (recall section 6-7). The basic DAC takes the form shown in figure 13-13. Since the relative amplitudes of the bit current generators must exactly match the bit weights of the digital signal, a DAC designed for signals of one code cannot be used with signals that are coded differently. A complete DAC is shown in figure 13-14. The input latch samples the digital data source at appropriate times and holds the data in parallel digital form for a

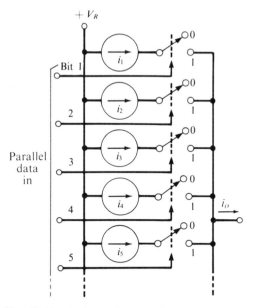

**Fig. 13-13.** Basic DAC. Each bit of the input data word controls a switch that either connects a current generator to or disconnects it from the output. When a switch is closed by a 1-level input bit, a current proportional to the weight value of its input bit is generated. The generated currents are then summed to produce a total output current $i_o$ proportional to the numerical value of the input word.

**Fig. 13-14.** Complete DAC. The basic DAC of figure 13-13 is combined with an optional input register and an output summing amplifier to make a complete DAC.

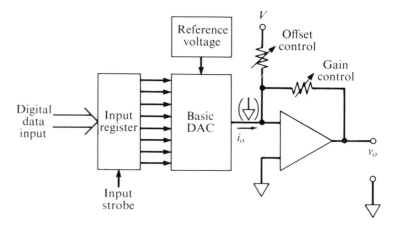

**Table 13-1.** DAC output for eight-bit binary-coded input.

| Binary | Fraction of full scale | $v_o$ for 10-V full scale |
|---|---|---|
| 11111111 | 255/256 | 9.961 V |
| 10000000 | 1/2 | 5.0 |
| 01000000 | 1/4 | 2.5 |
| 00100000 | 1/8 | 1.25 |
| 00010000 | 1/16 | 0.625 |
| 00001000 | 1/32 | 0.312 |
| 00000100 | 1/64 | 0.156 |
| 00000010 | 1/128 | 0.078 |
| 00000001 | 1/256 | 0.039 |
| 00000000 | 0 | 0.0 |

**Table 13-2.** DAC output for eight-bit BCD input.

| BCD | Percent full scale | $v_o$ for 10-V full scale |
|---|---|---|
| 10011001 | 99 | 9.90 V |
| 01010000 | 50 | 5.0 |
| 01000000 | 40 | 4.0 |
| 00100000 | 20 | 2.0 |
| 00010000 | 10 | 1.0 |
| 00001001 | 9 | 0.9 |
| 00000100 | 4 | 0.4 |
| 00000010 | 2 | 0.2 |
| 00000001 | 1 | 0.1 |
| 00000000 | 0 | 0.0 |

steady input to the DAC. In a microcomputer system the latch may be part of the I/O port. The reference voltage is a voltage source used to supply power to the current generators. In many designs the output current per bit weight is proportional to the reference voltage. The current summing is almost always achieved by using a current follower circuit to provide a virtual-common summing point for the current generators and to convert the current signal to a proportional voltage. Offset and gain controls can be added to adjust the overall number-to-voltage transfer function.

The following discussion of DAC devices and their operation begins with a description of the most commonly used digital input codes. Then several of the most popular methods for current generation and switching are introduced. A discussion of the limitations of DACs in the light of manufacturers' specified characteristics serves to summarize this section.

## Digital Input Codes

Commercial DACs are available for binary and 1 2 4 8 BCD data. The DAC input code may be unsigned, in which the sign of the input number is assumed to be positive and the entire range of output voltage or current is unipolar, or it may be signed, in which both positive and negative numbers are included and the output signal is bipolar.

**Unsigned codes.** The relations between the input word and the output signal for an eight-bit binary DAC and an eight-bit BCD DAC are given in tables 13-1 and 13-2. The weighting factor for each bit is clearly seen in these tables. Looking at the binary table, we see that the most significant bit (MSB) has a weight of one-half full scale; the next most significant bit, one-fourth full scale; and so on. With an eight-bit word, the least significant bit (LSB) has a weight of $1/256$ full scale, or 39 mV for 10-V full scale. This value corresponds to the smallest increment, or the **resolution**, of the DAC. Note that the maximum output voltage is the full-scale voltage less the value of the LSB. For a ten-bit binary converter, the LSB has a weight of $1/1024$ or 9.77 mV/10 V. The BCD code uses the bit positions less efficiently, as we have seen. For an eight-bit BCD converter, the LSB (and thus the resolution) is 0.1 V/10 V, about 2.5 times that of the binary DAC.

**Signed codes.**   When a digital word is converted to an analog domain signal, it is generally desired to have the output signal polarity indicate the sign of the number. A **bipolar DAC** accepts a signed input code and produces a bipolar output. There are several signed codes in current use. All of them use the MSB to indicate the sign. Four of the most common signed binary codes are illustrated in table 13-3 for four-bit (three bits for magnitude plus one bit for sign) numbers. The **sign-and-magnitude code** described in section 12-2 is a direct solution to the problem. The MSB gives the sign ($0 = +$, $1 = -$) and the remaining bits the absolute magnitude in straight binary code. The fact that there are two conditions that result in a zero output can be awkward in some applications. Note that no significant bits are lost by using the MSB as a sign bit. Although the sign-and-magnitude code is not very popular for binary-coded signals, it is almost universally adopted for bipolar BCD signals.

Another simple signed code derives from the concept of dividing the succession of states of a binary number into two halves with the higher valued half (MSB = 1) positive and the lower valued half (MSB = 0) negative. The number at one-half full scale ($1000 \cdots 0$) is arbitrarily assigned the zero value. The result, the **offset binary code**, is very easy to implement with the DAC of figure 13-14 by setting the offset to one-half full-scale so that when the basic DAC output is 0, the op amp output is negative and half of full scale; when the DAC output is half of full-scale, the op amp output is 0; and so on.

The signed code most often used in computers is the two's complement code (see section 12-2) because of its convenience for computation. A comparison of the two's complement and offset binary codes reveals that they are the same except for the state of the MSB, which is exactly the opposite for the two cases. Thus a DAC designed for offset binary can be converted to two's complement coding by inverting the MSB between the data source and the DAC input.

## DAC Circuits

Conceptually the simplest form of the DAC is the weighted resistor DAC shown in figure 13-15. Resistance values for a binary-coded converter are

**Table 13-3.**   Signed binary codes.

| Number | Sign and magnitude | Offset binary | Two's complement |
|---|---|---|---|
| +7 | 0 1 1 1 | 1 1 1 1 | 0 1 1 1 |
| +6 | 0 1 1 0 | 1 1 1 0 | 0 1 1 0 |
| +5 | 0 1 0 1 | 1 1 0 1 | 0 1 0 1 |
| +4 | 0 1 0 0 | 1 1 0 0 | 0 1 0 0 |
| +3 | 0 0 1 1 | 1 0 1 1 | 0 0 1 1 |
| +2 | 0 0 1 0 | 1 0 1 0 | 0 0 1 0 |
| +1 | 0 0 0 1 | 1 0 0 1 | 0 0 0 1 |
| +0 | 0 0 0 0 | 1 0 0 0 | 0 0 0 0 |
| −0 | 1 0 0 0 | 1 0 0 0 | 0 0 0 0 |
| −1 | 1 0 0 1 | 0 1 1 1 | 1 1 1 1 |
| −2 | 1 0 1 0 | 0 1 1 0 | 1 1 1 0 |
| −3 | 1 0 1 1 | 0 1 0 1 | 1 1 0 1 |
| −4 | 1 0 0 0 | 0 1 0 0 | 1 1 0 0 |
| −5 | 1 1 0 1 | 0 0 1 1 | 1 0 1 1 |
| −6 | 1 1 1 0 | 0 0 1 0 | 1 0 1 0 |
| −7 | 1 1 1 1 | 0 0 0 1 | 1 0 0 1 |
| −8 | | 0 0 0 0 | 1 0 0 0 |

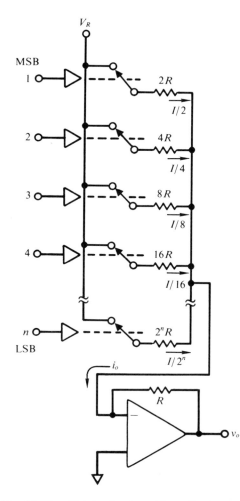

**Fig. 13-15.** Weighted resistor binary DAC. Precision resistors in series with the reference voltage $V_R$ produce the current generators, which are controlled by analog switches driven in response to the logic levels at the digital inputs. Each current generator produces an appropriate fraction of the full-scale current $I$, and the currents are summed by the op amp current follower.

shown. The magnitude of the current generated by each bit is directly proportional to the reference voltage $V_R$ and inversely proportional to the resistance. If $R$ is the resistance necessary to generate the full-scale current $I$ and the MSB is to generate a current of $I/2$, its resistance should be $2R$. The resistance for bit 2 is $4R$ for a current of $I/4$, and in general, for the $n$th bit the resistance is $2^n R$. The output voltage of the current follower can be written as

$$v_o = -IR \left( \frac{a_1}{2} + \frac{a_2}{4} + \frac{a_3}{8} + \frac{a_4}{16} + \cdots + \frac{a_n}{2^n} \right) \qquad (13\text{-}1)$$

where $I = V_R/R$ and $a_1$ is the logic level of bit 1 (0 or 1), $a_2$ the level of bit 2, etc. A $V_R$ of 10 V and an $I$ of 2 mA are common values. For these values, $R$ is 5 k$\Omega$. The resistor in the MSB generator is 10 k$\Omega$ and in the LSB generator, $2^n \times 5$ k$\Omega$. For an 8-bit converter, the LSB resistor is 1.28 M$\Omega$, and for a 12-bit converter, 20.48 M$\Omega$. The large range of resistance values required seriously limits the usefulness of this simple circuit. Not only are resistance tolerances hard to maintain over this range, but also the analog switches must have very low ON resistances and very high OFF resistances.

The switch and resistance requirements are eased greatly by the circuit of figure 13-16. As the diagram shows, the range of resistances required is considerably reduced. To calculate the series attenuator resistor required, consider the group of four generators to be a voltage generator with an equivalent source resistance of $(16/15)R$. (Note that the constant source resistance would not exist if the resistors were not always connected to common or $V_R$.) If bits 5–8 represent the next most significant digit, the series attenuator resistor can be chosen to give one-tenth the current from the most significant group. The result of that calculation, $9 \times (16/15)R$, is $9.6R$. Similarly, another group of four bits can be added to make a 12-bit binary or three-digit BCD DAC as shown.

**Ladder circuits.**   Another technique for reducing the required range of resistance values still more is the **ladder network** of resistors shown in figure 13-17. The expression for $i_o$ can best be found by considering the current sources one at a time. If only the MSB is 1, $i_1 = V_R/2R = I/2$. If only bit 2 is 1, a current $i_2$ is generated which splits at $N2$, part going to $N3$ and part going to $N1$ to be summed. The net resistance to common of the $R$-$2R$ network upward from point $N2$ is $2R$. This is a characteristic of the network. The resistance downward from $N2$ to common is $R$. The combined resistance to common at $N2$ is thus $2R/3$. The current $i_2$ is then

$$i_2 = V_R/(2R + 2R/3) = 3V_R/8R$$

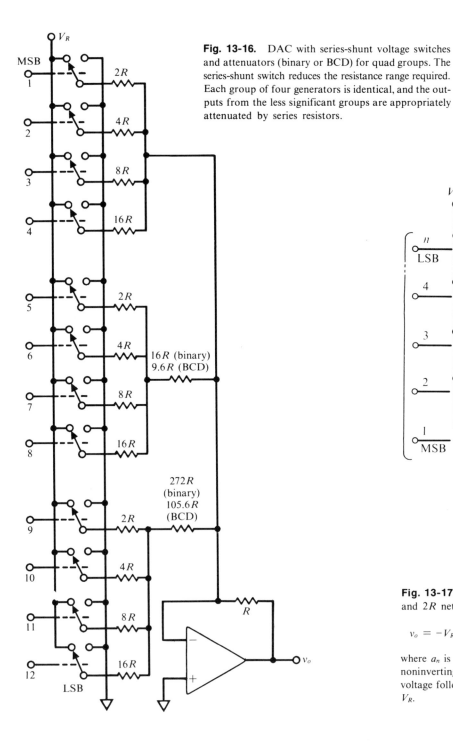

**Fig. 13-16.** DAC with series-shunt voltage switches and attenuators (binary or BCD) for quad groups. The series-shunt switch reduces the resistance range required. Each group of four generators is identical, and the outputs from the less significant groups are appropriately attenuated by series resistors.

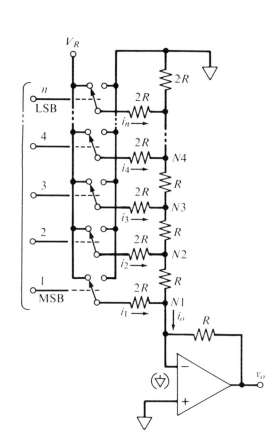

**Fig. 13-17.** Ladder network binary DAC. This $R$ and $2R$ network produces an output voltage

$$v_o = -V_R \left( \frac{a_1}{2} + \frac{a_2}{4} + \frac{a_3}{8} + \frac{a_4}{16} + \cdots + \frac{a_n}{2^n} \right)$$

where $a_n$ is the logic level of the $n$th input bit. For a noninverting output, point $N1$ could be connected to a voltage follower input to provide a full-scale output of $V_R$.

The current $i_2$ is split, two-thirds going to $N1$ and one-third to $N2$. Thus, the current from bit 2 to be summed is

$$\text{bit 2 current} = \frac{2}{3}\left(\frac{3V_R}{8R}\right) = \frac{V_R}{4R} = \frac{I}{4}$$

If only bit 3 is 1, the resistance from $N3$ is $2R$ upward and $5R/3$ downward. The current $i_3 = 11V_R/32R$. This current splits at $N3$ with $3V_R/16R$ going toward $N2$. At $N2$, this current splits again with two-thirds $V_R/8R$ ($I/8$), going to $N1$ to be summed. The remaining bits can be solved similarly to demonstrate that $i_o$ and $v_o$ for this DAC are the same as equation 13-1. The ladder network is very popular for binary DACs, particularly hybrids and ICs, because it requires only two resistance values. Note that the ladder network will not work without the series-shunt form of switching.

**Current switching.**    The switches in the DACs discussed above are applied in the **voltage-switching mode.** That is, they are used to determine what voltage is applied to the current-generating resistance network. The voltage changes in the switched circuits and the voltage across the open switch contacts approach $V_R$ V. This requires large switch drive voltages and reduces switching speed. Since currents are being summed, the series-shunt **current mode switch** can be used to advantage. A current-switching ladder network DAC is shown in figure 13-18. The currents $i_1$–$i_n$ are constant. The value of $i_1$ is clearly $V_R/2R = I/2$. The resistance to common at $N2$ is the $2R$ that generates $i_2$ in parallel with the $2R$ of the network above $N2$. The voltage at $N2$ is thus $V_R/2$, resulting in a current $i_2 = I/4$. Similarly, the voltage at $N3$ is $V_R/4$, and the current $i_3 = I/8$, and so on up the ladder. The current-switching technique is most often found in high-speed, IC designs.

**Summing and output amplifiers.**    The DAC output amplifier serves several purposes: summing the currents, converting the current to a voltage signal, and offsetting unipolar DACs for bipolar codes. As the switching speeds of DACs have increased, the output op amp response has become the limiting factor in DAC response speed. For this reason many DACs now have connections that allow the substitution of a faster external amplifier if desired. To obtain the fastest response, $i_o$ is connected directly to a load resistor. The load resistor is chosen to obtain a full-scale voltage of 0.1 V or less. For $I = 2$ mA, a load resistor of 50 $\Omega$ is ideal. If the DAC is to drive a high-speed scope display, the $i_o$ output is connected to the scope input with a terminated 50-$\Omega$ coaxial cable, which serves as both connection and load.

**Multiplying DACs.**    A multiplying DAC is one for which the reference voltage is supplied externally and for which the output current is accurately proportional to the value of $V_R$ applied. From equation 13-1, the DAC

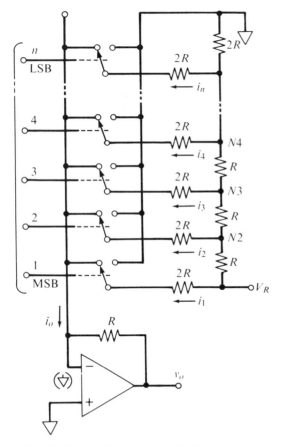

**Fig. 13-18.**  Ladder network DAC with current switching. The switches direct the generated currents to the op amp summing point or to common. All three contacts of each DPST switch are at the common voltage. Since there is no voltage change and no contact voltage, the switch drive signal can be small and the switching fast.

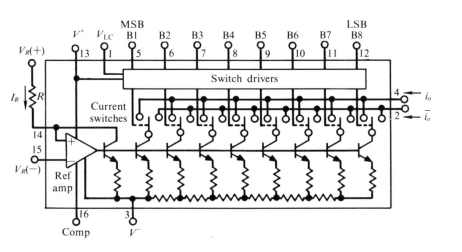

**Fig. 13-19.** Monolithic eight-bit DAC (DAC-08). This multiplying DAC has internal switches and an internal $R/2R$ ladder network. A reference current $I_R = [V_R(+) - V_R(-)]/R$ is applied to the reference amplifier which controls the base voltages of the transistors such that the current through the ladder, $i_{o(max)}$ is $I_R$ $(255/256)$. The current switches determine whether each bit position's share of that current appears at the $i_o$ or the $\bar{i}_o$ output. Thus the current at the complementary output $\bar{i}_o$ is $i_{o(max)} - i_o$.

output voltage is equal to the product of $V_R$ and the input digital number. The multiplying DAC is thus a multiplier with one digital and one analog domain input and an analog domain output. It is very convenient for the digital control of analog signals, controlled-gain amplifiers, and so on. A full four-quadrant multiplying DAC accepts a signed digital data code and a value of $V_R$ of either polarity and produces an output signal of appropriate polarity.

## IC Digital-to-Analog Converters

Integrated circuit DACs are available in a variety of internal designs. Many of the simpler (and cheaper) DACs are multiplying converters that require an external reference. Most simple IC DACs also require an external output op amp for current summing and current-to-voltage conversion. Typical of the IC multiplying DACs is the eight-bit DAC-08 illustrated in figure 13-19. This DAC features complementary current outputs $i_o$ and $\bar{i}_o$ and settling times of $\sim 100$ ns.

A somewhat more complicated eight-bit IC DAC is illustrated in figure 13-20. This system has a built-in reference, an internal latch, and an internal summing amplifier. Provision is made for adding an external offset voltage to accommodate signed as well as unsigned codes. Because of the input latches and internal circuitry, this DAC is directly compatible with most microprocessor systems with a minimum of external circuitry. Some of the critical characteristics and limitations of DAC devices are described in table 13-4 in the terms used by manufacturers in their specifications.

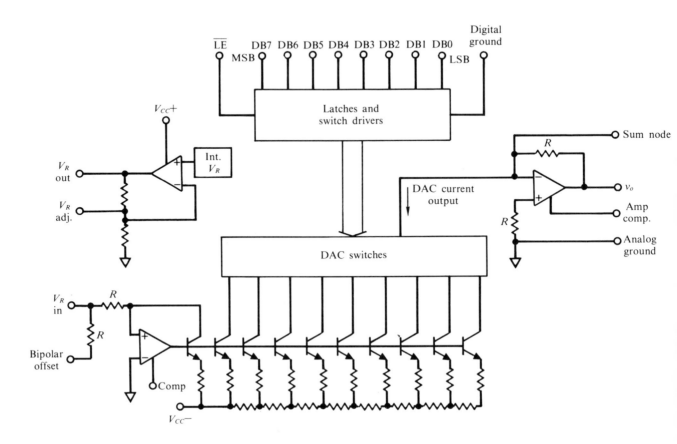

**Fig. 13-20.** Monolithic eight-bit, microprocessor-compatible DAC (NE 5018). This DAC has an internal reference (5 V nominal), an eight-bit latch, and an internal output amplifier in addition to the ladder and switching network. The DAC settling time is ~2 μs.

**Table 13-4.** DAC characteristics

---

*Resolution.*   The fraction of the full-scale range represented by the smallest possible change of the input number (1 part in $2^8$ for an eight-bit DAC, 1 part in $2^{10}$ for a ten-bit DAC, etc.).

*Accuracy.*   The difference between the expected and measured output voltage (or current) in terms of the change caused by changing the LSB. Most converters are specified to be accurate to at least $\pm 1/2$ LSB.

*Linearity.*   The nearness of the plot of analog output vs. increasing digital code to a straight line. Linearity does not imply absolute accuracy since the line may not go through the origin or the full-scale value.

*Monotonicity.*   A DAC is nonmonotonic if there is a momentary reversal in the expected direction of change. For instance, if when the code is increased 1 LSB, the output decreases 1/4 LSB instead and then increases 2 1/4 LSB on the next increment, the DAC is not monotonic.

**Table 13-4.** DAC characteristics (continued)

*Zero offset.* The output for an input code of zero. This is generally within one-half LSB for unipolar DACs, but for bipolar DACs an error in the offset circuit can cause considerable zero offset.

*Stability.* The constancy of the full-scale output with age and with variations in temperature and power supply.

*Settling time.* The time required for the output signal to settle within one-half LSB of its final value after a given change in input code (usually full scale). Settling time is generally limited by the output amplifier response.

*Glitches.* Transients or spikes that appear at the output when new digital data are applied. They can be removed by using a fast sample-and-hold circuit at the output, putting it in the hold mode only during the glitch-producing transition period.

## 13-4 Analog-to-Digital Converters

An **analog-to-digital converter (ADC)** is a circuit that converts an analog domain signal (current, voltage, or charge) to a digital signal that encodes a number proportional to the analog magnitude. Three distinctly different classes of converters exist; integrating ADCs, digital servo ADCs, and flash ADCs. We have encountered the integrating ADC in chapter 9, where the voltage-to-frequency converter, the dual-slope converter and the charge-balance converter were discussed. These converters are based on measuring the number of charge units contained in the analog signal over a period of time. The flash ADC, discussed at the end of this section, uses parallel comparators and a decoder to digitize the signal. It is by far the fastest of all ADCs.

In the **digital servo ADC** a DAC is used to generate an analog signal for comparison with the signal to be digitized. The digital input to the DAC is changed in a direction determined by the results of the comparison until the input signal and the DAC output are equal as shown in figure 13-21. The ADC is called a digital servo because it is a feedback system in which the feedback information is in the digital domain. There are several types of digital servo ADCs based on different digital registers and different methods for adjusting their contents. The three predominate types of digital servo ADCs, the staircase, the tracking, and the successive approximation converters are described below.

## Staircase ADC

The simplest digital servo converter is the **staircase ADC**. The register is simply a binary counter that counts pulses from a clock. Each pulse increases

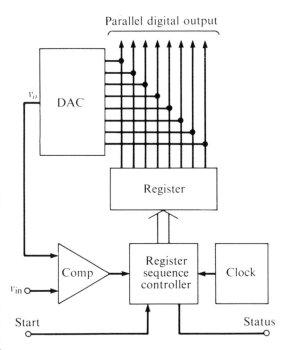

**Fig. 13-21.** Digital servo ADC. On application of a start signal, the register sequence controller alters the contents of the register until the DAC output is within one LSB of the analog input voltage. The comparator determines whether the number in the register will be increased or decreased, and the clock determines the rate of change. The status (end-of-conversion) line provides an appropriate logic-level signal when the converter is busy and a logic-level transition when the conversion is complete.

the DAC output by one LSB increment as shown in figure 13-22. The "staircase" appearance of the DAC output waveform gives the converter its name. The comparator output controls the counting gate and is used to indicate that the converter is busy. The staircase converter is quite simple, but it is also the slowest of the digital servo types. If the analog input voltage is full scale, the counter must count from all 0's to all 1's. This requires $2^n$-1 clock pulses. With a ten-bit DAC and a 10-MHz clock, for example, the conversion time for a full-scale analog input would be greater than 100 $\mu$s. This time can be prohibitively slow where high-speed conversions are desired.

The staircase ADC, like all ADCs, is subject to an uncertainty known as the **quantizing error**. Each step in the staircase output waveform corresponds to a counter increment of one LSB. Each step is said to correspond to one quantum, shown as $Q$ in figure 13-22. The total number of output states or quanta is $2^n$. The quantizing interval $Q$ represents the smallest analog difference that the converter can resolve. For the staircase ADC, it is apparent that the converter always rounds up to the quantizing level for which $v_o$ just exceeds $v_{in}$. It is thus possible for the output to be nearly one LSB in error for an input signal that just exceeds one quantizing level. In order to make the quantizing error symmetrical ($\pm 1/2$ LSB), the DAC output is normally offset by a voltage corresponding to $1/2$ LSB.

## Tracking ADC

The true digital servo should be able to adjust the output in increments in either direction in order to follow, or "track," the input quantity changes.

**Fig. 13-22.**   Staircase ADC. The conversion cycle begins when a start pulse clears the binary counter. Since $v_o$ is then less than $v_{in}$, the comparator goes to 1 and opens the counting gate. Counts are accumulated until $v_o$ just exceeds $v_{in}$. At this time the comparator goes to 0 and closes the counting gate. The parallel digital output of the counter is thus the digital equivalent of the analog input voltage.

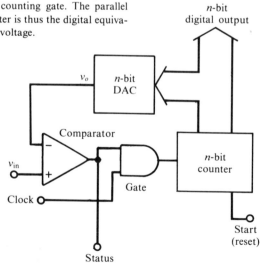

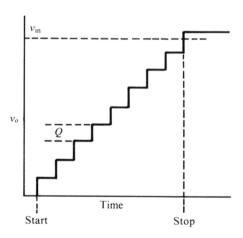

This can be achieved by using an up-down counter as the register in the block diagram of figure 13-21. The **tracking ADC** differs from the staircase ADC of figure 13-22 in that the clock is connected directly to the up-down counter input and the comparator output controls the count direction. The tracking ADC is also very simple. On the initial conversion, if the counter starts at zero, $2^n - 1$ clock cycles are required for full-scale output. However, the tracking ADC can follow small variations in the analog signal within a few clock cycles. It has the advantages of a continuous output and rapid updating at the clock rate. If a ten-bit DAC with a 10-V output range is used with a 10-MHz clock, the converter can follow input voltage changes as fast as $10^5$ V/s.

When a tracking ADC is used to convert a steady input signal, the digital output alternates, or "waffles," between the two adjacent quantizing levels that span the theoretically correct output value. The quantizing error is very noticeable in the tracking ADC because the "hunting," or alternation in the LSB is observable on repeated conversions; the staircase converter, on the other hand, always rounds up to the next higher quantizing level and provides a steady digital output. The maximum error occurs when the input signal is halfway between two quantizing levels. In this case the error would be $\pm 1/2$ the value of the LSB. In converter specifications, this maximum quantizing error of $\pm 1/2$ LSB is typically given.

## Successive Approximation ADC

The major difference among types of digital servo ADCs is in the method by which the number in the register is adjusted to give a DAC output equal to the input. In the staircase ADC, each numerical code is tried in order until the correct one appears. This is a simple but not very efficient way of searching for the right number. A much more efficient way is to divide the range into a small number of fields (usually two, but sometimes four) and to identify the field with the desired number in it. Then that field is divided into smaller segments, and so on until the final result is determined. This procedure is illustrated in figure 13-23. The first test, in effect, tests whether $v_{in}$ is in the upper or lower half of the range. Next the upper half of the range is divided in half and tested by making the next most significant bit a 1. Next the quarter between 1/2 and 3/4 is tested. Then the eighth between 5/8 and 3/4 is tested, and so on. The block diagram of a typical **successive approximation converter** is shown in figure 13-24.

The successive approximation procedure requires only one clock cycle per bit of conversion. The conversion time $t_c$ is constant and given by $t_c = n/f$, where $n$ is the number of bits in the converter and $f$ is the clock frequency. A ten-bit converter with a 10-MHz clock can complete a conversion every microsecond. However, this requires a DAC and comparator that

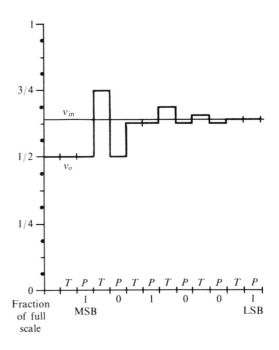

**Fig. 13-23.** Successive approximation search. At the start of the conversion the register is set to one-half scale (1000 . . . ). This gives a half-scale output for $v_o$ during the first test period (marked $T$), and the comparator indicates whether $v_o$ is too high or too low. Since the test shows $v_o$ to be too low, the 1 in the MSB is retained during the posting interval $P$. Next the upper half of the range is divided into two and tested by making the next most significant bit 1. This test shows $v_o$ to be too high, so the 1 is replaced by a 0 during the second posting period. The process continues until the LSB of the converter has been posted.

**Fig. 13-24.** Successive approximation ADC. A conversion cycle begins with a Start pulse, which sets the output of the two-registers to half scale (1000 . . . 0). The output of the storage register is converted by the DAC to give an analog comparison voltage $v_o$. If $v_o < v_{in}$, the 1 in the MSB of the storage register is retained, and the shift register shifts its 1 to the next most significant bit. If $v_o > v_{in}$, the logic programmer resets the MSB of the storage register. Each bit is then tested in succession until the LSB has been tested with the comparator output indicating whether to retain or reset the storage register test bit. An *end of convert* (EOC) pulse signifies that the conversion is complete.

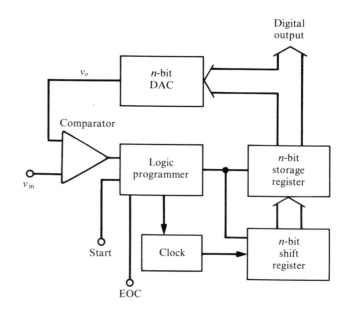

can settle to within one-half LSB from a half-scale step input in 0.1 $\mu$s, and indeed such high-speed converters with conversion times in the microsecond range are available. More common and less expensive, however, are 10-bit and 12-bit converters with conversion times of 4–30 $\mu$s—still very fast. One requirement of the successive approximation converter is that the analog input voltage remain absolutely constant during the conversion time. If it does not, errors in the most significant bit tests can occur. Analog sample-and-hold circuits, described in the next section, are usually employed to acquire the voltage to be converted and to hold it constant during conversion.

The successive approximation converter is subject, of course, to the quantizing error of $\pm 1/2$ LSB. Successive approximation converters are available in monolithic IC packages as well as in hybrid (combination of IC and discrete components) form. Successive approximation registers, which contain the logic programmer, the shift register, and the storage register are also available in IC form for use with an external DAC and comparator. A serial output is available on some successive-approximation ADCs. The serial information is obtained from the comparator output during the posting interval. Note from the waveform of figure 13-23 that the sign of the comparator output follows the posted bit value.

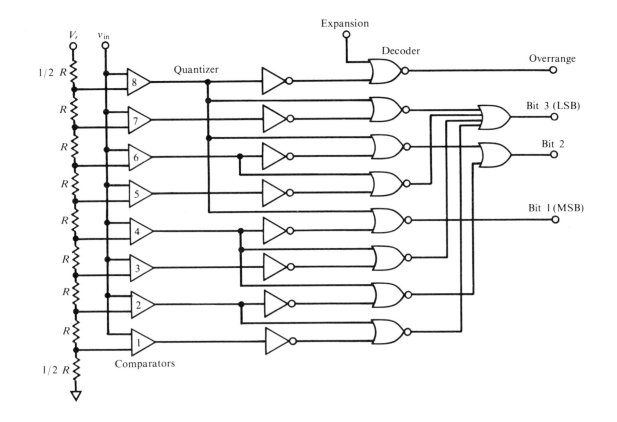

**Fig. 13-25.** Flash ADC. Eight comparators and a decoder provide a three-bit conversion plus overrange indication. The reference voltages for each comparator come from the divider on $V_r$. The quantizing level $Q$ is $Q = V_r/8$. The threshold for comparator 1 is $Q/2$ or $V_r/16$; that for comparator 2 is $3Q/2$ or $3V_r/16$, and so on.

## Flash ADC

Conversion rates in the 1–50 MHz range can be achieved at the expense of resolution by the **flash ADC** as shown in figure 13-25. The flash, or parallel, ADC consists of a bank of $2^n - 1$ comparators that perform the quantizing and decoding logic where $n$ is the number of bits in the conversion. The comparators all have different thresholds set by the reference voltage and its divider. The decoder logic produces a normal binary code from the seven output lines plus the overrange indication. The high speed of the flash converter results from the simultaneous comparison of all output levels. Very rapid decoding can be achieved with Schottky TTL or ECL gates. Thus the flash converter is essentially a continuous converter and needs no strobe to begin conversion. The major disadvantage of the flash ADC is the need for $2^n$ comparators for $n$ bits of conversion. For an 8-bit converter, 256 comparators are needed. Despite this staggering number, TRW, Inc. produces an 8-bit flash converter (model TDC-1007J) that converts in 33 ns.

## 13-5   Sample-and-Hold Circuits

Sample-and-hold circuits, introduced in chapter 6 where high-speed analog sampling was discussed, are used with analog-to-digital converters to hold the signal constant during conversion. The sample-and-hold thus captures and freezes the analog input signal prior to conversion and defines the instant of acquisition. This section considers several variations on the basic sample-and-hold circuit and discusses their major deviations from ideality. Examples of IC sample-and-hold circuits are presented.

### Sample-and-Hold Basics

The sample-and-hold function is generally accomplished by charging a capacitor with the analog signal during the sampling interval and then isolating the capacitor during the hold period with a high input impedance amplifier. Sample-and-hold circuits that spend most of the time following the input signal changes and only a small fraction of the time in the hold mode are often called **track-and-hold** circuits. Switching from the sample or track mode to the hold mode is normally accomplished by a fast analog switch operated from logic-level control signals.

The simplest sample-and-hold circuit uses a voltage follower for isolation purposes as shown in figure 13-26. From this circuit, some of the deviations from ideal behavior of sample-and-hold circuits can be recognized and characterized in terms of the two states of the circuit and the transitions from one state to the other. First, in order for the follower to track the input signal reliably, the time constant $RC$ and the amplifier response time in the sample mode must be short compared to the rate of change of the input signal. For this reason $R$ is quite low, and an input voltage follower is used to unload the analog signal source. Other characteristics of concern during the sampling process are the following: **offset**, an

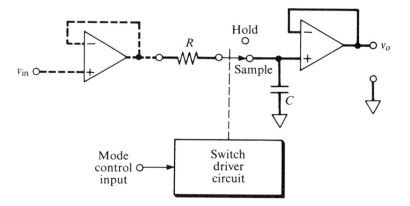

**Fig. 13-26.** Sample-and-hold circuit. In the sample mode the input signal charges capacitor $C$ through resistor $R$; in the hold mode the voltage on the capacitor is maintained (held) at the follower input. The input follower is often used to buffer the analog signal source from the low input impedance of the $RC$ network.

output voltage at zero input voltage; **gain error**, the deviation of the stated output-to-input voltage ratio (unity for the follower); and **settling time**, the time required for the output voltage to come within a given percent of its final value after an instantaneous change of some specified fraction in the input voltage. The sampling characteristics are clearly determined by the operational amplifiers used in the voltage follower circuits and the values of $C$ and $R_{ON}$ for the switch. Second, the transition from the sample mode to the hold mode includes errors of the following types: **aperture time**, the total time between the hold command and the actual opening of the hold switch, including the average delay and delay uncertainty where aperture time is frequently as small as 10–100 ns; **switching offset**, the change in voltage on $C$ caused by charge loss or gain during switching and its settling time. These errors have to do with the switch and drive circuits and the response speed of the output follower amplifier. Third, during the hold mode the problem of output drift is of greatest concern. The rate of change in the output voltage, in volts per second, is called the **droop**. It is caused by finite currents at the amplifier input or through the sampling switch which cause the voltage across $C$ to change with time. This drift in the hold voltage can be minimized by using an amplifier with very low input current and a switch with very high OFF resistance. Using a larger capacitance $C$ reduces drift, but it also increases the $RC$ time constant and decreases the tracking rate. Finally, in the transition from hold to sample, the **acquisition time**, the minimum sampling time to acquire the input voltage to within a given accuracy, is important. In most cases the acquisition time is the same as settling time in the sample mode. The critical characteristics in actual applications are the minimum sampling time (given by the acquisition time), the sampling error (given by the offsets, gain error, droop, and overall analog accuracy in percent of full-scale signal), and the sampling time uncertainty (the variation in, or unknown portion of, the aperture time).

## Other Sample-and-Hold Circuits

Accuracy in sampling can be improved at the expense of response speed by variations of the basic sample-and-hold circuit of figure 13-26. The closed-loop sample-and-hold shown in figure 13-27 is essentially the same as the basic circuit except that the sampling amplifier is included in the feedback loop of the input voltage follower. Amplifier 1 compares the output voltage to the input voltage and then charges capacitor $C$ until the error is reduced essentially to zero. This error-correcting feedback eliminates offset and common mode errors.

Another closed-loop sample-and-hold circuit that uses an integrator as the output amplifier is shown in figure 13-28. The advantage of this circuit over the error-correcting feedback circuit of figure 13-27 is that all three

**Fig. 13-27.** Closed-loop sample-and-hold. Op amp 1 is both an input buffer and an error-correcting amplifier. In the sample mode, current from op amp 1 charges the hold capacitor until the output $v_o$ equals the input $v_{in}$. In the hold mode, op amp 2 is a voltage follower, which maintains the output voltage at the sampled value. The diodes provide a stabilizing feedback loop for op amp 1 when the switch is opened. Resistor $R$ allows $v_o$ and the diode feedback voltage to differ during this time.

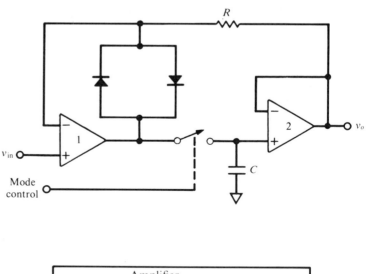

**Fig. 13-28.** Feedback sample-and-hold circuit using an integrator. The integrator keeps the output voltage exactly equal to the input voltage. The input amplifier may be either an op amp or an operational transconductance amplifier (OTA). The latter provides an output current proportional to its input voltage to charge the integration capacitor.

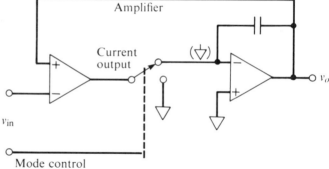

switch contacts are held at the common voltage in both modes. Because the voltage across the switch is essentially zero, switch leakage is greatly reduced. For all of the closed-loop configurations the acquisition time is the time for the entire circuit to settle on the input voltage, not just the capacitor voltage as in figure 13-26.

The sample-and-hold circuits of figures 13-26 to 13-28 are all available in convenient circuit modules or in hybrid or monolithic IC form with characteristics to suit a variety of applications. Acquisition times as low as 25 ns and overall accuracies better than 0.002% are readily obtained. The monolithic IC sample-and-hold circuits, such as the example shown in figure 13-29, require an external hold capacitor. To obtain high accuracy the hold capacitor should be carefully selected for high insulation resistance and low dielectric absorption. **Dielectric absorption** occurs because the dielectric inside the capacitor does not polarize instantaneously. As a result not all the

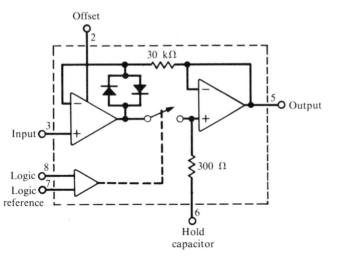

**Fig. 13-29.** Monolithic IC sample-and-hold. This circuit of the LF 398 type is an IC version of the circuit of figure 13-27. It has an acquisition time of less than 10 μs, a gain accuracy of 0.002%, and a droop rate as low as 5 mV/min with a 1-μF capacitor.

energy stored in the capacitor can be recovered instantaneously when the capacitor is discharged. In the sample-and-hold, if the capacitor has been in the hold mode for some time and then acquires a new voltage and holds it, the voltage on the capacitor tends to "creep" toward the original value. For this reason, it is best to operate the sample-and-hold circuit in the hold mode for as short a time as feasible and to use a high-quality polypropylene, polystyrene, or Teflon capacitor. These capacitors have dielectric absorption figures of less than 0.05%. Capacitors with Mylar, ceramic, or even polycarbonate dielectrics are usually unsatisfactory. (See appendix B.) Most hybrid and modular sample-and-holds include an internal hold capacitor.

In applications with high-resolution ADCs (12- to 16-bit converters), the sample-and-hold device should be carefully selected since its characteristics can easily limit the conversion accuracy.

## 13-6    Analog I/O Systems

The systems used for analog input and output to and from the microcomputer vary tremendously in complexity and versatility. At one extreme is an application in which the analog I/O tasks are invariant, such as a data acquisition system for a gas chromatograph. Since input signal levels are approximately known, the timing requirements are known, and the data rate is either fixed or variable only within a limited range, the system can be structured to meet this specific need. At the other extreme is the laboratory analog peripheral for a general-purpose microcomputer. Here the analog I/O system should have provision for encountering signals of widely varying levels. The timing should have built-in flexibility and the data rates should be widely variable. The analog input should include multiple input channels, an input amplifier, a sample-and-hold and an ADC. The analog output should contain DACs and output amplifiers. This latter analog I/O system must have enough hardware and software flexibility to handle most of the needs of a general-purpose system.

This section begins by considering some of the options available for interfacing DACs and ADCs to microcomputers. These simple converter systems and interfaces often suffice and are most economical for applications where great versatility is not needed. Next the more versatile "data acquisition system" available in modular, hybrid, and IC packages is considered. Finally, the highly versatile analog I/O boards produced by several manufacturers for a variety of computers are discussed.

## Converter System Interfacing

There are several options for interfacing DACs and ADCs to the computer. Besides the obvious ones involved in setting up the interface (whether to use programmed I/O, or DMA, basic I/O, or handshaking I/O, flags, or interrupts), there are many choices involved in the selection of the converter system and the I/O port structure. For fixed applications systems, a nonprogrammable port connected to a simple ADC or DAC may suffice. Or, if an extra I/O port is available in the system, the converter may be connected to that port. Alternatively, there are converter systems with built in I/O ports. Thus, it may be more economical to buy a microprocessor-compatible converter system and connect it directly to the computer bus. With an ADC for data acquisition, still other options exist—such as whether to time the acquisitions with a software timing loop or a hardware real-time clock. This choice is often dictated by the timing accuracy requirements as discussed in section 13-2.

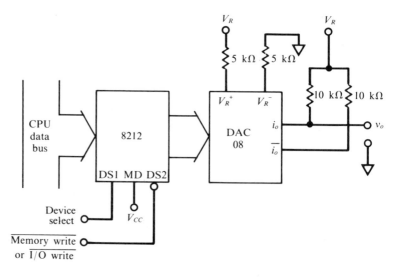

**Fig. 13-30.**   Interfacing the DAC-08 through an output port. Here the DAC-08 is connected as an offset binary DAC. The DAC is continuously updated whenever the port is selected by the microcomputer.

A simple DAC interface to the CPU data bus through an 8212 type output port is shown in figure 13-30. Here the DAC is connected to continuously convert the output of the 8212 buffer register, which is updated whenever the port is software selected. The interface to a DAC with an internal latch register, such as the type in figure 13-20, is quite simple as shown in figure 13-31. Since the DAC provides not only the internal latch but also the output amplifier and internal reference voltage, very few external components are needed.

These two interfacing approaches are also illustrated for two ADCs in figure 13-32. In figure 13-32a, a simple eight-bit ADC is shown interfaced through an input port. The Start convert signal is supplied by software, and the EOC (end of conversion) strobes the data into the 8212 latch. This either causes a program interrupt or sets a flag that is checked by the processor. The service routine selects the port and reads the data onto the bus.

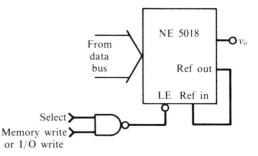

**Fig. 13-31.**  Interfacing the NE 5018 DAC. In this simple interface the internal latch is enabled whenever the device is selected.

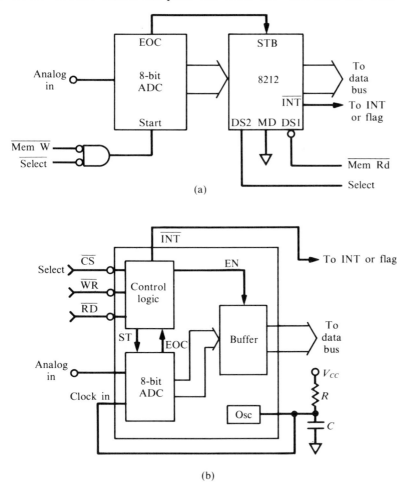

(a)

(b)

**Fig. 13-32.**  Two ADC interfaces. In (a) an eight-bit ADC is interfaced through an 8212 input port. In (b) a microprocessor compatible ADC of the ADC0801 type is shown. The ADC0801 converter (National) shown in (b) completes a conversion in 100 $\mu$s.

Figure 13-32b illustrates the use of an eight-bit microprocessor compatible ADC. The device shown has internal control logic, an on-chip clock generator, and an output buffer register in addition to the successive approximation converter. Conversions are started when the Memory write (or I/O write) line goes LO and the appropriate address occurs. At the end of conversion, $\overline{\text{INT}}$ goes LO, alerting the processor that the converter has finished its cycle. The service routine then selects the device through the $\overline{\text{RD}}$ and $\overline{\text{CS}}$ lines. This enables the output buffer and puts the data onto the data bus.

For timed data acquisition a software clock could be used with both converters in figure 13-32 to generate the appropriate delay between reading the data and starting the next conversion. For more accurate timing one of the real-time clocks discussed in section 13-2 could be used to generate the appropriate delay. In either case the next conversion would not begin until after the delay time had expired. If the real-time clock were used to interrupt the processor, the microcomputer could be free to do other tasks during the delay period.

Interfacing higher-resolution ADCs and DACs to an eight-bit microcomputer data bus is also readily achieved. A 12-bit ADC can be interfaced using two 8-bit I/O ports or by using 12 data lines in a programmable I/O port. Input and output words of greater than eight-bit lengths are transferred in two or more successive bytes. Thus the use of high-resolution DACs or ADCs does not alone justify the selection of a CPU with a 16-bit data bus.

## Data Acquisition Systems

Another viable option for analog inputs is to purchase a complete data acquisition system. Such systems are available in modular and hybrid form from several analog manufacturers; complete IC data acquisition systems are also available. The modular and hybrid data acquisition systems usually include an input multiplexer, an input instrumentation amplifier or programmable gain amplifier, a sample-and-hold circuit, an ADC, timing and control logic, and tristate output buffers for microcomputer bus compatibility. In some of the systems the internal connections between devices can be changed by the user so that it is possible to bypass the input amplifier or sample-and-hold circuit when these are not needed.

A typical data acquisition system is shown in figure 13-33. The input multiplexer allows 16 single-ended analog signals to be converted either randomly or sequentially. All of the control circuitry needed to operate the sample-and-hold and ADC is contained within the package. There are three sets of tristate output buffers that can be enabled separately. This allows data to be put on an eight-bit microcomputer data bus in two steps.

Even though the cost of a data acquisition system may be several hundred dollars, purchasing one is an excellent way to obtain a versatile analog input peripheral without investing a great deal of design time. Costs of such

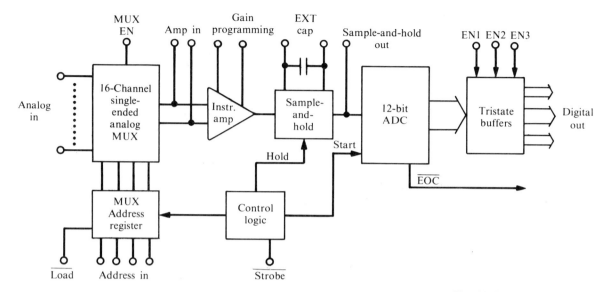

**Fig. 13-33.** Data acquisition system. This hybrid data acquisition system (Datel HDAS-16) is also available in an eight-channel differential input system (HDAS-8). A HI→LO transition at $\overline{\text{Strobe}}$ initiates acquisition and conversion. Total acquisition and conversion time is ~20 µs, which gives a throughput rate of 50 kHz.

peripherals may also be declining as several IC data acquisition systems have been announced. For example, the National Semiconductor ADC 0816 single-chip data acquisition system features a 16-channel multiplexer, an 8-bit ADC, and a tristate output buffer. The user must supply the ADC clock and any analog signal conditioning. For the latter the multiplexer is not internally connected to the ADC. This allows the user to insert an amplifier or a sample-and-hold circuit. This 40-pin IC is easily interfaced to most microcomputer systems because of its latched and decoded multiplexer address inputs and its latched tristate output buffer. Data acquisition modules without tristate output buffers usually require an input port to interface them to the microcomputer data bus.

## Analog Input/Output Boards

Many analog suppliers and some microcomputer suppliers now feature plug-in analog I/O boards for specific computer systems. These may be input boards, output boards or combination boards. The great virtue of these systems is that they literally need only be plugged into the CPU bus. Since these boards are general-purpose analog I/O systems, the potential purchaser must usually trade off the ease of interfacing against the extra expense and complexity of these boards. For example, many applications do not require the great versatility afforded by these analog I/O boards. Another consideration with complete plug-in systems is the problem of software development. Because of the great versatility of the analog I/O system, the programming may be much more complicated for a specific application than the comparable programming for less versatile hardware. On the other hand, the board functions may be compatible with available software.

Most analog I/O boards include input multiplexers, an input amplifier, a sample-and-hold circuit, one or more output DACs, and interfacing circuitry. Some include on-board RAM and ROM and real-time clocks. Typical of a real-time analog I/O board is the Analog Devices RTI-1200 board for Intel SBC-80 Single Board Computers shown in block diagram form in figure 13-34. The input multiplexer allows 16 channels of single-ended analog input or 8 differential inputs. A programmable gain amplifier has software selectable gains of one, two, four, and eight, effectively expanding the dynamic range of the 12-bit ADC to $2^{15}$. A fast (90 ns aperture time)

**Fig. 13-34.** Analog I/O board. This Analog Devices RTI-1200 board is a real-time analog I/O system for the Intel family of single-board computers.

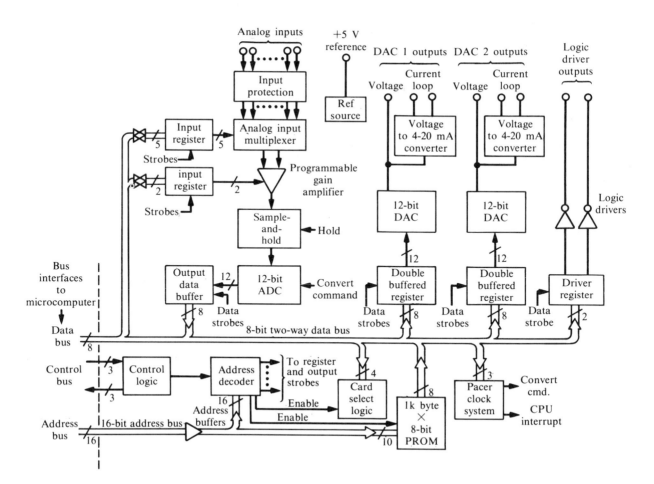

sample-and-hold circuit precedes the ADC. The board has provision for two optional 12-bit DACs driven by double buffered registers. It includes a 1K × 8 PROM socket that can be used to accommodate a PROM to perform calibrations, linearity tests, and other tasks. Two real-time clocks are provided; one an *RC* oscillator and the other a crystal-controlled clock. The clock pulses can be used to generate very accurately spaced analog-to-digital conversions or to perform a time-of-day clock function. This board operates on memory mapped I/O. The ADC EOC signal sets a bit in a status word and can be used to request a program interrupt.

Prices for analog I/O boards range from several hundred dollars to over one thousand dollars depending upon the options selected. This does not include software, but some manufacturers provide excellent examples of data acquisition programs in their users' manuals.

## 13-7 Converter System Applications

This section describes some examples of the use of converter systems in applications other than data acquisition. The acquisition of data is treated in chapter 14. Included here are the digital waveform generator, the Touch-Tone generator, the no-droop sample-and-hold, and the digital oscilloscope.

### Digital Waveform Generator

A waveform generator is a generator that produces a particular waveform at regular intervals. A waveform is digitally encoded by converting it to a digital word at regular intervals. Conversely, if a succession of digital words representing the waveform is applied to a DAC, the waveform is reproduced at the DAC output. This concept is most simply illustrated in figure 13-35a. The ramp function shown is digitally encoded as a sequence of digital words, each word one increment larger than the preceding one. Such a sequence is generated by a counter. If a triangular waveform is desired instead, an up-down counter can be used with a logic circuit that reverses the direction of the count when the counter reaches all 1's or all 0's. This technique is particularly well suited to slow ramp generation because the output is essentially drift free. The output waveform frequency is $f_c/2^n$, which is $f_c/1024$ for a ten-bit counter. Thus the sweep rate is changed by changing the clock frequency.

A ROM or RAM can be inserted between the counter and the DAC to provide virtually any waveform. The counter sequences the memory address, and the memory contents at each address are read out to the DAC. The waveform produced is whatever sequence was stored in the memory. If a sine function ROM is used, a sinusoidal output waveform results. If a RAM is used, any encoded waveform read into the RAM can be repeated indefinitely

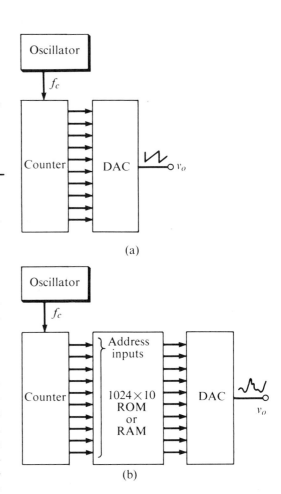

(a)

(b)

**Fig. 13-35.** Digital waveform generators, (a) ramp or triangle and (b) any function. In (a) the counter is incremented by an oscillator, and its parallel output is connected to a DAC. A ramp (staircase actually) voltage is generated, and the waveform is repeated each time the counter cycles through its modulus. Virtually any waveform can be generated by using a RAM or ROM between the counter and the DAC as in (b).

at the DAC output. The waveform repetition rate is determined by the clock frequency. With a microcomputer an extremely versatile waveform generator can be constructed. The computer can calculate the code required for almost any waveform desired at the DAC output.

## Touch-Tone Generator

The circuit that generates the pair of tones in the Touch-Tone telephone dialing system (recall section 9-7) is an ingenious combination of digital, time, and analog techniques. The block diagram of an IC dual-tone generator is shown in figure 13-36. The keyboard logic is arranged so that depressing a single key connects one of the $R$ (row) inputs with one of the $C$ (column) inputs. The keyboard logic detects the depressed key and sends appropriate frequency division codes to the rate multiplier-counters. The counters provide a cycle of addresses to the ROMs that drive the DACs with numerical values that approximate a sine wave. A single crystal oscillator provides the time base. The frequencies of the generated tones are very accurate.

**Fig. 13-36.** Block diagram of Mostek type 5085 dual-tone generator. The tones are generated by digital sine-wave generators composed of the counters, ROMs and DACs. The two tones are combined in the summing amplifier. The counters are actually rate multipliers controlled by the keyboard logic.

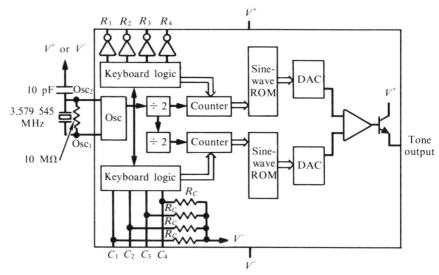

## No-Droop Sample-and-Hold Circuit

All the analog sample-and-hold circuits described in section 13-5 hold the analog data as a charge stored on a capacitor. As a result of capacitor charge leakage, the held voltage will eventually "droop." Data stored in a digital register, on the other hand, can be held indefinitely without loss or change of information. Thus a no-droop or infinite-hold sample-and-hold circuit can be achieved by converting the analog quantity to a digital signal for storage in a register and converting the register contents to a continuous analog output signal. Such a circuit is shown in figure 13-37. If track-and-hold operation is required, the analog-to-digital conversion cycle can be made repetitive by using the ADC status output to trigger the sample-and-hold command input. The register contents and DAC output are then updated at the maximum acquisition-conversion rate. To go into the hold mode, the gate in the automatic cycling connection is closed.

The accuracy of the circuit is determined by the gain accuracy of the analog sample-and-hold circuit and by the conversion accuracies of the ADC and DAC. The aperture time is determined by the analog sample-and-hold circuit alone; the output settling time is determined by the total time for both conversion processes. For very fast converters this can be less than one microsecond. In the track-and-hold mode the aperture time uncertainty is equal to the conversion cycle time. The signal is converted to the digital domain solely to take advantage of the nonvolatile storage available in that domain.

**Fig. 13-37.** No-droop sample-and-hold. At the desired sample time, a transition at the analog sample-and-hold control shifts the circuit to the hold mode and triggers the ADC after a short delay. At the end of the conversion the ADC status (EOC) output enables a latch to store the digital word. A DAC at the output of the latch provides a continuous analog output of the latch contents.

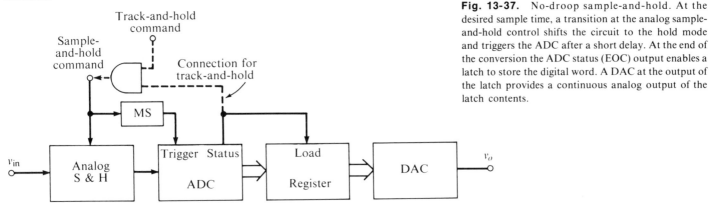

## Digital Storage Oscilloscope

A storage oscilloscope captures and holds a waveform obtained in a single sweep for long-term display on a CRT. Analog storage scopes use special screen phosphors or a charged mesh behind the screen to maintain the luminance of the display. The digital storage oscilloscope converts the input data into a parallel digital signal that is stored in semiconductor memory and then converted back to the analog domain for display on a conventional CRT. It is a combination of a data acquisition system, a no-droop sample-and-hold circuit and a waveform generator. In addition, some digital scopes can digitally process the required data to provide waveform analysis, signal averaging, frequency spectrum calculation and display, coordinate transformation, or other functions.

A block diagram of a basic digital storage scope is shown in figure 13-38. Most digital storage scopes provide numerical display of waveform amplitudes and time values. One or more sensors (user adjustable $x$ and $y$ coordinate positions on the display) permit the user to select the portion of the waveform to be numerically quantified or analyzed. This eliminates the errors caused by the limited screen resolution and the necessity for the operator to interpolate between screen markers. The data display rate is independent of the data acquisition rate. This allows horizontal expansion of the trace after acquisition, an operation impossible to achieve on a transient signal with an analog scope.

Digital storage oscilloscopes allow the retention of acquired data indefinitely, whereas the data stored with an analog scope fades with time and cannot be easily retrieved for readout to an external recorder or computer. The digital storage scope, however, is not yet as fast as an analog scope. A bandwidth of 10 MHz is considered fast for the digital scopes, while 400-MHz analog scopes have been available for some time. Equivalent time sampling can extend the upper frequency range on a digital scope for repetitive signals. With an analog storage scope, the brightness of the trace depends on the writing rate. This makes it difficult to display and store with uniform brightness a slow waveform with very fast transitions. Increasing the trace intensity to display the rapid transitions can lead to blooming and fading of the trace on the slower portions. The digital storage scope, of course, does not depend on the display intensity adjustment for effective storage of the data.

**Fig. 13-38.**  Block diagram of a digital storage oscilloscope. The analog input signal is digitized and stored in a solid-state memory at a rate determined by the time base settings. The memory content is then read out in order to the $y$ DAC for display on the CRT or (more slowly) for the external recorder output. The readout can be repeated indefinitely. The $x$-axis deflection signal is obtained by analog conversion of the memory address. Data acquisition in the digital oscilloscope is synchronized with the input signal waveform by using the trigger signal to control the memory address counter. If the counter is running continuously before the trigger and proceeds less than a full count after the trigger, the memory will contain a pre-trigger portion of the waveform. This unique ability to trigger *after* the event can be very useful.

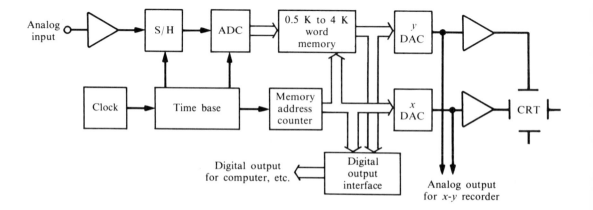

## Suggested Experiments

### 1. Parallel I/O ports.

Investigate a programmable I/O (PIO) interface device connected to the microcomputer bus. Input parallel digital information to the computer from a switch register, store the data in memory, and output the data through the port to binary indicator lights. Study various I/O modes such as basic I/O, handshaking I/O, and interrupt driven I/O. Investigate the various programmable port functions.

### 2. Serial I/O port.

Connect an oscilloscope to a serial I/O port. Write a short program to send a character entered from the keyboard repetitively to the serial port. Observe the oscilloscope display for the start and stop bits and the ASCII character. Repeat for several other characters.

### 3. Real-time clocks.

Investigate an LSI counter-timer chip interfaced to the microcomputer bus. Use the counter-timer as an events counter to interrupt the processor after a given number of events has occurred. Use a manual push button as the event generator. Connect an accurate oscillator to the counter-timer chip, and program it to provide a software selected time delay. Connect the counter-timer as a programmable monostable and as a square-wave generator. Use it as a divide-by-$N$ counter.

### 4. Digital-to-analog converters.

Use a switch register to provide digital information to an eight-bit monolithic DAC. Observe the analog output on a digital multimeter. Plot the analog output against the digital input. Determine the linearity and monotonicity of the DAC. Wire the DAC to give bipolar outputs by adding an output offset circuit. Investigate the offset binary code, the sign-and-magnitude code, and the two's complement code.

### 5. Analog-to-digital converters.

Investigate a staircase ADC by connecting an eight-bit up-counter, an eight-bit DAC, and a comparator in a digital servo system. Connect a voltage reference source as the ADC input and binary indicator lights as the output. Obtain the digital output vs. the analog input for several input voltages. Determine if the ADC rounds up or down. Make a tracking ADC by changing the counter to an up-down counter and using the comparator output to determine the count direction. Note the waffling between adjacent digital values and the ability of this ADC to track input signal changes. Investigate the successive approximation principle by constructing a manual successive approximation ADC. Then use a monolithic eight-bit successive approximation converter. Determine the conversion time and the output-input relationship.

### 6. Sample-and-hold circuits.

Wire a simple sample-and-hold circuit from an op amp, an analog switch, and an $RC$ circuit. Measure its acquisition time, settling time, gain error, tracking frequency, and droop rate. Obtain an IC or hybrid sample-and-hold and characterize it. Use the sample-and-hold as the input to a successive approximation converter. Sample and convert various points on a time-varying repetitive input signal.

### 7. DAC and ADC interfacing.

Interface an eight-bit DAC to the microcomputer via a parallel I/O port. Write a program to generate a voltage ramp or a triangular waveform (repetitive). Display the DAC output on an oscilloscope or on a strip chart recorder. Write a program to carry out a software-based successive approximation conversion using only a DAC and a comparator. Use it to convert several analog voltages, and determine its speed limitations. Interface an eight-bit successive approximation ADC through a parallel port, and obtain several conversions. Write a program to obtain the average of 1000 analog-to-digital conversions.

### 8. Data acquisition system.

Connect a sample-and-hold circuit and an ADC to a parallel input port. Write a program to acquire data from a time-varying analog input signal. Store the data, and use a DAC to output the acquired data to a strip chart recorder on a slow time scale.

## Questions and Problems

**1.** Show how a nonprogrammable I/O port such as the 8212 port of figure 13-2 could be used to interface eight LEDs to the CPU data bus. Use basic I/O in your interface.

**2.** Draw a block diagram for an interface between an eight-bit switch register and the CPU data bus using a nonprogrammable port in a handshaking I/O mode. Show a flow chart of a simple program to read the switch register contents.

**3.** Discuss the hardware and software tradeoffs involved in a decision whether to use a nonprogrammable I/O port or a PIO chip. Discuss specific cases in which one type of port has advantages over the other.

**4.** It is desired to use a microcomputer output port to turn on a 120-V ac lamp. Describe two methods by which this could be accomplished, and give specific circuit diagrams of the interface.

**5.** Parallel data transmission may be prohibitively expensive when an I/O device is located some distance from the microcomputer. Serial transmission techniques are therefore often used for interfacing remote devices even though these devices may be inherently parallel I/O devices. Consider, for example, the interfacing of a remote ADC to a microcomputer with a serial I/O port. Discuss what hardware and software would be needed to accomplish a serial link between the computer and the remote ADC. What data rate limitations would be encountered with a 12-bit converter and a UART with a maximum baud rate of 40 000?

**6.** Describe how a 500-ms clock output could be obtained from the hardware clock of figure 13-11 with a 1.00-MHz crystal oscillator as the frequency reference. What scaler output would be selected, and what would the counter modulus have to be?

**7.** What scaler output and counter modulus are needed to obtain a 10-s clock output from a hardware clock (similar to the one in figure 13-11) that derives its basic frequency from the 60-Hz power line?

**8.** The table below gives some of the relationships between bipolar binary codes in terms of the instructions for the interconversion. Complete the table.

| to \ To convert from | Sign magnitude | Two's complement | Offset binary |
|---|---|---|---|
| Sign magnitude | No change | If MSB = 1, complement other bits, add 00 ... 01 | |
| Two's complement | | No change | |
| Offset binary | | | No change |

**9.** Make a table listing the resolution obtainable from 8-, 10-, 12-, and 16-bit binary-coded DACs. Assume full scale is 10 V. Compare in the same table the resolution from 8-, 12-, and 16-bit BCD-coded DACs.

**10.** In the simple weighted resistor binary DAC of figure 13-15, $n = 8$, $V_R = 10$ V and $R = 5$ kΩ. What are the values of $v_o$ for the following digital inputs? a) 11111111, b) 10000000, c) 01001001, d) 00000001, e) 00001000

**11.** Repeat problem 10 for $V_R = 5.12$ V.

**12.** Design a weighted resistor (see fig. 13-15), 3-decade (12-bit) BCD DAC to produce a 10-V full-scale output. The output op amp is a current-to-voltage converter with a feedback resistance of 10 kΩ. Give the values of all the resistors needed. Would the DAC of figure 13-16 be more advantageous than the simple, weighted resistor DAC? Why?

**13.** A ten-bit bipolar DAC has a 10-V full-scale output. Because of resistor tolerances, drift in component values, etc., the output of the DAC could be in error by as much as $\Delta V$. (a) How large can $\Delta V$ be before the LSB of the DAC is no longer significant? (b) Repeat for 16-, 12- and 8-bit DACs.

**14.** The weighted resistor DAC of figure 13-15 is a 12-bit converter ($n = 12$). It is desired that the output error not exceed half the change in $v_o$ corresponding to a change in the LSB. (a) If only the resistor corresponding to the MSB is in error, what percentage error in the resistance can be tolerated? (b) If only the LSB resistor is in error, what percentage change in the resistance can be tolerated?

**15.** (a) For the DAC of figure 13-16, verify that for a straight binary eight-bit converter, the series resistor should be $16R$. (b) Verify that for a two-digit BCD DAC the series resistor should be $9.6R$.

**16.** For the ladder network DAC of figure 13-17, show that at any node the resistance is $2R$ in the direction of common, $2R$ in the direction of virtual common, and $2R$ in the direction of the switch.

**17.** For the ladder network DAC of figure 13-17, show that the total current drawn from the reference source $V_R$ is a constant value independent of the value of the digital input.

**18.** The binary ladder network DAC of figure 13-17 can be made noninverting if a voltage follower is substituted for the output current-to-voltage converter. Show that if only the MSB switch is closed, $v_o$ (the voltage at $N1$) is $V_R/2$. Also show that

$$v_o = V_R \left( \frac{a_1}{2} + \frac{a_2}{4} + \frac{a_3}{8} + \cdots + \frac{a_n}{2^n} \right)$$

where $a_n$ is the logic level of the $n$th input bit.

**19.** For the noninverting binary ladder DAC, considered in problem 18, bit 3 *only* is a logic 1; all other bits are logic 0's. Find the voltages at nodes $N3$, $N2$, and $N1$ in terms of $V_R$ and the resistance value $R$.

**20.** Explain how a multiplying DAC could be used as a programmable attenuator for an analog signal.

**21.** The simple staircase ADC of figure 13-22 always rounds up; the maximum error is thus one LSB. The error can be made symmetrical at $\pm 1/2$ LSB by offsetting the DAC output by half the analog voltage corresponding to a change of one LSB. (a) To build an eight-bit ADC with full-scale input range 0 to 10 V, what offset voltage should be used to assure that the maximum quantizing error is $\pm 1/2$ LSB? (b) Plot the DAC output voltage $v_o$ vs. time (as in fig. 13-23) for an input voltage of 5.0 V.

**22.** Suppose that an up-down counter is used in the ADC of figure 13-22 to make a tracking ADC. For an eight-bit ADC with a full-scale range of 0–10 V, plot the DAC output vs. time for an input voltage of 3.5 V.

**23.** Three ADCs are used to convert the same analog voltage signals. The ADCs are all eight-bit, 10-V full-scale converters. One is a staircase ADC, one is a tracking ADC, and the last is a successive approximation ADC. (a) If all three ADCs are clocked by a 1.00-MHz clock and all have initial outputs corresponding to 0 V inputs, compare the conversion times required when the analog input is changed to 7.498 V. What will the outputs of the three converters read for successive conversions of the 7.498 V input signal? (They will not be the same.) (b) Suppose the analog input is changed from 7.498 V to 9.900 V. Compare the conversion time of the three converters and the outputs for successive conversions of the 9.900-V signal.

**24.** A bipolar ADC used with offset binary code has a $\pm 5$ V full-scale range and 12-bit resolution. (a) What digital output corresponds to $-5$ V? (b) What analog voltage change is necessary to cause a one LSB change in the digital output? (c) What analog voltage input corresponds to a positive full scale output $(11 \ldots 11)$?

**25.** The basic sample-and-hold circuit of figure 13-26 is specified to have a droop rate in the hold mode of 200 mV/s with an external hold capacitor of 100 pF. (a) What is the droop rate if $C$ is changed to 5000 pF? (b) What is the droop rate for a $0.01-\mu F$ hold capacitor?

**26.** A sample-and-hold circuit has a sampling time of 70 ns and is used under ordinary operating temperatures (0 to $+100°C$). What type of hold capacitor would give the best performance? Why would a ceramic capacitor likely be unsatisfactory?

**27.** Challenge question: It is desired that the error introduced by a sample-and-hold circuit not exceed 0.01% for a 10-V peak-to-peak sine-wave input signal. If the only error is a 50-ns aperture time, what is the maximum allowable input frequency?

**28.** Compare the integrating sample-and-hold circuits of figure 13-28 to the basic sample-and-hold circuit of figure 13-26. Include accuracy, acquisition time, and droop rate in the comparison.

**29.** In multichannel data acquisition systems, various elements of the acquisition chain may be shared by multiplexing with multiple input sources. For example, a data acquisition system such as that shown in figure 13-33 shares a single amplifier, a single sample-and-hold, and a single ADC among multiple channels. In a second arrangement the multiplexer could be placed after the input amplifier so that each input has a separate amplifier. In a third arrangement, each channel might have a separate amplifier, sample-and-hold, and ADC. The ADC outputs could be digitally multiplexed and transmitted to the processor. Discuss the advantages and disadvantages of these three types of multichannel data acquisition systems. Consider cost, dynamic range limitations, needed multiplexer quality, accuracy, and ability to handle remote data sources in your discussion.

# Chapter 14

# Optimized Measurement and Control Systems

All measurement and control systems suffer from uncertainties that make them behave in less-than-ideal fashion. In implementing a measurement or control function, the experimenter is therefore usually concerned with achieving the optimum balance of speed, efficiency, and accuracy. The use of microprocessors as control elements in instrumentation systems is making possible a new breed of intelligent, self-optimizing systems that have tremendous potential for improving many operations.

Any systematic effort at optimization rests upon the accuracy and precision with which digital data are acquired because nothing can overcome erroneous information about the quantity that is measured or controlled. For that reason this section begins with a discussion of the sampling operation, the first step in most microcomputer-based measurement and control systems. The general methods for improving the precision of measurements are then discussed. Since measurement imprecision is often the result of electrical noise, the general properties that can be used to distinguish signals and noise are basic to any signal-to-noise enhancement methods. Among the signal-processing techniques discussed are analog and digital integration, digital filtering, lock-in amplification, boxcar integration, multichannel averaging, and correlation methods. Measurement of the rates of random events and rates of change are considered in a separate section because signal-to-noise ratio considerations for these measurements are somewhat special.

The microprocessor has opened new horizons in the area of automated control systems. The section on control introduces the principles and provides many illustrative examples of practical control systems. Open- and closed-loop control systems are considered. Self-optimizing systems in which the control system can adapt to changing system dynamics are next considered. This chapter concludes with some thoughts on the future of microcomputer automation.

# 14-1  Sampling of Time-Varying Quantities

The process of sampling a continuous (analog) signal that varies with time is an integral part of many electronic measurement and control systems. Sampling oscilloscopes, digital oscilloscopes, boxcar integrators, multichannel averagers and computer data acquisition systems are but a few of the electronic instruments that use the sampling operation. Sampling is also inherent in all analog-to-digital conversion techniques. In every case, the samples acquired are all that remain of the original waveform. Thus, the accuracy of representing the continuous signal depends upon obtaining accurate samples.

In this section we shall see that a finite number of samples can accurately describe a continuous waveform if the waveform is of limited bandwidth. Such a band-limited signal can then be sampled exactly if samples are taken at more than some minimum rate specified by the sampling theorem. The **sampling rate**, or frequency, is the reciprocal of the **sampling interval**, or the time between samples. The **sampling duration**, the total time over which samples are taken, is shown to be important for certain types of signals. The **aperture time** is the time over which the analog signal is averaged during sampling and is an important sampling parameter. In a sampling operation associated with an analog-to-digital conversion, the quantization process must also be considered. The time associated with quantizing the sample is called the **quantizing time**. The total time for an analog-to-digital conversion, which is the combination of the aperture time and the quantizing time, is called the **digitization time**. Each of these steps in the sampling process influences the accuracy with which the sampled data represent the original signal.

## Band-Limited Signals

Electrical signals almost always vary with time. A time-varying signal can be characterized by its **frequency spectrum**. We have encountered the frequency spectrum of a signal in chapter 2 where the harmonic composition of a square wave is discussed. Knowing the frequency composition of a signal is important if operations (such as sampling) are to be performed on the signal without loss of information. Also, as will be seen later in this chapter, differences in the frequency composition of signals and noise can be exploited to improve measurement precision.

The analysis of the frequency composition of a waveform is called **Fourier analysis**. In Fourier analysis an amplitude-time waveform is transformed into its spectrum, which is the amplitude of each frequency plotted against frequency (see note 14-1). These two representations of a waveform are often called the **time domain** and **frequency-domain** representations (see note 14-2).

**Note 14-1. Fourier Transform Analysis.**
Every amplitude-time waveform $v(t)$ has a related frequency spectrum $F(\omega)$. The two functions $v(t)$ and $F(\omega)$ are called Fourier transform pairs and are mathematically related by the Fourier integral

$$F(\omega) = \int_{-\infty}^{\infty} v(t)[\cos \omega t - j \sin \omega t]\,dt$$

where $j$ is the complex operator $\sqrt{-1}$.

**Note 14-2. Fourier Domains and Data Domains.**
The Fourier domain (time or frequency) refers to the manner in which the information in an entire waveform is encoded. It should not be confused with the manner in which a single data point is encoded (data domain).

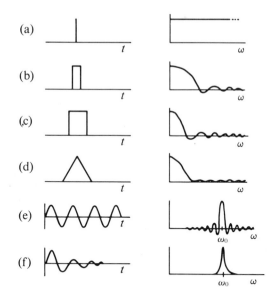

**Fig. 14-1.** Pictorial Fourier transform pairs. An infinitely sharp amplitude-time signal (a) has equal amplitudes at all frequency (white spectrum). For the rectangular pulses in (b) and (c), the frequency spectrum has the form of the function $(\sin x)/x$. As the pulse widens, the frequency spectrum narrows. The triangular pulse (d) gives a frequency spectrum of functional form $(\sin^2 x)/x^2$. A finite duration sine wave (e) and an exponentially decaying sine wave (f) are also shown.

Several pictorial Fourier transform pairs are illustrated in figure 14-1. In many cases, the frequency spectrum of a waveform is plotted as the amplitude density (in volts per hertz) vs. frequency (in hertz) to avoid mathematical problems in calculating the Fourier integral. Such a plot is called an **amplitude spectrum**. In some cases the power density (in watts per hertz) is plotted against frequency (in hertz). Such a plot is called a **power density spectrum** or simply a **power spectrum**.

A **band-limited signal** has an amplitude spectrum that is zero everywhere except in a particular frequency range. Limited bandwidth signals can result from purposely filtering a broadband signal or from bandwidth limitations of transducers, amplifiers, or other system components. Band-limited signals fall into two classes: dc signals and ac signals.

**Direct-current signals.**    A direct-current signal is one in which the current is always in the same direction and the magnitude of the current is constant over the period of interest. However, no signal can be constant indefinitely. Consider the output of a thermocouple used to monitor temperature. A typical plot of the transducer output voltage against time is shown in figure 14-2 along with the signal power density that results from Fourier analysis of the waveform. The signal frequency composition at frequencies higher than 0

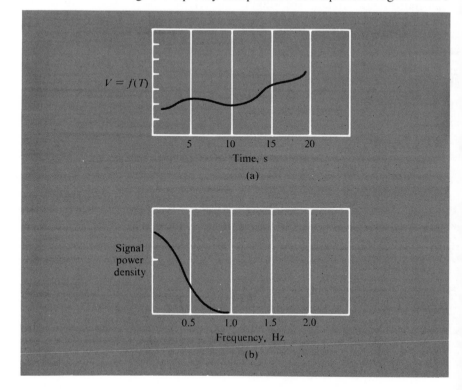

**Fig. 14-2.** Output voltage vs. time (a) and power spectrum (b) for a thermocouple. Note that most of the signal power is at or near 0 Hz (dc) but that some information is present at higher frequencies.

Hz (dc) may arise from actual temperature changes or from changes in the thermocouple transfer characteristics with time. If the temperature changes were of interest, it would be important that the system bandwidth be several hertz to avoid loss of information. If the short-time fluctuations were considered unimportant or undesirable, the signal bandwidth could be further narrowed by a low pass filter. In either case any electronic system for amplifying, modifying, and displaying such a signal must have a low-frequency response that extends to 0 Hz.

Similar spectra arise from other transducers that are usually considered to produce dc outputs. Because all signals have some bandwidth, a general definition of a **dc signal** is one whose power is concentrated in a band of frequencies near 0 Hz like the thermocouple signal of figure 14-2. For such signals the power is proportional to the square of the dc current or voltage.

**Alternating-current signals.**  An ac signal can also be usefully characterized by its power spectrum. In contrast to a dc signal, the power density in an ac signal occurs at frequencies higher than 0 Hz. Often dc signals are converted to ac signals by modulation, as described in chapter 8, in order to perform amplification and other operations at higher frequencies. Some typical ac signals and their power spectra are shown in figure 14-3. Such ac signals may, by their nature, be band-limited, or they may be intentionally

**Fig. 14-3.**  Ac signals and their power spectra. The audio-range signal (a) is a broadband signal as shown by its frequency spectrum. The chopped signal (b) contains odd harmonics as does a square wave. The peak signal (c) is similar to a dc signal, but the signal power may extend to very high frequencies.

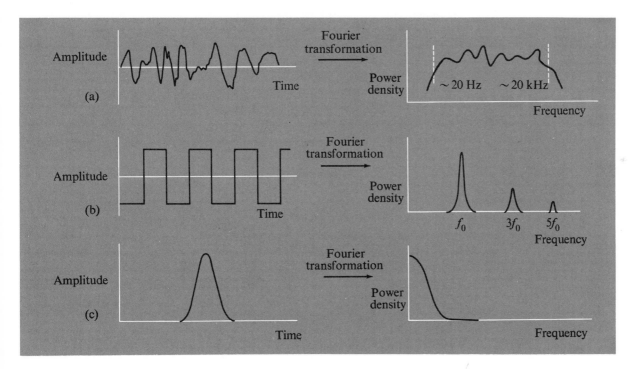

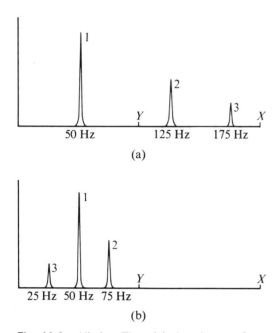

(a)

(b)

**Fig. 14-4.** Aliasing. The original analog waveform contains components at 50 Hz, 125 Hz and 175 Hz (a). If the sampling rate is 400 Hz, the Nyquist frequency is 200 Hz (point $X$), and no aliasing occurs. If a sampling rate of 200 Hz is chosen (point $Y$), the components above 100 Hz are undersampled and produce low-frequency aliases as in (b).

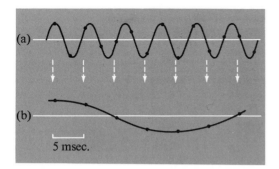

**Fig. 14-5.** Aliasing of 175 Hz to 25 Hz. In (a) a 175-Hz sine wave is shown. The points indicate samples taken at 400 Hz. These samples would accurately describe the original signal. If every other point is dropped out to simulate a 200-Hz sampling rate and the remaining points are connected, the 25-Hz alias results.

restricted in bandwidth by filters. For example, most of the information from the chopped signal in figure 14-3(b) is at the fundamental frequency $f_0$, and a bandpass filter or tuned amplifier could be used to increase the signal strength in this narrow frequency interval. In any case, as will be shown below, it is important to have a band-limited signal in order to carry out the sampling operation with high accuracy.

## Sampling Theorem

In sampling a continuous signal, we would expect samples taken at more closely spaced intervals to describe the original signal more accurately than samples taken at larger intervals. An exact description of a signal with unrestricted variation can be obtained only when the interval between samples approaches zero. For band-limited signals, however, there is a finite sampling rate that is sufficient to include all the information in the signal.

The Nyquist sampling theorem provides the quantitative basis for the rate at which samples must be taken, based on the bandwidth of the signal. The **sampling theorem** states that if a band-limited dc signal is sampled at a rate that is *twice the highest frequency component* in the signal, the sample values exactly describe the original signal. If the sampling rate is $1/\Delta t$ (where $\Delta t$ is the sampling interval), the signal must have no frequency components at frequencies greater than $1/(2\Delta t)$. The critical frequency $1/(2\Delta t)$ is called the **Nyquist frequency**. A signal with Fourier components extending from 0 to 200 Hz, for example, should be sampled at a rate of at least 400 samples per second, or every 2.5 ms. In order to ensure accurate sampling, the signal should be band-limited by an appropriate input filter prior to the sampling step. It is important to point out that sampling rates considerably higher than the Nyquist criterion are often used to ensure adequate sampling. As a rule of thumb, a sampling rate of ten times the Nyquist frequency is often employed.

If the sampling theorem is not followed and the sampling rate is too low, two kinds of errors result. First, the information in the signal at frequencies above the Nyquist frequency is lost, and, second, the undersampled high frequencies show up as spurious low frequencies. This latter error, known as **aliasing**, is illustrated in figure 14-4. As a result of undersampling at 200 Hz, the high-frequency information is lost and spurious low-frequency components are added to the signal as shown in figure 14-4b. The 50-Hz component is still properly sampled, but the 125- and 175-Hz components have aliases at 75 and 25 Hz. The way in which aliasing comes about can be appreciated by the simple example shown in figure 14-5. A familiar example of aliasing is the appearance of the slow rotation of the wheels of a rapidly moving stage coach in western movies because the frequency of spoke rotation is undersampled by the framing of the movie camera.

It is interesting to point out with respect to figure 14-4 that if no frequency components below 100 Hz had been present in the original signal, the undersampling, or fold-over, would not have been serious. The aliased high frequencies would not have overlapped with any other signal, and the position of fold-over could be accurately predicted. This points out a more general statement of the Nyquist sampling theorem that applies to band-limited ac signals as well as to band-limited dc signals. *A signal or waveform sampled at twice its bandwidth will be accurately sampled.* Thus, if all frequency components of a signal are located in a 100-Hz bandwidth, a 200-Hz sampling rate is adequate, no matter where the 100-Hz bandwidth is located along the frequency axis. This can considerably reduce the sampling rate for narrow bandwidth ac signals. However, since aliases may be generated in the process, the reconstruction of the signal from its aliases may be quite complicated. Again, in practice, sampling rates much higher than the Nyquist rate are often used.

With many signals, such as peak-shaped signals that contain Fourier components from dc out to very high frequencies, compromises must be made in choosing the sampling interval. Usually these signals are band-limited by a low pass filter giving minimal distortion and are sampled at 2–10 times the filter bandpass. Another factor that influences the choice of the sampling interval is the presence of noise at discrete frequencies. For example, if 60-Hz noise is a problem, taking samples every 16.67 ms, with a constant phase relation to the line frequency, is an excellent way of discriminating against the 60-Hz interference.

The total time over which samples are taken, the sampling duration, can also cause errors in the sampled signal. Consider the sampling of the peak-shaped signal shown in figure 14-6. Even if the sampling is done at the appropriate rate, an error occurs because of loss of signal information beyond the truncation points. This **truncation error** is often small because the information loss is low or can be made so simply by taking a few more points. Signals with long tails such as Lorentzian peaks give more problems than Gaussian or exponentially decaying signals.

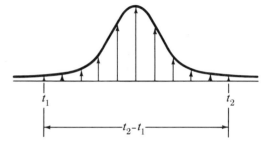

**Fig. 14-6.** Sampling duration. Sampling is begun at a time $t_1$ and continued until time $t_2$. The total sampling duration $t_d$ is $t_2 - t_1$. Information outside this total time is lost.

## Aperture Time

A sample of a continuous signal cannot be acquired instantaneously. Instead the signal must be observed for a finite length of time called the **aperture time**. The signal is effectively smoothed or averaged over the aperture time. In a sample-and-hold ADC system this averaging arises because the sampling switch follows a band-limited filter or amplifier and the switch cannot open instantaneously. The filtering effect of the switch and band-limiting circuitry can, of course, cause distortion of the higher-frequency information.

Of more direct concern is the uncertainty in the aperture time, or aperture time jitter. The signal that opens the sampling switch has often passed through several logic circuits, and each logic-level threshold has an associated uncertainty, or jitter, as described for the comparator in chapter 9. The error caused by the aperture time uncertainty is proportional to the rate of change of the signal and the magnitude of the uncertainty as shown in figure 14-7 for a sine-wave signal. To minimize this error the aperture time uncertainty should be as small as possible. For a signal changing at a rate of 1 V/$\mu$s, a 10-ns aperture jitter causes a 0.1% amplitude error. To reduce the error to 0.01% requires that the aperture uncertainty be reduced to 1 ns. The aperture time uncertainty should also be a small fraction of the sampling interval in order to obtain accurate sampling. Fortunately many modern sample-and-hold circuits have aperture time uncertainties in the nanosecond to subnanosecond range.

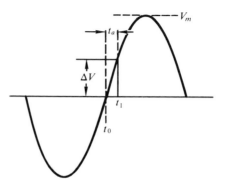

**Fig. 14-7.** Aperture uncertainty. The maximum error occurs when the sine wave crosses zero since its rate of change is maximum there. An upper limit estimate of the error ($\Delta V$) is $\Delta V = V_m \sin(2\pi f t_a)$ where $V_m$ is the amplitude of the sine wave, $f$ is its frequency, and $t_a$ is the aperture uncertainty, $t_1 - t_0$.

## Digitization Time

The above discussion of sampling rate and sampling duration is directly applicable to analog-to-digital conversion of a continuous waveform. However, more must be said about aperture time. An ADC cannot digitize a signal instantaneously. The **digitization time** is the interval between the start of the sampling operation and the appearance of the digitized signal. Digitization time is made up of the aperture time and the quantization time. In ADCs with input sample-and-hold circuits like a sample-and-hold input successive approximation converter, these two times are quite separate. With dual-slope ADCs the sample is integrated for a fixed period (the aperture time) and then quantized during the integration of the reference voltage. With some ADCs, however, it is difficult to distinguish the aperture and quantizing time. With a voltage-to-frequency converter ADC, the aperture and quantization times can be considered equal, because sampling and quantization occur continuously throughout the integration time.

The sample-and-hold input successive approximation ADC provides a short aperture time (often 25 ns or less), small aperture time uncertainty, and a distinct and relatively short quantization time (often only a few micro-seconds). This allows rapid sampling and digitization of analog waveforms. The successive approximation ADC can also be used for longer sampling intervals. The effective aperture time can be established by an input low pass filter, or a specific number of fast acquisitions can be averaged to define the aperture time. This latter approach has the potential advantage of increasing the effective number of bits in the converter. This effect of digital averaging is described in section 14-3.

## 14-2  Signal-to-Noise Ratio Enhancement by Bandwidth Reduction

In modern measurement and control systems it is increasingly necessary to measure weak electrical signals in the presence of noise. As sources and detectors are improved and weaker physical effects are utilized to provide information, the problem of discriminating between an information-conveying signal and extraneous, unwanted noise components becomes increasingly difficult. Fortunately, several elegant hardware- and computer-based techniques have been developed to aid measurement where the signal-to-noise ratio is quite small. This section and the following one are devoted to exploring the principles of these signal-to-noise ratio (S/N) enhancement techniques.

In general, signals and noise can be distinguished by their frequency characteristics and by the time of occurrence or phase coherence of their frequency components. This section considers S/N enhancement by bandwidth reduction (frequency discrimination) methods. Among the techniques included are low pass filtering, hardware and software averaging (integration), and digital filtering in the time domain (smoothing) and in the frequency domain. Waveform correlation techniques, which take advantage of the phase coherence of repetitive signals, are the subject of section 14-3.

## Signals and Noise

Electrical signals consist of a desirable signal component, which is related to some quantity of interest, and an undesirable component, which is termed noise. **Electrical noise** may thus be defined as any part of the observed electrical signal that is unwanted. Implicit in this definition is the concept that what is considered noise in one situation may be useful information under other conditions, and vice versa. Thus, developing techniques to discriminate against noise requires knowledge of the general properties of electrical signals and noise.

**Noise spectra.**  A **noise power density spectrum**, analogous to a signal power density spectrum, is a plot of noise power density (in watts per hertz)

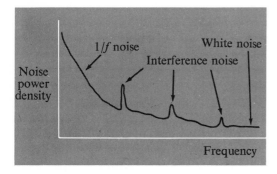

**Fig. 14-8.** Combined noise power density spectrum. At high frequencies white noise predominates; at low frequencies one-over-$f$ $1/f$ noise predominates. Interference noise has components at discrete frequencies.

**Note 14-3.   Johnson Noise.**

Because Johnson noise is due to thermal motion, the magnitude of the noise voltage increases with the temperature $T$. Thermal noise also increases as the resistance $R$ of the component increases. Since the noise power density is equal at all frequencies, the total noise voltage observed across a resistor depends upon the range of frequencies that the measurement system passes, i.e., the system bandwidth $\Delta f$. The quantitative relationship describing the rms noise voltage, $\bar{v}_{rms}$, is known as the Nyquist relation and is

$$\bar{v}_{rms} = (4k/TR\Delta f)^{1/2},$$

where $k$ is Boltzmann's constant.

**Note 14-4.   Shot Noise.**

The rms shot noise current $\bar{i}_{rms}$ due to an average current $i$ observed during a time interval $t$ is

$$\bar{i}_{rms} = (Q_e i/t)^{1/2}$$

where $Q_e$ is the charge on an electron. This shot noise equation, or Schottky equation, can also be expressed in terms of bandwidth because the bandwidth equivalent to an observation time $t$ is $\Delta f = 1/(2t)$. Thus, the rms shot noise current can also be written

$$\bar{i}_{rms} = (2Q_e i\Delta f)^{1/2}$$

against frequency. Noise sources give rise to three distinct types of power spectra, which are illustrated in figure 14-8 (see also note 9-1). **White noise** has an essentially flat power spectrum and can be considered a mixture of all frequencies with random amplitudes and phase angles. Low-frequency **one-over-$f$ ($1/f$) noise,** also called **flicker** or **pink noise,** has a spectrum in which the power density increases approximately with the reciprocal of the frequency at low frequencies. Such a spectrum is typical of the low-frequency drifts common in transducers, amplifiers, and analog components. Discrete frequency **interference noise** often arises from the 60-Hz power lines, radio transmitters, motors, and nearby oscillators, and the power density spectrum has peaks at the fundamental and harmonics of these frequencies. In a real electronic system all three types of noise are likely to be encountered.

From this description of power density spectra, one important distinction between signals and noise is clear. The distributions of signal power and noise power as functions of frequency are quite different. In general, noise tends to have a rather broad and featureless power spectrum except for discrete frequency noise. In contrast, many signals have or can be made to have relatively narrow and well-defined frequency regions where most of the signal power occurs. These general differences in the frequency distribution form the basis of the S/N enhancement techniques described in this section.

**Noise sources.**   The total noise in an electronic system results from two distinct types of noise: fundamental noise and excess (nonfundamental) noise. **Fundamental noise** arises from the motion of discrete charges in electrical circuits and cannot be completely eliminated. **Excess noise** arises from imperfect instrumentation or nonideal component behavior and can in principle be reduced to insignificant levels by careful practice and instrument design. Noise is also introduced in the process of converting an analog signal into a digital representation. This type of noise is called **quantizing noise.**

The two most important types of fundamental noise are Johnson noise and shot noise. **Johnson noise,** also called **thermal noise,** is produced by the random motion of electrons in resistive elements because of thermal agitation. Johnson noise has a white power density spectrum; that is, its power density is independent of frequency. The Johnson noise voltage across a resistor increases with increasing temperature, resistance, and system bandwidth (see note 14-3). **Shot noise** results from the random movement of discrete charges across junctions. Examples include the flow of charges across semiconductor junctions or between cathode and anode in a vacuum tube or phototube. Shot noise also has a white power density spectrum. The shot noise current increases with increasing average current and system bandwidth (see note 14-4).

Any noise above and beyond Johnson or shot noise is considered excess noise. In contrast to fundamental noise, excess noise is almost always frequency dependent. Interference noise from the 60-Hz power lines and noise with a $1/f$ power spectrum are excess noise. The $1/f$ noise is often considered to be synonymous with drift. It can be introduced by long-term power-supply fluctuations, changes in component values, temperature drifts, and other sources whose exact nature is poorly understood.

Quantizing noise is the result of the finite resolution of an analog-to-digital conversion. It is usually thought of only in terms of ADCs, but this type of noise is present in any process that converts a continuous infinite-resolution signal to a finite number of digits. Thus quantizing noise can be introduced in the manual conversion of a strip chart recorder deflection or a meter scale position to a numerical value as well as in an electronic analog-to-digital converter. The quantizing noise in an ADC can be visualized by the process illustrated in figure 14-9. If the DAC output is compared to the ADC input, it is apparent that the quantization process adds noise to the original signal. If the quantizing interval of the converter is $Q$, the rms value of the quantizing noise is $Q/\sqrt{12}$.

**Note 14-5.  Mean-Square Noise.**
For dc signals the mean-square noise can be defined as the average squared deviation of the signal from its mean value:

mean-square noise $=$

$$\left[\frac{(S - S_1)^2 + (S - S_2)^2 + \cdots + (S - S_n)^2}{n}\right]$$

where $S$ is the mean values of the signal, $S_1$, $S_2$, ..., $S_n$ are its instantaneous values, and $n$ is the total number of values. The mean-square noise is also called the variance of the signal, and the rms noise is its standard deviation.

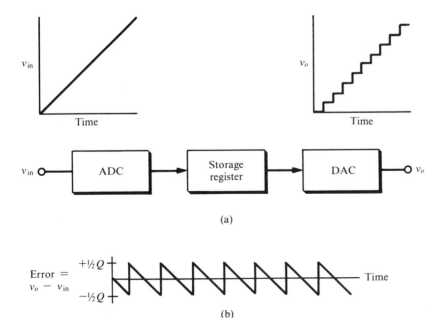

(a)

(b)

**Fig. 14-9.**  Quantizing noise. The circuit in (a) is used to reconstruct a slowly varying analog input signal by playing the ADC output back through a DAC. If the analog input is subtracted from the DAC output as shown in (b), the result is the quantizing noise. This noise has an average value of zero, a peak-to-peak value of $Q$ (the quantizing level) and an rms value of $Q/\sqrt{12}$.

**Signal-to-noise ratio.**    In most applications it is the total noise present that is of interest. If noise sources are completely independent, the total mean-square noise current of voltage $\overline{v_t}^2$ (see note 14-5) is the simple sum of the individual mean-square noise components (variances), $\overline{v}_{n1}^2$, $\overline{v}_{n2}^2$, ..., according to

$$\overline{v_t}^2 = \overline{v}_{n1}^2 + \overline{v}_{n2}^2 + \overline{v}_{n3}^2 + \ldots$$

The signal-to-noise ratio can be expressed as

$$S/N = \frac{\text{average signal}}{\text{rms noise}}$$

The S/N is also commonly expressed as a power ratio in decibels or as a voltage or current ratio in decibels.

For dc signals the signal-to-noise ratio is the reciprocal of the relative standard deviation of the measurement if electrical noise is the factor limiting measurement precision. For ac signals the relationship between S/N and precision is less straightforward. However, in many cases ac waveforms are converted to dc before they are displayed or digitized. In these situations, the S/N and the relative standard deviation are reciprocally related as they are for dc signals.

## Low Pass Filtering

Perhaps the most common method of enhancing the signal-to-noise ratio of a measurement is low pass filtering. Many signals of interest have major frequency components at dc (0 Hz) with bandwidths extending only a few hertz. In these cases a simple low pass filter can effectively limit the measurement system bandpass to that necessary to pass the signal frequencies. The characteristics of first- and higher-order active low pass filters are discussed in detail in chapter 8. It is simply necessary to choose the $RC$ time constant, and hence the bandwidth and phase shift characteristics, such that the signal frequencies are affected as little as desired. The improvement in signal-to-noise ratio by filtering comes at the expense of a decreased response time, which can lead to distortion of the signal. Thus, a compromise between the measurement precision (S/N) and the preservation of signal shape must be made.

The distortion in the signal can be accurately predicted if the **impulse response function** of the filter is known. The impulse response of a simple $RC$ low pass filter is illustrated in figure 14-10. A unit area pulse of decreasing pulsewidth produces, in the limit of zero pulsewidth, the response shown in figure 14-10c′. The impulse response function is important because any arbitrary input signal shape can be considered as the summation of a series of impulses. Thus the total response of the filter to an arbitrary signal is the superposition of the separate responses due to each impulse. This can be seen more clearly in figure 14-11. As the number of impulses becomes very large, the output response approaches the continuous response of figure 14-10b′.

The process of obtaining the superposition of a sequence of impulse responses is known as **convolution** (see note 14-6). The convolution of a peak-shaped signal with the impulse response of an $RC$ low pass filter is

**Note 14-6.   Convolution.**
Convolution is a multiplication-integration operation found widely in science. It relates the input and output of a system by the impulse response function. Mathematically, convolution involves multiplying the input signal $X(\tau)$ by a reflected and displaced version of the impulse response $I(t - \tau)$ and time-averaging or integrating the product as a function of the displacement $t$. The output $C(t)$ is obtained from the convolution integral

$$C(t) = \int_{-\infty}^{\infty} X(\tau) I(t - \tau) d\tau$$

where $\tau$ is a dummy variable of integration.

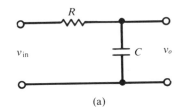

(a)

**Fig. 14-10.** Impulse response of $RC$ low pass filter. A pulse of height $1/\Delta t$ and width $\Delta t$(b) is applied to the low pass filter of (a). The output of the filter (b') rises exponentially towards $1/\Delta t$ during the pulse and decays exponentially after the pulse. The input pulse width is made shorter and shorter (at constant area) until the pulse width goes to zero (c). The result is the impulse response of the filter (c').

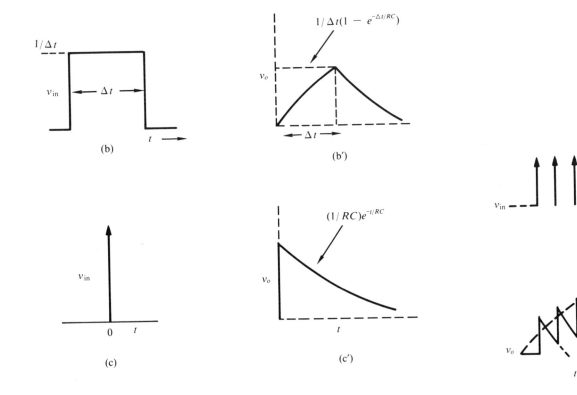

**Fig. 14-11.** $RC$ filter response to a rectangular pulse. The input pulse (a) is shown as a series of impulses. The filter output (b) is the superposition of the response towards each impulse.

illustrated in figure 14-12. When the convolution operation is carried out mathematically or graphically, the impulse response is reflected on the time axis, and the area of the product of the signal and the impulse response is evaluated for various relative displacements. The reflection is necessary to keep the time sense correct in that the early time edge of the response function must be applied first to the early time edge of the signal. In this way the previous values contribute to the current value with exponentially decreasing weight. The distortion shown in figure 14-12 includes altering the

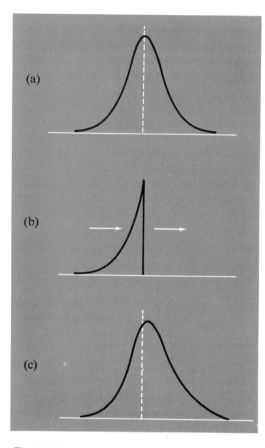

**Fig. 14-12.**   Distortion of a peak signal by *RC* filter. The distortion in the filtered signal (c) can be visualized by the convolution operation. The impulse response function (b) is reversed and moved to the right. At each displacement, the raw signal (a) and the response function are multiplied together, and the area of their product is taken. The resulting area vs. displacement is the distorted signal (c).

peak height, shifting the peak maximum, and skewing the peak with the generation of a trailing edge. These effects can be minimized by ensuring that the *RC* time constant is small relative to the time required to scan across the peak. The situation in figure 14-12 is for a Gaussian peak with a half-width of 1 s and a filter with 0.25-s time constant.

## Analog and Digital Integration

Integration is a widely used technique for S/N enhancement. Integration differs from *RC* filtering in that the response of an integrator to a rectangular pulse is a linear function of time. A linear integrator has a step function impulse response with the step lasting for the integration time. This gives the linear integrator constant weighting of all previous values in their contribution to the current value.

Linear integration of analog signals can be accomplished in several ways. Active and passive low pass filters approach linear behavior when the time constant is much larger than the integration time. An op amp integrator with timed switching, of course, makes an excellent linear integrator. Integrating ADCs such as the dual-slope and the voltage-to-frequency converter are also linear integrators. Another approach is to use digital integration in hardware with a fast adder following the analog-to-digital converter or in software by summing successive ADC values. Time-domain signals, such as pulse outputs from transducers, can be integrated by counting techniques.

**Integrating ADCs.**    An integrating ADC, using the dual-slope approach or voltage-to-frequency converter, is an excellent way to digitize analog signals if high conversion speed is unnecessary. A significant advantage of integrating ADCs is their ability to reject noise at certain frequencies. For the dual-slope converter, for example, the integral obtained during the signal integration period can be thought of as the sum of the integrals of the true signal and the noise. Since a sine wave has an average value of zero over one period, noise that has a period equal to the signal integration period has no effect on the output value. The frequency response of an integrating ADC with a 1.0-s integration time is illustrated in figure 14-13a. Note that the response function has nodes at 1.0 Hz, 2.0 Hz, etc. Thus noise signals of these frequencies are greatly attenutated. The ability to reject input signals at certain frequencies is called **normal mode rejection**. Normal mode rejection differs from common mode rejection in that a common mode signal is present at both inputs of a differential amplifier, whereas a normal mode signal is present at only one input.

Their excellent normal mode rejection is one reason that dual-slope converters are frequently used in digital voltmeters where the signal integration time is often made one period of the 60-Hz line frequency or 16.667 ms.

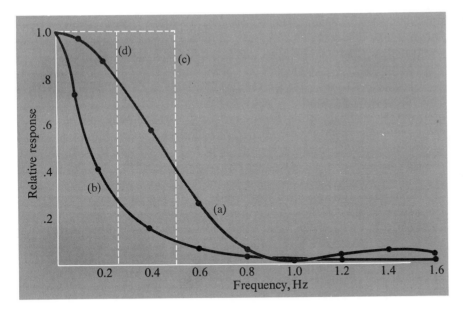

With this integration time the first node falls at 60 Hz. Since many discrete frequency noise sources can be traced ultimately to the power line, noise rejection at the power-line frequency is highly desirable. This filtering characteristic is also evident at any multiple of the power-line frequency and normal mode rejection ratios as large as 70 dB are practical.

Filters of differing frequency response characteristics are often compared on the basis of their equivalent bandwidths. The **equivalent bandwidth** $\Delta f$ is a rectangular bandwidth of area equal to that of the power spectrum of the filter. The equivalent bandwidth of a linear integrator of integration time $t$ is $\Delta f = 1/(2t)$. This is compared in figure 14-13 with the equivalent bandwidth of an *RC* low pass filter which is $\Delta f = 1/(4\,RC)$. Thus for a linear integrator to have a bandwidth equivalent to that of a *RC* low pass filter, the integration time must be twice the time constant of the filter ($t = 2\,RC$). For white noise and dc signals, the S/N improves with $(\Delta f)^{-1/2}$ or for an integrator the S/N is proportional to $t^{1/2}$.

Integration using integrating ADCs has two main limitations: the measurement bandwidth is based at 0 Hz (dc), which is the region of greatest $1/f$ noise, and the range of signals that can be integrated is limited to dc or slowly changing signals.

**Digital integration.**   Digital integration in hardware or software can provide excellent signal averaging characteristics. Many computer-based systems use multiple-point averaging of successive analog-to-digital conversions for this purpose. The number of points to be averaged, the total averaging time, and other variables can be program controlled. Many computer- or hardware-based digital averaging systems use the data acquisition scheme illustrated in figure 14-14. Here the ADC operates at nearly its maximum throughput rate. An input filter limits the bandwidth of the input signal to that necessary to provide accurate sampling. If the ADC had a conversion time of 8 $\mu$s and the sample-and-hold settling time was 1 $\mu$s, for example, the monostable might be set up for a 2-$\mu$s delay. The sampling rate of the data acquisition system would then be 100 kHz. The input filter would be chosen to limit the bandwidth to less than 50 kHz ($RC > 3.2$ $\mu$s). If no input signal changes faster than 1 ms were important, then it would be feasible to average one hundred data points.

Note that this approach provides essentially complete averaging over the 1-ms period even though the sample-and-hold aperture time may be as short as 50 ns because the input filter averages the raw input signal and the computer or hardware adder averages the analog-to-digital conversions. For slowly changing signals and white noise, the S/N should improve with the square root of the averaging time or with the square root of the number of points averaged. Thus averaging one hundred points should improve the S/N by a factor of ten over taking a single point.

Digital integration can also give rejection nodes for noise signals with periods equal to the averaging time. With a 1-ms averaging time, for example, rejection nodes should occur at 1 kHz and multiples of 1 kHz.

**Fig. 14-14.** Fast data acquisition system. An input $RC$ filter limits the bandwidth of the input signal. The end-of-convert signal $\overline{\text{EOC}}$ from the ADC puts the sample-and-hold in the hold mode. The ADC start-convert signal (ST) is delayed by a monostable multi-vibrator to allow the sample-and-hold to settle before conversion.

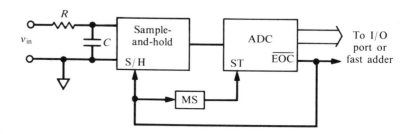

## Digital Filtering

Because filtering a signal can be considered as a process of weighting the data (convolution), filtering can readily be carried out in the digital domain. With computer systems virtually any weighting factor desired (filter function) can be achieved in software. Hardware systems can also perform the digital filtering operation. There are two general schemes by which the filtering operation can be performed: time-domain weighting and frequency-domain weighting.

In the time domain the digital filtering operation is often referred to as a data smoothing operation. The smoothing operation is normally carried out on data sampled at evenly spaced time intervals. One approach to smoothing is the *n*-point **moving averaging smooth**. Here *n* points are averaged to provide a smoothed data point that represents the central value, and the averaging function is moved along the time axis. For example, in a five-point moving average smooth, points 1 through 5 would be averaged to provide a smoothed data point replacing point 3 in the raw data array. Points 2 through 6 would then be averaged replacing point 4, etc.

In more complex smoothing operations, a polynomial can be used to approximate local sections of data, and the fitted polynomial then provides a smoothed value for the central point. Tables of weighting coefficients for a variety of smoothing functions including higher-order filtering operations are available. All of these smoothing techniques improve the S/N through bandwidth reduction. Unfortunately, since the smoothing function is only an approximation to each local section of data, the true signal can undergo distortion unless the smoothing parameters are carefully chosen.

In the frequency-domain approach, the Fourier transform of the signal is taken and applied to the desired frequency response of the filter. It was shown earlier that filtering can be considered to be the result of convolving the amplitude-time waveform of the signal with the impulse response function of the filter. An important Fourier transform theorem states that *convolution in the time domain is equivalent to multiplication in the frequency domain.* Hence the frequency domain digital filter is applied as illustrated in figure 14-15. Various fast Fourier transform algorithms are available to calculate the Fourier transform of the input digital signal. Essentially any desired frequency response curve can be set up, including many that are impossible to design with hardware. Filters with no phase shift, square cutoff filters, differentiating filters, and unique discrete frequency filters are all readily implemented. Filtering is implemented by multiplying the real output of the transformed input signal by the frequency response of the selected filter and regenerating the signal by inverse Fourier transformation.

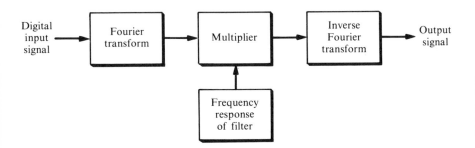

**Fig. 14-15.**  Digital filtering in the frequency domain. The Fourier transform of the input signal is computed and multiplied by the desired frequency response at all pertinent frequencies. The inverse Fourier transform is then computed to give the filtered output.

A simple example that demonstrates the effectiveness of this approach is shown in figure 14-16. Here the desired filter response is a low pass filter with an abrupt cutoff and no phase shift, characteristics that are impossible to achieve with analog filters. Note that the noise level of the filtered signal is much lower than that of the original signal. Analogous reduction of the noise level with analog techniques would have been difficult without distortion of the signal. Distortions can result in digital filtering if signal information is also attenuated by the filter. However, since the frequency-domain representation of the input signal can be displayed, it is often quite simple to choose the filter bandwidth to avoid distortions.

It is important to note that digital filtering in the frequency domain is exactly analogous to smoothing operations in the time domain. The smoothing approach is less complex computationally, but the Fourier transform approach allows easier visualization of the filtering operation.

**Fig. 14-16.** Digital filtering. The Fourier transform of the input signal (a) produces the real output shown in (b). This is multiplied by the desired filter response (c) to yield the modified real output (d). The inverse Fourier transform is then calculated to regenerate the filtered signal (e).

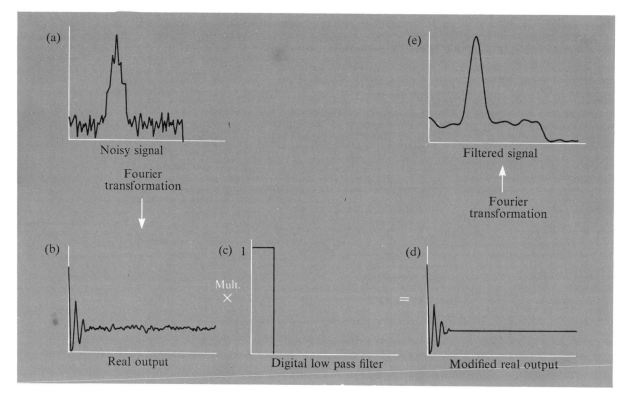

## 14-3 Correlation Techniques for Signal-to-Noise Enhancement

The preceding section emphasized the signal-to-noise enhancement techniques that depend on the different frequency characteristics of signal information and noise power. A second important difference between the frequency components of signals and noise is their phase relation. The frequency components that make up a signal are, in general, phase related; noise frequencies, on the other hand, typically are not related in phase to the signal frequencies or, for that matter, to other noise frequencies.

These two distinguishing properties of noise and signal frequencies (relative distribution and phase relation) are illustrated in figure 14-17 for two

**Fig. 14-17.** Noisy peak signals (a) and their amplitude (b) and phase spectra (c). The signal in column II has a higher noise level than that in column I. Note from the amplitude spectra (b) that most of the signal frequency components are located in a narrow band near 0 Hz and that the higher frequencies are primarily due to noise. In (c) most of the lower frequency components of both noisy signals are seen to be in phase (0 phase), while the higher frequency components are random in phase.

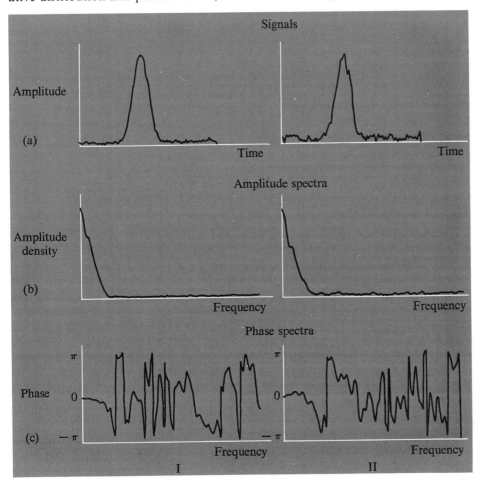

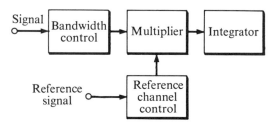

**Fig. 14-18.** Generalized block diagram of an S/N enhancement technique. The noisy signal is amplified with a band-limited amplifier and multiplied by a reference signal. Finally, the multiplier output is time-averaged or integrated.

**Note 14-7. Correlation.**
Correlation with continuous functions can be described by the following integral:

$$R_{xy}(\tau) = \lim_{T \to \infty} \frac{1}{2T} \int_{-T}^{+T} x(t)\, y(t - \tau)\, dt$$

Here $R_{xy}(\tau)$ represents the correlation function of the two signals $x(t)$ and $y(t)$ over the interval $-T$ to $+T$, and $\tau$ is the relative displacement. Many correlations are carried out on sampled waveforms where correlation is described by

$$R_{xy}(n\Delta t) = \sum_{t} x(t)\, y(t - n\Delta t)$$

Here $\Delta t$ is the sampling interval, and $n\Delta t$ is the relative displacement.

noisy peak-shaped signals. From the amplitude spectra, it is clear that attenuation of the higher frequencies by filtering would significantly improve the signal-to-noise ratio. In addition, the phase spectra show that both noisy signals have in-phase components at low frequencies. The similarity of the phase spectra in the low-frequency region indicates that the phases of the frequency components responsible for the peak are essentially the same in these two noisy signals. It is also important to note that the two phase spectra are quite dissimilar in the higher-frequency region where random noise is expected to predominate.

This phase coherence of the signal frequency components can be used to enhance the signal-to-noise ratio *if the signal is or can be made repetitive.* For example, if the amplitude-time waveforms in figure 14-17a could be obtained repetitively (by repetitive scanning or by repeated triggering of the signal initiation process) and the signals obtained on each repetition were added together, the frequency components that make up the peak would add in phase while the noise frequencies would add randomly and tend to cancel out. The S/N enhancement techniques presented in this section all depend on this basic principle to discriminate between signals and noise.

The section begins with a brief discussion of correlation because it is the correlation of two signals that can provide the necessary phase discrimination. Then several important S/N enhancement techniques are described. These include lock-in amplification, boxcar integration, multichannel signal averaging and waveform correlation techniques.

## Introduction to Correlation

A generalized block diagram of a signal-to-noise enhancement technique is shown in figure 14-18. The bandwidth control step discriminates against noise on the basis of its frequency distribution, and the multiplication-integration step discriminates against noise on the basis of the predictable time behavior of the signal information (phase coherence). The multiplication-integration operation is best described in terms of correlation. **Correlation** involves multiplying one signal by a delayed version of a second signal and integrating or time-averaging the product. Evaluating this time-averaged product over a range of relative displacements or delays, generates a correlation pattern that is a function of the relative displacement (see note 14-7).

Two general types of correlation are commonly distinguished. If the two signals are different, the process is called **cross-correlation**; if they are the same signal, it is **auto-correlation**. The correlation of two functions is very similar to their convolution (see note 14-8). Correlation of two signals is equivalent to multiplication of their frequency spectra as illustrated in figure 14-19. The Fourier transform of a rectangular pulse is the $(\sin x)/x$ function, and the product of the two transforms is $(\sin^2 x)/x^2$. This product of two

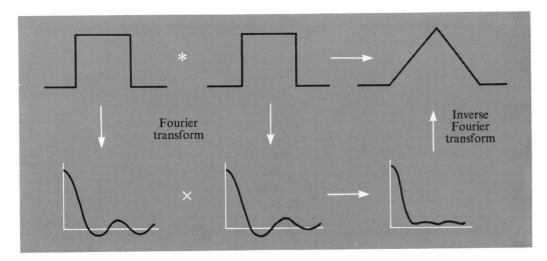

spectra is often called a **cross spectrum**. Inverse Fourier transformation yields the triangular auto-correlation function. The effects of a particular correlation operation are often more easily discerned by thinking in terms of multiplying spectra rather than correlating waveforms.

The multiplication-integration operation (correlation) provides discrimination between phase-related signal components and randomly phased noise components. Some of the techniques described below involve only a simplified correlation operation in the sense that the correlation function is evaluated at only a single relative displacement. In general a measurement technique is referred to specifically as a correlation technique when the complete correlation function of two waveforms is evaluated.

**Fig. 14-19.** Fourier transform representation of correlation. The asterisk * indicates correlation and $\times$ multiplication. The auto-correlation of a rectangular signal yields a triangular waveform (top path). The same result can be obtained by multiplying the Fourier transforms together and taking the inverse transform of the product (bottom path).

## Lock-In Amplification

The lock-in amplifier introduced in chapter 8 is an example of a S/N enhancement system that uses a cross-correlation technique. The basic steps in a complete lock-in amplifier system include modulation, selective amplification (often tuned amplification), synchronous demodulation, and low pass filtering. The demodulation step of the lock-in amplifier provides the phase discrimination. Synchronous demodulation can be carried out using a four quadrant multiplier as shown in figure 14-20. The result of the synchronous demodulation is a full-wave rectified output of those signal components of the same frequency *and phase* as the reference. The final step in the recovery of the signal is to send the multiplier output through a low pass filter. This step simply decreases the fluctuations of the synchronously rectified carrier and produces an output voltage proportional to the amplitude of the carrier wave. Note that the cross-correlation operation is present in the

**Note 14-8. Correlation and Convolution.**
Correlation and convolution both involve the multiplication-integration operation. In convolution the impulse response function must be reversed from left to right before carrying out the multiplication-integration operation. For the correlation depicted in figure 14-9, the scanning function is symmetrical, and thus correlation and convolution give equivalent results.

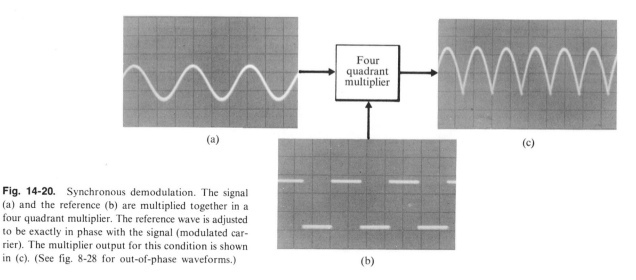

(a)

(b)

(c)

**Fig. 14-20.** Synchronous demodulation. The signal (a) and the reference (b) are multiplied together in a four quadrant multiplier. The reference wave is adjusted to be exactly in phase with the signal (modulated carrier). The multiplier output for this condition is shown in (c). (See fig. 8-28 for out-of-phase waveforms.)

synchronous demodulation and low pass filtering step. In this case, the cross-correlation is carried out at only one relative displacement ($\tau = 0$).

The performance of a lock-in amplifier is shown in figure 14-21. Despite the complete "burial" of the signal in noise, the output dc level has high S/N, and the amplitude of the observed sine wave can be easily measured.

Hardware lock-in amplifiers are available from several manufacturers. It is also possible to simulate lock-in amplifier performance in *real time* with an interfaced microcomputer system. The reference waveform can be used to trigger data acquisition of the carrier wave. If the frequency of the carrier is not too high, say 1 kHz or less, several samples can be acquired during each half-cycle of the carrier. Alternate half-cycles are then added together with the correct polarity so as to carry out synchronous demodulation. The values for several hundred cycles are then averaged to provide a "low pass filtered" output via a DAC.

## Boxcar Integration

The boxcar integrator is a versatile gated integrator for measuring repetitive signals. The boxcar technique involves gating out a particular section of a waveform and integrating successive gated signals to improve the signal-to-noise ratio. It is particularly useful for measuring repetitive short pulse signals and signals that have a slow repetition rate or low duty cycle. A block diagram of a boxcar integrator is shown in figure 14-22.

The analog gating operation can be thought of as a multiplication operation in which the input waveform is multiplied by a normalized rectangular pulse (amplitude = 1). The analog gating and integration operation is then a form of cross-correlation in which the signal waveform is

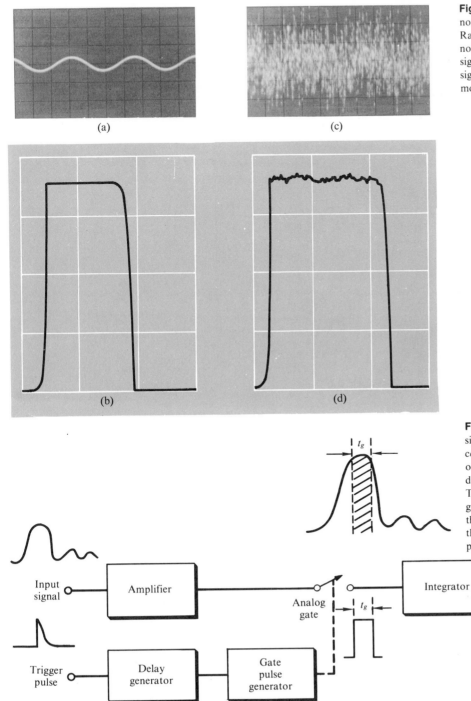

(a)

(c)

(b)

(d)

**Fig. 14-21.** Performance of a lock-in amplifier. The noise-free input in (a) is recovered as a dc level in (b). Random noise added to the signal in (a) produces the noisy input in (c). The resulting dc output for the noisy signal is shown in (d). The rise and fall in the output signals (b) and (d) result from the application and removal of the input signal.

**Fig. 14-22.** Boxcar integrator. The amplified input signal is gated to an integrator for a short gate time $t_g$ controlled by the gate pulse generator. The gate pulse occurs at a fixed time delay relative to the trigger signal derived from the start of the repetitive input signal. The same time slice of the input waveform can be integrated for multiple repetitions to improve the S/N, or the delay generator can slowly scan the gate time across the input waveform. This allows examination of multiple sections of the waveform or of the entire waveform.

**Fig. 14-23.** Boxcar integrator performance. A repetitive exponential decay (a) is applied to the scanning boxcar and the output (b) is obtained on a recorder. The same signal obscured in noise (c) gives the output shown in (d). Complete recovery is obtained by plotting amplitude against delay time as shown in (e) for the noisy signal of (c).

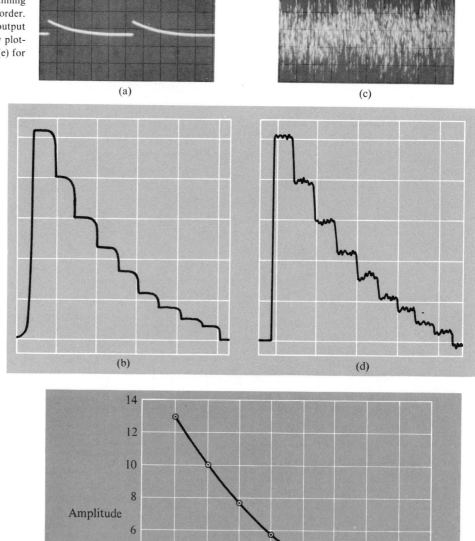

cross-correlated with a rectangular gating pulse at a relative displacement $\tau$ set by the delay generator. The boxcar gate delay can also be slowly scanned in time so that the cross-correlation is carried out across the complete waveform. With a sufficiently narrow gate pulse and a relatively slow sweep on the delay time, the shape of a pulse as brief as 2 ns can be recorded on a strip chart recorder. Gate pulse widths as low as 100 ps are available on commercial boxcar integrators.

The key step in the boxcar integrator system is the analog gating. This step is a sampling operation, and all the criteria for accurate sampling discussed in section 14-1 must be satisfied. The gate pulse width $t_g$ is the aperture time of the sampling operation. If the gate pulse is scanned across the waveform, the increments in the delay time between samples must be small enough to satisfy the Nyquist sampling theorem.

The noise discrimination capability of the boxcar integrator is quite impressive as is illustrated in figure 14-23. Note that the signal-to-noise ratio in the recovered waveform is substantially higher than that of the noisy signal.

The boxcar integrator function is also readily implemented with a computer data acquisition system as illustrated in figure 14-24. In the example

**Fig. 14-24.** Computer boxcar integrator. A programmable timer is used to generate a variable delay between the trigger signal and the acquisition of the sample. Through program control a specified number of samples can be taken at any given delay, the delay can be scanned at the desired rate, and the time increment between points can be varied.

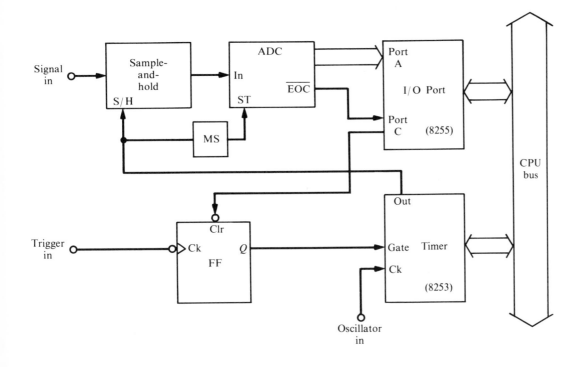

shown, the time of sampling is controlled by a real-time clock. Multiple-point averaging and scanning are readily accomplished with quite simple programming. A computer boxcar integrator system, however, cannot achieve the very fast aperture times of a hardware system.

The boxcar integrator is thus a very versatile measurement system. It can be thought of as a lock-in amplifier for pulsed signals. As a gated integrator, the boxcar is a single-channel averager for repetitive waveforms. In its scanning mode, it is a cross-correlation computer. The boxcar integrator can measure very fast waveforms and narrow pulses. However, for many signals multichannel averagers offer significantly faster measurement times because hundreds of channels can be averaged simultaneously. Also more general cross-correlation analysis requires more complex cross-correlations than those between a simple rectangular pulse and a signal waveform.

## Multichannel Averaging

In many experimental measurements it is necessary to recover a complete repetitive signal from a noisy waveform. A **multichannel averager** acquires a large number of evenly spaced samples across the complete waveform on *each* repetition instead of one sample per repetition as in the boxcar integrator. The samples from successive repetitions are averaged or integrated as in the boxcar integrator. However, since the multichannel averager takes $N$ samples across a waveform, the complete waveform can be recovered $N$ times as fast. As with any averaging or integrating technique, the signal builds up as the number of scans, and the noise as the square root of the number of scans. Thus the S/N improves as $N^{1/2}$.

Most multichannel averagers are digital instruments based on the general block diagram of figure 14-25. Each repetition of the waveform is digitized at a desired number of points, and these values are added to previous values and stored in memory. The general structure of the digital multichannel averager may be based on a stand-alone hardware system, or it may be just a program in a microcomputer system. Hardware averagers are usually faster than the software-based averagers, but multichannel averagers are not, in general, extremely fast. Sampling rates are normally less than 200 kHz for hardware averagers and are often less than 50 kHz for computer-based systems. Typical input circuitry includes a fast sample-and-hold and a fast analog-to-digital converter. Precision can be traded off for sampling rate if desired.

If averaging is not required, considerably faster sampling rates can be achieved. Instruments called **transient recorders** can digitize and store a complete single trace of a waveform. Sampling rates for transient recorders can be as high as 10–50 MHz with six- to eight-bit resolution. The transient recorder block diagram is similar to that shown in figure 14-25 except that the adder is not used.

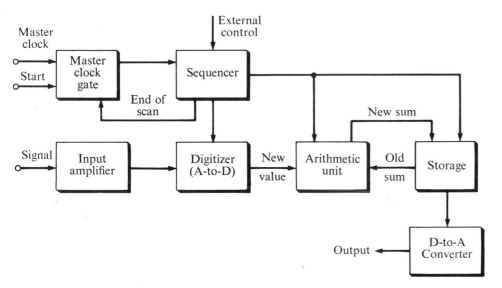

**Fig. 14-25.** Multichannel averager block diagram. The start pulse that initiates the repetition of the signal waveform gates a master clock to a sequencer (hardware or computer) that controls the sampling, digitization, addition, and storage of the new values. A DAC provides an analog output for display.

In all multichannel averagers the sampling operation described previously is an extremely important step. All the criteria for accurate sampling with respect to sampling rate, sampling duration, aperture time, and quantizing time must be satisfied.

Multichannel averaging can achieve impressive S/N enhancement for certain types of signals as illustrated in figure 14-26. One hundred scans should improve the S/N by a factor of ten compared to a single scan. Quantizing noise can limit the measurement precision when the ADC does not have enough resolution or, more often, when two large signals are subtracted to yield a small result. If the noise in the system (ADC and signal) is much less than the quantization interval $Q$, quantizing noise cannot be reduced by averaging. This is true because the noise is round-off error, which is identically reproduced on each scan. However, a level of system noise of $Q$ or greater randomizes the quantizing error and allows quantization noise reduction by averaging. In cases where the total system noise is very small a S/N enhancement can be realized by adding random noise to the original signal prior to multichannel averaging. As long as the random noise is not too intense, its final level can be reduced to an acceptable point by averaging.

## Waveform Correlation Techniques

The basic correlation operation has been present in all the S/N enhancement techniques discussed in this section. However, except for the scanning boxcar integrator, the correlation was evaluated at only one relative displacement and even then with only a rectangular reference pulse. Cross-correlation and auto-correlation of complete waveforms are, not surprisingly, also very useful S/N enhancement techniques.

**Fig. 14-26.** Effects of multichannel averaging. One scan of a noisy waveform is shown in (a); the result of 100 scans is shown in (b).

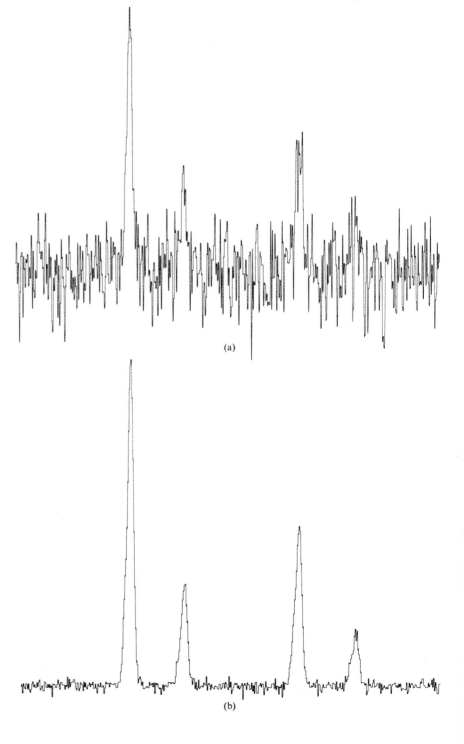

(a)

(b)

In cross-correlation it is useful to distinguish the signal waveform and the reference waveform and to think in terms of the reference waveform moving with respect to the signal waveform. For example, in the scanning boxcar integrator the rectangular gate pulse is the reference waveform that is shifted or slid across the signal. In more general cross-correlations any desired shape could be used for the reference waveform. The optimal shape of the reference cross-correlation waveform for maximum S/N enhancement has been thoroughly investigated. For repetitive single pulse signals, a noise-free version of the signal itself provides a close-to-optimal reference pulse shape. An example of the cross-correlation operation for S/N enhancement is shown in figure 14-27.

The goal of cross-correlation with pulse waveforms need not be simply S/N enhancement. Useful modifications of the signal waveform such as differentiation and resolution enhancement can be carried out by cross-correlation with bipolar pulses. In a general sense, cross-correlations carried out with complex reference waveforms can be considered a measure of the degree of similarity between the signal and the reference waveforms. The

**Fig. 14-27.** S/N enhancement by cross-correlation. The noisy signal on the left is cross-correlated (indicated by *) with a noise-free version of itself, resulting in the smoothed signal on the right. Note that some distortion in signal shape is introduced as the signals have been broadened and rounded by the cross-correlation. The noise level in signal (b) greatly exceeds that in (a).

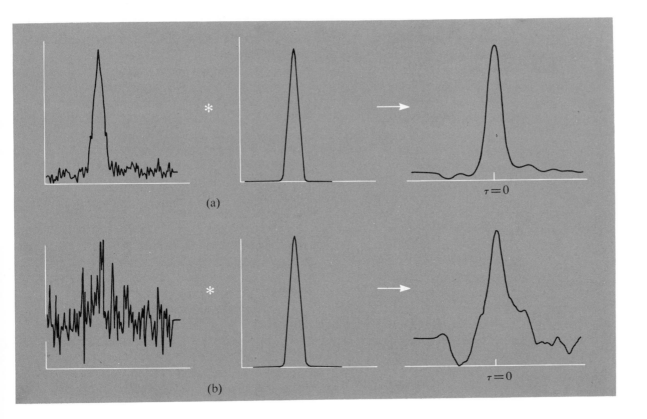

value of the cross-correlation function at zero displacement ($\tau = 0$) is the sum or time average of the product of the two waveforms, and its magnitude is a measure of the common features shared by the two waveforms. As such, cross-correlation is highly useful in automated signal and pattern recognition systems. As with simple pulse cross-correlations, more complex correlations can be carried out under conditions of poor S/N.

Correlation carried out between a signal waveform and a time-shifted version of the same signal waveform is called auto-correlation. This type of correlation can be useful in recovering a periodic signal from noise when no reference waveform is available. The signal becomes its own reference. However, any phase information in the original periodic signal is lost in the auto-correlation operation. The auto-correlation patterns of several periodic waveforms are illustrated in figure 14-28. Note that the pattern for random noise is quite different from that of the signals shown. Thus if a periodic waveform is present in a noisy signal, its auto-correlation pattern persists after that for the noise has decayed away.

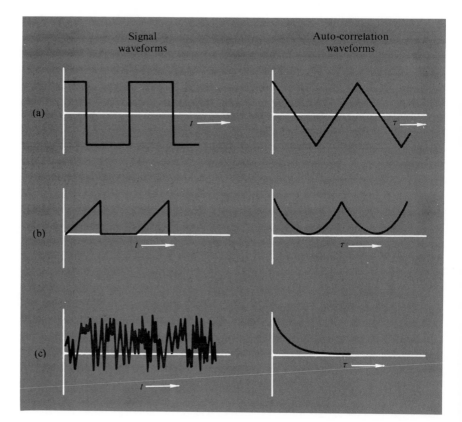

**Fig. 14-28.**   Auto-correlation patterns of several waveforms. When auto-correlated, random noise of limited bandwidth (c) gives rise to an exponential decay. Periodic signals (a) and (b) give patterns that persist after the noise has decayed.

Although a few manufacturers supply stand-alone correlation instrumentation, it is not widely available at this time. However, correlation analysis can be readily performed with computer-based systems. This provides great flexibility in pretreating the data, in carrying out the correlation operation, and in displaying the data for easier interpretation. The correlation may be carried out on stored waveforms by multiplication of the Fourier transforms of the waveforms and inverse transformation or by time-averaging the product of the two waveforms as a function of displacement.

## 14-4  Rate Measurements

It is often necessary to measure the rates of processes or events, and some aspects of rate measurements are distinct from direct measurements of magnitude. Rate measurements divide into the two categories of this section—measuring the rate of occurrence of discrete events, and determining the rate of change of some physical or chemical quantity from a measurement of the rate of change of a related electrical quantity.

## Rates of Randomly Spaced Events

The rate of a regularly recurring event is just its frequency. The measurement of frequency has been discussed in chapters 4 and 9; in general, frequency measurement is accomplished by a counter that accumulates the number of events or cycles occurring during a specific time interval. It is helpful here to remember that a frequency measurement is an integrating measurement. All events during the gate time are counted and the result, events per time, is the average rate of events over that time. Many events of interest in nature occur at irregular intervals—radioactive decay of unstable nuclei and the arrival of photons from an ordinary light source, for example. These events occur randomly in time, and this irregularity of occurrence affects the measurement process.

At low rates of incidence, photons, decay particles, and ions are often detected by conversion to emitted electrons followed by electron multiplication. The electron multiplier is the charge-amplification dynode chain used in the photomultiplier tube of figure 4-31. Each incident particle or photon that causes electron emission produces a cascade of $10^5$ to $10^8$ electrons at the last dynode. This burst of electrons has a charge of $10^{-11}$ to $10^{-14}$ C, which causes a current pulse in the connection to the last dynode. All the cascading electrons produced by a single particle arrive at the last dynode within a time span of 3–10 ns. The peak pulse current (pulse charge divided by pulse width) is thus in the range of 1 $\mu$A to 3 mA, a readily detectable pulse amplitude. Two systems are used to measure rates with electron-multiplying detectors: analog circuits that measure the average current produced by the

**Fig. 14-29.** Analog and digital techniques for measuring the event rate at an electron multiplier detector. In the analog circuit (a) a current follower converts the electron multiplier current to a proportional output voltage. The current is equal to the charge per second, which is the number of pulses per second $N_p$ times the electron multiplier gain $\overline{G}$ times the electron charge $Q_e$. The capacitor $C$ produces a low pass filter with an $RC$ time constant much longer than the interval between pulses. The digital circuit uses a high-frequency amplifier and comparator to produce a logic pulse for each current pulse. The counter reading $N$ is the number of pulses that occur over the gate time interval $t$.

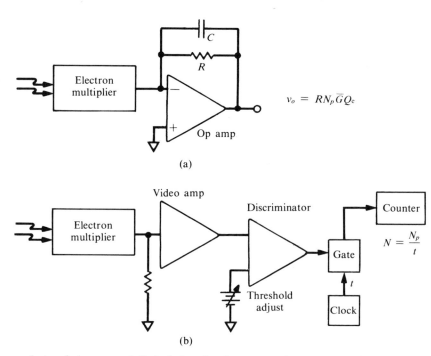

(a)

(b)

### Note 14-9.  Derivation of Standard Deviation of Analog Random Count Rate Output Signal.

The standard deviation $\sigma_p$ for a number $N_p$ of random events is $\sqrt{N_p}$.

$$\sigma_p = N_p^{1/2}$$

This standard deviation can be expressed in terms of the current signal by multiplying the above deviation by the average current contributed by each pulse.

$$\sigma_i = N_p^{1/2}\,\frac{\overline{G}Q_e}{t} = \left(\frac{N_p\overline{G}^2 Q_e^2}{t^2}\right)^{1/2}$$

where $t$ is the time over which the current is averaged. Since $N_p\overline{G}Q_e/t$ is equal to the average current $i$,

$$\sigma_i = \left(\frac{i\overline{G}Q_e}{t}\right)^{1/2}$$

The standard deviation of the current follower output voltage is obtained by multiplying $\sigma_i$ by the feedback resistor value $R$:

$$\sigma_v = \left(\frac{i\overline{G}Q_e R^2}{t}\right)^{1/2}$$

If an $RC$ filter is used instead of an integration time $t$, the equivalent integration time of the filter $(2RC)$ can be substituted:

$$\sigma_v = \left(\frac{iGQ_e R^2}{2RC}\right)^{1/2} = \left(\frac{iR\overline{G}Q_e}{2C}\right)^{1/2}$$

packets of charge and digital circuits that count the number of current pulses in a given time. The digital technique is called **pulse counting** or, in optical systems, **photon counting**. Both these techniques are illustrated in figure 14-29.

The randomness of the pulse times causes an uncertainty in the measured rate because there is a variation in the number of events that occur in a given period. The larger the number of pulses measured, the smaller the relative error in the measurement. For the digital technique the standard deviation $\sigma$ for the measurement is equal to the square root of the count value, $\sqrt{N_p}$. In the case of the analog measurement, the standard deviation of the current follower output voltage $\sigma_v$ is $(iR\overline{G}Q_e/2C)^{1/2}$, where $i$ is the average output current of the electron multiplier. (For the derivation of this expression, see note 14-9). The relative standard deviation $\sigma_v/v_o$ of the analog output increases as the signal decreases as can be seen by dividing the expression for $\sigma_v$ by $v_o = iR$ to obtain

$$\frac{\sigma_v}{v_o} = \left(\frac{\overline{G}Q_e}{2iRC}\right)^{1/2}$$

This same expression shows, as expected, that the relative standard deviation improves as the $RC$ filter time constant increases.

At low count rates, the pulse counting technique has a number of distinct advantages. First, at very low count rates, the average analog current is extremely low. (At 1000 pulses per second, the current is $10^{-8}$ to $10^{-11}$ A depending on $\overline{G}$.) Such low currents require a large (and therefore noisy) feedback resistor in the current follower, a long time constant to smooth out the pulses, and a low-drift amplifier to minimize $1/f$ noise. Other error sources are seen from the counting circuit waveforms shown in figure 14-30. At low count rates the current from the low-level pulses can become significant in the analog technique as can the pulse-to-pulse charge variation. Conversely, with the pulse counting technique long integration times can be employed without drift to obtain high measurement precision even for very low count rates.

The pulse counting technique would be ideal if it were not for its problems at high count rates. These problems are due to an inability to distinguish pulses that occur at nearly the same time. The **resolving time**, or **dead time**, of the counting system is defined as the minimum distinguishable time interval between two pulses. The dead time for most systems is longer than the pulse width from the electron multiplier. It results from pulse broadening in the amplifier, the limited rise and fall times of the discriminator, and the maximum clock rate of the counter. The dead time for a good counting system is about 10 ns. If more than one pulse occurs in any 10-ns interval, **pulse pile-up** has occurred, causing a count error. The probability $P_n$ that $n$ pulses will occur within the resolving time $t_d$ of a pulse counter is given by the Poisson distribution as

$$P_n = \frac{(\overline{R}t_d)^n}{n!} \exp(-\overline{R}t_d)$$

where $\overline{R}$ is the average pulse rate. If no pulse or one pulse arrives during $t_d$, no pulse overlap occurs, but if two or more arrive during $t_d$, count loss occurs. The probability $P_l$ that overlap will occur is the total probability (unity) less each of the other probabilities.

$$P_l = 1 - P_0 - P_1 = 1 - \exp(-\overline{R}t_d) - \overline{R}t_d \exp(-\overline{R}t_d)$$

Figure 14-31 shows the error due to pulse pile-up as a function of frequency for two values of $t_d$. Note that a counting system with a $t_d$ of 10 ns has a count error of 0.1% at an average count frequency of 4.5 MHz. This same counter could count regularly spaced events up to 100 MHz without error! The average current at a pulse rate of 900 kHz, where the pulse counting error begins to be significant for a 50-ns dead time, is between 0.014 and 14 $\mu$A, depending on the multiplier gain. Since this current level is easily measured with a simple op amp circuit, the widest range random event rate measuring system would combine a pulse counter for low rates and a simple analog circuit for average currents above $10^{-8}$ A.

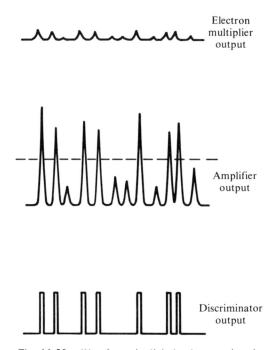

**Fig. 14-30.** Waveforms in digital pulse counting circuit. The amplified signal from the electron multiplier shows low-amplitude noise pulses that arise from electron emissions unrelated to the measured events that start part way down the dynode chain. The heights of pulses from true events are unequal because of variations in multiplier gain from pulse to pulse. When the discriminator level is appropriately set, multiplier noise and gain variations do not affect the pulse counting measurement.

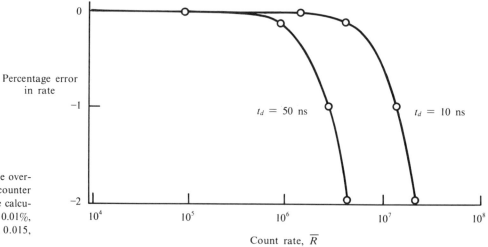

**Fig. 14-31.** Pulse counting error due to pulse overlap. $\overline{R}$ is the average count rate, and $t_d$ is the counter dead time. Errors for other values of $t_d$ can be calculated from the values of $\overline{R}t_d$ for 2%, 1%, 0.1%, 0.01%, and 0.001% error; these are 0.22, 0.15, 0.045, 0.015, and 0.005, respectively.

Commercial pulse counting systems sometimes offer several additional operating modes. One is to measure rate by determining the time required to achieve a predetermined count value. This mode maintains a constant standard deviation for each measurement. If the source of the random events is chopped, a kind of lock-in detection can be used by controlling the counting mode with the chopper drive. Counts obtained during a "source-OFF" state can be subtracted from the counts obtained during an equal duration ON state. An up-down counter is convenient for this application. If this cycle is repeated many times without clearing the counter, the noise in the ON and OFF rates averages out, and the difference between the ON and OFF counts accumulates.

## Rates of Change

The rate of change of a signal is an important quantity in many measurement and control systems. Some applications of rate measurement systems (differentiators) are illustrated in figure 14-32. The derivative signal in figure 14-32a provides a sharp transition from positive to negative voltage when the input slope reverses. This can be a very useful control signal as described in the next section. In figure 14-32b, the derivative is used to minimize the effects of a variable background signal. The second derivative shows very little change except when the information signal is present. In figure 14-32c, derivatives are used to locate the inflection point of a sigmoid curve. Either the first or the second derivative can be used as a control signal to cause some action to occur exactly at the time of the inflection. Rate measurements are also widely used in rate studies of chemical reactions and in velocity and acceleration measurements in physics.

Unfortunately rate measurements are often imprecise because of noise. The large rates of change common to noise pulses can mask the rate information about the signal. Because of the substantial influence of noise on differentiator outputs, the number of proposed rate measurement techniques is only slightly lower than the number of investigators who use rate-of-change information.

The simple op amp differentiator presented in chapter 5 is usually unsatisfactory unless its frequency response is highly degraded by putting a capacitor in parallel with the feedback resistor and adding an input resistor. This makes the differentiator an integrator (low pass filter) at high frequencies. One very good technique for generating reliable rate-of-change signals is based on integration of two successive segments of the input signal and subtraction of the resulting areas. The integration and subtraction operation can be implemented with an op amp integrator and a switchable inverter or with a voltage-to-frequency converter and an up-down counter. Subtraction of successive dual-slope conversions in a computer-based system is also used for slow rates of change.

In computer-based systems derivatives are often obtained with differentiating filters. These can be applied in the time domain or in the frequency domain. A variety of multi-point derivative smoothing functions have been proposed for these operations. All are similar to the frequency-domain differentiation illustrated in figure 14-33. The derivative theorem of Fourier transformation states that multiplication of the Fourier transform of a signal by a linear ramp is equivalent to taking its first derivative. The ramp shown in figure 14-33b turns around and rolls off at high frequencies to suppress noise. The frequency response of this differentiating filter is exactly analogous to that of an analog differentiator with high-frequency roll-off. An excellent approach for obtaining initial rates of change in computerized chemical rate measurements is to use a differentiating filter to locate the time window where a constant slope occurs. Then the raw data in that window are fit by a linear least-squares procedure, and the slope of the resulting straight line is obtained.

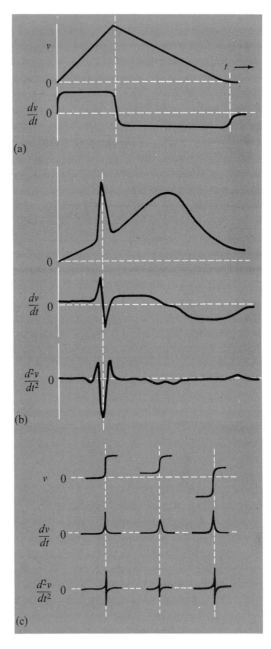

**Fig. 14-32.** Applications of derivatives. In (a) the derivative is proportional to the positive or negative slope of the input. In (b) the first or second derivative is used to locate a peak on a sloping baseline or to enhance resolution. In (c) the first and second derivatives of sigmoid curves are seen to be independent of the dc level.

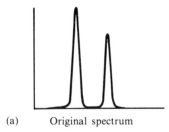
(a)    Original spectrum

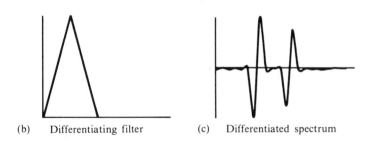
(b)    Differentiating filter          (c)    Differentiated spectrum

**Fig. 14-33.** Application of a derivative filter. The Fourier transform of the spectrum (a) is multiplied by the differentiating filter function (see text) (b) and the inverse transform taken. The resulting differentiated spectrum is shown in (c). This is analogous to a derivative smoothing operation in the time domain.

**Fig. 14-34.** Block diagram of generalized control system. In response to a control setting or input signal, the controller actuates a control element such as a motor, valve, or switch so that the desired action occurs in the controlled process, which might be heating, illumination, rotation, etc. The result of this action can be monitored by sensing the process variable $PV$ and producing a signal related to the controlled quantity. In a closed-loop system, this signal is compared with a set point signal (the signal for the desired value of $PV$), and the controller acts to reduce the difference between these signals to zero. Signal conditioning is generally used to make the sensor output linear so that the conditioned signal is proportional to the process variable.

## 14-5    Control

We have seen that the microcomputer is basically a device for performing a programmable sequence of operations. The ability to vary the program, the availability of conditional branch and loop sequences, and the speed of execution of the programmed operations make the microcomputer an extremely versatile and powerful general-purpose sequencer. This sequencing capability can be applied to the control of external devices and processes through interfaces that effectively make such controls part of the CPUs instruction set. In this way the computer can alter signal pathways, open or close valves, control power to motors, heaters, and lamps, generate test signals, and perform other tasks at times and in sequences determined by the operating program.

·The various control functions can be either open-loop or closed-loop as illustrated in figure 14-34. In **open-loop control**, the setting of the controller results in the desired change in the process variable. For example, in the control of a cooling fan in a piece of equipment, the control element is a switch that is actuated by the controller. The manipulated quantity is the power to the fan, and the process variable is the movement of air through the

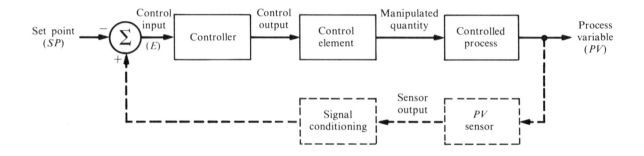

equipment or perhaps the temperature of some part of the equipment. It is assumed that when actuated, the switch closes, the fan is powered, and the air moves at a rate that keeps the equipment at a safe temperature. In **closed-loop control** the process variable $PV$ is monitored by a sensor. The value of $PV$ is then compared with the desired value $SP$ to produce an error signal $E$. The controller responds to the error signal by changing the manipulated quantity to bring $PV$ closer to $SP$. A home furnace is a closed-loop system. The thermostat compares the room air temperature with the set value and closes a switch when the temperature is below the set value. The controller responds to the closed switch by activating the burner to perform the process of heating the air. The heating process continues until the set point temperature is reached.

In this section the application of microprocessors in control systems is explored. The discussion includes interfaces for control, control system dynamics and modes, and examples of practical control systems.

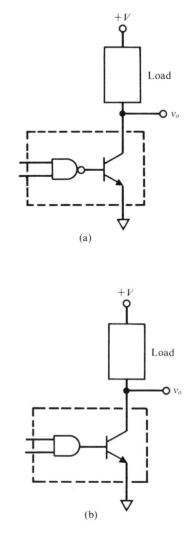

(a)

## Digital Drivers

In order for a microcomputer to function as the controller in an automated system, it must be interfaced to the required control elements. This is generally accomplished with a parallel output port. Since each bit in the latch of an eight-bit parallel port is independently controllable by the word written into it, eight different devices could be turned on and off by that one port. The interface to the device is completed by a **digital driver** circuit that can control power to the device in response to the logic level at its input. One of the simplest digital drivers is the open-collector gate shown in figure 14-35. The gate output is the collector of a transistor driven by the gate logic. The output is thus a logic-controlled switch for positive current to common. The voltage and current in the switched circuit are limited by the characteristics of the output transistor. The limits for some representative TTL devices are given in table 14-1.

Many low-voltage dc devices can be controlled by the open-collector buffer and driver just by connecting them in the load position shown in figure 14-35. Such loads include small incandescent lamps and LEDs (with appropriate current-limiting resistor in series). If the device is inductive, such as a relay coil, solenoid, or motor, the circuit of figure 14-36 should be used. The supply voltage $V$ is chosen to match the requirements of the load, with 0.2 to 0.7 V allowed for the drop across the on transistor. Normal relays and solenoids require that the current in the coil remain on as long as the armature is to remain in the ON position. This uses power during the ON time and the device automatically reverts to OFF when the system power is turned off. A **latching relay** or **latching solenoid** remains in its ON position once put there by a momentary current. To return it to the OFF state, a

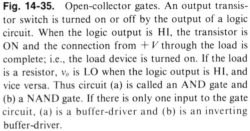

(b)

**Fig. 14-35.** Open-collector gates. An output transistor switch is turned on or off by the output of a logic circuit. When the logic output is HI, the transistor is ON and the connection from $+V$ through the load is complete; i.e., the load device is turned on. If the load is a resistor, $v_o$ is LO when the logic output is HI, and vice versa. Thus circuit (a) is called an AND gate and (b) a NAND gate. If there is only one input to the gate circuit, (a) is a buffer-driver and (b) is an inverting buffer-driver.

**Table 14-1.** Output capabilities of open-collector gates.

| Type | Function (B/D = buffer-driver) | +V(max), V | $I_L$(max), mA |
|------|-------------------------------|-----------|---------------|
| 7401, 03 | Quad NAND | 5.5 | 16 |
| 7405 | Hex inverter | 5.5 | 16 |
| 7406/7416 | Hex inverter B/D | 30/15 | 40 |
| 7407/7417 | Hex B/D | 30/15 | 40 |
| 7409 | Quad AND | 5.5 | 16 |
| MC 75461 | Dual AND B/D | 30 | 300 |
| MC 75462 | Dual NAND B/D | 30 | 300 |

momentary current is applied to an opposing coil, or current is removed from a holding coil.

The relay shown in figure 14-36 is an example of a control device that provides electrical isolation between the controller and the control element. There need be no direct connection between the digital driver and the circuit controlled by the relay contacts. This can protect the controller from accidental damage due to high voltages or power in the controlled circuits. The maximum controlled circuit power is limited only by the relay contact ratings. A device that provides isolation between two logic-level systems is the **opto-isolator**. An input gate is used to control an LED that is optically coupled to a phototransistor as in figure 6-17c. The phototransistor controls the input state of a separately powered output gate. Opto-isolators are very useful when digital information must be exchanged between two systems where the commons may not be at the same voltage. Circuits using ac line voltage can be controlled by a relay or a logic-activated zero-crossing switch. This device, described in figure 7-25, is often combined with an opto-isolator input to provide complete electrical separation between the logic and power-line circuits.

The above drivers all provide simple two-level, or ON-OFF control. Additional levels of control can be achieved by using more bits of the output port and some means of converting the digital output word into a multi-level control signal. For control in the analog domain, a DAC is generally used. A DAC is not designed to provide significant power at its output.

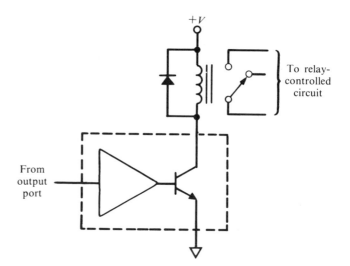

**Fig. 14-36.** Driving an inductive load with an open-collector output. The diode protects the transistor from the voltage spike produced when the current through the coil is turned off.

Therefore a DAC output that is to control a load such as a dc motor or heater requires a power amplifier. A simple power amplifier can be made from an op amp and booster as shown in figure 14-37. Incorporating the booster amplifier in the follower-with-gain feedback loop as shown maintains precise control over the booster output voltage. The booster amplifiers available from op amp manufacturers can provide many watts of output power to a load. For still greater power the DAC output can be used to control a voltage-programmable power supply (see section 7-8).

Another kind of driver that can provide a variable control output is a pulsed drive output. An ON-OFF driver is used to provide an output drive pulse that adds a single increment to the manipulated quantity. The total for the manipulated quantity or the rate of change of that quantity is equal to the number or rate of incrementing pulses applied. A primary example of pulsed control is the **stepper motor**. The shaft of a stepper motor rotates an exact fraction of a turn for each pulse applied to the motor windings. This is actually accomplished by a multipoled permanent magnet rotor surrounded by fixed electromagnetic poles driven in a four-step sequence. Between steps the rotor is "locked" in place by electromagnetic fields that attract the rotor poles. Upon the application of a step, electromagnetic fields one-fourth of a rotor pole position away are energized, and the rotor is attracted in that direction. On the next step, the rotor moves another quarter of a pole position, and so on. A motor with a 12-pole rotor completes one revolution in exactly 48 steps. The coil energizing sequence in a four-coil (unipolar)

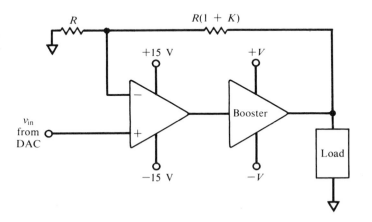

**Fig. 14-37.** Power amplifier for driving a load from a DAC. The booster amplifier acts as a higher current (and sometimes higher voltage) output stage for the op amp. Because it is included in the gain-controlling feedback loop, the circuit acts as a high output voltage follower with a gain of $K$.

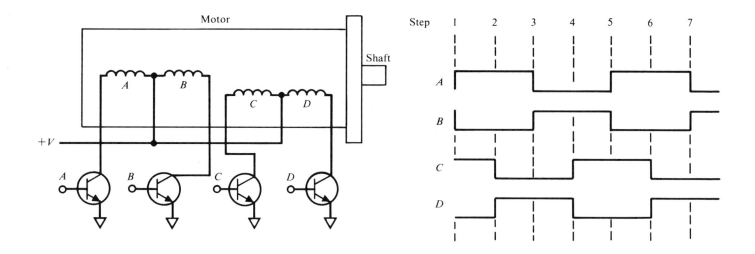

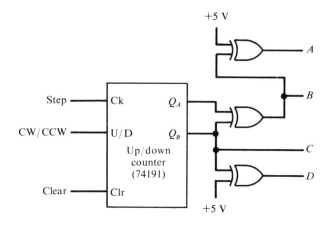

**Fig. 14-38.** Stepper motor coils and drive sequence. When coils *A* and *C* are energized, they attract a N pole of the permanent magnet rotor. Coils *B* and *D* attract the S poles since their current is in the opposite direction. As the coils are sequentially energized, the rotor turns in steps. If the pulse sequence is reversed, the rotor turns the other direction.

motor is illustrated in figure 14-38. The pulse sequence can be generated by software using four bits of an output port or by clocking a counter and logic circuit to produce the desired sequence as shown in figure 14-39.

Incremental or pulsed drivers can be used to generate increments of charge, heat, light, or many other quantities, and the controller can act to adjust the increment rate to obtain the desired value of the process variable (voltage, temperature, illumination, etc.). Pulsed control can be economical and precise (cf. the switching regulator power supply of fig. 3-25), and under microcomputer control it has the additional advantages of being simple and inherently digital.

**Fig. 14-39.** Stepper motor drive sequence generator. The first two stages of a binary up-down counter are used to provide a bidirectional four-step sequence. Exclusive-OR gates decode each of the four states into the appropriate combinations of drive signals as shown in figure 14-38.

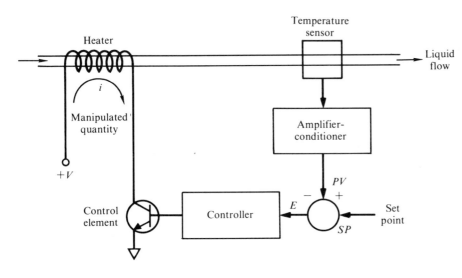

**Fig. 14-40.** Temperature controller for flow system. The current in the heater is controlled to produce the set point temperature in the flowing liquid. Response delays in the heater and sensor and dead time in the flow system between the heater and sensor contribute to the dynamics of the control system. The ideal controller would compensate for undesirable dynamic characteristics of other components to produce a stable, responsive, and accurate control system.

## Control System Dynamics

A typical control system is shown in figure 14-40. Every element in the control loop affects the system response to changes in some variables (flow rate, set point, or input temperature) and thus contributes to the system dynamics. There is **lag** in the heater since the temperature of the heating element and pipe cannot change instantly with changes in heater current. Then there is a delay called **dead time** between the heating of a portion of the liquid and the transport of that portion to the sensor. The sensor, too, has lag, since its temperature cannot instantly follow the changes in liquid temperature. The controller can be designed to respond in any of several ways to an error input. The current applied may be fully on, proportional to $E$, proportional to the rate of change of $E$, proportional to the accumulated error, or it may respond in some other way. The controller's response to the error signal is called its **control law** or **control algorithm**. The choice of control law affects the system dynamics; the controller can either compensate for or exaggerate the effects of the dynamic characteristics of the other elements.

The simplest control algorithm is **ON-OFF control**. In this the manipulated quantity is simply on or off depending on whether $PV$ is greater or less than $SP$. The dynamics of ON-OFF control are illustrated in figure 14-41. The lag and dead time in the system result in overshooting the set point from both directions. Cycling and overshoot are characteristic of ON-OFF control. If the controller has fast response and negligible ON-OFF gap, the amplitude and frequency of the $PV$ variation are not affected by the controller but by all other lags and delays in the system.

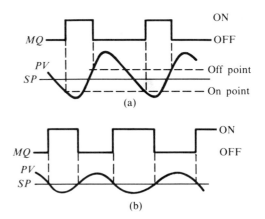

**Fig. 14-41.** ON-OFF control dynamics. In (a) the turn-on and turn-off values for $PV$ are separated. When $PV$ drops below the ON point, the controller turns on the manipulated quantity $MQ$. The decrease in $PV$ continues, however, due to lags and delays. The manipulated quantity is turned off when $PV$ exceeds the OFF point, and overshoot is again produced by the delay. Eliminating the gap between ON and OFF points as in (b) reduces the variation in $PV$ and increases the control cycle frequency.

**Fig. 14-42.** Response of proportional controller to step change in $E$. Two cases are illustrated; loop gain $K = 0.5$ (solid lines) and $K = 1$ (dashed lines). The system has a finite dead time, which is longer than the lags. The step change in $E$ affects the process immediately according to the gain. This change does not appear at $PV$ until after the dead time. An increase in $PV$ results in a decrease in $E$ and the process. This decreases $PV$ one lag time later and increases $E$ again. For $K < 1$, a steady value is reached with less than half the error corrected. For $K = 1$, a continuous oscillation results, and if $K > 1$, the amplitude of the oscillations increases.

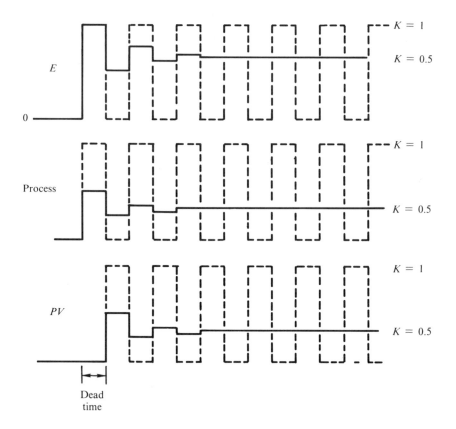

Dead time

**Note 14-10.   Derivation of Error for Proportional Controller.**

The error signal $E$ is the difference between $SP$ and $PV$:

$$E = SP - PV$$

Since $SP$ is constant, a change in $E$ results only from a change in $PV$:

$$\Delta E = -\Delta PV$$

A change in $PV$ results from the controller action $(K\Delta E)$, and the perturbation on the system $PV'$ in terms of the effect it would have on $PV$ is uncompensated.

$$\Delta PV = K\Delta E + PV'$$

Substituting to eliminate $\Delta E$,

$$\Delta PV = -K\Delta PV + PV'$$

The ratio of the change in $PV$ to the perturbation on $PV$ is then

$$\frac{\Delta PV}{PV'} = \frac{1}{1 + K}$$

A more sophisticated control algorithm is **proportional control**. In this strategy the manipulated quantity is set at a value proportional to the error magnitude $E$. Thus as $PV$ approaches $SP$, $E$ decreases, the drive quantity decreases, and $PV$ changes more slowly. This increasingly gradual approach to $SP$ can eliminate the overshoot and cycling problems of ON-OFF control. The dynamic response of such a control system depends upon the overall gain, dead time, and lag in the control loop. The **loop gain** $K$ is the change in $PV$ that results from a unit change in $E$ when the loop is opened by disconnecting $PV$ from the summing unit. (In fig. 14-40, $K < 1$ if a value of $E$ equivalent to a 10° temperature error results in a temperature change of less than 10°.) The response dynamics of a proportional control system are shown in figure 14-42 for $K \leq 1$. The perturbation in $E$ could have come from a change in $SP$ or from a change in the process conditions. As the dynamics show, a loop gain of less than one results in imperfect compensation while a loop gain of one or more produces oscillations with a period of twice the dead time. The fraction of the perturbation that is uncompensated is $1/(1 + K)$ which is 2/3 for $K = 0.5$, 1/2 for $K = 1$ and approaches zero as $K$ approaches infinity (see note 14-10). It is clear that the dead time in the system prevents the use of a large enough value of $K$ to provide accurate control. In some systems this is not observed because the lags in the process and detector responses are much longer than the dead time. For such systems the loop gain is very low at the frequency at which dead time oscillations would occur. Accurate control is achieved on a longer time scale by the much higher loop gain at low-frequencies.

A controller can be tailored to give close to optimum response for a system by adding averaging and rate terms to the control algorithm. The controller response function is then

$$MV = K_P E + K_I \int E dt + K_D \frac{dE}{dt}$$

where $MV$ is the manipulated variable. This control algorithm is called **PID controller** for the three terms (*proportional, integral, and derivative*) in its response function. The integral or averaging term is particularly useful in oscillation-prone systems where $K_P$ must be kept low. A correction term proportional to the integral of all past values of $E$ is applied to change $MV$. As long as an error exists, the integral term operates to reduce it. The integral term is good, therefore, for long-term accuracy. The derivative or rate term is designed to aid the response speed on the basis of giving the controller a stronger response to sudden changes in the error signal. The proportionality constants for the three terms must be carefully chosen in a given system to provide quick and accurate response to perturbations as well as good stability against oscillations.

One advantage of using a microprocessor for the controller is that once it is interfaced to the sensor and control element, any useful algorithm can be implemented or adjusted by simply changing the program. Even more complex algorithms that respond to combinations of sensors are very practical. Another advantange is that, given the time scale of many processes, one microprocessor can often manage several control loops. Subsystem control can be economically implemented so as to optimize sections of a process independently. For example, the mixing ratios of fuel components could be controlled by a loop separate from the one that controls the rate of consumption of mixed fuel. In turn, the fuel consumption might be only one of several controlled aspects of an overall process. Each subsystem can be optimized individually, and overall control is greatly simplified because each of the perturbing variables is separately controlled.

## Practical Control Systems

Modern electronic devices and techniques offer a variety of solutions to practical problems in measurement and control. The systems discussed in this section illustrate the kinds of options available and their relative merits.

**Temperature controller.**  Two approaches to a computer-controlled temperature regulator are shown in figure 14-43. In the first, the computer is not part of the control loop but provides the set point value. This controller

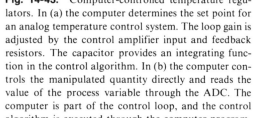

**Fig. 14-43.** Computer-controlled temperature regulators. In (a) the computer determines the set point for an analog temperature control system. The loop gain is adjusted by the control amplifier input and feedback resistors. The capacitor provides an integrating function in the control algorithm. In (b) the computer controls the manipulated quantity directly and reads the value of the process variable through the ADC. The computer is part of the control loop, and the control algorithm is executed through the computer program.

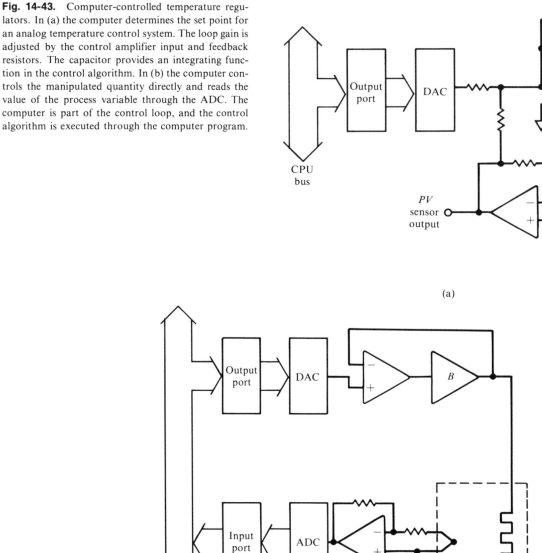

assumes that the *SP* command is correctly executed by the analog controller. The *PV* sensor output could be interfaced through an ADC so the computer could check for reliable operation. Alternatively, a comparator could be used to check the difference between the DAC and *PV* sensor outputs and raise a flag in the I/O port if the difference is excessive. In any case, the computer's attention is only needed to vary *SP* or to respond to control failure. The control system of figure 14-43b depends on the computer to maintain the correct value for the heater voltage. This it does by reading the value for *PV* through the interfaced ADC and calculating the appropriate response. The advantages of including the computer in the control loop are: the control algorithm is readily changed without altering the system hardware, continuous information about the process dynamics is available to the computer, and the computer detection of control error is inherent in the system. The disadvantage is that the computer must execute the control loop subroutine often enough to maintain adequate control.

These two approaches to temperature control illustrate the **hardware-software trade-off** in interface and control system design. Operations performed in the interface hardware increase the hardware complexity, but they relieve the computer of the need to provide frequent or extended attention to the interfaced task.

**DC motor controller.** A dc motor controller with position sensing is shown in figure 14-44. An encoded wheel is attached to the motor shaft, and its clear and opaque sections are sensed by opto-interrupters as in figure 4-33. In this case two interrupters, separated by half a segment (or 1½, 2½, etc.), provide both speed and direction information. An edge-triggered *D* flip-flop decodes the direction information as shown and an up-down counter keeps track of the net rotations from the reference point when the counter

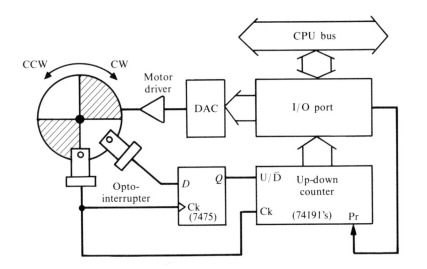

**Fig. 14-44.** Dc motor controller with position encoder. The motor speed and direction are controlled by the CPU through the DAC. Clockwise (CW) rotations cause the counter to increment, and counter clockwise rotations (CCW) reduce the count. For clockwise rotation, the LO→HI (dark→light) trigger at the *D* FF occurs when *D* is HI, so *Q* is HI (count up). For counterclockwise rotation, *D* is LO (dark) when the LO→HI trigger occurs, so *Q* is low and the counter reverses. The encoding wheel shown produces two counts per rotation. The position of whatever the motor is driving can be read from the counter contents within half a revolution of the motor shaft.

was reset. The resolution of the controller (counts per revolution) can be increased by using a wheel with more sectors. If a binary counter is used, the count is in 2's complement notation. With the interface shown, the CPU can control the motor to turn its load to any desired position as determined by reading the counter. The control algorithm can include slowing the motor as the counter approaches the desired value, approaching the desired value from the same direction to eliminate gear backlash effects, automatic resetting upon encountering an absolute load position index (a separate, single-point detector on the motor-driven load), or other actions. The counting function could be done by software, in which case the input interface would be one bit for the $U/\overline{D}$ indicator and a flag for the count (Ck) signal. If the motor control interface were an ON-OFF pulse-width or pulse-rate controller, the interface would be extremely simple.

Another example of dc motor control is the motor-speed controller in figure 14-45. The control loop is identical to the phase-locked loop controller of figure 9-24, except that the voltage-controlled oscillator is replaced by a

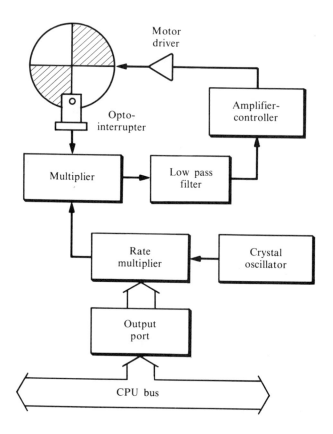

**Fig. 14-45.** Dc motor-speed controller. The output frequency from the opto-interrupter is compared with a CPU-controlled fraction of the frequency of a crystal oscillator. The result of that comparison determines the control signal to the motor driver. The control loop is of the phase-lock loop type and involves both analog and frequency domain signals.

motor and opto-interrupter. The control loop signal is in the frequency domain at the sensor output and is converted to an analog signal by the multiplier phase comparator. The computer can provide the speed set point as shown, or it could take over the multiplier, filter, and controller functions in the control loop. In a completely computer-controlled system, the computer real-time clock would provide the frequency reference, and the only interface would be the motor drive control and a flag input from the opto-interrupter. The choice between software and hardware control would be based on the need for a simple interface or a sophisticated control algorithm and the load on the CPU from other tasks.

The recognition that motor speed and frequency are both rates and therefore analogous quantities allows the techniques of frequency control to be applied to the control of motor speed. In solving a new design problem, the identification of an already solved analogous problem can provide a direct solution.

**Light-intensity controller.**    The control of light intensity is used in figure 14-46 to illustrate the desirability of deriving the feedback control signal from the quantity that is actually to be controlled. Controlling the voltage across a lamp certainly produces a more constant illumination than operating the lamp from an uncontrolled supply. However, this approach assumes that the power input is constant (the lamp resistance is constant) and the relationship of power input to light output is constant (that is, there are no aging effects in the lamp). Sampling of the light intensity with a photodiode as in figure 14-46b allows a control loop to keep that quantity constant at the set point value. Such a feedback light control is often used in spectrophotometry, and long-term stability is improved. In such systems several of the operations are performed on the light before it impinges on the sample. In the illustrative system of figure 14-46c, the light is focused on a monochromator entrance slit, a particular portion of the spectrum is selected, and the light is passed through a cell containing the sample material in a solvent. Even with constant illumination, the light intensity at the sample varies with movement of the filament, changes in the monochromator wavelength setting, and solvent. Measuring the light intensity just before it illuminates the sample allows the controller to compensate for all but the solvent effect. In other words, all effects within the control loop are compensated. The effect of the solvent is equalized in the feedback beam by a solvent-only reference cell. The constant beam intensity greatly aids this system in accurately determining the sample absorbance $A_s$, the log of the ratio of the sample and reference intensities.

**Dedicated microprocessor controllers.**    In the above discussion, the advantages of using the CPU as the controller in feedback control systems are weighed against the task burden this adds to the CPU. The development

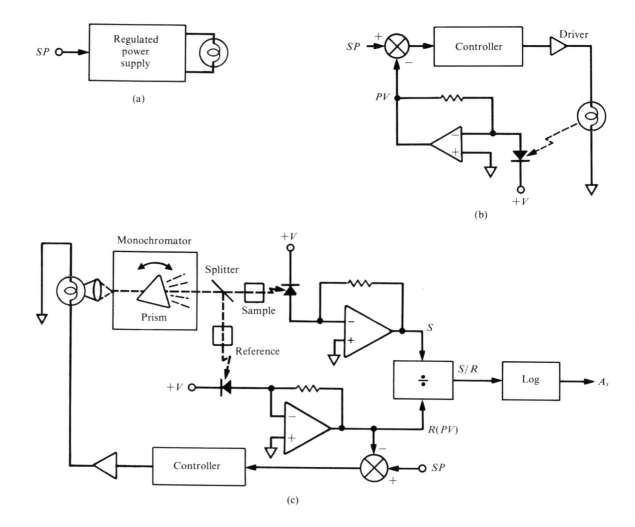

**Fig. 14-46.** Light intensity control. The regulated power supply (a) controls the voltage (or current) to the lamp. Adding a light detector (b) controls the actual light output rather than the power input to the lamp. In a spectrophotometer (c), many other factors affect that portion of the light used in the measurement. To keep that portion constant, the feedback quantity should be sampled as close to the point of use as possible.

of inexpensive microprocessors and memory has made it economical, in many instances, to design a hardware controller using a dedicated microprocessor. In other words, the main microcomputer is interfaced to a controller that incorporates another microprocessor to achieve the control function. Thus the hardware simplicity and software flexibility are obtained without burdening the main CPU with the control task. Some microprocessors such as the Intel 8048 have been designed for general-purpose peripheral control; others have been tailored to specific tasks for which there is a substantial market.

LSI controllers that include microprocessors are available for control of CRT displays, floppy disc drives, and stepper motors. The control algorithms are often in ROM, which is part of the LSI controller IC. Except for the device handlers, these "smart" control chips are all that is needed to interface the main CPU bus with the controlled device.

The development of microprocessor control systems for applications in home heating, cooling, cooking, and lighting systems to optimize efficiency without loss of effectiveness could affect great savings in our energy resources. The improvements resulting from sophisticated, computer-controlled systems for automobile engines are already widely recognized. Industrial processes could similarly be improved through the adoption of these increasingly inexpensive and effective control techniques. Because of the now widespread awareness of the values of conservation, dramatic advances in control system applications are to be expected in the near future.

## 14-6   Self-Optimizing Systems

Instruments, machines, and peripheral devices are increasingly referred to as being "smart" or having local intelligence. Usually this means that there is a CPU controlling the device operation, but it can also mean that the instrument is adaptive—that is, it does not continue to follow blindly a routine that is inappropriate for the circumstances. One of the most exciting areas of modern electronic technology is in the development of intelligent systems. Electronic **adaptive control systems** provide a level of automation beyond mere control. Through the speed and precision of modern electronics, adaptive systems can greatly extend the capabilities of virtually any operation or measurement. In this section, examples of four types of "optimizing" control systems are described: an input amplifier that automatically adjusts to the best amplification, some techniques that compensate for imperfect system components, a system that alters its measurement strategy in response to changing signal conditions, and a system that searches for the particular combination of variables that best achieves a given objective.

### Autoranging

One relatively simple operation that a smart controller can accomplish is to keep a measuring instrument in range. In a voltage measurement system, for example, a programmable gain amplifier is often used between the signal input and the ADC. When the ADC indicates an overrange, the amplifier gain is reduced. A digital comparator monitors the ADC output. If the conversion falls below the comparison value, the amplifier gain is increased. Note that the resulting gain setting is part of the conversion information. In a digital meter the autoranging circuit keeps the decimal point in the display in the correct place. In a computer system the gain control lines must be interfaced to an input port to complete the information about the signal amplitude. A computer-controlled autoranging system is shown in figure 14-47. The gains of the three integrating amplifiers provide a 64-fold range control, and the control of the integration time can extend that range by orders of magnitude. With a twelve-bit converter, the steps of $2^3$ in amplifier gain ensure that nine bits of the ADC are used to give a minimum resolution of one part in $2^9 = 1:512$.

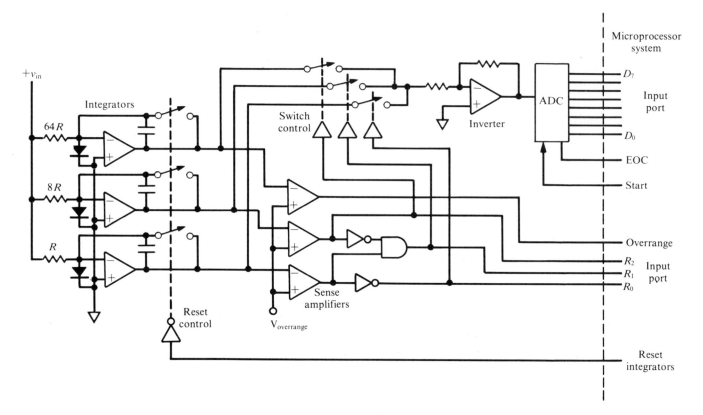

**Fig. 14-47.** Autoranging integrating amplifier for a microcomputer-based data acquisition system. The sense amplifiers detect integrator overrange and, through the logic, connect the most appropriate integrator to the ADC input. The computer triggers the conversion at the end of the integration time. The three simultaneous integrations with different gain greatly reduce the possibility of a lost data point due to overranging.

## Autocalibration

A smart controller can be used to test for and compensate for drift in device characteristics. For example, a principal limitation in high-resolution ADCs and DACs is the problem of keeping the drift in offset and gain to less than the value of the LSB. A controller can be used to test for drift in these parameters and make necessary adjustments. A DAC designed on this principle is shown in figure 14-48. Note that low-resolution control is sufficient to calibrate a high-resolution system because only a few of the least significant bits are affected by the drift.

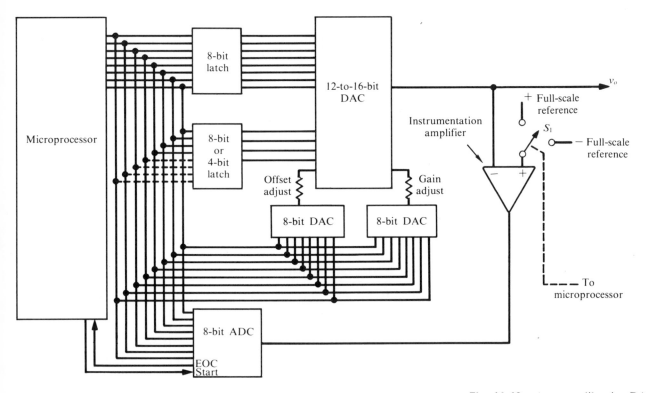

**Fig. 14-48.** An autocalibrating DAC. The microprocessor periodically compares the DAC output extremes with positive and negative reference values. The error is read from the eight-bit ADC output and used by the computer to determine the needed change in gain or offset. These are applied by altering the values in the low-resolution gain or offset DACs.

## Autocompensation

Most traditional measurement systems attempt to minimize the change in any conditions that might cause an error in the measured value. Even so, the dominant source of error in many measurements is the variation in quantities that affect the measurement system characteristics. For example, if temperature is a factor in a strain gauge or a pH measurement, its effects could be compensated by measuring the temperature accurately, experimentally determining the effect of temperature on the measured quantity, and applying an appropriate compensating correction to each measured result. A rough temperature control could be used to limit the degree of compensation needed.

Compensation can also be used for known errors in generated signals. For example, a charge pulse generator can be made by gating a controlled current pulse for a controlled time. The charge, which is the current-time product, can readily be made accurate to within 1% for various combinations of current and time. However, to obtain 0.1% accuracy of charge generation is much more difficult. In one example, the error in the charge was quite constant for each combination of current and time. This is an example of **fixed pattern noise**. It is common in array detectors and other multiple element devices. The computer controller was then used to determine a table of errors for each $i \times t$ combination by measuring the charge delivered to a known load. The computer used this table to correct the charge error, and a better than tenfold improvement in accuracy was obtained without any change in the hardware.

In another system, autoranging was used to maintain the maximum resolution from an ADC. However, the large dynamic range required large differences in the values of the gain resistors, and constant adjustment was required to eliminate apparent "steps" in the measured value that occurred at some range changes. The problem was cured by software that measured the step caused by each range change and corrected all subsequent values by that amount.

## Optimization of Strategy

As shown in the previous section, the best choice of control algorithm depends greatly on the characteristics of the system being controlled. Thus, if conditions in the system change considerably, the algorithm may not only be suboptimal; it may produce an unstable system. There are two solutions to such a problem. One is to make a controller with sufficiently low gain and response time that it is stable under all expected conditions (and optimum under none). The other is to analyze the system's response to changes in the driven variable and to adjust the control algorithm as conditions change in the system. This approach is widely used in the flight control systems of commercial airliners.

Adaptive control is a useful concept in measurement as well. The optimum measurement strategy often changes with signal conditions. An example is in instruments that measure the spectrum of the light emitted from a sample or a star. If the light source is weak, photon counting might be used. Waiting at each setting of the grating or prism until a set number of counts have accumulated would give a constant standard deviation for each value. However, if the source does not emit light at some wavelengths in the spectrum, the integration time of the system is longest at the wavelengths of no interest. An adaptive strategy could be used to determine first whether there is enough light to be of interest and then either to wait for the desired number of counts or to move to the next part of the spectrum. If the counting (integration) is to continue for a long time, the count rate may drift during this period because of slow variations in the system (background counts, amplifier gain, and so on). In such cases an adaptive measurement strategy could anticipate the need for long integration times based on low signal levels and switch to a synchronous (lock-in) detection mode to eliminate possible drifts. The implementation of adaptable strategies allows the operator to choose between speed and accuracy in a given measurement and then to have the chosen goal implemented in the most efficient way.

## Simplex Optimization

In most measurement and control systems, the final result obtained may be influenced by several experimental variables, and it is often desirable to find the set of conditions that yields the best results. If the variables do not interact, each factor can be investigated in turn to determine the optimum set of conditions. In general, however, experimental variables do interact with each other, and the single factor approach does not always readily yield the optimum. The **simplex optimization** method provides a systematic search for an optimum set of conditions even when variables are interactive. In the simplex approach several experimental variables are changed simultaneously until the optimum response is obtained. The method is well suited for computer-controlled systems because response measurements, calculations, and variable adjustments are required in real time.

The simplex is a geometric figure defined by a number of points equal to the number of factors (variables) plus one. A two-dimensional simplex is a triangle and a three dimensional simplex is a tetrahedron. In practice the simplex is moved across the response surface by a prescribed set of rules until it reaches the optimum response or undergoes failure.

A major problem in any optimization procedure is defining the goal of the optimization and choosing the variables to be optimized. In a spectroscopic experiment, for example, many different optimization goals can be defined. Some worthwhile goals are highest measurement precision (signal-to-noise), highest measurement accuracy, highest signal, best signal-to-background ratio, highest resolution, etc. Among the many variables influencing the measurement are monochromator slit width, photomultiplier voltage, current-to-voltage converter gain and bandwidth, and any variables that influence the light output. Obviously the choice of the optimization goal and the specific variables to be optimized greatly influences the set of conditions obtained.

Let us assume that a chemical reaction is being optimized for best yield. For simplicity, the variables considered are the temperature of the reaction mixture and the pH of the mixture. Let us also assume that a computer can control these variables and monitor the yield after each change in conditions. To illustrate the initial simplex and its moves toward the optimum, consider the contour map of figure 14-49a. Three evaluations of yield vs. temperature and pH define the initial simplex. The simplex moves by discarding the

**Fig. 14-49.** Two-dimensional simplex for optimization of product yield. The lines represent equal product yields in a chemical reaction as a function of temperature and pH. Points A, B and C in (a) represent the vertices of the initial simplex. The point of lowest response (point A) is discarded and reflected across the face of the remaining points generating a new simplex, BCD in (b). The lowest response of the BCD simplex (point C) is now discarded and reflected to give point E. This process continues until the optimum response has been found.

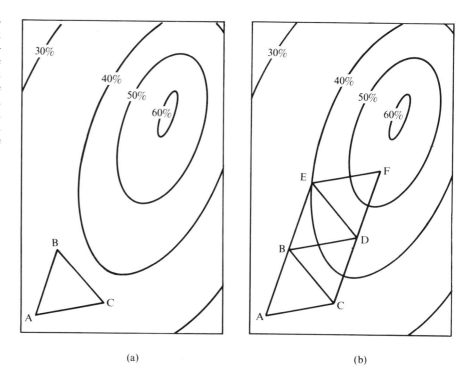

(a)                  (b)

vertex with the worst response and replacing it with a response obtained at the mirror image position across the face of the remaining points. Thus, in figure 14-49 point A is discarded and reflected across the BC face to generate a new set of variables for which the response is evaluated. This new vertex is indicated as point D in figure 14-49b. Points B, C and D now form a new simplex. If the new point in the second simplex (point D) had the worst response, it would not be discarded as in the initial simplex for this would only regenerate the ABC simplex. Instead the second worst response would be rejected, and the process continued. Boundaries can be assigned to the independent variables to limit the simplex to an appropriate region. Modifications to the simplex procedure allow expansion and contraction of the simplex which can accelerate movement and cause the simplex to adapt to particular response surfaces.

Simplex optimization is particularly well suited to computer-controlled experiments where several interactive variables are present. It has been applied to such problems as optimizing magnetic field homogeneity in nuclear magnetic response spectroscopy, optimizing signal intensity vs. spatial observation window in flame spectroscopy as well as optimizing yields in synthetic chemical procedures. As more and more instrumental systems become intelligent, simplex procedures are certain to become more commonplace.

## 14-7 Conclusions

From the many examples presented in this text, it is apparent that scientific and process control instrumentation and measurement techniques can be considerably improved by modern microelectronic concepts and technology. There have been dramatic improvements in the past few years, and we should witness even greater developments during the 1980s. It seems inevitable that nearly all scientific research and development laboratories will become highly automated during this decade, and consumer products will be influenced increasingly by the ubiquitous microprocessor.

Perhaps this will be the decade of super laboratory robots. If so, we would be wise to gain some insight into their potential impact on an already revolutionary scientific era. How "intelligent" and versatile will these robots be? Will they provide an economical work force? Will they work "intelligently" on both research and routine projects? How will they affect scientific and manpower requirements? These questions, of course, were not answered by the information in this text because of their speculative nature. However, we hope that your study of the concepts and techniques presented will spur your interest and enable you to envision the potential future impact of microelectronics on all of our lives. Only with vision will we know how best to proceed.

# Suggested Experiments

### 1.   Sampling and aliasing.
Connect a sample-and-hold circuit and a successive approximation ADC to a parallel input port of a microcomputer. Write a program to acquire data at a sampling rate of ~10 kHz. Connect the sine-wave output from a function generator to the sample-and-hold input. Frequencies less than 5 kHz should be sampled correctly. Run the data acquisition program with sine-wave inputs of 1 kHz, 2.5 kHz, 5 kHz, 7.5 kHz, and 9 kHz. Plot the sampled data sets, and determine whether aliasing has occurred.

### 2.   Signals and noise.
Connect a function generator to the data acquisition system of the previous experiments. Acquire a sine-wave signal at an appropriate sampling rate. Use a fast Fourier transform routine to obtain the amplitude spectrum of the signal. Repeat for other waveshapes. Connect a white noise generator to the data acquisition system. Limit the bandwidth of the noise to less than one-half the sampling rate. Obtain its amplitude spectrum. Connect the noise generator and a function generator to a summing amplifier to obtain a noisy sine-wave signal. Obtain the amplitude spectrum of the noisy signal.

### 3.   Analog and digital integration.
Connect the noise generator and a dc signal to a summing amplifier, and use the noisy signal as the input to the data acquisition system. Use a program that acquires and averages an operator-selected number of data points. Find the average values, the standard deviations, and the signal-to-noise ratios for summing 1, 4, 16, 32, 64 and 128 conversions. Connect a dual-slope ADC to a parallel input port. Measure the amplitude of a dc signal combined with a periodic "noise" source (the sine-wave output of the function generator). Set the sine-wave generator frequency so that the ADC integration time is an exact multiple of the noise period. Obtain the average and standard deviation of ten conversions. Repeat with the sine-wave generator frequency set away from the rejection node.

### 4.   Digital filtering.
Have the computer generate a Gaussian-shaped peak with random noise, and study the effect of smoothing parameters on noise reduction and signal distortion. Use a moving-average smooth and various polynomial smoothing functions. Develop a frequency-domain low pass digital filter with a sharp cutoff, and apply it to the Fourier transform of the Gaussian peak. Try different filter functions.

### 5.   Boxcar integrator.
Study a computer-controlled scanning boxcar integrator with a programmable timer to vary the delay time. Use the boxcar as a gated integrator to improve the S/N of the noisy repetitive pulsed signal. Then use the boxcar in its scanning mode to acquire a noisy sine-wave signal. Vary the number of points averaged and the scanning time.

### 6.   Multichannel averager.
Use a data acquisition system to acquire a transient signal repetitively for a program-controlled number of scans. Investigate the influence of quantizing noise by measurements on a ramp signal with a very small slope. Obtain results on a single ramp and on multiple repetitive ramps. Add a small amount of noise to the signal and repeat.

### 7.   Auto- and cross-correlation.
Cross correlate a noisy Gaussian peak-shaped signal with a noise-free Gaussian peak. Investigate the effect of parameters (peak height, half-width) on noise reduction and signal distortion. Auto-correlate several waveforms (square wave, triangular wave, sine wave) with and without added noise.

### 8.   Rates of random events.
Use a hardware or software random pulse rate generator as a counting source. Relate the standard deviation of the frequency measured to the gate time used and the number of counts produced. Compare the result with the expected relationship.

### 9.   Measuring rate of change.
Use a differentiator to measure the rate of change of a signal from a triangular wave generator or from a DAC that is driven from a regularly incremented counter. Compare the results of the analog rate measurement with a computer differentiation of the same waveforms.

### 10.   Open-collector driver.
Interface an open-collector driver IC to a microcomputer through an I/O port. Use the open-collector driver to control several types of loads such as an LED indicator, a relay, and a small dc motor.

### 11.   Temperature controller.
Attach a thermistor to a 500-Ω, 0.5-W resistor. Connect the resistor to the output of an op amp or booster with at least 20-mA output current capacity. Arrange a bridge circuit for the thermistor to obtain a voltage related to the temperature. Design and test an analog and a computer-based circuit to control the temperature of the resistor.

### 12.   Motor-speed controller.
Attach a slotted disc to the shaft of a small dc motor, and use an opto-interrupter to monitor the disc position. Arrange to control the motor through an output port, and read the opto-interrupter through an input port. Implement a program that will maintain a constant motor speed by adjusting the width of regularly spaced motor-drive pulses.

# Questions and Problems

**1.** A waveform is composed of 10-Hz, 30-Hz, and 50-Hz frequency components. The waveform is to be sampled at equal intervals. (a) What is the minimum sampling rate that will provide an accurate representation of the waveform? (b) Use a diagram to show what aliases (if any) would be generated for sampling rates of 25 Hz, 70 Hz, and 120 Hz.

**2.** The sampling operation produces a waveform that is pulse-amplitude modulated. If the sampling rate is appropriate, the original waveform can be recovered by passing the samples through a low pass filter. A band-limited dc signal has a maximum frequency content of 20 Hz. A simple $RC$ low pass filter after the sample-and-hold circuit is to be used to reconstruct the original waveform. Specify the appropriate sampling frequency, and give values of $R$ and $C$ such that the recovered signal suffers negligible distortion.

**3.** A female vocalist (with overtones to 5 kHz) sings into a microphone whose output is sampled and recorded on tape. (a) Determine the minimum sampling rate required, and show appropriate values for an $RC$ low pass filter used to reconstruct the signal. (b) How many samples would have to be taken for a 4-minute recording? (c) What would the vocalist sound like if half the samples were lost?

**4.** Challenge question: Several interesting properties of waveforms and their spectra can be illustrated by calculating the Fourier transform of a sinusoidal waveform of finite length. If a function is a real, even time function [$f(-t) = f(t)$], its transform is purely real (no imaginary component) and can be calculated from

$$F(f) = 2 \int_0^t f(t) \cos (2\pi ft)dt$$

where $F(f)$ is the amplitude spectrum of waveform $f(t)$. (a) Use this equation and standard integral tables to calculate the amplitude spectrum of a cosine wave $f(t) = \cos(2\pi f't)$. (b) Plot the spectrum for $f' = 100$ Hz and $t = 0.1$ and 1.0 s. (c) Plot the square of the amplitude spectrum (power spectrum) for $f' = 100$ Hz and $t = 0.1$ s. (d) What happens to the power spectrum when $t$ gets large? What happens when $t$ gets small?

**5.** Calculate the rms Johnson noise voltage for a 100-M$\Omega$ resistor at 300 K and a system bandwidth of 10 Hz.

**6.** The rms noise strength of a signal recorded by an oscilloscope or a strip chart recorder is usually estimated to be one-fifth the peak-to-peak fluctuation level of the signal. Consider the definition of the rms value and the standard deviation of a signal and explain why the above rule of thumb is statistically sound.

**7.** Measurements of a fluctuating dc voltage gave the following results:

| | | |
|---|---|---|
| 6.19 V | 5.66 | 5.82 |
| 5.90 | 5.81 | 5.91 |
| 5.68 | 5.83 | 6.05 |
| 5.92 | 5.99 | 5.68 |
| 5.36 | 6.00 | 5.69 |

(a) Using the normal definition of standard deviation, calculate the standard deviation, the relative standard deviation, and the S/N. (b) Plot the values, and use one-fifth the peak-to-peak fluctuation as the estimate of the standard deviation. Compare the S/N obtained with that in (a).

**8.** The current output of a photomultiplier tube had a noise component due solely to shot noise. (a) If the average current was $5 \times 10^{-9}$ A and the measurement system bandwidth was $\Delta f = 5$ Hz, calculate the rms shot noise current and the S/N. (b) A change in the light level striking the photomultiplier increased the average current to $5 \times 10^{-7}$ A. Recalculate the shot noise current and the S/N.

**9.** A forward biased diode operated with a constant current through it is sometimes used as a white noise generator. (a) Calculate the rms shot noise current in a diode operated at 5 A into the summing point of a current-to-voltage converter of 250 kHz bandwidth. (b) What feedback resistance would be needed for the rms noise voltage at the current follower output to be 500 mV? (c) What would be the Johnson noise voltage across the feedback resistor at 25°C?

**10.** A photodiode is operated into an op amp current follower with a feedback resistor of 1 M$\Omega$. At what photodiode current will the rms shot noise voltage equal the rms Johnson noise voltage if the temperature is 25°C?

**11.** For white noise the signal-to-noise ratio at the output of an analog integrator should improve as $t^{1/2}$, where $t$ is the integration time. For long integration times (greater than a few seconds), the S/N improvement with $t$ is usually not as great as expected. Discuss why this is true. Does digital integration suffer from the same limitations?

**12.** Ten measurements of a dc signal with a computer-based system gave a relative standard deviation of 5.5%. How many repetitive measurements of the same signal should be averaged to get 0.1% relative standard deviation?

**13.** Sketch the function that results from the convolution of the two functions in figure 14-50. Label the time and amplitude axes appropriately. Note that the resulting function is similar to the slit function in spectroscopy for unequal exit and entrance slits.

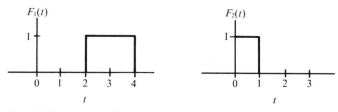

**Fig. 14-50.**   Problem 13.

**14.**   A linear ramp of 1-s duration, reaches relative amplitude 2 at the end of the 1-s sweep and resets to zero amplitude at this time. It is applied to an integrator whose impulse response is a unit height rectangular pulse of 2-s duration. Sketch the function that results from convolution of the ramp with the integrator impulse response function.

**15.**   The impulse response of an $RC$ filter is given in figure 14-10, and that of a linear integrator is given in problem 14 above. Another very useful time response is the step function response. Use the impulse response functions to sketch the step response of the $RC$ filter and the linear integrator. Show the appropriate graphical convolution.

**16.**   Consider the impulse response function of the linear integrator. What frequency response would a digital filter require to simulate integrator behavior? Sketch the response. How does the frequency response of a differentiator differ from that of an integrator?

**17.**   The following signal-to-noise enhancement techniques are available in a laboratory: lock-in amplification, boxcar integration, multichannel averaging. (a) For each technique describe how the S/N enhancement occurs. (b) Give the types of signals for which each technique is applicable.

**18.**   Which of the S/N enhancement techniques in problem 17 would be most suited for (a) measurement of the fluorescence induced by a 5-ns laser pulse; (b) measurement of the phase shift between a sinusoidal stimulation and the response; and (c) repetitive scan measurements of a weak nuclear magnetic resonance signal?

**19.**   Discuss why phase information is lost in auto-correlation techniques. Consider problem 4 in your answer.

**20.**   A fast oscilloscope is to be used to observe the current pulses from a photomultiplier tube. Assume that the electron multiplier has a gain of $10^6$ and produces a pulse 10 ns long. A short piece of coaxial cable connects the PMT anode and the oscilloscope input. The scope input is internally terminated with a 50-$\Omega$ resistor. Calculate the expected pulse amplitude (in mV).

**21.**   What is the standard deviation and the relative standard deviation of a random count measurement of 4258 counts? How many counts are required to achieve a relative standard deviation of 0.1%?

**22.**   What is the minimum error due to pulse overlap for a count rate of 200 kHz and a dead time of 20 ns? What is the maximum count rate if the error is to be below 0.1%?

**23.**   Describe the difference between and the sources and effects of lag and dead time in the manual adjustment of water temperature for (a) taking a shower and (b) taking a bath.

**24.**   Choose a control system with which you are familiar (such as riding a bicycle), and identify the functions and quantities in the generalized control system of figure 14-34.

**25.**   The input voltage of an op amp must never be allowed to exceed the range bounded by the amplifier power supply voltages. If the booster amplifier in figure 14-37 is powered by $+$ and $-$ 50 V supplies, what is the minimum gain the follower with gain can safely have?

**26.**   In the computer-controlled system of figure 14-43, a finite time is required for the computer to acquire the $PV$ value, compute the desired control response, and apply that through the DAC. Is this delay dead time or lag? Why? Explain the problem that exists if the delay through the computer is large compared to the lag in the heater and sensor. Describe a possible solution.

**27.**   Using sketches, describe a control system that maintains a constant level of light from a source by adjusting the power of the lamp to keep the pulse rate from a photomultiplier tube detector at the desired level. Comment on the effect of the count statistics in this control application.

# Grounding and Shielding

# Appendix A

The quality of electronic measurement and control systems often depends directly on the care taken by the designer and the user in minimizing unwanted noise and pickup. Interference noise problems are often difficult to assess, and their elimination remains something of an art; however, following the basic guidelines developed in this appendix when designing and interconnecting system components can solve a large fraction of all noise problems and eliminate much of the frustration and time involved in tracking down an interference problem that appears during use.

## Grounding

Voltage is not an absolute quantity but is the potential difference between two points. In order to establish and maintain reproducible and safe voltages in a circuit, a stable reference point from which all voltages are measured must be established. This single stable reference point is called the circuit common. When a circuit is linked to other circuits in a measurement system, the commons of the circuits are often connected together to provide the same common for the entire system. The circuit or system common may also be connected to the universal common, earth ground, by connection to a ground rod, water pipe, or power-line common (see notes 3-2 and 5-1). The term "ground" has come to be a general term for the system common whether it is connected to earth ground or not. Thus the verb "to ground" may mean either to connect to common or to earth ground. The more specific terms common and earth ground are preferable, but since they lack convenient verb forms, they are often not used. Compound terms such as ground loops and ground plane also apply to a system common as well as to an earth ground. Thus whenever the term ground is used, the context must be studied to determine which is meant.

**Safety grounds.** For safety reasons the chassis of electronic equipment must be grounded. If it is not, stray impedances between a voltage source and the chassis and between the chassis and ground can cause a fraction of the source voltage to appear on the chassis presenting a shock hazard. An even more dangerous situation could arise if the insulation between the ac power line and the chassis were to break down and the ac line were to

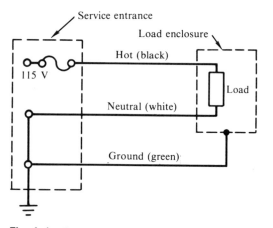

**Fig. A-1.** Standard three-wire U.S. 115-V ac power distribution. The hot wire (black) is fused, and any load current is returned through the neutral wire (white). Only the ground wire (green) carries current during a fault and then only until the fuse or a circuit breaker interrupts the circuit.

**Fig. A-2.** Ground loops. In (a) the signal common (1) is at a different voltage than the amplifier common (2). This gives rise to a ground loop and an erroneous signal. Ground loops can be eliminated by establishing a single common point as in (b).

contact the chassis. If the ac line can be fused, the chassis is at the line voltage and capable of supplying an amount of current limited only by the fuse. Grounding the chassis, however, causes the fuse to blow when a breakdown of the power-line insulation occurs, removing the danger. Safety grounds are always connected to earth ground; signal commons may or may not be earth grounded.

In the United States the National Electrical Code requires the three-wire 115-V ac power distribution system illustrated in figure A-1. Enclosures should be connected to the safety ground (green) since it carries no current and is always at ground potential. The neutral and safety ground are connected together only at the point where the ac power enters the building.

**Signal commons.**   When signal commons are considered, it is important to remember that no conductor is perfect—that is, all conductors have inductance and resistance—and that two physically separate commons or earth grounds are seldom, if ever, at the same voltage.

The commons of low-frequency analog circuits are usually best interconnected at a single point. Use of a single point common to eliminate ground loops is shown in figure A-2 for an op amp inverting amplifier. A ground loop is particularly troublesome if the two commons are unstable with respect to each other. It is important that the connections to the single common point have very low resistance and high current carrying capacity so that ohmic drops along the connections are minimized. Typically a large copper wire or foil is used. This is particularly important when several connections are made to a single common and when some of the connections are long, as they would be when the signal source must be remote from the measurement circuits. Even so, at radio frequencies the resistance is increased by the "skin effect," and inductive reactance can be very large.

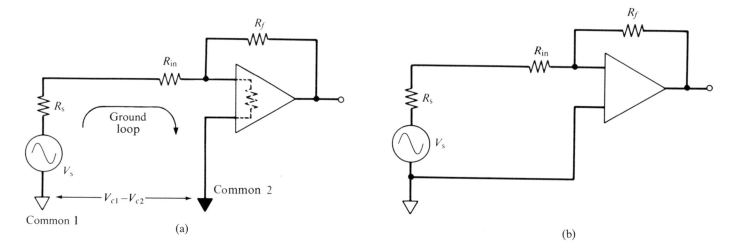

(a)                                                                    (b)

When several circuits share a single point common, a series or parallel connection can be made as illustrated in figure A-3. In both cases the resistance to common must be very low since ultimately a single conductor must carry the sum of all the currents from every component in the system. It may in fact become impractical to have a single common point because the current carrying capacity cannot be provided. In this case it may be safer to have several stable common points and tolerate some ground loops. This sort of compromise is often necessary in solving the grounding problems associated with large installations and buildings, such as a computer center, or when the circuitry is subjected to interference that may cause large currents, such as that from electrical storms.

In many laboratory measurement situations, it is not possible or practical to have a single common point, particularly if the signal source is remote from the measurement system. In these cases, it is advantageous to use a differential or instrumentation amplifier as discussed in section 8-1. Even though a potential difference exists between the signal common and the amplifier common, the erroneous signals generated by ground loops are common mode and are rejected by the difference amplifier (see figure 8-1). Therefore it is unnecessary for the two common voltages to be stable with respect to each other. The input impedance of the difference amplifier should be large compared to the source impedance in order to keep the lines identical and thus retain high common mode rejection.

Single point common systems suffer serious limitations at high frequencies (>1 MHz) because the inductive reactance of lengthy conductors to common can be large. Therefore, high-frequency circuits, including digital circuits, use a multipoint common system, as illustrated in figure A-4, to minimize the impedance. Such multipoint commons should not be used at low frequencies (<1 MHz) because the common currents from all circuits go through the same impedance of the ground plane and cross couple.

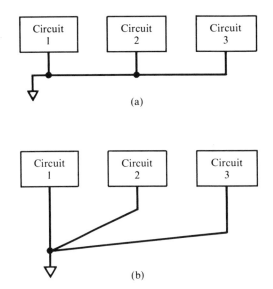

**Fig. A-3.** Single point commons. The series connection (a) is simpler from a wiring standpoint than the parallel connection (b). However, the parallel connection is less noisy because currents to common from the different circuits do not cross couple.

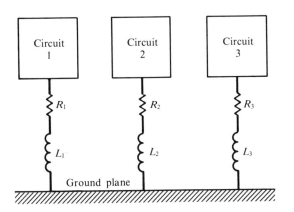

**Fig. A-4.** Multipoint common system for high frequencies. Circuits are connected to the closest low-impedance ground plane. Because a ground plane can be made with very low inductance, its impedance can be low compared to that of separate conductors connected to a single point. Each connection to the ground plane should be kept as short as possible (only a fraction of an inch in very high frequency circuits).

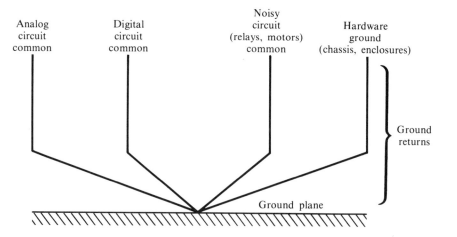

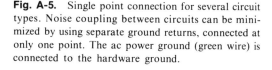

**Fig. A-5.** Single point connection for several circuit types. Noise coupling between circuits can be minimized by using separate ground returns, connected at only one point. The ac power ground (green wire) is connected to the hardware ground.

A high quality common is necessary on circuit boards that contain digital ICs because of their rapid switching speeds. Either a low-impedance ground bus or a ground plane covering a large percentage of the board (60% or more) is satisfactory. If power buses are used, inductances can be kept low by making the bus as wide as possible (0.1 in. or more). Unused gate inputs should be connected to common or through a series resistor to $V_{CC}$, whichever is appropriate.

Most complex systems require separate ground returns (connections to common) for circuits of widely different power and noise levels. Thus several low-level analog circuits can share a common ground return that is different from the digital ground return. Electromechanical components that tend to generate noise, like relays and motors, can share a ground return line separate from the signal grounds as illustrated in figure A-5. Connecting the separate ground returns at a single point greatly minimizes most low-frequency grounding problems.

**Power-supply decoupling.** Since the dc power supply and its distribution system are usually shared by a variety of circuits, noise generated by one circuit can couple to other circuits through the power system. To minimize this coupling, it is good practice to place a capacitor across the power bus at each circuit. Decoupling analog circuits is very important since any noise on the supply lines can add noise to the signal, influence amplifier gains, and possibly cause oscillations.

Digital circuits with totem-pole output stages are particularly strong sources of power supply voltage transients. When a totem-pole gate switches states, there is a short period of time in which both output transistors are on and a low impedance path exists between the supply and ground. This can result in power-supply current spikes as large as 100 mA each time the gate changes state. A high-speed decoupling capacitor, typically 0.01 to 0.1 $\mu$F, at the IC package acts as an extra current source and helps prevent supply voltage transients during switching. Decoupling capacitors (10–100 $\mu$F) should also be used on each printed circuit board at the point of power entrance to the board. For capacitor characteristics, see appendix B. Additional information on the avoidance of transmission-line effects with TTL circuits is given in table 11-3 of chapter 11.

## Shielding

Another very important method for minimizing the pickup of unwanted interference noise is shielding. Shielding involves surrounding signal-carrying wires, components, circuits, or complete systems with a conducting material that is connected to common. Although shielding of cables is stressed here, the same principles apply to shielding amplifiers, components, or systems.

**Shielded cables.**    There are several different kinds of shielded cables available for external signal connections and critical internal connections. Coaxial cable (see fig. A-6) is useful from zero frequency (dc) to several hundred megahertz since its characteristic impedance is uniform over this range (see sect. 11-4). The coaxial cable provides good protection against other electric fields and capacitive pickup if the shield is grounded. However, the shield is part of the signal path, and grounding on both ends can cause ground loops. The triax cable shown in figure A-7 provides an additional outer copper braid that is insulated from the signal conductors and acts as a true shield. Triax cables are, unfortunately, expensive and rather awkward to use.

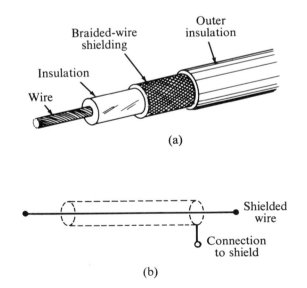

(a)

(b)

**Fig. A-6.** Coaxial cable, construction (a) and schematic (b). The braided shield provides excellent electric field protection but only fair magnetic shielding. The shield should be terminated uniformly by a coaxial connector (BNC, UHF, or type N) to avoid concentration of current on one side of the shield.

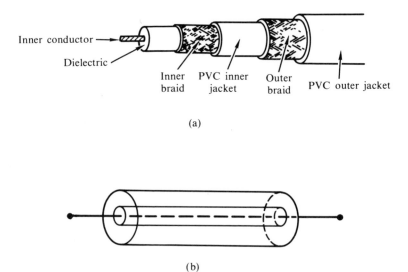

(a)

(b)

**Fig. A-7.**  Triax cable construction (a) and schematic (b). The inner conductor is surrounded by an inner braided shield as in normal coaxial cable and by an outer shield. The outer braid is grounded and bypasses both ground-loop currents and capacitive pickup.

Shielded twisted pairs (see fig. A-8) have characteristics similar to triax cable, but they are normally limited to signal frequencies from dc to 1 MHz. The inner conductors carry the signal current, and the shield carries the noise current. Twisting the two signal-carrying wires provides cancellation of randomly induced noise pickup and protection against capacitive coupling and magnetic fields. Any noise current in the shield couples equally to the two conductors by mutual inductance, and the induced voltages thus cancel.

**Grounding of cable shields.**    A very important question that invariably arises when shielded cables are used is where to ground the cable shield so as to avoid ground loops and capacitive pickup. For low-frequency circuits where single point commons are used, the shield should be grounded at only one point. The exact placement of the shield-to-ground connection depends upon whether the signal source or the receiver is grounded.

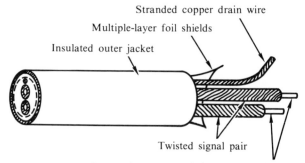

**Fig. A-8.**  Shielded twisted-pair cable. The foil shield provides better coverage and more effective magnetic shielding than a braided shield but less flexibility and durability.

For illustrative purposes it will be assumed that a low-frequency (<1 MHz) signal source is being connected to an amplifier. If the signal source has one grounded lead, but neither amplifier input is grounded, the shield should always be connected to the source common even if this is not at earth ground. Such a connection is shown in figure A-9a for a coaxial cable and in figure A-9b for a twisted-pair cable. The signal common is the only connection point that produces no noise voltage between the amplifier input terminals.

If the signal source is floating but one amplifier lead is grounded, the only shield connection that precludes a noise voltage between the amplifier input terminals is the amplifier common as shown in figure A-9c and d. If the signal circuit is grounded both at the source and at the amplifier, a ground loop results, and the amount of noise pickup depends on the susceptibility of the loop to electric and magnetic fields. The preferred connection of the shield ground in this case is shown in figure A-9e and f. In each case, grounding the shield at both ends forces some of the ground-loop current through the shield rather than the center conductor or twisted wire pair. To provide better noise immunity in this case, it is necessary to break the ground loop by an isolation transformer, an optical coupler, or a differential amplifier as discussed below.

**Fig. A-9.** Preferred connections of cable shields for grounded source (a,b), grounded amplifier (c,d), and grounded source and amplifier (e,f).

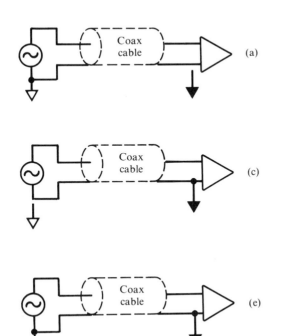
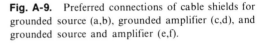
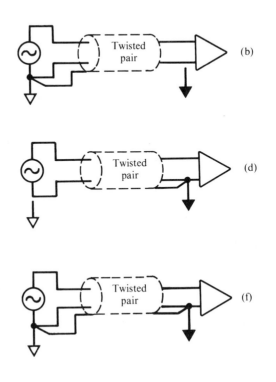

## Isolation

Ground loops can be broken by transformer coupling between amplifier stages. For ac signals below about 5 MHz a simple isolation transformer as shown in figure A-10 effectively isolates the input circuitry from the output. Optical isolators as discussed in chapter 6 can provide excellent isolation between circuits. Optical isolators contain an LED and a phototransistor or photodiode in the same package. They are especially useful for isolation of digital circuits where linearity through the coupler is unnecessary.

Many analog manufacturers produce isolation amplifiers that are transformer or optically coupled. Isolation amplifiers can be used as simply as op amps and yet provide excellent ohmic isolation between the signal source and the output. Transformer-coupled isolation amplifiers use modulation techniques so that their low-frequency response extends to dc. Many contain dc-dc converters so that the input amplifier power supply is also isolated from the output stage supply. In applications where gain accuracy and linearity are most important, transformer-coupled amplifiers provide excellent characteristics; optically coupled amplifiers provide higher speed.

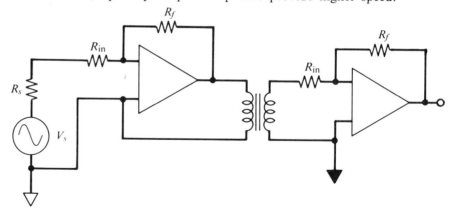

**Fig. A-10.** Isolation transformer for ac signals. The two amplifier stages are transformer coupled to provide excellent isolation between the input and output circuits.

## RF Shielding

High-frequency interference in circuits is frequently referred to as **RF (radio-frequency) interference**. Many sources of RF interference can be found in laboratory environments. Spark sources, flash lamps, and gaseous discharges for lasers are but a few. RF interference can be quite serious, rendering many digital circuits completely inoperable.

Enclosing the sensitive circuit in a metal shield and using shielded cable can provide RF shielding. A conductor that has a high surface area (mesh or braid) makes an excellent RF ground. The shield should be terminated at both ends as is a signal cable for high-frequency signals (see sect. 11-4). For best shielding two separate shields such as are provided by triax cable should be used.

# Guarding

Many available differential amplifiers and measurement devices have an additional shield called a **guard shield**. The guard shield, which surrounds the amplifier or meter, should be held at a potential that prevents current through any unbalanced source impedance. This eliminates the differential input noise voltage. The rule to follow when connecting a guard shield is to ensure that no common mode current occurs in any of the input resistances. This usually means that the guard should be connected to the source terminal with the lowest impedance to common. In addition a shield around a high-gain amplifier should always be connected to the amplifier common.

Consider, for example, the measurement of the $IR$ drop across resistor $R_s$ with a guarded digital voltmeter as shown in the circuit of figure A-11. Connecting the guard to the low-impedance source terminal, eliminates the noise current in the input circuitry of the meter.

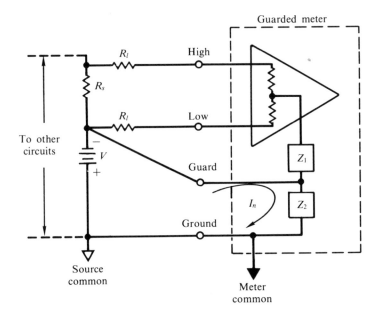

**Fig. A-11.** Use of a guarded meter. Connect the guard to the low-impedance terminal of the source to avoid noise currents at the meter input. If the guard is connected to the source common or meter common, any noise in voltage source $V$ or any induced noise causes a noise current in the low meter lead. If the guard is connected to the low meter terminal, a noise current is produced in the line resistance $R_l$.

# Appendix B

# Components

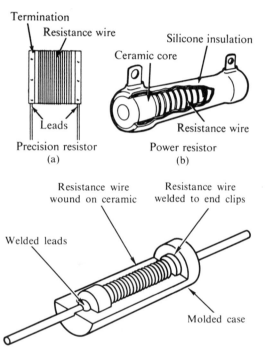

Fig. B-1. Wire-wound resistors, (a) high precision, (b) high power, with radial leads, (c) medium power or precision with axial leads.

**Fig. B-2.** Carbon composition resistor.

## Resistors

Commercial resistors are wire wound, carbon composition, or film. As the name suggests, **wire-wound resistors** are made by winding resistance wire around an insulating form and providing contact at each end. Three types of wire-wound resistors are shown in figure B-1. Wire-wound resistors can have high precision and high power dissipation capabilities. A **carbon composition resistor** is made by forming a mixture of graphite powder, silica, and a binder into a solid cylinder pressed between two conductors as shown in figure B-2. The resistance is changed by varying the ratio of carbon to silica in the resistive element. **Film resistors** are formed by depositing a thin conducting film on a cylindrical insulating ceramic substrate. Contact is then made to each end. For precision resistors, a spiral groove can be cut through the film to increase the resistance as shown in figure B-3. The automated cutter ceases when the desired resistance is attained. The conducting film in a film resistor can be metal, metal oxide, carbon, or cermet, which is a mixture of glass and metal alloys that is glazed onto an alumina substrate. The cermet film is relatively thick and very durable.

**Tolerances.** The tolerance of a resistor is the expected agreement of its actual and nominal resistances when purchased, expressed as a percentage of the nominal value. Resistor tolerances can be divided into four categories:

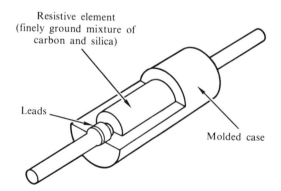

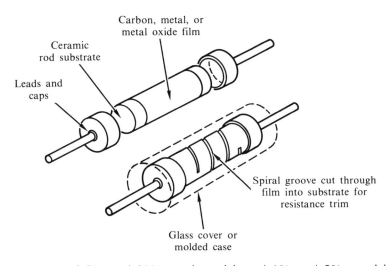

**Fig. B-3.** Film resistor.

general-purpose, $\pm5\%$ to $\pm20\%$; semi-precision, $\pm1\%$ to $\pm5\%$; precision, $\pm0.1\%$ to $\pm1\%$; and ultra-precision, $\pm0.01\%$ to $\pm0.1\%$. Carbon composition resistors are semi-precision. Metal-film resistors can achieve accuracies in the precision range, but only wire-wound resistors are capable of ultra-precision applications.

General-purpose resistors are available in $\pm5$, $\pm10$, and $\pm20\%$ tolerances. The available values of these resistors are shown in table B-1. The resistor values listed increase in increments of approximately 5%. This accounts for the nonuniform intervals and the lack of round numbers. Note that the two significant digits are the same in each decade. The resistances of general-purpose resistors are marked on them in a code of colored bands as shown in figure B-4. Each number from 0 to 9 has been assigned a color as

**Fig. B-4.** General-purpose resistor color code.

| Color | Figure | Multiplier | Tolerance |
|-------|--------|-----------|-----------|
| Black | 0 | 1 | — |
| Brown | 1 | 10 | $\pm1\%$ |
| Red | 2 | 100 | $\pm2\%$ |
| Orange | 3 | 1000 | — |
| Yellow | 4 | $10^4$ | — |
| Green | 5 | $10^5$ | $\pm0.5\%$ |
| Blue | 6 | $10^6$ | $\pm0.25\%$ |
| Violet | 7 | — | $\pm0.1\%$ |
| Gray | 8 | — | $\pm0.05\%$ |
| White | 9 | — | — |
| Gold | — | 0.1 | $\pm5\%$ |
| Silver | — | — | $\pm10\%$ |
| No band | — | — | $\pm20\%$ |

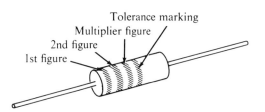

**Table B-1.**  General-purpose resistor values. ±10% and ±20% are in bold. All values available in ±5% tolerance.

| | | | | | | | |
|---|---|---|---|---|---|---|---|
| **1.0** | **10** | **100** | **1000** | **10 k** | **100 k** | **1.0 M** | **10 M** |
| 1.1 | 11 | 110 | 1100 | 11 k | 110 k | 1.1 M | 11 M |
| **1.2** | **12** | **120** | **1200** | **12 k** | **120 k** | **1.2 M** | **12 M** |
| 1.3 | 13 | 130 | 1300 | 13 k | 130 k | 1.3 M | 13 M |
| **1.5** | **15** | **150** | **1500** | **15 k** | **150 k** | **1.5 M** | **15 M** |
| 1.6 | 16 | 160 | 1600 | 16 k | 160 k | 1.6 M | 16 M |
| **1.8** | **18** | **180** | **1800** | **18 k** | **180 k** | **1.8 M** | **18 M** |
| 2.0 | 20 | 200 | 2000 | 20 k | 200 k | 2.0 M | 20 M |
| **2.2** | **22** | **220** | **2200** | **22 k** | **220 k** | **2.2 M** | **22 M** |
| 2.4 | 24 | 240 | 2400 | 24 k | 240 k | 2.4 M | |
| **2.7** | **27** | **270** | **2700** | **27 k** | **270 k** | **2.7 M** | |
| 3.0 | 30 | 300 | 3000 | 30 k | 300 k | 3.0 M | |
| **3.3** | **33** | **330** | **3300** | **33 k** | **330 k** | **3.3 M** | |
| 3.6 | 36 | 360 | 3600 | 36 k | 360 k | 3.6 M | |
| **3.9** | **39** | **390** | **3900** | **39 k** | **390 k** | **3.9 M** | |
| 4.3 | 43 | 430 | 4300 | 43 k | 430 k | 4.3 M | |
| **4.7** | **47** | **470** | **4700** | **47 k** | **470 k** | **4.7 M** | |
| 5.1 | 51 | 510 | 5100 | 51 k | 510 k | 5.1 M | |
| **5.6** | **56** | **560** | **5600** | **56 k** | **560 k** | **5.6 M** | |
| 6.2 | 62 | 620 | 6200 | 62 k | 620 k | 6.2 M | |
| **6.8** | **68** | **680** | **6800** | **68 k** | **680 k** | **6.8 M** | |
| 7.5 | 75 | 750 | 7500 | 75 k | 750 k | 7.5 M | |
| **8.2** | **82** | **820** | **8200** | **82 k** | **820 k** | **8.2 M** | |
| 9.1 | 91 | 910 | 9100 | 91 k | 910 k | 9.1 M | |

shown. Note that from 2 to 7 the colors follow the spectrum. The color of the band closest to the end of the resistor represents the first figure of the resistance; the second band, the second figure; and the third band, the power of 10 by which to multiply the first two figures to get the total resistance. Thus a resistor coded with yellow, violet, and red bands is $47 \times 100\ \Omega$ or 4.7 kΩ. Blue, gray, green is $68 \times 10^5\ \Omega$, or 6.8 MΩ; and green, blue, black is 56 Ω. For resistances between 1 and 10 Ω, gold is used for the third band. Thus orange, white, gold is 3.9 Ω.

Precision resistors have standard values that are within 1% or less of each other. This results in many more values per decade as shown in table B-2. The color code markings for precision resistors are shown in figure B-5. Four bands are used to give three significant digits and a multiplier. The wide band at the far end of the resistor indicates a precision resistor and gives the tolerance according to the table shown. The resistance of precision resistors is most often printed on the body along with alphanumeric code for other data about the resistor. The resistance is generally given as the digits and multiplier, just as in the color code. Thus 103 is $10 \times 10^3\ \Omega = 10.0\ k\Omega$ and 3322 is $332 \times 10^2\ \Omega = 33.2\ k\Omega$.

Factors that can change the value of a resistor in application include temperature variation, hours operated at high temperature, soldering, shock, overload, and moisture. In close tolerance situations, these should be considered. Manufacturer's installation and handling recommendations should be followed, and excessive electrical, thermal, or mechanical stress should be avoided. The temperature coefficient of resistance may be quite significant. This is generally given as $\alpha_R$, the relative change in $R$ per degree Celsius from the value at 25°C. For a composition resistor, $\alpha_R$ may be 0.005, which means that the resistance would change 0.5% per degree Celsius. A resistor that is 10 kΩ at 25°C might be 11.25 kΩ at 50°C. Metal-film and wire-wound resistors have the lowest temperature coefficients of resistance.

**Power.**    A resistor converts electrical power to heat. The amount of power thus converted can be calculated from

| Tolerance code | |
|---|---|
| Black | — |
| Brown | ±1% |
| Red | ±2% |
| Orange | — |
| Yellow | — |
| Green | ±0.5% |
| Blue | ±0.25% |
| Violet | ±0.1% |
| Gray | ±0.05% |
| White | — |
| Gold | ±5% |
| Silver | ±10% |

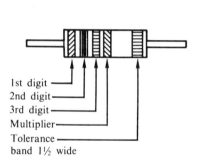

1st digit
2nd digit
3rd digit
Multiplier
Tolerance
band 1½ wide

**Fig. B-5.**  Color code position for precision resistors.

$$P = IV, \qquad P = I^2 R, \qquad \text{and} \quad P = V^2 / R$$

where $P$ is the power in watts, and $I$ and $V$ are the current in and the voltage across the resistance $R$. When power is dissipated in a resistor, the resistor heats up until its rate of heat loss to the surroundings equals the rate of electrical heat conversion. Resistors differ in their ability to stand high temperatures and to lose heat to the surroundings. Increasing the size of the resistive element aids in both respects. The approximate sizes of composition resistors in four power ratings are shown in figure B-6. For dissipation greater than 2 W, power resistors are used. Rather low gauge resistance wire is wound on a ceramic form which withstands high temperatures. Power resistors should be mounted so that they can readily dissipate heat to the air or the instrument case.

Table B-3 summarizes the ratings of power dissipation, size, tolerance, temperature coefficient, and maximum voltage for a number of common resistor types. Metal film resistors are seen to withstand higher temperatures than composition and other film types. The rating is the maximum temperature of the air surrounding the resistor in which it can dissipate its rated power.

**Series and parallel resistors.** The resistance $R_s$ of $N$ resistors in series is $R_s = R_1 + R_2 + R_3 + \cdots + R_N$. The actual value of the resistance can differ from the nominal value by the error tolerance which is generally a constant percentage of the nominal value. The error tolerance of resistors in series is then a combination of the individual error tolerances. If the resistances and tolerances are all equal, then the tolerance of the combination is the square root of the sum of the squares of the individual tolerances since the variances are additive. If the resistances are unequal, the tolerance of the largest $R$ often dominates. If all the tolerances are an equal percentage, the tolerances of all smaller resistors may be negligible. The power dissipated in resistors in series is in direct proportion to the resistance since $P = I^2 R$, and the current is the same in each. Awareness of these facts can allow use of lower accuracy and lower power for the lower valued resistors in a series circuit.

**Table B-2.** Precision resistor values. Values are given for only one decade. All other decades have the same significant digits. ±1% values are in bold. All values available in ±0.1%, ±0.25%, and ±0.5%.

| | | | | | | | |
|---|---|---|---|---|---|---|---|
| **1.00** | **1.33** | **1.78** | **2.37** | **3.16** | **4.22** | **5.62** | **7.50** |
| 1.01 | 1.35 | 1.80 | 2.40 | 3.20 | 4.27 | 5.69 | 7.59 |
| **1.02** | **1.37** | **1.82** | **2.43** | **3.24** | **4.32** | **5.76** | **7.68** |
| 1.04 | 1.38 | 1.84 | 2.46 | 3.28 | 4.37 | 5.83 | 7.77 |
| **1.05** | **1.40** | **1.87** | **2.49** | **3.32** | **4.42** | **5.90** | **7.87** |
| 1.06 | 1.42 | 1.89 | 2.52 | 3.36 | 4.48 | 5.97 | 7.96 |
| **1.07** | **1.43** | **1.91** | **2.55** | **3.40** | **4.53** | **6.04** | **8.06** |
| 1.09 | 1.45 | 1.93 | 2.58 | 3.44 | 4.59 | 6.12 | 8.16 |
| **1.10** | **1.47** | **1.96** | **2.61** | **3.48** | **4.64** | **6.19** | **8.25** |
| 1.11 | 1.49 | 1.98 | 2.64 | 3.52 | 4.70 | 6.26 | 8.35 |
| **1.13** | **1.50** | **2.00** | **2.67** | **3.57** | **4.75** | **6.34** | **8.45** |
| 1.14 | 1.52 | 2.03 | 2.71 | 3.61 | 4.81 | 6.42 | 8.56 |
| **1.15** | **1.54** | **2.05** | **2.74** | **3.65** | **4.87** | **6.49** | **8.66** |
| 1.17 | 1.56 | 2.08 | 2.77 | 3.70 | 4.93 | 6.57 | 8.76 |
| **1.18** | **1.58** | **2.10** | **2.80** | **3.74** | **4.99** | **6.65** | **8.87** |
| 1.20 | 1.60 | 2.13 | 2.84 | 3.79 | 5.05 | 6.73 | 8.98 |
| **1.21** | **1.62** | **2.15** | **2.87** | **3.83** | **5.11** | **6.81** | **9.09** |
| 1.23 | 1.64 | 2.18 | 2.91 | 3.88 | 5.17 | 6.90 | 9.20 |
| **1.24** | **1.65** | **2.21** | **2.94** | **3.92** | **5.23** | **6.98** | **9.31** |
| 1.26 | 1.67 | 2.23 | 2.98 | 3.97 | 5.30 | 7.06 | 9.42 |
| **1.27** | **1.69** | **2.26** | **3.01** | **4.02** | **5.36** | **7.15** | **9.53** |
| 1.29 | 1.72 | 2.29 | 3.05 | 4.07 | 5.42 | 7.23 | 9.65 |
| **1.30** | **1.74** | **2.32** | **3.09** | **4.12** | **5.49** | **7.32** | **9.76** |
| 1.32 | 1.76 | 2.34 | 3.12 | 4.17 | 5.56 | 7.41 | 9.88 |

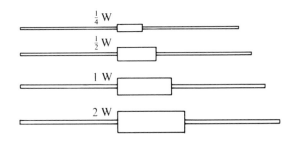

**Fig. B-6.** Composition resistors of various power ratings drawn to actual size.

**Table B-3.** Resistor properties.

| Resistive material | Power rating at $T_A = P_D$, W | Ambient temperature, $T_A$ °C | Body dimensions, in. $X_L$ | $D$ | Resistance range, ohms | Temperature coefficient, $\alpha_R \times 10^6$, $K^{-1}$ (Note 2) | Resistance tolerance, ±% | Maximum working voltage, $V_M$, V |
|---|---|---|---|---|---|---|---|---|
| Metal film .......... | 1/20 | 125 | 0.150 | 0.065 | 10 to $2 \times 10^5$ | −100 to +25 | 0.1, 0.5, 1 | |
| Deposited carbon film | 1/10 | 70 | 0.250 | 0.090 | 1 to $4 \times 10^5$ | −250 to +500 | 1, 2, 5, 10 | |
| Metal film .......... | 1/10 | 125 | 0.250 | 0.110 | 10 to $3 \times 10^5$ | −100 to +25 | 0.1, 0.5, 1 | |
| Carbon composition . | 1/8 | 70 | 0.145 | 0.062 | 2.7 to $1 \times 10^8$ | −5000 to +5000 | 5, 10, 20 | |
| Deposited carbon film | 1/8 | 70 | 0.375 | 0.125 | 1 to $3 \times 10^6$ | −1500 to −250 | 1, 2, 5, 10 | 150 |
| Cermet film........ | 1/8 | 70 | 0.150 | 0.065 | 4.7 to $1.5 \times 10^5$ | −200 to +100 | 1, 2, 5 | |
| Metal film .......... | 1/8 | 125 | 0.375 | 0.125 | 25 to $1.5 \times 10^6$ | −100 to +25 | 0.1, 0.5, 1 | |
| Wire-wound precision | 1/8 | 125 | 0.375 | 0.125 | 1 to $3 \times 10^5$ | +20 | 0.05, 0.1, 0.5, 1 | |
| Carbon composition . | 1/4 | 70 | 0.250 | 0.090 | 2.7 to $1 \times 10^8$ | −5000 to +5000 | 5, 10, 20 | 250 |
| Deposited carbon film | 1/4 | 70 | 0.625 | 0.188 | 1 to $5 \times 10^6$ | −1500 to −250 | 1, 2, 5, 10 | 250 |
| Cermet film........ | 1/4 | 70 | 0.250 | 0.090 | 10 to $1.6 \times 10^5$ | −200 to +100 | 1, 2, 5 | |
| Metal film .......... | 1/4 | 125 | 0.625 | 0.188 | 25 to $3 \times 10^6$ | −100 to +25 | 0.1, 0.5, 1 | |
| Wire-wound precision | 1/4 | 125 | 0.750 | 0.250 | 10 to $4.5 \times 10^5$ | +10 | 0.01, 0.05, 0.1, 1 | |
| Carbon composition . | 1/2 | 70 | 0.375 | 0.140 | 1 to $1 \times 10^8$ | −5000 to +5000 | 5, 10, 20 | 350 |
| Deposited carbon film | 1/2 | 70 | 0.750 | 0.250 | 1 to $1 \times 10^7$ | −1500 to −250 | 1, 2, 5, 10 | 350 |
| Cermet film........ | 1/2 | 70 | 0.375 | 0.140 | 4.3 to $1 \times 10^6$ | −200 to +100 | 1, 2, 5 | |
| Metal film .......... | 1/2 | 125 | 0.750 | 0.250 | 10 to $4 \times 10^6$ | −100 to +25 | 0.1, 0.5, 1 | 700 |
| Wire-wound precision | 1/2 | 125 | 1.000 | 0.375 | 10 to $1.2 \times 10^6$ | +10 | 0.01, 0.05, 0.1, 1 | |
| Carbon composition . | 1 | 70 | 0.562 | 0.225 | 2.7 to $1 \times 10^8$ | −5000 to +5000 | 5, 10, 20 | 500 |
| Deposited carbon film | 1 | 70 | 1.062 | 0.375 | 1 to $1.5 \times 10^7$ | −1500 to −250 | 1, 2, 5, 10 | 500 |
| Cermet film........ | 1 | 70 | 0.562 | 0.190 | 10 to $1 \times 10^6$ | −200 to +100 | 1, 2, 5 | |
| Metal film .......... | 1 | 125 | 1.062 | 0.375 | 25 to $4 \times 10^6$ | −100 to +25 | 0.1, 0.5, 1 | 1000 |
| Wire-wound power .. | 1 | 25 | 0.250 | 0.085 | 0.5 to $1 \times 10^3$ | +20 to +100 | 0.1, 0.5, 1 | |
| Carbon composition . | 2 | 70 | 0.688 | 0.318 | 10 to $1 \times 10^8$ | −5000 to +5000 | 5, 10, 20 | 1500 |
| Deposited carbon film | 2 | 70 | 2.188 | 0.375 | 2 to $1 \times 10^8$ | −1500 to −250 | 1, 2 | 1500 |
| Cermet film........ | 2 | 70 | 0.688 | 0.318 | 10 to $1.5 \times 10^6$ | −200 to +100 | 1, 2, 5 | |
| Metal film .......... | 2 | 125 | 2.188 | 0.375 | 100 to $6 \times 10^6$ | −100 to +25 | 0.1, 0.5, 1 | |
| Wire-wound power .. | 2 | 25 | 0.406 | 0.094 | 0.5 to $2.5 \times 10^3$ | +20 to +100 | 0.1, 0.5, 1 | |
| Cermet film........ | 2–115 | 25 | | | 10 to $1 \times 10^6$ | −500 to +500 | 1, 2, 5, 10 | |
| Wire-wound power .. | 2–250 | 25 | | | 0.1 to $2.7 \times 10^5$ | +50 | 0.1, 0.5, 1, 3 | |

*Notes.*

1. Reprinted with permission from L.J. Giacoletto, *Electronics Designers' Handbook*, 2nd ed., McGraw-Hill Book Co., NY, 1977, p. 3–38.

2. $\alpha_R = \dfrac{1}{R} \dfrac{\delta R}{\delta T}\bigg|_{T=25°C}$

The combined resistance of $N$ parallel resistors is

$$R_p = \frac{1}{1/R_1 + 1/R_2 + 1/R_3 + \cdots 1/R_N}$$

In the parallel case, the tolerance of the resistor of lowest resistance has the largest influence on the combined tolerance. The power dissipated in the resistors is in inverse proportion to their value since $P = V^2/R$ and the same voltage is applied to all.

**Noise.**  Two types of electrical noise are generated in resistors, Johnson noise and current noise. Johnson noise is caused by the random thermal motion of charge carriers in a conductor and is unavoidable. The rms amplitude of Johnson noise is

$$\overline{v}_{\text{rms}} = (4k\,TR\,\Delta f)^{1/2}$$

where $k$ is Boltzmann's constant, $T$ is the temperature in $°K$, $R$ is the resistance and $\Delta f$ is the bandwidth of the system influenced by the noise. This noise is clearly reduced at lower temperatures, resistance, and bandwidth.

  **Current noise** is a low-frequency noise caused by a current in a non-homogeneous conductor such as the carbon composition material. Like $1/f$ noise, current noise is inversely proportional to frequency. Current noise is considerable in carbon composition resistors where it is generally the dominant noise source. For low-noise applications, metal film or wire-wound resistors should be used. The noise in these resistors approaches the theoretical lower limit of the Johnson noise equation.

**Frequency characteristics.**  A practical resistor is not a pure resistance. The inductance of its leads or windings and the capacitance of its end terminals result in the approximate equivalent circuit shown in figure B-7. The consequence of the reactances is that the overall impedance is a function of the frequency of the current or voltage applied. The capacitance and inductance associated with the resistance varies considerably from one type of resistor to another. The carbon composition resistor has negligible inductance and a capacitance of 0.25 to 0.5 pF. The 3-dB point for a parallel $RC$ circuit is $f_0 = 1/(2\pi RC)$. If the capacitance does not depend on the value of $R$, $Rf_0$ is a constant.

$$Rf_0 = 1/(2\pi C)$$

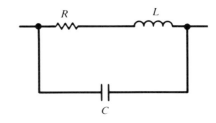

**Fig. B-7.**  Equivalent circuit of a practical resistor.

For a typical carbon composition resistor, the value of $Rf_0$ is about 0.5 MHz·MΩ. This means that the effective resistance of a 1-MΩ resistor is "down 3 dB"—that is, it is 707 kΩ—at 0.5 MHz. At 10 MHz, the highest value of $R$ that can be used with less than 30% error is 50 kΩ. Carbon and metal-film resistors have an $Rf_0$ constant of generally 3–5 MHz·MΩ. The slight series inductance increases the impedance at high frequencies and partially compensates for the capacitive decrease. Wire-wound resistors can have very high values of series inductance because of the coil-type winding of the resistance wire. Ultra-high precision wire-wound resistors are thus often truly precise only for dc circuits. Some wire-wound resistors use noninductive windings (two parallel coils wound in opposite directions). This can extend the useful frequency range of wire-wound resistors into the low audio-frequency range.

**Variable resistors.**  Some resistors are made to have an adjustable or variable resistance by exposing the resistive element so that a movable contact, or wiper, can contact the element at any point along the resistor. Such resistors are called rheostats or potentiometers. The value of the total or maximum resistance is usually stamped on the case. The resistive element is usually either a coil of resistance wire or a strip of resistive film. Several types of potentiometers are shown in figure B-8. The word **rheostat** is used for high wattage potentiometers, and a **trimmer** is a potentiometer used for an adjustment that is required only occasionally.

Potentiometers are made of the same materials as fixed resistors and so have similar qualities. In addition, several other characteristics are applied. The linearity of a potentiometer often affects its accuracy in a given application. The deviation of the linear relationship between output resistance and degree of rotation is usually given as the maximum resistance deviation from a straight line relative to the total (end-to-end) resistance. Potentiometers may require anywhere from 270° to 25 complete turns to traverse the entire resistance range. Their lifetime is measured in the number of cycles (end-to-end rotations) they are designed to endure. Trimmers have a low cycle lifetime (in the hundreds); some cermet and wire-wound potentiometers give very long service rated in hundreds of thousands of revolutions.

The power rating of a potentiometer determines the power it can dissipate. This assumes the entire element is involved in the dissipation. If the current is passing through only part, as between the wiper and one end, the power rating must be reduced proportionately. A safe approach is to calculate the maximum current for the full resistance at the rated power and then not exceed this same current through any part of the resistance. For example, a 1-W, 10-k$\Omega$ potentiometer can withstand a current of $I = \sqrt{P/R} = \sqrt{10^{-4}} = 10^{-2}$ A. Thus the current through any part should be limited to 10 mA or less.

## Conductors

Metallic conductors are used to connect electronic components in the desired circuit. The assumption of negligible resistance and reactance in the connecting wires and contacts is often valid. However, conductors do have finite resistances, and therefore limited current-carrying capacity. Conducting wires also exhibit a small inductance and capacitance that can be significant at high frequencies. Two types of conductors are considered in this section, wires and printed circuit foil.

**Wire.**    Metallic wire is an essential part of all electronic systems. Its low resistance can provide a nearly ideal electrical connection between components. Wire is also used to fabricate inductors, transformers, and wire-wound resistors. The usual electrical hookup wire is made of copper because of its very high conductivity and good flexibility at moderate costs. The copper wire is often plated or "tinned" with a thin layer of another metal

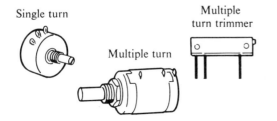

Single turn

Multiple turn

Multiple turn trimmer

**Fig. B-8.**    Variable resistors.

such as silver or tin to make it easier to solder other components to it and then covered with an insulator, usually plastic. Copper wire is available in many diameters, called **gauges**. The larger the diameter, the lower the resistance per unit length. Table B-4 gives the resistances of several sizes of copper wire. A conductor made of several collected small wires is called **stranded wire**. It offers improved flexibility and is less likely to break under repeated flexing. To provide some perspective, 22 gauge wire is the normal hookup wire size, 18 gauge is used for household lamp cords, and 12 and 14 gauge are used for house wiring. The current values given in parentheses in table B-4 would bring the temperature of the wire to 100°C if the wire were bundled or enclosed and the ambient temperature were 57°C (135°F).

A single wire of diameter $d$ cm that is $h$ cm above a ground plane exhibits a capacitance, an inductance, and therefore a characteristic impedance (see chap. 11 on transmission lines) according to the following relationships:

$$C = \frac{24.12}{\log (4h/d)} \ (pF/m)$$

$$L = 0.46 \log \frac{4h}{d} \ (\mu H/m)$$

$$Z_0 = 138 \log \frac{4h}{d} \ (\Omega)$$

**Printed circuit foil.** Most component-to-component connections these days are accomplished by patterns of copper foil on an insulating epoxy circuit board. This construction technique is simple and reliable since all the connections are correctly completed once the components have been properly soldered in place on the board. Most printed circuit (PC) boards are 1/16 or 3/32 in. (1.59 or 2.38 mm) thick. The copper foil pattern may be on one or both sides. The copper foil is very thin, from 0.001 to 0.0025 in. (0.04 to 0.1 mm), but it is often coated with a thicker layer of tin/lead solder. The width of the conducting foil trace between two contacts determines the maximum current that should pass in that connection as shown in table B-5. Exceeding the maximum recommended current can cause overheating and destruction of the circuit board. The traces on the PC boards are often quite close together. The minimum practical spacing between traces depends upon the maximum difference in voltage of the traces. Table B-6 gives the minimum spacing recommended for various voltages.

In the design of the PC board pattern, it is often necessary to make a connection between the patterns on the two sides of the board by a feedthrough. In some boards, the holes are copper plated after drilling. These plated-through boards automatically connect top and bottom patterns wherever a hole is drilled. Only one side of a plated-through board need be

**Table B-4.** Copper wire characteristics.

| A.W.G.* | Number of strands | Diameter per strand, mm | Resistance, $\Omega/M$ | Current capacity, A |
|---------|-------------------|-------------------------|------------------------|---------------------|
| 30 | 1 | 0.255 | 0.346 | 0.144 |
| 24 | 1 | 0.511 | 0.0804 | |
| 24 | 7 | 0.022 | 0.0804 | 0.577 |
| 22 | 1 | 0.644 | 0.0501 | |
| 22 | 7 | 0.025 | 0.053 | 0.918 (5) |
| 20 | 1 | 0.812 | 0.0316 | |
| 20 | 7 | 0.032 | 0.0332 | 1.46 (7.5) |
| 18 | 1 | 1.024 | 0.0198 | |
| 18 | 7 | 0.040 | 0.0203 | 2.32 (10) |
| 16 | 1 | 1.291 | 0.0125 | |
| 16 | 19 | 0.029 | 0.0130 | 3.69 (13) |
| 14 | 1 | 1.628 | 0.0078 | 5.87 (17) |
| 12 | 1 | 2.053 | 0.0049 | 9.33 (23) |

*American Wire Gauge (A.W.G.) is a means of specifying relative wire diameter. The lower the A.W.G. number, the larger the diameter and the lower the resistance per unit length.

**Table B-5.** Recommended maximum currents for various printed circuit trace widths.*

| Trace width, in. | 0.015 | 0.03 | 0.08 | 0.12 | 0.156 |
|------------------|-------|------|------|------|-------|
| Maximum current, A | 1 | 2 | 3 | 4 | 5 |

*Assumes a 0.00125-in. foil, continuous current, and adequate ventilation.

**Table B-6.** Minimum spacing between printed circuit traces for various maximum voltages.

| Voltage, V | 5 | 25 | 50 | 120 | 300 |
|------------|-----|------|------|------|------|
| Minimum separation, in. | 0.010 | 0.013 | 0.025 | 0.060 | 0.120 |

soldered when mounting (and demounting) components. If the board is not plated-through, component leads or wires soldered on both sides of the board make the top-to-bottom connections.

High-frequency or low-noise circuit boards often use a nearly solid foil on one side of the board as a ground plane. This limits the connecting traces (other than ground connections) to the other side. The use of a ground plane increases shielding and reduces crosstalk between the traces (see app. A). The trace and ground plane combination also produces a transmission line effect for the connection with capacitance, inductance, and a characteristic impedance.

## Capacitors

A capacitor consists of two conductors separated by an insulator. A pictorial representation of a simple capacitor is shown in figure B-9 as two metal plates of area $a$ separated by a distance $d$. The capacitance is directly proportional to the overlapping area of the plates and inversely proportional to the distance between them. Thus, $C = \epsilon a / d$ where $\epsilon$ is the proportionality constant. If a vacuum occupies the space between the capacitor plates, the value of $\epsilon$ is $8.854 \times 10^{-12}$ farads per meter ($\text{Fm}^{-1}$). This constant is called $\epsilon_0$, the permittivity of free space. If an insulating material with polarizable molecules (a dielectric) is placed between the plates, the molecules of the dielectric tend to align under the influence of an applied voltage. This process takes energy from the field and increases the capacitance (the charge required to attain a particular voltage). The ratio of the capacitance with the dielectric to that with a vacuum is the **dielectric constant** $K_d$ for that material. Table B-7 lists the $K_d$ values for a number of commonly used dielectrics. The capacitance of a parallel plate capacitor with a dielectric insulator is

$$C = \epsilon_0 K_d a / d$$

**Capacitor types.** Practical capacitors are made with various combinations of conductors and dielectrics. Families of capacitors are based on the type of dielectric employed—film such as paper or plastic, rigid material such as mica or ceramic, metal oxide such as $Al_2O_3$ or $Ta_2O_5$, and fluid such as oil or gas.

**Film capacitors** are made by rolling two strips of foil and dielectric into a cylinder as shown in figure B-10. Then the unit is sealed with wax or plastic. The larger the area of the foil and the thinner the dielectric film, the higher the capacitance. However, if too high a voltage is applied to the capacitor, the dielectric breaks down, and a discharge occurs between the two foils. Thus capacitors are rated by both capacitance and breakdown voltage. Higher breakdown voltages require thicker dielectric films and thus a larger size for a given capacitance. Tubular film capacitors are made with

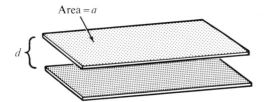

**Fig. B-9.**  Pictorial representation of a capacitor.

**Table B-7.**  Dielectric constants for common materials.

| Material | $K_d$ |
|---|---|
| Vacuum | 1.0000 |
| Air | 1.0001 |
| Paper (impregnated) | 3.7 |
| Polyester | 3.0–4.5 |
| Polystyrene | 2.5 |
| Polycarbonate | 3.2 |
| Polypropylene | 2.1 |
| Polysulfone | 3.1 |
| Teflon | 2.1 |
| Mica | 5.4 |
| Glass | 4.5–9.1 |
| Quartz | 3.8 |
| Steatite | 6.0 |
| Titanium dioxide | 80–120 |
| Barium titanate | 200–16 000 |
| Aluminum oxide | 8.4 |
| Tantalum oxide | 27.6 |
| Oil | 2.2 |

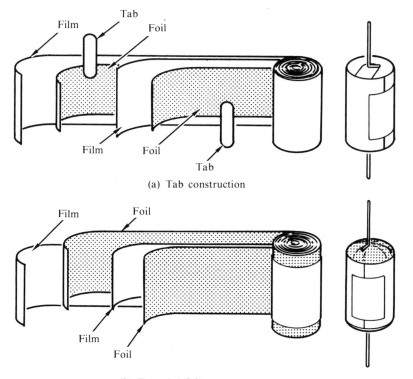

(a) Tab construction

(b) Extended foil construction

**Fig. B-10.** Construction of tubular film capacitors. The tab construction (a) is more economical, but current must pass through the coiled foil to get to the capacitance of the inner windings. The coiled path adds inductance to the device. This is avoided by extending the foils (b) to opposite sides to make contact along the long edge of the foil. When finished, one foil is the outermost conductor, and the other is shielded by it.

films of impregnated paper or of various plastics such as polyester, polystyrene, polycarbonate, polypropylene, polysulfone, and polytetrafluoroethylene (Teflon). Paper has good high-voltage characteristics but is being replaced by polyester. The plastic dielectrics differ in dielectric constant, breakdown voltage, resistance, and degree of signal loss (see Frequency Characteristics, below). Polycarbonate combines high quality with small size. The premium film capacitors are made with Teflon.

The foil in the film capacitor can be replaced by vapor-depositing a metal coating on one side of the film. Such metallized film capacitors are very compact since the metal coating is much thinner than the foil. Because the current-carrying capability of the conductor is also greatly reduced, metallized film capacitors should not be used in situations where high ac currents are encountered. Metallized capacitors are more resistant to breakdown since an arc occurring at a weak spot in the dielectric vaporizes the metal, breaking the connection to the area.

Capacitors made from rigid dielectrics such as mica, glass, or ceramic are stacked rather than rolled as shown in figure B-11. Metallized mica is also used. When glass is used, 0.001-in. films of glass are interleaved with

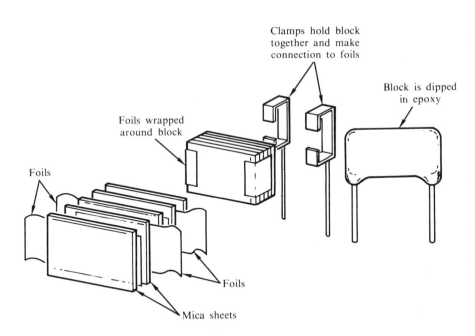

Clamps hold block together and make connection to foils

Block is dipped in epoxy

Foils wrapped around block

Foils

Foils

Mica sheets

**Fig. B-11.** Construction of mica capacitor.

metal foil. After the contacts are attached, the assembly is surrounded with glass and fired to produce an extremely durable device. Glass capacitors are stable to very high temperatures and are quite insensitive to shock. Ceramic dielectric sheets are often stacked with conducting layers of silver paste. The stack is then fired and encapsulated. The very high $K_d$ of some ceramics makes these extremely compact and often single-thickness capacitors in disc or tube shape such as shown in figure B-12 are practical. Manufacturers distinguish two types of ceramic capacitors. Type I capacitors are made from ceramic that is largely titanium dioxide ($TiO_2$). They have reasonable temperature stability and good high-frequency characteristics. Type II capacitors are made largely from barium titanate ($BaTiO_3$). Their capacitance is highly dependent on temperature and voltage, and they have poor high-frequency characteristics. The extremely compact Type II ceramics thus have limited application.

**Electrolytic capacitors** are made of aluminum foil, tantalum foil, or tantalum sponge with a surface that has an anodic formation of metal oxide film. The anodized metal foil is in an electrolytic solution. The oxide film is the dielectric between the metal and the conducting solution. Because the dielectric is so thin, a high capacitance can be obtained in a small space. Most electrolytic capacitors must be used in a circuit where the polarity is always in one direction. If the polarity is reversed, the oxide is reduced destroying the dielectric, and gas evolved at the electrode contacting the solution can cause the capacitor to explode. Some electrolytic capacitors are made of two anodized metal electrodes connected by the electrolyte. Such capacitors can be used with bipolar signals or connected either way in the circuit. The fact that the oxide film is a relatively low resistance dielectric results in significant leakage current. The oxide in aluminum electrolytics tends to deteriorate with time even when not used. Some tantalum electrolytics use a dry electrolyte and can thus be more completely encapsulated. Electrolytic capacitors are generally used for filtering applications where a leaky dielectric can be tolerated and where large capacitance in a small space is essential.

Oil, air, and gas dielectric capacitors are used for specialized applications, generally not electronic in nature. Various types of oil capacitors are used for heavy ac currents such as in circuits with large line-operated motors, for heavy dc filtering such as in a power supply for arc welding, and for high-energy discharges such as in a flash lamp supply. Air and gas dielectrics are used where very high precision and stability or very high breakdown voltage are required.

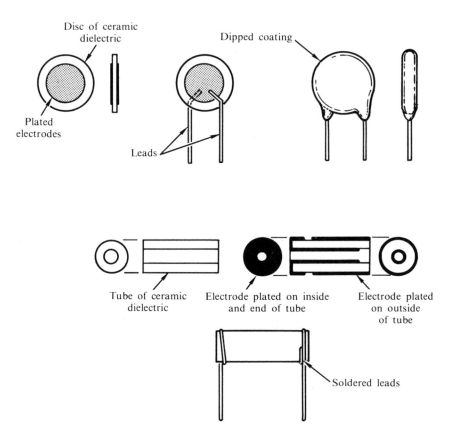

**Fig. B-12.**  Disc and tubular ceramic capacitors.

**Values and tolerance.**   Capacitors can be obtained in various tolerance ratings from $\pm 20\%$ to $\pm 0.5\%$. Because of dimensional changes, capacitors have a high temperature coefficient of capacitance. Among film capacitors, polycarbonate and Teflon have the lowest temperature dependence and polyester has the highest. Mica and glass capacitors with temperature coefficients of less than 100 ppm/°C can be obtained. Among ceramics, only very low $K_d$ dielectric materials have reasonable temperature stability for critical applications.

Humidity can also affect capacitance values by as much as 20% in paper capacitors. Teflon and polypropylene have the lowest water absorption among film types, and polyester is relatively high. Mica and glass capacitors are unaffected by humidity.

The capacitance of an ideal capacitor is not affected by, the voltage applied. This is generally true of all types of capacitors except electrolytics and Type II ceramics. Fairly serious distortion can result when these latter types are used for coupling or in high pass or low pass filters.

Capacitor values range from a few tenths of a picofarad to thousands of microfarads. The capacitance value is generally marked on the body of the capacitor. If the value is followed by MF, $\mu$F, or MFD, or if the value marked is less than one, the value is in microfarads (examples: 0.1 MFD, .22 MF, .01 $\pm 20\%$). If the value is greater than one and microfarads are not indicated, the value is in picofarads. This value can be expressed directly in pF (examples: 7.5, 27, 470, 1600) or in a three- or four-digit code. The code digits are followed by a letter that gives the tolerance according to table B-8. As in precision resistor marking, the first digits in the code are the significant digits, and the last digit is a multiplier. Thus capacitors marked 220 K, 471 J, and 1003 F have values of 22 pF $\pm 10\%$, 470 pF $\pm 5\%$ and 0.1 $\mu$F $\pm 1\%$. The contact to the outside foil of tubular film capacitors is indicated by a band encircling that end of the capacitor. For lowest noise this end should be connected to the circuit point with the lower impedance to common. Some mica and tubular ceramic capacitors are color coded as shown in figure B-13.

A capacitor does not hold a charge indefinitely because the dielectric is never a perfect insulator. Capacitors are rated for **leakage**, the conduction through the dielectric, by the leakage resistance-capacitance product in $M\Omega \cdot \mu$F. The lowest leakage capacitors are those of Teflon or polystyrene film, which have $10^6$ $M\Omega \cdot \mu$F resistance at 25°C. (A 0.01-$\mu$ capacitor has a leakage resistance of $10^{15}$ $\Omega$.) Other dielectrics have less than one tenth this resistance. High temperature increases leakage. The resistance of polystyrene capacitors falls to $10^5$ $M\Omega \cdot \mu$F at 85°C. Electrolytic capacitors show high leakage.

Some of the charge on a capacitor cannot be recovered immediately. This is due to **dielectric absorption** in which long-term polarizations or interfacial effects absorb energy that cannot be quickly returned. Thus a charged

**Table B-8.**   Tolerance code for capacitors.

| Letter | Tolerance |
|--------|-----------|
| M | $\pm 20\%$ |
| K | $\pm 10\%$ |
| J | $\pm 5\%$ |
| G | $\pm 2\%$ |
| F | $\pm 1\%$ |
| E | $\pm 0.5\%$ |

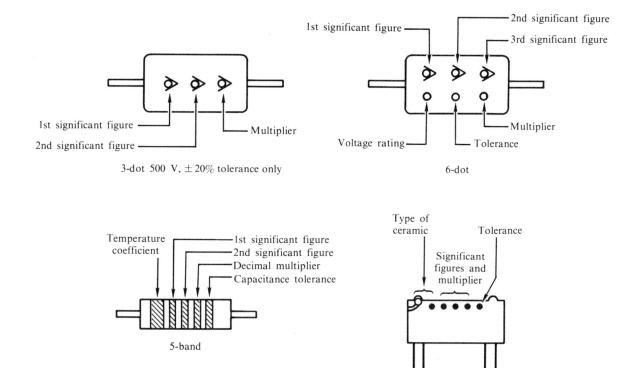

3-dot 500 V, ±20% tolerance only

6-dot

5-band

6-dot

Color code for mica and ceramic capacitors

| | | | Capacitance tolerance | | Temp. coeif. ppm/ °C | |
|---|---|---|---|---|---|---|
| Color | Significant figure | Multiplier | More than 10 pF | Less than 10 pF | | Voltage rating |
| Black | 0 | 1 | ±20% | 2.0 pF | 0 | |
| Brown | 1 | 10 | ±1% | | 30 | 100 |
| Red | 2 | 100 | ±2% | | 80 | 200 |
| Orange | 3 | 1 000 | | | 150 | 300 |
| Yellow | 4 | 10 000 | | | 220 | 400 |
| Green | 5 | | ±5% | 0.5 pF | 330 | 500 |
| Blue | 6 | | | | 470 | 600 |
| Violet | 7 | | | | 750 | 700 |
| Gray | 8 | 0.01 | | 0.25 pF | 30 | 800 |
| White | 9 | 0.1 | ±10% | 1.0 pF | 500 | 900 |
| Gold | | | ±0.5% | | | 1000 |
| Silver | | | ±10% | | | 2000 |

**Fig. B-13.**   Color code for mica and ceramic capacitors.

capacitor can be momentarily discharged to zero volts but have a small fraction of its original voltage on it several moments later. Dielectric absorption is measured as the fraction of the charge that is not available in a short time. Teflon and polystyrene have the lowest values at about 0.02–0.05%. Polypropylene is 0.1%, polycarbonate is 0.5%, and polyester and paper are much worse. Some applications such as analog integrators require capacitors with high charge accuracy. For such applications, low leakage and low dielectric absorption are essential.

**Series and parallel capacitance.** The capacitance of parallel capacitors is additive so that

$$C_p = C_1 + C_2 + C_3 + \cdots + C_N$$

When capacitors are in series, their reactances (inversely proportional to capacitance) are additive. Thus $X_{C_s} = X_{C_1} + X_{C_2} + \cdots + X_{C_N}$. When $1/(\omega C)$ is substituted for $X_C$ and the $\omega$s are cancelled,

$$C_s = \frac{1}{1/C_1 + 1/C_2 + \cdots + 1/C_N}$$

When unequal capacitances are in parallel, the largest values influence the combined capacitance most, but for capacitors in series the combined capacitance is always less than the smallest value. The breakdown voltage of parallel capacitors must all be at least equal to the applied voltage; in series combinations of capacitors, the total applied voltage is divided among the capacitors in inverse proportion to their value. Two 0.01-$\mu$F, 200-V capacitors in series would give a combined capacitance of 0.005 $\mu$F and a breakdown voltage of 400 V.

**Frequency characteristics.** The equivalent circuit of a practical capacitor is shown in figure B-14. Assume that $R_p$ is so much larger than $X_C$ that it can be neglected. The series resistance $R_s$ includes the lead resistance, but the power loss on charging and discharging due to polarizing the dielectric is generally much larger. This loss is proportional to the current and shows up as an effective series resistance. The inductance is due to the capacitor leads and foil. The magnitude of the total impedance of the capacitor is not then simply $X_C$; it is $Z = [R_s^2 + (X_C - X_L)^2]^{1/2}$. A vector plot of the complex impedance is shown in figure B-15. The effect of $R_s$ is indicated by three different measures as shown by the equations in figure B-15. The **power factor** PF is the ratio of the resistive loss to the total impedance. The **dissipation factor** DF is the ratio of the resistance to the net reactance. The ideal PF and DF are zero, and the ideal $Q$ is infinite. Because the dissipation factor varies somewhat with frequency, the value at 1 kHz is generally specified. The lowest values of DF for film capacitors are $2 \times 10^{-4}$ for Teflon, $3 \times 10^{-4}$ for polypropylene, and $5 \times 10^{-4}$ for polystyrene. Polyester and

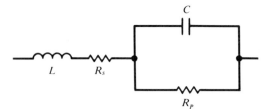

**Fig. B-14.** Equivalent circuit of a practical capacitor. $R_p$ is the leakage resistance, and $R_s$ includes lead resistance and losses in the polarization of the dielectric.

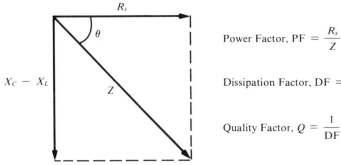

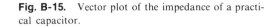

Power Factor, $\text{PF} = \dfrac{R_s}{Z} = \cos\theta$

Dissipation Factor, $\text{DF} = \dfrac{R_s}{X_C - X_L} = \cot\theta$

Quality Factor, $Q = \dfrac{1}{\text{DF}}$

**Fig. B-15.** Vector plot of the impedance of a practical capacitor.

Type I ceramics have DF values of about 0.005. At high frequencies (>10 k Hz for a 0.01-$\mu$F capacitor), where the value of $X_C$ becomes quite low, the value of DF increases rapidly.

The impedance of an ideal capacitor decreases linearly as the frequency increases. Plots of the impedance against frequency for some actual capacitors are shown in figure B-16. The impedance decreases with increasing frequency as expected until the increasing impedance due to the inductance is of comparable value. At higher frequencies the impedance increases, and the device actually exhibits a net inductance at these frequencies. The frequency at which the capacitive and inductive reactances are equal is the resonant frequency for the capacitor. Obviously, a capacitor should not be used above its resonant frequency if capacitive reactance is desired. Note that the resonant frequency increases with decreasing nominal capacitance for a tubular film capacitor. For very high frequency applications, rigid dielectric capacitors are recommended. The stacked construction eliminates much of the inductance. Type I ceramics have negligible frequency effects past 100 MHz.

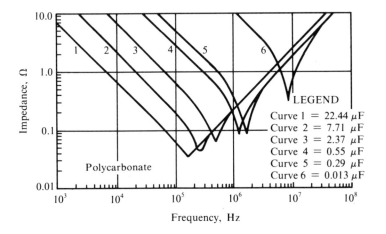

**Fig. B-16.** Impedance vs. frequency curves for several values of metallized polycarbonate capacitors.

**Variable capacitors.**    Adjustable capacitors are available with values up to a few hundred picofarads. They are made by adjusting the fraction of the area of the conducting plates that overlaps to produce the capacitance. The geometries used are rotating semicircles and sliding cylinders. The dielectrics are air, mica, and ceramic. Some variable capacitors are trimmers, designed to be adjusted only occasionally; others, like radio tuning capacitors, are made for frequent use. Varactor diodes, pn junction diodes in which the capacitance across the junction depends on the reverse bias voltage, are finding increasing use as adjustable capacitors. An adjustable resistor or a DAC is used to vary the diode bias voltage. The characteristics of several types of variable capacitors are summarized in table B-9.

**Table B-9.**    Characteristics of selected variable capacitors.

| Type | Capacitance range, pF | dc working voltage | Typical $Q$ at 1 MHz | Maximum temperature,°C | Relative cost |
|---|---|---|---|---|---|
| Trimmer: | | | | | |
| ceramic............ | 7–45 | 500 | 500 | 85 | medium |
| quartz.............. | 0.5–30 | 1250 | 1500 | 125 | high |
| mica .............. | 7.5–50 | 200 | 250 | 70 | medium |
|  | 1400–3000 | | | | |
| film............... | 5–50 | 100 | | | low |
| air ................ | 9–145 | 500 | 1500 | 85 | medium |
| Multigang air: | | | | | |
| general purpose ..... | 15–100 | 500 | 250 | 125 | medium |
|  | 350–550 | | | | |
| precision ........... | 25–115 | 250 | 2000 | 125 | high |
|  | 100–1150 | | | | |
| Voltage-controlled diode | 2–6 | 20–100 | 750 | 125 | medium |
|  | 30–100 | | | | |

Reprinted with permission, from L.J. Giacoletto, *Electronics Designers' Handbook*, 2nd ed., McGraw-Hill Book Co., NY, 1977, p. 3–37.

## Inductors and Transformers

**Inductors.**    Inductors are never pure inductances because there is always some resistance in and some capacitance between the coil windings. When choosing an inductor (occasionally called a choke) for a specific application, it is necessary to consider the value of the inductance, the dc resistance of the coil, the current-carrying capacity of the coil windings, the breakdown voltage between the coil and the frame, and the frequency range in which the coil is designed to operate.

Inductors are available with inductance values ranging from several hundred henrys down to a few microhenrys. To obtain a very high inductance it is necessary to have a coil of many turns. The inductance can be further increased by winding the coil on a closed-loop iron or ferrite core. To obtain as pure an inductance as possible, the dc resistance of the windings should be reduced to a minimum. This can be done by increasing the wire size, which, of course, increases the size of the choke. The size of the wire also determines the current-handling capacity of the choke since the work done in forcing a current through a resistance is converted to heat in the resistance. Magnetic losses in an iron core also account for some heating, and this heating restricts any choke to a certain safe operating current. The windings of the coil must be insulated from the frame as well as from each other. Heavier insulation, which necessarily makes the choke more bulky, is used in applications where there will be a high voltage between the frame and the winding. The losses sustained in the iron core increase as the frequency increases. At about 15 kHz they become so large that the iron core must be abandoned. This results in coils of reduced coupling efficiency, but fortunately very large inductance values are not used at high frequencies. The iron-core chokes are restricted to low-frequency applications, but ferrite-core chokes are good at high frequencies. Several practical inductors are illustrated in figure B-17. In a variable inductor the magnetic coupling between the windings of the coil or the effective number of turns is varied in order to change the total inductance value. The magnetic coupling can be changed by varying the orientation of part of the coil winding, or the effective number of turns can be varied by positioning a silver-plated brass core.

Another type of variable choke is called the swinging choke. The inductance of this choke increases with decreasing current. This type of choke is used to advantage in certain power-supply circuits.

Large inductors, rated in henries, are used principally in power applications. The frequency in these circuits is relatively low, generally 60 Hz or low multiples thereof. In high-frequency circuits, such as those found in FM radios and television sets, very small inductors (of the order of microhenries) are frequently used.

**Transformers.**   A transformer is two or more inductors arranged so that the coils are inductively coupled to each other. As in simple inductors, the coupling can be through air, ferrite, or iron depending on the size of the inductance and its desired frequency of operation. Two kinds of transformers are most commonly available as standard products. High-frequency pulse transformers are used for coupling ac signals while isolating the dc levels of primary and secondary circuits. Power transformers are used to provide various 60-Hz voltages from the power lines. Power transformers now usually have two primary windings. This allows them to be used with either 115-V or 230-V power as shown in figure B-18. A single power transformer can provide several voltages for different circuit needs by including multiple secondaries in the same transformer.

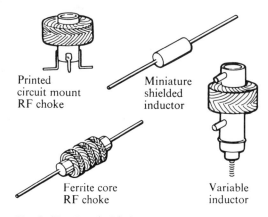

**Fig. B-17.**   Practical inductors.

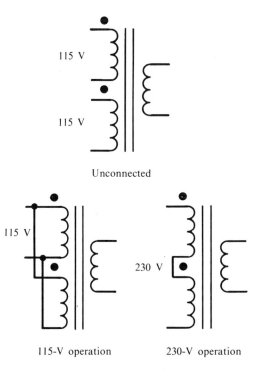

**Fig. B-18.**   Connection of dual-primary power transformers for 115-V or 230-V power sources.

# Appendix C     Manufacturer's Specifications

This appendix presents specification sheets for selected semiconductor devices and integrated circuits. The specifications for a transistor, a diode, and several ICs are presented as examples of the manufacturer's literature. This appendix begins with a list of major semiconductor and IC manufacturers and their addresses.

## Semiconductor and IC Manufacturers

Applications information, data books, and specification sheets for particular devices can be obtained from the following companies:

Advanced Micro Devices, Inc. (AMD), 901 Thompson Place, Sunnyvale, CA 94086

American Microsystems Inc (AMI), 3800 Homestead Road, Santa Clara, CA 95051

Analog Devices (AD), P.O. Box 280, Route 1 Industrial Park, Norwood, MA 02062

Burr-Brown Research Corp., International Airport Industrial Park, Tucson, AZ 85734

Datel-Intersil, 11 Cabot Blvd., Mansfield, MA 02048

EMM/SEMI, 2000 W. 14th Street, Tempe, AZ 85281

Exar Integrated Systems, 750 Palomar Avenue, Sunnyvale, CA 94086

Fairchild Semiconductor, 464 Ellis Street, Mountain View, CA 94042

Fujitsu America, Inc., 2945 Kifer Road, Santa Clara, CA 95051

General Electric Co. (GE), Schenectady, NY 13201

General Instrument Corp. (GI), 600 West John Street, Hicksville, NY 11802

Harris Semiconductor, P.O. Box 883, Melbourne, FL 32901

Hewlett-Packard (HP), 11000 Wolfe Road, Cupertino, CA 95014

Hitachi America, Ltd., 707 W. Algonquin Road, Arlington Heights, IL 60005

Hughes Aircraft, 500 Superior Avenue, Newport Beach, CA 92662

Intech/Function Modules, Inc., 282 Brokaw Road, Santa Clara, CA 95050

Intel Corp., 3065 Bowers Avenue, Santa Clara, CA 95051

Intersil Inc., 10900 N. Tantau Avenue, Cupertino, CA 95014

ITT Semiconductors, 500 Broadway, Lawrence, MA 01841

Lambda Semiconductor, 121 International Drive, Corpus Christi, TX 78410

Mitel Semiconductor Inc., P.O. Box 13089, Danata, Ottawa, Canada K2K 1X3

Monolithic Memories, 1165 East Arques Avenue, Sunnyvale, CA 94086

Mostek Corp., 1215 West Crosby Road, Carrollton, TX 75006

Motorola Semiconductor Products, 5005 East McDowell Road, Phoenix, AZ 85008

National Semiconductor, 2900 Semiconductor Drive, Santa Clara, CA 95051

NEC Electron, Inc., 3120 Central Expressway, Santa Clara, CA 95051

OKI Semiconductor, 1333 Lawrence Expressway, Santa Clara, CA 95051

Plessey Semiconductors, 1641 Kaiser, Irvine, CA 92714

Precision Monolithics, Inc. (PMI), 1500 Space Park Drive, Santa Clara, CA 95050

Pro-Log Corp., 2411 Garden Road, Monterey, CA 93940

Raytheon Semiconductor, 350 Ellis Street, Mountain View, CA 94042

RCA Solid State Div., Box 3200, Somerville, NJ 08876

Reticon Corp., 345 Potrero, Sunnyvale, CA 94086

Rockwell Microelectronics Div., P.O. Box 3669, 3310 Miraloma Avenue, Anaheim, CA 92803

Signetics Corp., 811 East Arques Avenue, Sunnyvale, CA 94086

Silicon General, Inc., 11651 Monarch Street, Garden Grove, CA 92641

Siliconix, Inc., 2201 Laurelwood Road, Santa Clara, CA 95054

Solitron Devices, Inc., 8808 Balboa Avenue, San Diego, CA 92123

Sprague Electric Co., 115 Northeast Cutoff, Worcester, MA 01606

Synertek, Inc., 3001 Stender Way, Santa Clara, CA 95051

Teledyne Philbrick, Allied Drive at Route 128, Dedham, MA 02026

Teledyne Semiconductor, 1300 Terra Bella Avenue, Mountain View, CA 94043

Texas Instruments, Inc. (TI), P.O. Box 225012, Dallas, TX 75265

Toshiba America, Inc., 2151 Michelson Drive, Irvine, CA 92715

TRW LSI Products, P.O. Box 1125, Redondo Beach, CA 90278

Western Digital Corp., 3128 Red Hill Avenue, Newport Beach, CA 92663

Zilog, Inc., 10340 Bubb Road, Cupertino, CA 95014

## IC Logos

Manufacturers of ICs use different logos and symbols on their products for ease in identification. For the newcomer, however, these markings are often not self-explanatory. Therefore, the following list has been compiled to aid in identifying the manufacturers of IC devices.

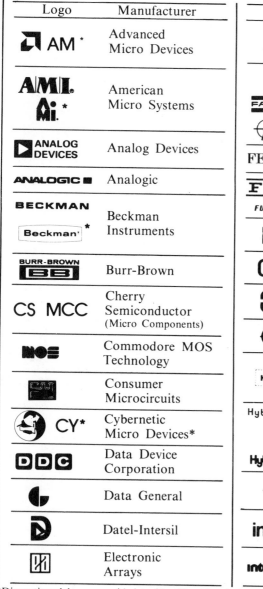

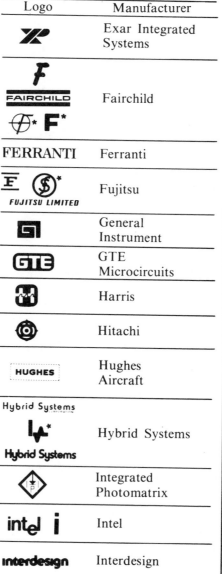

| Logo | Manufacturer |
|---|---|
| AM * | Advanced Micro Devices |
| AMI. Ai. * | American Micro Systems |
| ANALOG DEVICES | Analog Devices |
| ANALOGIC | Analogic |
| BECKMAN Beckman' * | Beckman Instruments |
| BURR-BROWN BB | Burr-Brown |
| CS MCC | Cherry Semiconductor (Micro Components) |
| MOS | Commodore MOS Technology |
| | Consumer Microcircuits |
| CY* | Cybernetic Micro Devices* |
| DDC | Data Device Corporation |
| | Data General |
| D | Datel-Intersil |
| | Electronic Arrays |

| Logo | Manufacturer |
|---|---|
| | Exar Integrated Systems |
| F FAIRCHILD F* F* | Fairchild |
| FERRANTI | Ferranti |
| F FUJITSU LIMITED * | Fujitsu |
| GI | General Instrument |
| GTE | GTE Microcircuits |
| | Harris |
| | Hitachi |
| HUGHES | Hughes Aircraft |
| Hybrid Systems * Hybrid Systems | Hybrid Systems |
| P | Integrated Photomatrix |
| intel i | Intel |
| interdesign | Interdesign |

| Logo | Manufacturer |
|---|---|
| INTERSIL AMS * | Intersil |
| ITT Semiconductors | ITT Semiconductors |
| LSI | LSI Computer Systems |
| λ | Lambda |
| ML | Master Logic |
| | Micro Networks |
| Mii | Micropac Industries |
| M | Micro Power Systems |
| | Mitel Semiconductor |
| | Mitsubishi |
| MMI * | Monolithic Memories |
| MOSTEK | Mostek |
| | Motorola |
| NS * | National Semiconductor |

*Discontinued logos provided to identify older devices.

Reprinted with permission from IC Master, United Technical Publications, Inc., Garden City, NY, 1981.

| Logo | Manufacturer |
|------|--------------|
| **NCR** | NCR |
| *NEC* | NEC |
| *Nitron* / **Ni** * NC * | Nitron |
| **NORTEC** | Nortec |
| ◇OKI® | OKI |
| **OEi** / OEI * | Optical Electronics |
| ▲ MATSUSHITA | Panasonic (Matsushita) |
| **PLESSEY** | Plessey |
| **PMI** / PMI * | Precision Monolithics |
| **RCA** Solid State | RCA |
| RAY | Raytheon Semiconductor |
| **RETICON**® | Reticon |
| Rockwell | Rockwell (Collins) |

| Logo | Manufacturer |
|------|--------------|
| SGS / ATES * | SGS-ATES Semiconductor |
| ⬡ | Sanyo |
| **signetics** S * | Signetics |
| SG Silicon General | Silicon General |
| SSi | Silicon Systems |
| B | Siliconix |
| SS | Solid State Scientific |
| S S * | Solitron |
| ② SPRAGUE | Sprague Electric |
| ⌇ | Standard Microsystems |
| ⚡ | Supertex |
| S | Synertec |
| TELEDYNE CRYSTALONICS | Teledyne Crystalonics |

| Logo | Manufacturer |
|------|--------------|
| TELEDYNE PHILBRICK | Teledyne Philbrick |
| ⬤ | Teledyne Semiconductor |
| TELE FUN KEN  TFK | Telefunken |
| ti | Texas Instruments |
| ◆ | Thomson-CSF Components |
| **TMX** | TMX |
| *Toshiba* | Toshiba |
| *TRW* | TRW |
| ⬒ | Universal Semiconductor |
| ∅ | Western Digital |
| **Zilog** | Zilog |

*Discontinued logos provided to identify older devices.

Reprinted with permission from IC Master, United Technical Publications, Inc., Garden City, NY, 1981.

## Specification Sheets

Each manufacturer of semiconductor and IC devices produces specification sheets that describe in detail the characteristics of their products. Although the specific formats of the specifications vary from manufacturer to manufacturer, most contain a description of the device, a set of absolute maximum ratings, the recommended operating conditions for the device, the electrical and switching characteristics for the device, and applications data. Where applicable, waveform information and graphical data concerning the variation of parameters are also shown.

The device descriptions often include lead or pin assignments, function tables, if applicable, and information concerning the types of applications for which the device is intended. The absolute maximum ratings specify maximum values for supply voltage, input voltage, operating temperature, and so on. If these maximum ratings are exceeded, the device can be destroyed. The recommended operating conditions give the values of supply voltage, input or output voltages or currents, and temperature that are recommended for the device under normal circumstances. The electrical and switching characteristics contain detailed information about the device. Usually minimum, maximum, and typical values for various static characteristics and speed parameters are given. The applications information often includes typical circuits in which the device can be used.

In the pages that follow, specification sheets for selected digital and linear ICs, transistors, and diodes are presented. The wealth of information available from each manufacturer is apparent in these selected examples.

## TYPES 1N914, 1N914A, 1N914B, 1N915, 1N916, 1N916A, 1N916B, 1N917
## SILICON SWITCHING DIODES
BULLETIN NO. DL-S 7311954, MARCH 1973

**FAST SWITCHING DIODES**

● **Rugged Double-Plug Construction**

**Electrical Equivalents**

1N914 . . . 1N4148 . . . 1N4531
1N914A . . . 1N4446
1N914B . . . 1N4448
1N916 . . . 1N4149
1N916A . . . 1N4447
1N916B . . . 1N4449

**mechanical data**

Double-plug construction affords integral positive contacts by means of a thermal compression bond. Moisture-free stability is ensured through hermetic sealing. The coefficients of thermal expansion of the glass case and the dumet plugs are closely matched to allow extreme temperature excursions. Hot-solder-dipped leads are standard.

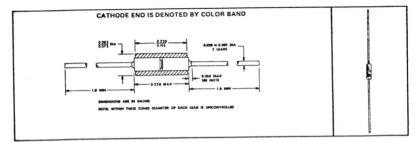

CATHODE END IS DENOTED BY COLOR BAND

**absolute maximum ratings at specified free-air temperature**

| | | 1N914 1N914A 1N914B | 1N915 | 1N916 1N916A 1N916B | 1N917 | UNIT |
|---|---|---|---|---|---|---|
| Working Peak Reverse Voltage from −65°C to 150°C | | 75* | 50* | 75* | 30* | V |
| Average Rectified Forward Current (See Note 1) | at (or below) 25°C | 75* | 75* | 75* | 50* | mA |
| | at 150°C | 10* | 10* | 10* | 10* | |
| Peak Surge Current, 1 Second at 25°C (See Note 2) | | 500* | 500 | 500* | 300 | mA |
| Continuous Power Dissipation at (or below) 25°C (See Note 3) | | 250* | 250 | 250* | 250 | mW |
| Operating Free-Air Temperature Range | | −65 to 175 | | | | °C |
| Storage Temperature Range | | −65 to 200* | | | | °C |
| Lead Temperature 1/16 Inch from Case for 10 Seconds | | 300 | | | | °C |

NOTES: 1. These values may be applied continuously under a single-phase 60-Hz half-sine-wave operation with resistive load.
2. These values apply for a one-second square-wave pulse with the devices at nonoperating thermal equilibrium immediately prior to the surge.
3. Derate linearly to 175°C free-air temperature at the rate of 1.67 mW/°C.

*JEDEC registered data

## TEXAS INSTRUMENTS
INCORPORATED
POST OFFICE BOX 5012 • DALLAS, TEXAS 75222

1

Reprinted with permission from *The Transistor and Diode Data Book for Design Engineers*, Texas Instruments, Inc., Dallas, TX, 1973, p. 10–19.

## TYPES 1N914, 1N914A, 1N914B, 1N915, 1N916, 1N916A, 1N916B, 1N917
## SILICON SWITCHING DIODES

### 1N914 SERIES AND 1N915

*electrical characteristics at 25°C free-air temperature (unless otherwise noted)

| PARAMETER | | TEST CONDITIONS | | 1N914 MIN | 1N914 MAX | 1N914A MIN | 1N914A MAX | 1N914B MIN | 1N914B MAX | 1N915 MIN | 1N915 MAX | UNIT |
|---|---|---|---|---|---|---|---|---|---|---|---|---|
| $V_{(BR)}$ | Reverse Breakdown Voltage | $I_R = 100\,\mu A$ | | 100 | | 100 | | 100 | | 65 | | V |
| $I_R$ | Static Reverse Current | $V_R = 10\,V$ | | | | | | | | | 25 | nA |
| | | $V_R = 20\,V$ | | | 25 | | 25 | | 25 | | | |
| | | $V_R = 20\,V$, | $T_A = 100°C$ | | | | | | 3 | | 5 | |
| | | $V_R = 20\,V$, | $T_A = 150°C$ | | 50 | | 50 | | 50 | | | $\mu A$ |
| | | $V_R = 50\,V$ | | | | | | | | | 5 | |
| | | $V_R = 75\,V$ | | | 5 | | 5 | | 5 | | | |
| $V_F$ | Static Forward Voltage | $I_F = 5\,mA$ | | | | | | 0.62 | 0.72 | 0.6 | 0.73 | |
| | | $I_F = 10\,mA$ | | | 1 | | | | | | | |
| | | $I_F = 20\,mA$ | See Note 4 | | | | 1 | | | | | V |
| | | $I_F = 50\,mA$ | | | | | | | | | 1 | |
| | | $I_F = 100\,mA$ | | | | | | | 1 | | | |
| $C_T$ | Total Capacitance | $V_R = 0$, | $f = 1\,MHz$ | | 4 | | 4 | | 4 | | 4 | pF |

### 1N916 SERIES AND 1N917

*electrical characteristics at 25°C free-air temperature (unless otherwise noted)

| PARAMETER | | TEST CONDITIONS | | 1N916 MIN | 1N916 MAX | 1N916A MIN | 1N916A MAX | 1N916B MIN | 1N916B MAX | 1N917 MIN | 1N917 MAX | UNIT |
|---|---|---|---|---|---|---|---|---|---|---|---|---|
| $V_{(BR)}$ | Reverse Breakdown Voltage | $I_R = 100\,\mu A$ | | 100 | | 100 | | 100 | | 40 | | V |
| $I_R$ | Static Reverse Current | $V_R = 10\,V$ | | | | | | | | | 50 | nA |
| | | $V_R = 20\,V$ | | | 25 | | 25 | | 25 | | | |
| | | $V_R = 20\,V$, | $T_A = 100°C$ | | | | | | 3 | | 25 | |
| | | $V_R = 20\,V$, | $T_A = 150°C$ | | 50 | | 50 | | 50 | | | $\mu A$ |
| | | $V_R = 75\,V$ | | | 5 | | 5 | | 5 | | | |
| $V_F$ | Static Forward Voltage | $I_F = 0.25\,mA$ | | | | | | | | | 0.64 | |
| | | $I_F = 1.5\,mA$ | | | | | | | | | 0.74 | |
| | | $I_F = 3.5\,mA$ | | | | | | | | | 0.83 | |
| | | $I_F = 5\,mA$ | | | | | | 0.63 | 0.73 | | | V |
| | | $I_F = 10\,mA$ | | | 1 | | | | | | 1 | |
| | | $I_F = 20\,mA$ | See Note 4 | | | | 1 | | | | | |
| | | $I_F = 30\,mA$ | | | | | | | 1 | | | |
| $C_T$ | Total Capacitance | $V_R = 0$, | $f = 1\,MHz$ | | 2 | | 2 | | 2 | | 2.5 | pF |

NOTE 4: These parameters must be measured using pulse techniques. $t_w = 300\,\mu s$, duty cycle ≤ 2%.

*JEDEC registered data

**TEXAS INSTRUMENTS**
INCORPORATED
POST OFFICE BOX 5012 · DALLAS, TEXAS 75222

Reprinted with permission from *The Transistor and Diode Data Book for Design Engineers,* Texas Instruments, Inc., Dallas, TX, 1973, p. 10–20.

## TYPES 1N914, 1N914A, 1N914B, 1N915, 1N916, 1N916A, 1N916B, 1N917
## SILICON SWITCHING DIODES

**operating characteristics at 25°C free-air temperature**

| PARAMETER | | TEST CONDITIONS | 1N914 1N914A 1N914B 1N916 1N916A 1N916B | | 1N915 | | 1N917 | | UNIT |
|---|---|---|---|---|---|---|---|---|---|
| | | | MIN | MAX | MIN | MAX | MIN | MAX | |
| $t_{rr}$ | Reverse Recovery Time | $I_F = 10$ mA,  $I_{RM} = 10$ mA, $i_{rr} = 1$ mA, $R_L = 100$ Ω,  See Figure 1 (Condition 1) | | 8 | | 10* | | 3* | ns |
| | | $I_F = 10$ mA,  $V_R = 6$ V,  $i_{rr} = 1$ mA, $R_L = 100$ Ω,  See Figure 1 (Condition 2) | | 4* | | | | | ns |
| $V_{FM(rec)}$ | Forward Recovery Voltage | $I_F = 50$ mA,  $R_L = 50$ Ω,  See Figure 2 | | 2.5* | | | | | V |
| $\eta_r$ | Rectification Efficiency | $V_r = 2$ V,  $R_L = 5$ kΩ,  $C_L = 20$ pF, $Z_{source} = 50$ Ω,  f = 100 MHz | 45* | | | | | | % |

### PARAMETER MEASUREMENT INFORMATION

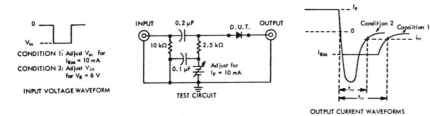

**CONDITION 1:** Adjust $V_{in}$ for $I_{RM} = 10$ mA
**CONDITION 2:** Adjust $V_{in}$ for $V_R = 6$ V

INPUT VOLTAGE WAVEFORM

TEST CIRCUIT

OUTPUT CURRENT WAVEFORMS

**FIGURE 1 — REVERSE RECOVERY TIME**

NOTES: a. The input pulse is supplied by a generator with the following characteristics: $Z_{out} = 50$ Ω, $t_r \leq 0.5$ ns, $t_w = 100$ ns.
b. Output waveforms are monitored on an oscilloscope with the following characteristics: $t_r \leq 0.6$ ns, $Z_{in} = 50$ Ω.

TEST CIRCUIT

VOLTAGE WAVEFORMS

**FIGURE 2 — FORWARD RECOVERY VOLTAGE**

NOTES: c. The input pulse is supplied by a generator with the following characteristics: $Z_{out} = 50$ Ω, $t_r \leq 30$ ns, $t_w = 100$ ns, PRR = 5 to 100 kHz.
d. The output waveform is monitored on an oscilloscope with the following characteristics: $t_r \leq 15$ ns, $R_{in} \geq 1$ MΩ, $C_{in} \leq 5$ pF.

\* JEDEC registered data

PRINTED IN U S A
TI cannot assume any responsibility for any circuits shown
or represent that they are free from patent infringement.
TEXAS INSTRUMENTS RESERVES THE RIGHT TO MAKE CHANGES AT ANY TIME
IN ORDER TO IMPROVE DESIGN AND TO SUPPLY THE BEST PRODUCT POSSIBLE.

**TEXAS INSTRUMENTS**
INCORPORATED
POST OFFICE BOX 5012 • DALLAS, TEXAS 75222

Reprinted with permission from *The Transistor and Diode Data Book for Design Engineers,*
Texas Instruments, Inc., Dallas, TX, 1973, p. 10–21.

# TYPES 2N2217 THRU 2N2222, 2N2218A, 2N2219A, 2N2221A, 2N2222A
## N-P-N SILICON TRANSISTORS
BULLETIN NO. DL-S 7311916, MARCH 1973

### DESIGNED FOR HIGH-SPEED, MEDIUM-POWER SWITCHING AND GENERAL PURPOSE AMPLIFIER APPLICATIONS

- $h_{FE}$ . . . Guaranteed from 100 $\mu$A to 500 mA
- High $f_T$ at 20 V, 20 mA . . . 300 MHz (2N2219A, 2N2222A)
  250 MHz (all others)
- 2N2218, 2N2221 for Complementary Use with 2N2904, 2N2906
- 2N2219, 2N2222 for Complementary Use with 2N2905, 2N2906

*mechanical data

Device types 2N2217, 2N2218, 2N2218A, 2N2219, and 2N2219A are in JEDEC TO-5 packages.
Device types 2N2220, 2N2221, 2N2221A, 2N2222, and 2N2222A are in JEDEC TO-18 packages.

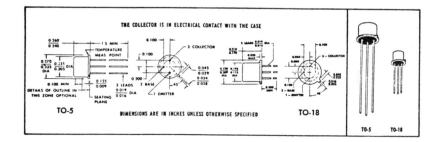

*absolute maximum ratings at 25°C free-air temperature (unless otherwise noted)

|  | 2N2217 2N2218 2N2219 | 2N2218A 2N2219A | 2N2220 2N2221 2N2222 | 2N2221A 2N2222A | UNIT |
|---|---|---|---|---|---|
| Collector-Base Voltage | 60 | 75 | 60 | 75 | V |
| Collector-Emitter Voltage (See Note 1) | 30 | 40 | 30 | 40 | V |
| Emitter-Base Voltage | 5 | 6 | 5 | 6 | V |
| Continuous Collector Current | 0.8 | 0.8 | 0.8 | 0.8 | A |
| Continuous Device Dissipation at (or below) 25°C Free-Air Temperature (See Notes 2 and 3) | 0.8 | 0.8 | 0.5 | 0.5 | W |
| Continuous Device Dissipation at (or below) 25°C Case Temperature (See Notes 4 and 5) | 3 | 3 | 1.8 | 1.8 | W |
| Operating Collector Junction Temperature Range | −65 to 175 | | | | °C |
| Storage Temperature Range | −65 to 200 | | | | °C |
| Lead Temperature 1/16 Inch from Case for 10 Seconds | 230 | | | | °C |

NOTES: 1. These values apply between 0 and 500 mA collector current when the base-emitter diode is open-circuited.
2. Derate 2N2217, 2N2218, 2N2218A, 2N2219, and 2N2219A linearly to 175°C free-air temperature at the rate of 5.33 mW/°C.
3. Derate 2N2220, 2N2221, 2N2221A, 2N2222, and 2N2222A linearly to 175°C free-air temperature at the rate of 3.33 mW/°C.
4. Derate 2N2217, 2N2218, 2N2218A, 2N2219, and 2N2219A linearly to 175°C case temperature at the rate of 20.0 mW/°C.
5. Derate 2N2220, 2N2221, 2N2221A, 2N2222, and 2N2222A linearly to 175°C case temperature at the rate of 12.0 mW/°C.

*JEDEC registered data. This data sheet contains all applicable registered data in effect at the time of publication.

USES CHIP N24

## TEXAS INSTRUMENTS
INCORPORATED
POST OFFICE BOX 5012 • DALLAS, TEXAS 75222

Reprinted with permission from *The Transistor and Diode Data Book for Design Engineers,* Texas Instruments, Inc., Dallas, TX, 1973, p. 4–93.

## TYPES 2N2217 THRU 2N2222, 2N2218A, 2N2219A, 2N2221A, 2N2222A
## N-P-N SILICON TRANSISTORS

### 2N2217 THRU 2N2222

*electrical characteristics at 25°C free-air temperature (unless otherwise noted)

| PARAMETER | | TEST CONDITIONS | TO-5 → 2N2217 TO-18 → 2N2220 | | TO-5 → 2N2218 TO-18 → 2N2221 | | TO-5 → 2N2219 TO-18 → 2N2222 | | UNIT |
|---|---|---|---|---|---|---|---|---|---|
| | | | MIN | MAX | MIN | MAX | MIN | MAX | |
| $V_{(BR)CBO}$ | Collector-Base Breakdown Voltage | $I_C = 10\ \mu A$, $I_E = 0$ | 60 | | 60 | | 60 | | V |
| $V_{(BR)CEO}$ | Collector-Emitter Breakdown Voltage | $I_C = 10\ mA$, $I_B = 0$, See Note 6 | 30 | | 30 | | 30 | | V |
| $V_{(BR)EBO}$ | Emitter-Base Breakdown Voltage | $I_E = 10\ \mu A$, $I_C = 0$ | 5 | | 5 | | 5 | | V |
| $I_{CBO}$ | Collector Cutoff Current | $V_{CB} = 50\ V$, $I_E = 0$ | | 10 | | 10 | | 10 | nA |
| | | $V_{CB} = 50\ V$, $I_E = 0$, $T_A = 150°C$ | | 10 | | 10 | | 10 | $\mu A$ |
| $I_{EBO}$ | Emitter Cutoff Current | $V_{EB} = 3\ V$, $I_C = 0$ | | 10 | | 10 | | 10 | nA |
| $h_{FE}$ | Static Forward Current Transfer Ratio | $V_{CE} = 10\ V$, $I_C = 100\ \mu A$ | | | 20 | | 35 | | |
| | | $V_{CE} = 10\ V$, $I_C = 1\ mA$ | 12 | | 25 | | 50 | | |
| | | $V_{CE} = 10\ V$, $I_C = 10\ mA$ | 17 | | 35 | | 75 | | |
| | | $V_{CE} = 10\ V$, $I_C = 150\ mA$  See Note 6 | 20 | 60 | 40 | 120 | 100 | 300 | |
| | | $V_{CE} = 10\ V$, $I_C = 500\ mA$ | | | 20 | | 30 | | |
| | | $V_{CE} = 1\ V$, $I_C = 150\ mA$ | 10 | | 20 | | 50 | | |
| $V_{BE}$ | Base-Emitter Voltage | $I_B = 15\ mA$, $I_C = 150\ mA$  See Note 6 | | 1.3 | | 1.3 | | 1.3 | V |
| | | $I_B = 50\ mA$, $I_C = 500\ mA$ | | | | 2.6 | | 2.6 | |
| $V_{CE(sat)}$ | Collector-Emitter Saturation Voltage | $I_B = 15\ mA$, $I_C = 150\ mA$  See Note 6 | | 0.4 | | 0.4 | | 0.4 | V |
| | | $I_B = 50\ mA$, $I_C = 500\ mA$ | | | | 1.6 | | 1.6 | |
| $|h_{fe}|$ | Small-Signal Common-Emitter Forward Current Transfer Ratio | $V_{CE} = 20\ V$, $I_C = 20\ mA$, $f = 100\ MHz$ | 2.5 | | 2.5 | | 2.5 | | |
| $f_T$ | Transition Frequency | $V_{CE} = 20\ V$, $I_C = 20\ mA$, See Note 7 | 250 | | 250 | | 250 | | MHz |
| $C_{obo}$ | Common-Base Open-Circuit Output Capacitance | $V_{CB} = 10\ V$, $I_E = 0$, $f = 1\ MHz$ | 8 | | 8 | | 8 | | pF |
| $h_{ie(real)}$ | Real Part of Small-Signal Common-Emitter Input Impedance | $V_{CE} = 20\ V$, $I_C = 20\ mA$, $f = 300\ MHz$ | 60 | | 60 | | 60 | | Ω |

NOTES:   6.  These parameters must be measured using pulse techniques. $t_w = 300\ \mu s$, duty cycle ≤ 2%.
7.  To obtain $f_T$, the $|h_{fe}|$ response with frequency is extrapolated at the rate of −6 dB per octave from $f = 100\ MHz$ to the frequency at which $|h_{fe}| = 1$.

switching characteristics at 25°C free-air temperature

| | PARAMETER | TEST CONDITIONS† | TYP | UNIT |
|---|---|---|---|---|
| $t_d$ | Delay Time | $V_{CC} = 30\ V$,      $I_C = 150\ mA$,  $I_{B(1)} = 15\ mA$, | 5 | ns |
| $t_r$ | Rise Time | $V_{BE(off)} = -0.5\ V$,                     See Figure 1 | 15 | ns |
| $t_s$ | Storage Time | $V_{CC} = 30\ V$,        $I_C = 150\ mA$,  $I_{B(1)} = 15\ mA$, | 190 | ns |
| $t_f$ | Fall Time | $I_{B(2)} = -15\ mA$,                     See Figure 2 | 23 | ns |

†Voltage and current values shown are nominal; exact values vary slightly with transistor parameters.

*JEDEC registered data

### TEXAS INSTRUMENTS
INCORPORATED
POST OFFICE BOX 5012 • DALLAS, TEXAS 75222

Reprinted with permission from *The Transistor and Diode Data Book for Design Engineers,* Texas Instruments, Inc., Dallas, TX, 1973, p. 4–94.

### TYPES 2N2217 THRU 2N2222, 2N2218A, 2N2219A, 2N2221A, 2N2222A
### N-P-N SILICON TRANSISTORS

**2N2218A, 2N2219A, 2N2221A, 2N2222A**

**•electrical characteristics at 25°C free-air temperature (unless otherwise noted)**

| PARAMETER | | TEST CONDITIONS | TO-5 → <br> TO-18 → | 2N2218A <br> 2N2221A | | 2N2219A <br> 2N2222A | | UNIT |
|---|---|---|---|---|---|---|---|---|
| | | | | MIN | MAX | MIN | MAX | |
| $V_{(BR)CBO}$ | Collector-Base Breakdown Voltage | $I_C = 10\,\mu A$, $I_E = 0$ | | 75 | | 75 | | V |
| $V_{(BR)CEO}$ | Collector-Emitter Breakdown Voltage | $I_C = 10\,mA$, $I_B = 0$, | See Note 6 | 40 | | 40 | | V |
| $V_{(BR)EBO}$ | Emitter-Base Breakdown Voltage | $I_E = 10\,\mu A$, $I_C = 0$ | | 6 | | 6 | | V |
| $I_{CBO}$ | Collector Cutoff Current | $V_{CB} = 60\,V$, $I_E = 0$ | | | 10 | | 10 | nA |
| | | $V_{CB} = 60\,V$, $I_E = 0$, | $T_A = 150°C$ | | 10 | | 10 | μA |
| $I_{CEV}$ | Collector Cutoff Current | $V_{CE} = 60\,V$, $V_{BE} = -3\,V$ | | | 10 | | 10 | nA |
| $I_{BEV}$ | Base Cutoff Current | $V_{CE} = 60\,V$, $V_{BE} = -3\,V$ | | | -20 | | -20 | nA |
| $I_{EBO}$ | Emitter Cutoff Current | $V_{EB} = 3\,V$, $I_C = 0$ | | | 10 | | 10 | nA |
| $h_{FE}$ | Static Forward Current <br> Transfer Ratio | $V_{CE} = 10\,V$, $I_C = 100\,\mu A$ | | 20 | | 35 | | |
| | | $V_{CE} = 10\,V$, $I_C = 1\,mA$ | | 25 | | 50 | | |
| | | $V_{CE} = 10\,V$, $I_C = 10\,mA$ | | 35 | | 75 | | |
| | | $V_{CE} = 10\,V$, $I_C = 150\,mA$ | See Note 6 | 40 | 120 | 100 | 300 | |
| | | $V_{CE} = 10\,V$, $I_C = 500\,mA$ | | 25 | | 40 | | |
| | | $V_{CE} = 1\,V$, $I_C = 150\,mA$ | | 20 | | 50 | | |
| | | $V_{CE} = 10\,V$, $I_C = 10\,mA$, <br> $T_A = -55°C$ | | 15 | | 35 | | |
| $V_{BE}$ | Base-Emitter Voltage | $I_B = 15\,mA$, $I_C = 150\,mA$ | See Note 6 | 0.6 | 1.2 | 0.6 | 1.2 | V |
| | | $I_B = 50\,mA$, $I_C = 500\,mA$ | | | 2 | | 2 | |
| $V_{CE(sat)}$ | Collector-Emitter Saturation Voltage | $I_B = 15\,mA$, $I_C = 150\,mA$ | See Note 6 | | 0.3 | | 0.3 | V |
| | | $I_B = 50\,mA$, $I_C = 500\,mA$ | | | 1 | | 1 | |
| $h_{ie}$ | Small-Signal Common-Emitter <br> Input Impedance | $V_{CE} = 10\,V$, $I_C = 1\,mA$ | | 1 | 3.5 | 2 | 8 | kΩ |
| | | $V_{CE} = 10\,V$, $I_C = 10\,mA$ | | 0.2 | 1 | 0.25 | 1.25 | |
| $h_{fe}$ | Small-Signal Forward Current <br> Transfer Ratio | $V_{CE} = 10\,V$, $I_C = 1\,mA$ | | 30 | 150 | 50 | 300 | |
| | | $V_{CE} = 10\,V$, $I_C = 10\,mA$ | $f = 1\,kHz$ | 50 | 300 | 75 | 375 | |
| $h_{re}$ | Small-Signal Common-Emitter <br> Reverse Voltage Transfer Ratio | $V_{CE} = 10\,V$, $I_C = 1\,mA$ | | | $5\times10^{-4}$ | | $8\times10^{-4}$ | |
| | | $V_{CE} = 10\,V$, $I_C = 10\,mA$ | | | $2.5\times10^{-4}$ | | $4\times10^{-4}$ | |
| $h_{oe}$ | Small-Signal Common-Emitter <br> Output Admittance | $V_{CE} = 10\,V$, $I_C = 1\,mA$ | | 3 | 15 | 5 | 35 | μmho |
| | | $V_{CE} = 10\,V$, $I_C = 10\,mA$ | | 10 | 100 | 25 | 200 | |
| $|h_{fe}|$ | Small-Signal Common-Emitter <br> Forward Current Transfer Ratio | $V_{CE} = 20\,V$, $I_C = 20\,mA$, | $f = 100\,MHz$ | 2.5 | | 3 | | |
| $f_T$ | Transition Frequency | $V_{CE} = 20\,V$, $I_C = 20\,mA$, | See Note 7 | 250 | | 300 | | MHz |
| $C_{obo}$ | Common-Base Open-Circuit <br> Output Capacitance | $V_{CB} = 10\,V$, $I_E = 0$, | $f = 100\,kHz$ | | 8 | | 8 | pF |
| $C_{ibo}$ | Common-Base Open-Circuit <br> Input Capacitance | $V_{EB} = 0.5\,V$, $I_C = 0$, | $f = 100\,kHz$ | | 25 | | 25 | pF |
| $h_{ie(real)}$ | Real Part of Small-Signal <br> Common-Emitter Input Impedance | $V_{CE} = 20\,V$, $I_C = 20\,mA$, | $f = 300\,MHz$ | | 60 | | 60 | Ω |
| $r_b'C_c$ | Collector-Base Time Constant | $V_{CE} = 20\,V$, $I_C = 20\,mA$, | $f = 31.8\,MHz$ | | 150 | | 150 | ps |

NOTES: 6. These parameters must be measured using pulse techniques. $t_w = 300\,\mu s$, duty cycle ≤ 2%.
7. To obtain $f_T$, the $|h_{fe}|$ response with frequency is extrapolated at the rate of −6 dB per octave from $f = 100\,MHz$ to the frequency at which $|h_{fe}| = 1$.
*JEDEC registered data

### TEXAS INSTRUMENTS
INCORPORATED
POST OFFICE BOX 5012 • DALLAS, TEXAS 75222

Reprinted with permission from *The Transistor and Diode Data Book for Design Engineers*, Texas Instruments, Inc., Dallas, TX, 1973, p. 4-95.

## TYPES 2N2217 THRU 2N2222, 2N2218A, 2N2219A, 2N2221A, 2N2222A
## N-P-N SILICON TRANSISTORS

**\*operating characteristics at 25°C free-air temperature**

| PARAMETER | | TEST CONDITIONS | TO-5 → | 2N2218A | 2N2219A | |
|---|---|---|---|---|---|---|
| | | | TO-18 → | 2N2221A | 2N2222A | UNIT |
| | | | | MAX | MAX | |
| F | Spot Noise Figure | $V_{CE}$ = 10 V,  $I_C$ = 100 μA,  $R_G$ = 1 kΩ, f = 1 kHz | | | 4 | dB |

**\*switching characteristics at 25°C free-air temperature**

| PARAMETER | | TEST CONDITIONS† | TO-5 → | 2N2218A | 2N2219A | |
|---|---|---|---|---|---|---|
| | | | TO-18 → | 2N2221A | 2N2222A | UNIT |
| | | | | MAX | MAX | |
| $t_d$ | Delay Time | $V_{CC}$ = 30 V,    $I_C$ = 150 mA,    $I_{B(1)}$ = 15 mA, $V_{BE(off)}$ = −0.5 V,    See Figure 1 | | 10 | 10 | ns |
| $t_r$ | Rise Time | | | 25 | 25 | ns |
| $τ_A$ | Active Region Time Constant‡ | | | 2.5 | 2.5 | ns |
| $t_s$ | Storage Time | $V_{CC}$ = 30 V,    $I_C$ = 150 mA,    $I_{B(1)}$ = 15 mA, $I_{B(2)}$ = −15 mA,    See Figure 2 | | 225 | 225 | ns |
| $t_f$ | Fall Time | | | 60 | 60 | ns |

†Voltage and current values shown are nominal; exact values vary slightly with transistor parameters.

‡Under the given conditions $τ_A$ is equal to $\frac{t_r}{10}$ .

### \*PARAMETER MEASUREMENT INFORMATION

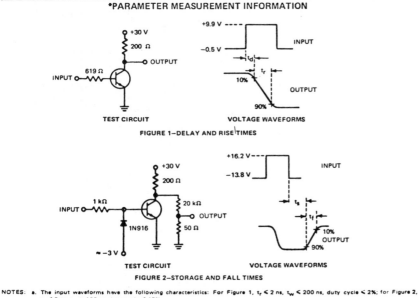

**FIGURE 1—DELAY AND RISE TIMES**

**FIGURE 2—STORAGE AND FALL TIMES**

NOTES:  a. The input waveforms have the following characteristics: For Figure 1, $t_r$ ≤ 2 ns, $t_w$ ≤ 200 ns, duty cycle ≤ 2%; for Figure 2, $t_f$ ≤ 5 ns, $t_w$ ≈ 100 μs, duty cycle ≤ 17%.

b. All waveforms are monitored on an oscilloscope with the following characteristics: $t_r$ ≤ 5 ns, $R_{in}$ ≥ 100 kΩ, $C_{in}$ ≤ 12 pF.

\*JEDEC registered data

PRINTED IN U.S.A

TI cannot assume any responsibility for any circuits shown
or represent that they are free from patent infringement.

TEXAS INSTRUMENTS RESERVES THE RIGHT TO MAKE CHANGES AT ANY TIME
IN ORDER TO IMPROVE DESIGN AND TO SUPPLY THE BEST PRODUCT POSSIBLE.

**TEXAS INSTRUMENTS**
INCORPORATED
POST OFFICE BOX 5012 · DALLAS, TEXAS 75222

Reprinted with permission from *The Transistor and Diode Data Book for Design Engineers*, 2nd ed., Texas Instruments, Inc. Dallas, TX, 1976, p. 4–96.

## POSITIVE-NAND GATES AND INVERTERS WITH TOTEM-POLE OUTPUTS

### recommended operating conditions

| | | SERIES 54 / SERIES 74 '00, '04, '10, '20, '30 | | | SERIES 54H / SERIES 74H 'H00, 'H04, 'H10, 'H20, 'H30 | | | SERIES 54L / SERIES 74L 'L00, 'L04, 'L10, 'L20, 'L30 | | | SERIES 54LS / SERIES 74LS 'LS00, 'LS04, 'LS10, 'LS20, 'LS30 | | | SERIES 54S / SERIES 74S 'S00, 'S04, 'S10, 'S20, 'S30, 'S133 | | | UNIT |
|---|---|---|---|---|---|---|---|---|---|---|---|---|---|---|---|---|---|
| | | MIN | NOM | MAX | MIN | NOM | MAX | MIN | NOM | MAX | MIN | NOM | MAX | MIN | NOM | MAX | |
| Supply voltage, Vcc | 54 Family | 4.5 | 5 | 5.5 | 4.5 | 5 | 5.5 | 4.5 | 5 | 5.5 | 4.5 | 5 | 5.5 | 4.5 | 5 | 5.5 | V |
| | 74 Family | 4.75 | 5 | 5.25 | 4.75 | 5 | 5.25 | 4.75 | 5 | 5.25 | 4.75 | 5 | 5.25 | 4.75 | 5 | 5.25 | |
| High-level output current, IOH | 54 Family | | | -400 | | | -500 | | | -100 | | | -400 | | | -1000 | µA |
| | 74 Family | | | -400 | | | -500 | | | -200 | | | -400 | | | -1000 | |
| Low-level output current, IOL | 54 Family | | | 16 | | | 20 | | | 2 | | | 4 | | | 20 | mA |
| | 74 Family | | | 16 | | | 20 | | | 3.6 | | | 8 | | | 20 | |
| Operating free-air temperature, TA | 54 Family | -55 | | 125 | -55 | | 125 | -55 | | 125 | -55 | | 125 | -55 | | 125 | °C |
| | 74 Family | 0 | | 70 | 0 | | 70 | 0 | | 70 | 0 | | 70 | 0 | | 70 | |

### electrical characteristics over recommended operating free-air temperature range (unless otherwise noted)

| PARAMETER | TEST FIGURE | TEST CONDITIONS[†] | SERIES 54 / SERIES 74 '00, '04, '10, '20, '30 | | | SERIES 54H / SERIES 74H 'H00, 'H04, 'H10, 'H20, 'H30 | | | SERIES 54L / SERIES 74L 'L00, 'L04, 'L10, 'L20, 'L30 | | | SERIES 54LS / SERIES 74LS 'LS00, 'LS04, 'LS10, 'LS20, 'LS30 | | | SERIES 54S / SERIES 74S 'S00, 'S04, 'S10, 'S20, 'S30, 'S133 | | | UNIT |
|---|---|---|---|---|---|---|---|---|---|---|---|---|---|---|---|---|---|---|
| | | | MIN | TYP‡ | MAX | MIN | TYP‡ | MAX | MIN | TYP‡ | MAX | MIN | TYP‡ | MAX | MIN | TYP‡ | MAX | |
| VIH  High-level input voltage | 1,2 | | 2 | | | 2 | | | 2 | | | 2 | | | 2 | | | V |
| VIL  Low-level input voltage | 1,2 | 54 Family | | | 0.8 | | | 0.8 | | | 0.7 | | | 0.7 | | | 0.8 | V |
| | | 74 Family | | | 0.8 | | | 0.8 | | | 0.7 | | | 0.8 | | | 0.8 | |
| VIK  Input clamp voltage | 3 | Vcc = MIN, II = § | | | -1.5 | | | -1.5 | | | | | | -1.5 | | | -1.2 | V |
| VOH  High-level output voltage | 1 | Vcc = MIN, VIH = VIL max, IOH = MAX | 54 Family | 2.4 | 3.4 | | 2.4 | 3.5 | | 2.4 | 3.3 | | 2.5 | 3.4 | | 2.5 | 3.4 | V |
| | | | 74 Family | 2.4 | 3.4 | | 2.4 | 3.5 | | 2.4 | 3.2 | | 2.7 | 3.4 | | 2.7 | 3.4 | |
| VOL  Low-level output voltage | 2 | Vcc = MIN, VIH = 2V, IOL = MAX | 54 Family | | 0.2 | 0.4 | | 0.2 | 0.4 | | 0.15 | 0.3 | | 0.25 | 0.4 | | | 0.5 | V |
| | | | 74 Family | | 0.2 | 0.4 | | 0.2 | 0.4 | | 0.2 | 0.4 | | 0.25 | 0.5 | | | 0.5 | |
| | | IOL = 4 mA  Series 74LS | | | | | | | | | | | | 0.4 | | | | |
| II  Input current at maximum input voltage | 4 | VI = 5.5 V / VI = 7 V | Vcc = MAX | | | 1 | | | 1 | | | 0.1 | | | 0.1 | | | 1 | mA |
| IIH  High-level input current | 4 | VIH = 2.4 V / VIH = 2.7 V | Vcc = MAX | | | 40 | | | 50 | | | 10 | | | 20 | | | 50 | µA |
| IIL  Low-level input current | 5 | VIL = 0.3 V / VIL = 0.4 V / VIL = 0.5 V | Vcc = MAX | | | -1.6 | | | -2 | | | -0.18 | | | -0.4 | | | -2 | mA |
| IOS  Short-circuit output current♦ | 6 | Vcc = MAX | 54 Family | -20 | | -55 | -40 | | -100 | -3 | | -15 | -20 | | -100 | -40 | | -100 | mA |
| | | | 74 Family | -18 | | -55 | -40 | | -100 | -3 | | -15 | -20 | | -100 | -40 | | -100 | |
| ICC  Supply current | 7 | Vcc = MAX | | | | | | | | | | | See table on next page | | | | | | mA |

[†] For conditions shown as MIN or MAX, use the appropriate value specified under recommended operating conditions.

[‡] All typical values are at Vcc = 5 V, TA = 25°C.

[§] II = -12 mA for SN54'/SN74', -8 mA for SN54H'/SN74H', and -18 mA for SN54LS'/SN74LS' and SN54S'/SN74S'.

[♦] Not more than one output should be shorted at a time, and for SN54H'/SN74H', SN54S'/SN74S', and SN54LS'/SN74LS', duration of short-circuit should not exceed 1 second.

TEXAS INSTRUMENTS
INCORPORATED
POST OFFICE BOX 5012 • DALLAS, TEXAS 75222

Reprinted with permission from *The TTL Data Book for Design Engineers*, 2nd ed., Texas Instruments, Inc., Dallas, TX, 1976, p. 6–2.

# POSITIVE-NAND GATES AND INVERTERS WITH TOTEM-POLE OUTPUTS

## supply current¶

| TYPE | $I_{CCH}$ (mA) Total with outputs high | | $I_{CCL}$ (mA) Total with outputs low | | $I_{CC}$ (mA) Average per gate (50% duty cycle) |
|---|---|---|---|---|---|
| | TYP | MAX | TYP | MAX | TYP |
| '00 | 4 | 8 | 12 | 22 | 2 |
| '04 | 6 | 12 | 18 | 33 | 2 |
| '10 | 3 | 6 | 9 | 16.5 | 2 |
| '20 | 2 | 4 | 6 | 11 | 2 |
| '30 | 1 | 2 | 3 | 6 | 2 |
| 'H00 | 10 | 16.8 | 26 | 40 | 4.5 |
| 'H04 | 16 | 26 | 40 | 58 | 4.5 |
| 'H10 | 7.5 | 12.6 | 19.5 | 30 | 4.5 |
| 'H20 | 5 | 8.4 | 13 | 20 | 4.5 |
| 'H30 | 2.5 | 4.2 | 6.5 | 10 | 4.5 |
| 'L00 | 0.44 | 0.8 | 1.16 | 2.04 | 0.20 |
| 'L04 | 0.66 | 1.2 | 1.74 | 3.06 | 0.20 |
| 'L10 | 0.33 | 0.6 | 0.87 | 1.53 | 0.20 |
| 'L20 | 0.22 | 0.4 | 0.58 | 1.02 | 0.20 |
| SN54L30 | 0.11 | 0.33 | 0.29 | 0.51 | 0.20 |
| SN74L30 | 0.11 | 0.2 | 0.29 | 0.51 | 0.20 |
| 'LS00 | 0.8 | 1.6 | 2.4 | 4.4 | 0.4 |
| 'LS04 | 1.2 | 2.4 | 3.6 | 6.6 | 0.4 |
| 'LS10 | 0.6 | 1.2 | 1.8 | 3.3 | 0.4 |
| 'LS20 | 0.4 | 0.8 | 1.2 | 2.2 | 0.4 |
| 'LS30 | 0.35 | 0.5 | 0.6 | 1.1 | 0.48 |
| 'S00 | 10 | 16 | 20 | 36 | 3.75 |
| 'S04 | 16 | 24 | 30 | 54 | 3.75 |
| 'S10 | 7.5 | 12 | 15 | 27 | 3.75 |
| 'S20 | 5 | 8 | 10 | 18 | 3.75 |
| 'S30 | 3 | 5 | 6.5 | 10 | 4.25 |
| 'S133 | 3 | 5 | 5.5 | 10 | 4.25 |

¶ Maximum values of $I_{CC}$ are over the recommended operating ranges of $V_{CC}$ and $T_A$; typical values are at $V_{CC} = 5$ V, $T_A = 25°C$.

## switching characteristics at $V_{CC} = 5$ V, $T_A = 25°C$

| TYPE | TEST CONDITIONS# | $t_{PLH}$ (ns) Propagation delay time, low-to-high-level output | | | $t_{PHL}$ (ns) Propagation delay time, high-to-low-level output | | |
|---|---|---|---|---|---|---|---|
| | | MIN | TYP | MAX | MIN | TYP | MAX |
| '00, '10 | $C_L = 15$ pF, $R_L = 400$ Ω | | 11 | 22 | | 7 | 15 |
| '04, '20 | | | 12 | 22 | | 8 | 15 |
| '30 | | · | 13 | .22 | | 8 | 15 |
| 'H00 | $C_L = 25$ pF, $R_L = 280$ Ω | | 5.9 | 10 | | 6.2 | 10 |
| 'H04 | | | 6 | 10 | | 6.5 | 10 |
| 'H10 | | | 5.9 | 10 | | 6.3 | 10 |
| 'H20 | | | 6 | 10 | | 7 | 10 |
| 'H30 | | | 6.8 | 10 | | 8.9 | 12 |
| 'L00, 'L04, 'L10, 'L20 | $C_L = 50$ pF, $R_L = 4$ kΩ | | 35 | 60 | | 31 | 60 |
| 'L30 | | | 35 | 60 | | 70 | 100 |
| 'LS00, 'LS04 | $C_L = 15$ pF, $R_L = 2$ kΩ | | 9 | 15 | | 10 | 15 |
| 'LS10, 'LS20 | | | 8 | 15 | | 13 | 20 |
| 'LS30 | | | | | | | |
| 'S00, 'S04 | $C_L = 15$ pF, $R_L = 280$ Ω | | 3 | 4.5 | | 3 | 5 |
| 'S10, 'S20 | $C_L = 50$ pF, $R_L = 280$ Ω | | 4.5 | | | 4.5 | |
| 'S30, 'S133 | $C_L = 50$ pF, $R_L = 280$ Ω | | 5.5 | | | 6.5 | 7 |

#Load circuits and voltage waveforms are shown on pages 3-10 and 3-11.

'S00, 'S04, 'S10, 'S20, 'S30, 'S133 CIRCUITS

'LS00, 'LS04, 'LS10, 'LS20, 'LS30 CIRCUITS
*The 12-kΩ resistor is not on 'LS30.

'H00, 'H04, 'H10, 'H20, 'H30 CIRCUITS

Resistor values shown are nominal and in ohms.

### schematics (each gate)

| CIRCUIT | R1 | R2 | R3 | R4 |
|---|---|---|---|---|
| '00, '04, '10, '20, '30 | 4 | 1.6 | 130 | 1 |
| 'L00, 'L04, 'L10, 'L20, 'L30 | 40 | 20 | 500 | 12 |

'00, '04, '10, '20, '30, 'L00, 'L04, 'L10, 'L20, 'L30, CIRCUITS

Input clamp diodes not on SN54L'/SN74L' circuits.

TEXAS INSTRUMENTS
INCORPORATED
POST OFFICE BOX 5012 · DALLAS, TEXAS 75222

Reprinted with permission from *The TTL Data Book for Design Engineers*, 2nd ed., Texas Instruments, Inc., Dallas, TX, 1976, p. 6-3.

## SERIES 54/74 FLIP-FLOPS

**recommended operating conditions**

| | SERIES 54/74 | '70 MIN | NOM | MAX | '72, '73, '76, '107 MIN | NOM | MAX | '74 MIN | NOM | MAX | '108 MIN | NOM | MAX | '110 MIN | NOM | MAX | '111 MIN | NOM | MAX | UNIT |
|---|---|---|---|---|---|---|---|---|---|---|---|---|---|---|---|---|---|---|---|---|
| Supply voltage, V_CC | Series 54 | 4.5 | 5 | 5.5 | 4.5 | 5 | 5.5 | 4.5 | 5 | 5.5 | 4.5 | 5 | 5.5 | 4.5 | 5 | 5.5 | 4.5 | 5 | 5.5 | V |
| | Series 74 | 4.75 | 5 | 5.25 | 4.75 | 5 | 5.25 | 4.75 | 5 | 5.25 | 4.75 | 5 | 5.25 | 4.75 | 5 | 5.25 | 4.75 | 5 | 5.25 | |
| High-level output current, I_OH | | | | -400 | | | -400 | | | -400 | | | -800 | | | -800 | | | -800 | µA |
| Low-level output current, I_OL | | | | 16 | | | 16 | | | 16 | | | 16 | | | 16 | | | 16 | mA |
| Pulse width, t_w | Clock high | 20 | | | 20 | | | 30 | | | 20 | | | 25 | | | 25 | | | ns |
| | Clock low | 30 | | | 47 | | | 37 | | | | | | 25 | | | 25 | | | |
| | Preset or clear low | 25 | | | 25 | | | 30 | | | 20 | | | 25 | | | 25 | | | |
| Input setup time, t_su | | 20† | | | 0† | | | 20† | | | 10† | | | 20† | | | 0† | | | ns |
| Input hold time, t_h | | 5† | | | 0† | | | 5† | | | 6† | | | 5† | | | 30† | | | ns |
| Operating free-air temperature, T_A | Series 54 | -55 | | 125 | -55 | | 125 | -55 | | 125 | -55 | | 125 | -55 | | 125 | -55 | | 125 | °C |
| | Series 74 | 0 | | 70 | 0 | | 70 | 0 | | 70 | 0 | | 70 | 0 | | 70 | 0 | | 70 | |

† The arrow indicates the edge of the clock pulse used for reference: ↑ for the rising edge, ↓ for the falling edge.

**electrical characteristics over recommended operating free-air temperature range (unless otherwise noted)**

| PARAMETER | | TEST CONDITIONS† | '70 MIN | TYP‡ | MAX | '72, '73, '76, '107 MIN | TYP‡ | MAX | '74 MIN | TYP‡ | MAX | '109 MIN | TYP‡ | MAX | '110 MIN | TYP‡ | MAX | '111 MIN | TYP‡ | MAX | UNIT |
|---|---|---|---|---|---|---|---|---|---|---|---|---|---|---|---|---|---|---|---|---|---|
| V_IH High-level input voltage | | | 2 | | | 2 | | | 2 | | | 2 | | | 2 | | | 2 | | | V |
| V_IL Low-level input voltage | | | | | 0.8 | | | 0.8 | | | 0.8 | | | 0.8 | | | 0.8 | | | 0.8 | V |
| V_IK Input clamp voltage | | V_CC = MIN, I_I = -12 mA | | | -1.5 | | | -1.5 | | | -1.5 | | | -1.5 | | | -1.5 | | | -1.5 | V |
| V_OH High-level output voltage | | V_CC = MIN, V_IH = 2 V, V_IL = 0.8 V, I_OH = MAX | 2.4 | 3.4 | | 2.4 | 3.4 | | 2.4 | 3.4 | | 2.4 | 3.4 | | 2.4 | 3.4 | | 2.4 | 3.4 | | V |
| V_OL Low-level output voltage | | V_CC = MIN, V_IH = 2 V, V_IL = 0.8 V, I_OL = 16 mA | | 0.2 | 0.4 | | 0.2 | 0.4 | | 0.2 | 0.4 | | 0.2 | 0.4 | | 0.2 | 0.4 | | 0.2 | 0.4 | V |
| I_I Input current at maximum input voltage | | V_CC = MAX, V_I = 5.5 V | | | 1 | | | 1 | | | 1 | | | 1 | | | 1 | | | 1 | mA |
| I_IH High-level input current | D, J, K, or K̄ | V_CC = MAX, V_I = 2.4 V | | | 40 | | | 40 | | | 40 | | | 40 | | | 40 | | | 40 | µA |
| | Clear | | | | 80 | | | 80 | | | 120 | | | 160 | | | 160 | | | 80 | |
| | Preset | | | | 80 | | | 80 | | | 80 | | | 80 | | | 160 | | | 80 | |
| | Clock | | | | 40 | | | 80 | | | 80 | | | 40 | | | 40 | | | 120 | |
| I_IL Low-level input current | D, J, K, or K̄ | V_CC = MAX, V_I = 0.4 V | | | -1.6 | | | -1.6 | | | -1.6 | | | -1.6 | | | -1.6 | | | -1.6 | mA |
| | Clear | | | | -3.2 | | | -3.2 | | | -3.2 | | | -4.8 | | | -3.2 | | | -3.2 | |
| | Preset | | | | -3.2 | | | -3.2 | | | -1.6 | | | -3.2 | | | -3.2 | | | -3.2 | |
| | Clock | | | | -1.6 | | | -3.2 | | | -3.2 | | | -1.6 | | | -1.6 | | | -4.8 | |
| I_OS Short-circuit output current* | Series 54 | V_CC = MAX | -20 | | -57 | -20 | | -57 | -20 | | -57 | -30 | | -85 | -20 | | -57 | -20 | | -57 | mA |
| | Series 74 | | -18 | | -57 | -18 | | -57 | -18 | | -57 | -30 | | -85 | -18 | | -57 | -18 | | -57 | |
| I_CC Supply current (Average per flip-flop) | | V_CC = MAX, See Note 1 | 13 | 26 | | 10 | 20 | | 8.5 | 15 | | 9 | 15 | | 20 | 34 | | 14 | 20.5 | | mA |

† For conditions shown as MIN or MAX, use the appropriate value specified under recommended operating conditions.
‡ All typical values are at V_CC = 5 V, T_A = 25°C.
*Not more than one output should be shorted at a time.
● Clear is tested with preset high and preset is tested with clear high.
NOTE 1: With all outputs open, I_CC is measured with the Q and Q̄ outputs high in turn. At the time of measurement, the clock input is at 4.5 V for the '70, '110, and '111; and

### TEXAS INSTRUMENTS
INCORPORATED
POST OFFICE BOX 5012 • DALLAS, TEXAS 75222

Reprinted with permission from *The TTL Data Book for Design Engineers*, 2nd ed., Texas Instruments, Inc., Dallas, TX, 1976, p. 6–46.

## SERIES 54/74 FLIP-FLOPS

### switching characteristics, VCC = 5 V, TA = 25°C

| PARAMETER | FROM (INPUT) | TO (OUTPUT) | TEST CONDITIONS | '70 MIN | '70 TYP | '70 MAX | '72,'73 '76,'107 MIN | '72,'73 '76,'107 TYP | '72,'73 '76,'107 MAX | '74 MIN | '74 TYP | '74 MAX | '109 MIN | '109 TYP | '109 MAX | '110 MIN | '110 TYP | '110 MAX | '111 MIN | '111 TYP | '111 MAX | UNIT |
|---|---|---|---|---|---|---|---|---|---|---|---|---|---|---|---|---|---|---|---|---|---|---|
| $f_{max}$ | | | | 20 | 36 | | 15 | 20 | | 15 | 25 | | 25 | 33 | | 20 | 25 | | 20 | 25 | | MHz |
| $t_{PLH}$ | Preset | Q | | | | 50 | | 16 | 25 | | | 25 | | 10 | 16 | | 12 | 20 | | 12 | 18 | ns |
| $t_{PHL}$ | (as applicable) | Q̄ | CL = 15 pF, | | | 50 | | 25 | 40 | | | 40 | | 23 | 35 | | 18 | 25 | | 21 | 30 | ns |
| $t_{PLH}$ | Clear | Q̄ | RL = 400 Ω, | | | 50 | | 16 | 25 | | | 25 | | 10 | 15 | | 12 | 20 | | 12 | 18 | ns |
| $t_{PHL}$ | (as applicable) | Q | See Note 2 | | | 50 | | 25 | 40 | | | 40 | | 17 | 25 | | 18 | 25 | | 21 | 30 | ns |
| $t_{PLH}$ | Clock | Q or Q̄ | | | 27 | 50 | | 16 | 25 | | 14 | 25 | | 10 | 16 | | 20 | 30 | | 12 | 17 | ns |
| $t_{PHL}$ | | | | | 18 | 50 | | 25 | 40 | | 20 | 40 | | 18 | 28 | | 13 | 20 | | 20 | 30 | ns |

¶ $f_{max}$ ≡ maximum clock frequency; $t_{PLH}$ ≡ propagation delay time, low-to-high-level output; $t_{PHL}$ ≡ propagation delay time, high-to-low-level output.
NOTE 2: Load circuit and voltage waveforms are shown on page 3-10.

### functional block diagrams

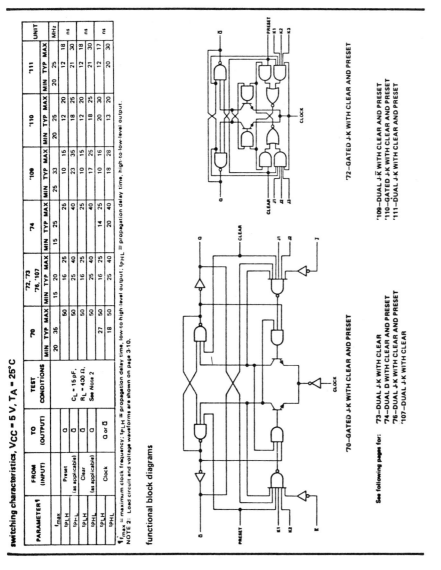

'70–GATED J-K WITH CLEAR AND PRESET

'72–GATED J-K̄ WITH CLEAR AND PRESET

See following pages for:  '73–DUAL J-K WITH CLEAR
'74–DUAL D WITH CLEAR AND PRESET
'76–DUAL J-K WITH CLEAR AND PRESET
'107–DUAL J-K WITH CLEAR

'109–DUAL J-K̄ WITH CLEAR AND PRESET
'110–GATED J-K WITH CLEAR AND PRESET
'111–DUAL J-K WITH CLEAR AND PRESET

TEXAS INSTRUMENTS
INCORPORATED
POST OFFICE BOX 5012 • DALLAS, TEXAS 75222

Reprinted with permission from *The TTL Data Book for Design Engineers*, 2nd ed., Texas Instruments, Inc., Dallas, TX, 1976, p. 6–47.

## TTL
## MSI

### TYPES SN5485, SN54L85, SN54LS85, SN54S85, SN7485, SN74L85, SN74LS85, SN74S85
### 4-BIT MAGNITUDE COMPARATORS
BULLETIN NO. DL-S 7611810, MARCH 1974 – REVISED OCTOBER 1976

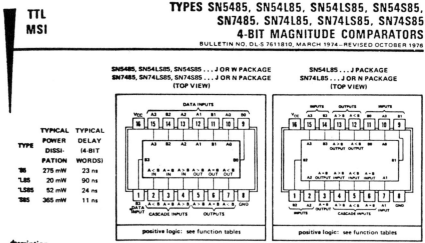

| TYPE | TYPICAL POWER DISSIPATION | TYPICAL DELAY (4-BIT WORDS) |
|------|------|------|
| '85 | 275 mW | 23 ns |
| 'L85 | 20 mW | 90 ns |
| 'LS85 | 52 mW | 24 ns |
| 'S85 | 365 mW | 11 ns |

### description

These four-bit magnitude comparators perform comparison of straight binary and straight BCD (8-4-2-1) codes. Three fully decoded decisions about two 4-bit words (A, B) are made and are externally available at three outputs. These devices are fully expandable to any number of bits without external gates. Words of greater length may be compared by connecting comparators in cascade. The $A > B$, $A < B$, and $A = B$ outputs of a stage handling less-significant bits are connected to the corresponding $A > B$, $A < B$, and $A = B$ inputs of the next stage handling more-significant bits. The stage handling the least-significant bits must have a high-level voltage applied to the $A = B$ input and in addition for the 'L85, low-level voltages applied to the $A > B$ and $A < B$ inputs. The cascading paths of the '85, 'LS85, and 'S85 are implemented with only a two-gate-level delay to reduce overall comparison times for long words. An alternate method of cascading which further reduces the comparison time is shown in the typical application data.

### FUNCTION TABLES

| COMPARING INPUTS | | | | CASCADING INPUTS | | | OUTPUTS | | |
|---|---|---|---|---|---|---|---|---|---|
| A3, B3 | A2, B2 | A1, B1 | A0, B0 | A > B | A < B | A = B | A > B | A < B | A = B |
| A3 > B3 | X | X | X | X | X | X | H | L | L |
| A3 < B3 | X | X | X | X | X | X | L | H | L |
| A3 = B3 | A2 > B2 | X | X | X | X | X | H | L | L |
| A3 = B3 | A2 < B2 | X | X | X | X | X | L | H | L |
| A3 = B3 | A2 = B2 | A1 > B1 | X | X | X | X | H | L | L |
| A3 = B3 | A2 = B2 | A1 < B1 | X | X | X | X | L | H | L |
| A3 = B3 | A2 = B2 | A1 = B1 | A0 > B0 | X | X | X | H | L | L |
| A3 = B3 | A2 = B2 | A1 = B1 | A0 < B0 | X | X | X | L | H | L |
| A3 = B3 | A2 = B2 | A1 = B1 | A0 = B0 | H | L | L | H | L | L |
| A3 = B3 | A2 = B2 | A1 = B1 | A0 = B0 | L | H | L | L | H | L |
| A3 = B3 | A2 = B2 | A1 = B1 | A0 = B0 | L | L | H | L | L | H |

'85, 'LS85, 'S85

| A3 = B3 | A2 = B2 | A1 = B1 | A0 = B0 | X | X | H | L | L | H |
| A3 = B3 | A2 = B2 | A1 = B1 | A0 = B0 | H | H | L | L | L | L |
| A3 = B3 | A2 = B2 | A1 = B1 | A0 = B0 | L | L | L | H | H | L |

'L85

| A3 = B3 | A2 = B2 | A1 = B1 | A0 = B0 | L | H | H | L | H | H |
| A3 = B3 | A2 = B2 | A1 = B1 | A0 = B0 | H | L | H | H | L | H |
| A3 = B3 | A2 = B2 | A1 = B1 | A0 = B0 | H | H | H | H | H | H |
| A3 = B3 | A2 = B2 | A1 = B1 | A0 = B0 | H | H | L | H | H | L |
| A3 = B3 | A2 = B2 | A1 = B1 | A0 = B0 | L | L | L | L | L | L |

H = high level, L = low level, X = irrelevant

## TEXAS INSTRUMENTS
INCORPORATED
POST OFFICE BOX 5012 • DALLAS, TEXAS 75222

Reprinted with permission from *The TTL Data Book for Design Engineers*, 2nd ed., Texas Instruments, Inc., Dallas, TX, 1976, p. 7–57.

**TYPES SN5485, SN54L85, SN54LS85, SN54S85,
SN7485, SN74L85, SN74LS85, SN74S85
4-BIT MAGNITUDE COMPARATORS**

**functional block diagrams**

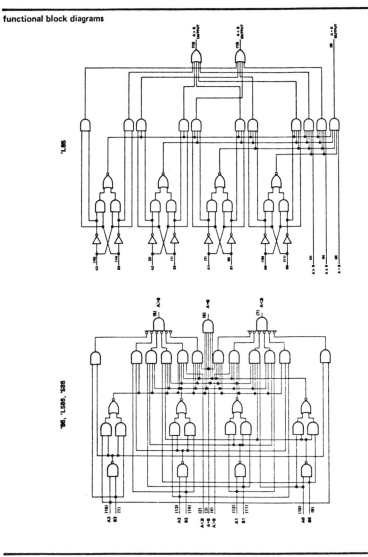

**TEXAS INSTRUMENTS**
INCORPORATED
POST OFFICE BOX 5012 · DALLAS, TEXAS 75222

Reprinted with permission from *The TTL Data Book for Design Engineers*, 2nd ed., Texas Instruments, Inc., Dallas, TX, 1976, p. 7–58.

TYPES SN5485, SN54L85, SN54LS85, SN54S85,
SN7485, SN74L85, SN74LS85, SN74S85
4-BIT MAGNITUDE COMPARATORS

schematics of inputs and outputs

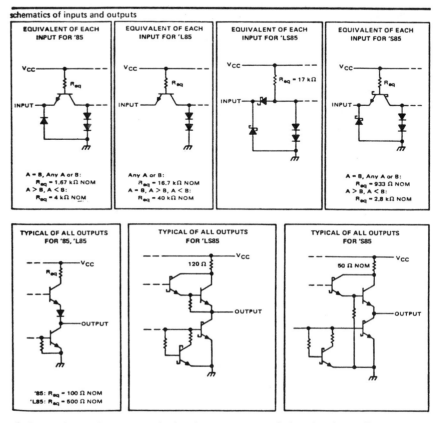

absolute maximum ratings over operating free-air temperature range (unless otherwise noted)

|  | SN54' SN54S' | SN54L' | SN54LS' | SN74' SN74S' | SN74L' | SN74LS' | UNIT |
|---|---|---|---|---|---|---|---|
| Supply voltage, $V_{CC}$ (see Note 1) | 7 | 8 | 7 | 7 | 8 | 7 | V |
| Input voltage (see Note 2) | 5.5 | 5.5 | 7 | 5.5 | 5.5 | 7 | V |
| Interemitter voltage (see Note 3) | 5.5 | | 5.5 | | | | V |
| Operating free-air temperature range | −55 to 125 | | | 0 to 70 | | | °C |
| Storage temperature range | −65 to 150 | | | −65 to 150 | | | °C |

NOTES:  1.  Voltage values, except interemitter voltage, are with respect to network ground terminal.
2.  Input voltages for 'L85 must be zero or positive with respect to network ground terminal.
3.  This is the voltage between two emitters of a multiple-emitter input transistor. This rating applies to each A input in conjunction with its respective B input of the '85 and 'S85.

TEXAS INSTRUMENTS
INCORPORATED
POST OFFICE BOX 5012 • DALLAS, TEXAS 75222

Reprinted with permission from *The TTL Data Book for Design Engineers*, 2nd ed., Texas Instruments, Inc., Dallas, TX, 1976, p. 7–59.

## TYPES SN5485, SN7485
## 4-BIT MAGNITUDE COMPARATORS

**recommended operating conditions**

| | SN5485 MIN | SN5485 NOM | SN5485 MAX | SN7485 MIN | SN7485 NOM | SN7485 MAX | UNIT |
|---|---|---|---|---|---|---|---|
| Supply voltage, $V_{CC}$ | 4.5 | 5 | 5.5 | 4.75 | 5 | 5.25 | V |
| High-level output current, $I_{OH}$ | | | −400 | | | −400 | µA |
| Low-level output current, $I_{OL}$ | | | 16 | | | 16 | mA |
| Operating free-air temperature, $T_A$ | −55 | | 125 | 0 | | 70 | °C |

**electrical characteristics over recommended operating free-air temperature range (unless otherwise noted)**

| | PARAMETER | | TEST CONDITIONS[†] | | MIN | TYP[‡] | MAX | UNIT |
|---|---|---|---|---|---|---|---|---|
| $V_{IH}$ | High-level input voltage | | | | 2 | | | V |
| $V_{IL}$ | Low-level input voltage | | | | | | 0.8 | V |
| $V_{IK}$ | Input clamp voltage | | $V_{CC}$ = MIN, | $I_I$ = −12 mA | | | −1.5 | V |
| $V_{OH}$ | High-level output voltage | | $V_{CC}$ = MIN, $V_{IL}$ = 0.8 V, | $V_{IH}$ = 2 V, $I_{OH}$ = −400 µA | 2.4 | 3.4 | | V |
| $V_{OL}$ | Low-level output voltage | | $V_{CC}$ = MIN, $V_{IL}$ = 0.8 V, | $V_{IH}$ = 2 V, $I_{OL}$ = 16 mA | | 0.2 | 0.4 | V |
| $I_I$ | Input current at maximum input voltage | | $V_{CC}$ = MAX, | $V_I$ = 5.5 V | | | 1 | mA |
| $I_{IH}$ | High-level input current | A < B, A > B inputs | $V_{CC}$ = MAX, | $V_I$ = 2.4 V | | | 40 | µA |
| | | all other inputs | | | | | 120 | |
| $I_{IL}$ | Low-level input current | A < B, A > B inputs | $V_{CC}$ = MAX, | $V_I$ = 0.4 V | | | −1.6 | mA |
| | | all other inputs | | | | | −4.8 | |
| $I_{OS}$ | Short-circuit output current[§] | SN5485 | $V_{CC}$ = MAX, $V_O$ = 0 | | −20 | | −55 | mA |
| | | SN7485 | | | −18 | | −55 | |
| $I_{CC}$ | Supply current | | $V_{CC}$ = MAX, See Note 4 | | | 55 | 88 | mA |

[†]For conditions shown as MIN or MAX, use the appropriate value specified under recommended operating conditions.
[‡]All typical values are at $V_{CC}$ = 5 V, $T_A$ = 25°C.
[§]Not more than one output should be shorted at a time.
NOTE 4: $I_{CC}$ is measured with outputs open, A = B grounded, and all other inputs at 4.5 V.

**switching characteristics, $V_{CC}$ = 5 V, $T_A$ = 25°C**

| PARAMETER[¶] | FROM INPUT | TO OUTPUT | NUMBER OF GATE LEVELS | TEST CONDITIONS | MIN | TYP | MAX | UNIT |
|---|---|---|---|---|---|---|---|---|
| $t_{PLH}$ | Any A or B data input | A < B, A > B | 1 | | | 7 | | ns |
| | | | 2 | | | 12 | | |
| | | | 3 | | | 17 | 26 | |
| | | A = B | 4 | | | 23 | 35 | |
| $t_{PHL}$ | Any A or B data input | A < B, A > B | 1 | | | 11 | | ns |
| | | | 2 | | | 15 | | |
| | | | 3 | $C_L$ = 15 pF, | | 20 | 30 | |
| | | A = B | 4 | $R_L$ = 400 Ω, See Note 5 | | 20 | 30 | |
| $t_{PLH}$ | A < B or A = B | A > B | 1 | | | 7 | 11 | ns |
| $t_{PHL}$ | A < B or A = B | A > B | 1 | | | 11 | 17 | ns |
| $t_{PLH}$ | A = B | A = B | 2 | | | 13 | 20 | ns |
| $t_{PHL}$ | A = B | A = B | 2 | | | 11 | 17 | ns |
| $t_{PLH}$ | A > B or A = B | A < B | 1 | | | 7 | 11 | ns |
| $t_{PHL}$ | A > B or A = B | A < B | 1 | | | 11 | 17 | ns |

[¶]$t_{PLH}$ = propagation delay time, low-to-high-level output
$t_{PHL}$ = propagation delay time, high-to-low-level output.
NOTE 5: Load circuit and voltage waveforms are shown on page 3-10.

### TEXAS INSTRUMENTS
INCORPORATED
POST OFFICE BOX 5012 • DALLAS, TEXAS 75222

Reprinted with permission from *The TTL Data Book for Design Engineers*, 2nd ed., Texas Instruments, Inc., Dallas, TX, 1976, p. 7–60.

## TYPES SN54LS85, SN74LS85
## 4-BIT MAGNITUDE COMPARATORS
REVISED OCTOBER 1976

**recommended operating conditions**

| | SN54LS85 | | | SN74LS85 | | | UNIT |
|---|---|---|---|---|---|---|---|
| | MIN | NOM | MAX | MIN | NOM | MAX | |
| Supply voltage, $V_{CC}$ | 4.5 | 5 | 5.5 | 4.75 | 5 | 5.25 | V |
| High-level output current, $I_{OH}$ | | | −400 | | | −400 | µA |
| Low-level output current, $I_{OL}$ | | | 4 | | | 8 | mA |
| Operating free-air temperature, $T_A$ | −55 | | 125 | 0 | | 70 | °C |

**electrical characteristics over recommended operating free-air temperature range (unless otherwise noted)**

| PARAMETER | | TEST CONDITIONS[†] | | SN54LS85 | | | SN74LS85 | | | UNIT |
|---|---|---|---|---|---|---|---|---|---|---|
| | | | | MIN | TYP[‡] | MAX | MIN | TYP[‡] | MAX | |
| $V_{IH}$ | High-level input voltage | | | 2 | | | 2 | | | V |
| $V_{IL}$ | Low-level input voltage | | | | | 0.7 | | | 0.8 | V |
| $V_{IK}$ | Input clamp voltage | $V_{CC}$ = MIN, | $I_I$ = −18 mA | | | −1.5 | | | −1.5 | V |
| $V_{OH}$ | High-level output voltage | $V_{CC}$ = MIN,   $V_{IH}$ = 2 V, $V_{IL}$ = $V_{IL}$ max,  $I_{OH}$ = −400 µA | | 2.5 | 3.4 | | 2.7 | 3.4 | | V |
| $V_{OL}$ | Low-level output voltage | $V_{CC}$ = MIN, $V_{IH}$ = 2 V, $V_{IL}$ = $V_{IL}$ max | $I_{OL}$ = 4 mA | 0.25 | 0.4 | | 0.25 | 0.4 | V |
| | | | $I_{OL}$ = 8 mA | | | | 0.35 | 0.5 | | |
| $I_I$ | Input current at maximum input voltage | A < B, A > B inputs | $V_{CC}$ = MAX,   $V_I$ = 7 V | | | 0.1 | | | 0.1 | mA |
| | | all other inputs | | | | 0.3 | | | 0.3 | |
| $I_{IH}$ | High-level input current | A < B, A > B inputs | $V_{CC}$ = MAX,   $V_I$ = 2.7 V | | | 20 | | | 20 | µA |
| | | all other inputs | | | | 60 | | | 60 | |
| $I_{IL}$ | Low-level input current | A < B, A > B inputs | $V_{CC}$ = MAX,   $V_I$ = 0.4 V | | | −0.4 | | | −0.4 | mA |
| | | all other inputs | | | | −1.2 | | | −1.2 | |
| $I_{OS}$ | Short-circuit output current [§] | $V_{CC}$ = MAX | | −20 | | −100 | −20 | | −100 | mA |
| $I_{CC}$ | Supply current | $V_{CC}$ = MAX,   See Note 4 | | | 10.4 | 20 | | 10.4 | 20 | mA |

[†]For conditions shown as MIN or MAX, use the appropriate value specified under recommended operating conditions.
[‡]All typical values are at $V_{CC}$ = 5 V, $T_A$ = 25°C.
[§]Not more than one output should be shorted at a time, and duration of the short-circuit should not exceed one second.
NOTE 4: $I_{CC}$ is measured with outputs open, A = B grounded, and all other inputs at 4.5 V.

**switching characteristics, $V_{CC}$ = 5 V, $T_A$ = 25°C**

| PARAMETER[¶] | FROM INPUT | TO OUTPUT | NUMBER OF GATE LEVELS | TEST CONDITIONS | MIN | TYP | MAX | UNIT |
|---|---|---|---|---|---|---|---|---|
| $t_{PLH}$ | Any A or B data input | A < B, A > B | 1 | | | 14 | | ns |
| | | | 2 | | | 19 | | |
| | | | 3 | | | 24 | 36 | |
| | | A = B | 4 | | | 27 | 45 | |
| $t_{PHL}$ | Any A or B data input | A < B, A > B | 1 | $C_L$ = 15 pF, $R_L$ = 2 kΩ, See Note 7 | | 11 | | ns |
| | | | 2 | | | 15 | | |
| | | | 3 | | | 20 | 30 | |
| | | A = B | 4 | | | 23 | 45 | |
| $t_{PLH}$ | A < B or A = B | A > B | 1 | | | 14 | 22 | ns |
| $t_{PHL}$ | A < B or A = B | A > B | 1 | | | 11 | 17 | ns |
| $t_{PLH}$ | A = B | A = B | 2 | | | 13 | 20 | ns |
| $t_{PHL}$ | A = B | A = B | 2 | | | 13 | 26 | ns |
| $t_{PLH}$ | A > B or A = B | A < B | 1 | | | 14 | 22 | ns |
| $t_{PHL}$ | A > B or A = B | A < B | 1 | | | 11 | 17 | ns |

[¶]$t_{PLH}$ = propagation delay time, low-to-high-level output
$t_{PHL}$ = propagation delay time, high-to-low-level output
NOTE 7:  Load circuit and voltage waveforms are shown on page 3-11.

**TEXAS INSTRUMENTS**
INCORPORATED
POST OFFICE BOX 5012 • DALLAS, TEXAS 75222

Reprinted with permission from *The TTL Data Book for Design Engineers*, 2nd ed., Texas Instruments, Inc., Dallas, TX, 1976, p. 7–62.

**TYPES SN5485, SN54L85, SN54LS85, SN54S85,**
**SN7485, SN74L85, SN74LS85, SN74S85**
**4-BIT MAGNITUDE COMPARATORS**

## TYPICAL APPLICATION DATA

### COMPARISON OF TWO N-BIT WORDS

This application demonstrates how these magnitude comparators can be cascaded to compare longer words. The example illustrated shows the comparison of two 24-bit words; however, the design is expandable to n-bits. As an example, one comparator can be used with five of the 24-bit comparators illustrated to expand the word length to 120-bits. Typical comparison times for various word lengths using the '85, 'L85, 'LS85, or 'S85 are:

| WORD LENGTH | NUMBER OF PKGS | '85 | 'L85 | 'LS85 | 'S85 |
|---|---|---|---|---|---|
| 1-4 bits | 1 | 23 ns | 90 ns | 24 ns | 11 ns |
| 5-24 bits | 2-6 | 46 ns | 180 ns | 48 ns | 22 ns |
| 25-120 bits | 8-31 | 69 ns | 270 ns | 72 ns | 33 ns |

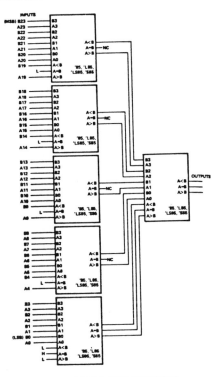

**COMPARISON OF TWO 24-BIT WORDS**

**TEXAS INSTRUMENTS**
INCORPORATED
POST OFFICE BOX 5012 • DALLAS, TEXAS 75222

Reprinted with permission from *The TTL Data Book for Design Engineers*, 2nd ed., Texas Instruments, Inc., Dallas, TX, 1976, p. 7–64.

**LINEAR
INTEGRATED CIRCUITS**

**TL081B, TL082B, TL084B
TYPES TL080 THRU TL084, TL080A THRU TL084A,
JFET-INPUT OPERATIONAL AMPLIFIERS**

BULLETIN NO. DL-S 7712484, JANUARY 1977

### 26 DEVICES COVER COMMERCIAL, INDUSTRIAL, AND MILITARY TEMPERATURE RANGES

- Low Power Consumption
- Wide Common-Mode and Differential Voltage Ranges
- Low Input Bias and Offset Currents
- Output Short-Circuit Protection
- High Input Impedance . . . JFET-Input Stage
- Internal Frequency Compensation (Except TL080, TL080A)
- Latch-Up-Free Operation
- High Slew Rate . . . 12 V/µs Typ

### description

The TL080 JFET-input operational amplifier family is designed to offer a wider selection than any previously developed operational amplifier family. Each of these JFET-input operational amplifiers incorporates well-matched, high-voltage JFET and bipolar transistors in a monolithic integrated circuit. The devices feature high slew rates, low input bias and offset currents, and low offset voltage temperature coefficient. Offset adjustment and external compensation options are available within the TL080 Family.

Device types with an "M" suffix are characterized for operation over the full military temperature range of −55°C to 125°C, those with an "I" suffix are characterized for operation from −25°C to 85°C, and those with a "C" suffix are characterized for operation from 0°C to 70°C.

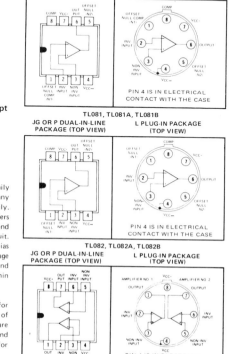

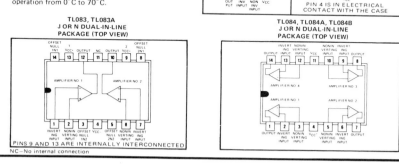

### TEXAS INSTRUMENTS
INCORPORATED

POST OFFICE BOX 5012 • DALLAS, TEXAS 75222

Reprinted with permission from Texas Instruments, Inc., Dallas, TX.

## TYPES TL080 THRU TL084, TL080A THRU TL084A, TL081B, TL082B, TL084B JFET-INPUT OPERATINAL AMPLIFIERS

schematic (each amplifier)

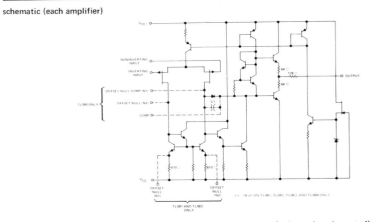

absolute maximum ratings over operating free-air temperature range (unless otherwise noted)

| | | TL08_M TL08_AM | TL08_I | TL08_C TL08_AC TL08_BC | UNIT |
|---|---|---|---|---|---|
| Supply voltage, V$_{CC}$ (see Note 1) | | 22 | 22 | 18 | V |
| Supply voltage, V$_{CC}$ (see Note 1) | | −22 | −22 | −18 | V |
| Differential input voltage (see Note 2) | | ±30 | ±30 | ±30 | V |
| Input voltage (see Notes 1 and 3) | | ±15 | ±15 | ±15 | V |
| Duration of output short circuit (see Note 4) | | Unlimited | Unlimited | Unlimited | |
| Continuous total dissipation at (or below) 25°C free-air | J,JG,N, or P Package | 680 | 680 | 680 | mW |
| temperature (see Note 5) | L Package | 625 | 625 | 625 | |
| Operating free-air temperature range | | −55 to 125 | −25 to 85 | 0 to 70 | °C |
| Storage temperature range | | −65 to 150 | −65 to 150 | −65 to 150 | °C |
| Lead temperature 1/16 inch from case for 60 seconds | N or P Package | 300 | 300 | 300 | °C |
| Lead temperature 1/16 inch from case for 10 seconds | N or P Package | 260 | 260 | 260 | °C |

NOTES: 1. All voltage values, except differential voltages, are with respect to the zero reference level (ground) of the supply voltages where the zero reference level is the midpoint between V$_{CC+}$ and V$_{CC−}$.
2. Differential voltages are at the noninverting input terminal with respect to the inverting input terminal.
3. The magnitude of the input voltage must never exceed the magnitude of the supply voltage or 15 volts, whichever is less.
4. The output may be shorted to ground or to either supply. Temperature and/or supply voltages must be limited to ensure that the dissipation rating is not exceeded.
5. For operation above 25°C free-air temperature, refer to Dissipation Derating Table.

DISSIPATION DERATING TABLE

| PACKAGE | POWER RATING | DERATING FACTOR | ABOVE T$_A$ |
|---|---|---|---|
| J | 680 mW | 8.2 mW/°C | 67°C |
| JG | 680 mW | 7.4 mW/°C | 58°C |
| L | 625 mW | 5.0 mW/°C | 25°C |
| N | 625 mW | 9.2 mW/°C | 76°C |
| P | 680 mW | 8.0 mW/°C | 65°C |

## TEXAS INSTRUMENTS
INCORPORATED
POST OFFICE BOX 5012 • DALLAS, TEXAS 75222

Reprinted with permission from Texas Instruments, Inc., Dallas, TX.

## TYPES TL080 THRU TL084, TL080A THRU TL084A, TL081B, TL082B, TL084B JFET-INPUT OPERATIONAL AMPLIFIERS

electrical characteristics, $V_{CC\pm} = \pm 15$ V

| PARAMETER | TEST CONDITIONS† | | TL08_M TL08_AM MIN TYP MAX | TL08_I MIN TYP MAX | TL08_C TL08_AC TL08_BC MIN TYP MAX | UNIT |
|---|---|---|---|---|---|---|
| $V_{IO}$ Input offset voltage | $R_S = 50\ \Omega$, $T_A = 25°C$ | TL08_ | 6 | 6 | 15 | mV |
| | | TL08_A‡ | 3 | | 6 | |
| | | '81B,'82B,'84B | | | 3 | |
| | $R_S = 50\ \Omega$, $T_A$ = full range | TL08_ | 9 | 9 | 20 | |
| | | TL08_A‡ | 5 | | 7.5 | |
| | | '81B,'82B,'84B | | | 5 | |
| $\alpha VIO$ Temperature coefficient of input offset voltage | $R_S = 50\ \Omega$, $T_A$ = full range | | 10 | 10 | 10 | $\mu V/°C$ |
| $I_{IO}$ Input offset current | $T_A = 25°C$ | TL08_ | 0.1 | 0.1 | 0.2 | nA |
| | | TL08_A‡ | 0.05 | | 0.1 | |
| | | '81B,'82B,'84B | | | 0.05 | |
| | $T_A$ = full range | TL08_ | 20 | 10 | 5 | |
| | | TL08_A‡ | 10 | | 3 | |
| | | '81B,'82B,'84B | | | 2 | |
| $I_{IB}$ Input bias current | $T_A = 25°C$ | TL08_ | 0.2 | 0.2 | 0.4 | nA |
| | | TL08_A‡ | 0.1 | | 0.2 | |
| | | '81B,'82B,'84B | | | 0.1 | |
| | $T_A$ = full range | TL08_ | 50 | 20 | 10 | |
| | | TL08_A‡ | 25 | | 5 | |
| | | '81B,'82B,'84B | | | 4 | |
| $V_{ICR}$ Common-mode input voltage range | $T_A = 25°C$ | TL08_ | +12 | +12 | +10 | V |
| | | TL08_A‡ | +12 | | +12 | |
| | | '81B,'82B,'84B | | | +12 | |
| $V_{OPP}$ Maximum peak-to-peak output voltage swing | $T_A = 25°C$, $R_L = 10\ k\Omega$ | | 24 27 | 24 27 | 24 27 | V |
| | $T_A$ = full range, $R_L \geq 10\ k\Omega$ | | 24 | 24 | 24 | |
| | $R_L \geq 2\ k\Omega$ | | 20 24 | 20 24 | 20 24 | |
| $A_{VD}$ Large-signal differential voltage amplification | $R_L \geq 2\ k\Omega$, $V_O = \pm 10$ V, $T_A = 25°C$ | TL08_ | 50 200 | 50 200 | 25 200 | V/mV |
| | | TL08_A‡ | 50 200 | | 50 200 | |
| | | '81B,'82B,'84B | | | 50 200 | |
| | $R_L \geq 2\ k\Omega$, $V_O = \pm 10$ V, $T_A$ = full range | TL08_ | 25 | 25 | 15 | |
| | | TL08_A‡ | 25 | | 25 | |
| | | '81B,'82B,'84B | | | 25 | |
| $B_1$ Unity-gain bandwidth | $T_A = 25°C$ | | 3 | 3 | 3 | MHz |
| $r_i$ Input resistance | $T_A = 25°C$ | | $10^{12}$ | $10^{12}$ | $10^{12}$ | $\Omega$ |
| CMRR Common-mode rejection ratio | $R_S \leq 10\ k\Omega$, $T_A = 25°C$ | TL08_ | 80 | 80 | 70 | dB |
| | | TL088_A‡ | 80 | | 80 | |
| | | '81B,'82B,'84B | | | 80 | |
| $k_{SVR}$ Supply voltage rejection ratio ($\Delta V_{CC\pm}/\Delta V_{IO}$) | $R_S \leq 10\ k\Omega$, $T_A = 25°C$ | TL08_ | 80 | 80 | 70 | dB |
| | | TL08_A‡ | 80 | | 80 | |
| | | '81B,'82B,'84B | | | 80 | |
| $V_n$ Equivalent input noise voltage | $A_{VD} = 100$, $R_S = 100\ \Omega$, $f = 1$ kHz, $T_A = 25°C$ | | 47 | 47 | 47 | $nV/\sqrt{Hz}$ |
| $I_{CC}$ Supply current (per amplifier) | No load, No signal, $T_A = 25°C$ | | 1.4 2.8 | 1.4 2.8 | 1.4 2.8 | mA |
| $V_{o1}/V_{o2}$ Channel separation | $A_{VD} = 100$, $T_A = 25°C$ | | 120 | 120 | 120 | dB |

† All characteristics are specified under open-loop conditions unless otherwise noted. Full range for $T_A$ is $-55°C$ to $125°C$ for TL08_M and TL08_AM, $-25°C$ to $85°C$ for TL08_I, and $0°C$ to $70°C$ for TL08_C, TL08_AC, and TL08_BC.

‡ Types TL080AM and TL083AM are not defined by this data sheet.

§ Input bias currents of a FET-input operational amplifier are normal junction reverse currents, which are temperature sensitive as shown in Figure 11. Pulse techniques must be used that will maintain the junction temperature as close to the ambient temperature as is possible.

### TEXAS INSTRUMENTS
INCORPORATED
POST OFFICE BOX 5012 • DALLAS, TEXAS 75222

Reprinted with permission from Texas Instruments, Inc., Dallas, TX.

## TYPES TL080 THRU TL084, TL080A THRU TL084A, TL081B, TL082B, TL084B
### JFET-INPUT OPERATINAL AMPLIFIERS

operating characteristics, $V_{CC} = \pm 15$ V, $T_A = 25°C$

| PARAMETER | | TEST CONDITIONS | | MIN | TYP | MAX | UNIT |
|---|---|---|---|---|---|---|---|
| SR | Slew rate at unity gain | $V_I = 10$ V, $C_L = 100$ pF, | $R_L = 2$ kΩ, See Figure 1 | | 1 2 | | V/µs |
| $t_r$ | Rise time | $V_I = 20$ mV, | $R_L = 2$ kΩ, | | 0.1 | | µs |
| | Overshoot factor | $C_L = 100$ pF, | See Figure 1 | | 10 % | | |

## PARAMETER MEASUREMENT INFORMATION

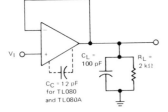

FIGURE 1—UNITY-GAIN AMPLIFIER

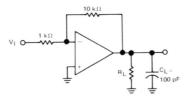

FIGURE 2—GAIN-OF-10 INVERTING AMPLIFIER

## INPUT OFFSET VOLTAGE NULL CIRCUITS

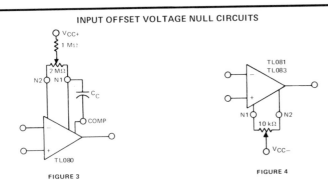

FIGURE 3

FIGURE 4

**TEXAS INSTRUMENTS**
INCORPORATED
POST OFFICE BOX 5012 • DALLAS, TEXAS 75222

Reprinted with permission from Texas Instruments, Inc., Dallas, TX.

## TYPES TL080 THRU TL084, TL080A THRU TL084A, TL081B, TL082B, TL084B JFET-INPUT OPERATIONAL AMPLIFIERS

### TYPICAL CHARACTERISTICS[†]

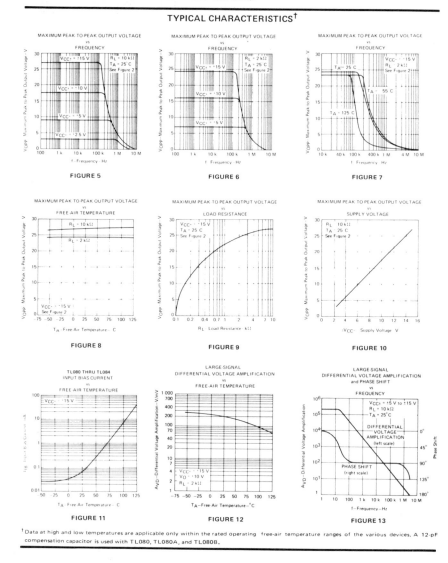

†Data at high and low temperatures are applicable only within the rated operating free-air temperature ranges of the various devices. A 12-pF compensation capacitor is used with TL080, TL080A, and TL080B.

## TEXAS INSTRUMENTS
INCORPORATED
POST OFFICE BOX 5012 • DALLAS, TEXAS 75222

Reprinted with permission from Texas Instruments, Inc., Dallas, TX.

## TYPES TL080 THRU TL084, TL080A THRU TL084A, TL081B, TL082B, TL084B
## JFET-INPUT OPERATINAL AMPLIFIERS

### TYPICAL CHARACTERISTICS[†]

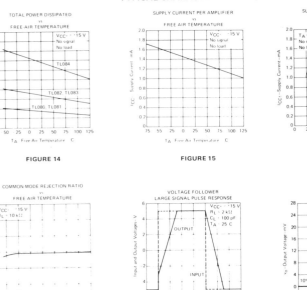

FIGURE 14

FIGURE 15

FIGURE 16

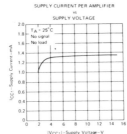

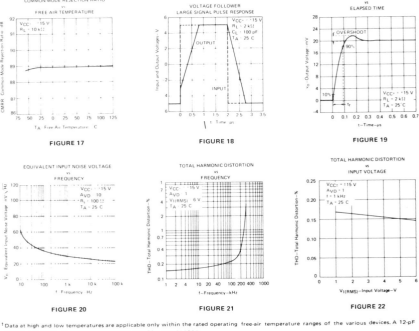

FIGURE 17

FIGURE 18

FIGURE 19

FIGURE 20

FIGURE 21

FIGURE 22

[†] Data at high and low temperatures are applicable only within the rated operating free-air temperature ranges of the various devices. A 12-pF compensation capacitor is used with TL080, TL080A, and TL080B.

**TEXAS INSTRUMENTS**
INCORPORATED
POST OFFICE BOX 5012 • DALLAS, TEXAS 75222

PRINTED IN U.S.A
TI cannot assume any responsibility for any circuits shown
or represent that they are free from patent infringement

TEXAS INSTRUMENTS RESERVES THE RIGHT TO MAKE CHANGES AT ANY TIME
IN ORDER TO IMPROVE DESIGN AND TO SUPPLY THE BEST PRODUCT POSSIBLE.

Reprinted with permission from Texas Instruments, Inc., Dallas, TX.

## 8-BIT μP-COMPATIBLE D/A CONVERTER

### NE5018

NE5018-F,N

### DESCRIPTION

The NE5018 is a complete 8-bit digital to analog converter subsystem on one monolithic chip. The data inputs have input latches, controlled by a latch enable pin. The data and latch enable inputs are ultra-low loading for easy interfacing with all logic systems. The latches appear transparent when the $\overline{LE}$ input is in the low state. When $\overline{LE}$ goes high, the input data present at the moment of transition is latched and retained until $\overline{LE}$ again goes low. This feature allows easy compatibility with most microprocessors.

The chip also comprises a stable voltage reference (5V nominal) and a high slew rate buffer amplifier. The voltage reference may be externally trimmed with a potentiometer for easy adjustment of full scale, while maintaining a low temperature co-efficient.

The output of the buffer amplifier may be offset so as to provide bipolar as well as unipolar operation.

### FEATURES

• 8-bit resolution
• Input latches
• Low-loading data inputs
• On-chip voltage reference
• Output buffer amplifier
• Accurate to ± 1/2 LSB
• Monotonic to 8 bits
• Amplifier and reference both short-circuit protected
• Compatible with 2650, 8080 and many other μP's.

### APPLICATIONS

• Precision 8-bit D/A converters
• A/D converters
• Programmable power supplies
• Test equipment
• Measuring instruments
• Analog-digital multiplication

### PIN CONFIGURATION

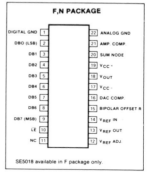

**F,N PACKAGE**

| Pin | | Pin | |
|---|---|---|---|
| DIGITAL GND | 1 | 22 | ANALOG GND |
| DB0 (LSB) | 2 | 21 | AMP. COMP. |
| DB1 | 3 | 20 | SUM NODE |
| DB2 | 4 | 19 | $V_{CC}+$ |
| DB3 | 5 | 18 | $V_{OUT}$ |
| DB4 | 6 | 17 | $V_{CC}-$ |
| DB5 | 7 | 16 | DAC COMP. |
| DB6 | 8 | 15 | BIPOLAR OFFSET R |
| DB7 (MSB) | 9 | 14 | $V_{REF\ IN}$ |
| $\overline{LE}$ | 10 | 13 | $V_{REF\ OUT}$ |
| NC | 11 | 12 | $V_{REF\ ADJ.}$ |

SE5018 available in F package only.

### ABSOLUTE MAXIMUM RATINGS

| PARAMETER | | RATING | UNIT |
|---|---|---|---|
| $V_{CC}+$ | Positive supply voltage | 18 | V |
| $V_{CC}-$ | Negative supply voltage | –18 | V |
| $V_{IN}$ | Logic input voltage | 0 to 18 | V |
| $V_{REFIN}$ | Voltage at $V_{REF}$ input | 12 | V |
| $V_{REFADJ}$ | Voltage at $V_{REF}$ adjust | 0 to $V_{REF}$ | V |
| $V_{SUM}$ | Voltage at sum node | 12 | V |
| $I_{REFSC}$ | Short-circuit current to ground at $V_{REF}$ OUT | Continuous | |
| $I_{OUTSC}$ | Short-circuit current to ground or either supply at $V_{OUT}$ | Continuous | |
| $I_{REF}$ | Reference input current | 5 | mA |
| $P_D$ | Power dissipation* | | |
| | -N package | 800 | mW |
| | -F package | 1000 | mW |
| $T_A$ | Operating temperature range | | |
| | SE5018 | –55 to +125 | °C |
| | NE5018 | 0 to +70 | °C |
| $T_{STG}$ | Storage temperature range | –65 to +150 | °C |
| $T_{SOLD}$ | Lead soldering temperature (10 seconds) | 300 | °C |

*NOTE

For N package, derate at 120°C/W above 35°C
For F package, derate at 75°C/W above 75°C

**signetics**

Reprinted with permission from Signetics Corp., Sunnyvale, CA.

## 8-BIT μP-COMPATIBLE D/A CONVERTER

**NE5018**

NE5018-F,N

### DC ELECTRICAL CHARACTERISTICS
$V_{CC}+$ = 15V, $V_{CC}-$ = -15V, SE5018. -55°C ≤ $T_A$ ≤ 125°C, NE5018. 0°C ≤ $T_A$ ≤ 70°C unless otherwise specified.

| | PARAMETER | TEST CONDITIONS | SE5018 | | | NE5018 | | | UNITS |
|---|---|---|---|---|---|---|---|---|---|
| | | | Min | Typ | Max | Min | Typ | Max | |
| $V_{CC}+$ | Positive supply voltage | | | 15 | | | 15 | | V |
| $V_{CC}-$ | Negative supply voltage | | | -15 | | | -15 | | V |
| | Resolution | | | 8 | | | 8 | | bits |
| | Relative accuracy | | | | ±0.19 | | | ±0.19 | % |
| $T_S$ | Settling time | To ± 1/2LSB, 10V step | | 2 | | | 2 | | μs |
| PSRR | Power supply | $V_{CC}+$  +12 to +18V | | ±1 | | | ±1 | | mV/V |
| | Rejection ratio | $V_{CC}-$  -12 to -18V | | | | | | | |
| $I_{CC}+$ | Positive supply current | $V_{CC}+$ = 15V | | 8 | | | 8 | | mA |
| $I_{CC}-$ | Negative supply current | $V_{CC}-$ = -15V | | -10 | | | -10 | | mA |
| $I_{IN}(0)$ | Logic "0" input current | $V_{IN}$ = 0V | | 5 | | | 5 | | μA |
| $V_{IN}(0)$ | Logic "0" input voltage | | | | 0.8 | | | 0.8 | V |
| $V_{IN}(1)$ | Logic "1" input voltage | | 2.0 | | | 2.0 | | | V |
| $T_{PWLE}$ | Latch enable pulse width | | | 400 | | | 400 | | ns |

### BLOCK DIAGRAM

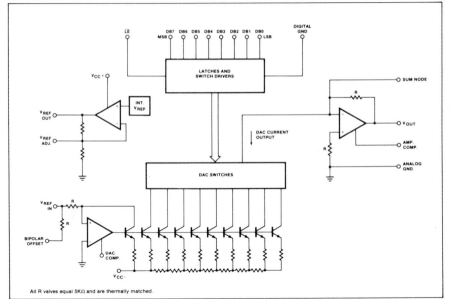

All R valves equal 5KΩ and are thermally matched.

signetics

---

Reprinted with permission from Signetics Corp., Sunnyvale, CA.

## 8-BIT μP-COMPATIBLE D/A CONVERTER

NE5018

NE5018-F,N

**EQUIVALENT SCHEMATIC**

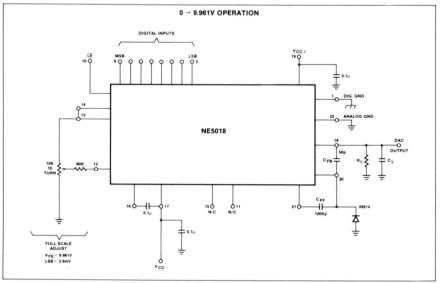

a subsidiary of **U.S. Philips Corporation**

Signetics Corporation
PO Box 9052
811 East Arques Avenue
Sunnyvale, California 94086
Telephone 408/739-7700

Printed in USA   Aug. 1977

Reprinted with permission from Signetics Corp., Sunnyvale, CA.

MAY 1976

**NATIONAL**

LF198/LF298/LF398 monolithic sample and hold circuits

## LF198/LF298/LF398 monolithic sample and hold circuits

### general description

The LF198/LF298/LF398 are monolithic sample and hold circuits which utilize BI-FET technology to obtain ultra-high dc accuracy with fast acquisition of signal and low droop rate. Operating as a unity gain follower, dc gain accuracy is 0.002% typical and acquisition time is as low as $6\mu s$ to 0.01%. A bipolar input stage is used to achieve low offset voltage and wide bandwidth. Input offset adjust is accomplished with a single pin and does not degrade input offset drift. The wide bandwidth allows the LF198 to be included inside the feedback loop of 1 MHz op amps without having stability problems. Input impedance of $10^{10}\Omega$ allows high source impedances to be used without degrading accuracy.

P-channel junction FET's are combined with bipolar devices in the output amplifier to give droop rates as low as 5 mV/min with a $1\mu F$ hold capacitor. The JFET's have much lower noise than MOS devices used in previous designs and do not exhibit high temperature instabilities. The overall design guarantees no feedthrough from input to output in the hold mode even for input signals equal to the supply voltages.

### features

- Operates from ±5V to ±18V supplies
- Less than $10\mu s$ acquisition time
- TTL, PMOS, CMOS compatible logic input
- 0.5 mV typical hold step at $C_h = 0.01\mu F$
- Low input offset
- 0.002% gain accuracy
- Low output noise in hold mode
- Input characteristics do not change during hold mode
- High supply rejection ratio in sample or hold
- Wide bandwidth

Logic inputs on the LF198 are fully differential with low input current, allowing direct connection to TTL, PMOS, and CMOS. Differential threshold is 1.4V. The LF198 will operate from ±5V to ±18V supplies. It is available in an 8-lead TO-5 package.

### functional diagram

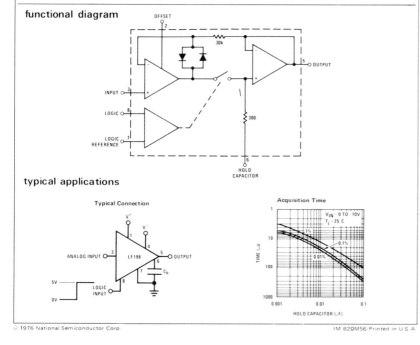

### typical applications

**Typical Connection**

**Acquisition Time**

© 1976 National Semiconductor Corp.

IM B20M56/Printed in U.S.A

---

Reprinted with permission from National Semiconductor Corp., Santa Clara, CA.

## absolute maximum ratings

| | | | |
|---|---|---|---|
| Supply Voltage | ±18V | Input Voltage | Equal to Supply Voltage |
| Power Dissipation (Package Limitation) (Note 1) | 500 mW | Logic To Logic Reference Differential Voltage | +7V, −30V |
| Operating Ambient Temperature Range | | (Note 2) | |
| LF198 | −55°C to +125°C | Output Short Circuit Duration | Indefinite |
| LF298 | −25°C to +85°C | Hold Capacitor Short Circuit Duration | 10 sec |
| LF398 | 0°C to +70°C | Lead Temperature (Soldering, 10 seconds) | 300°C |
| Storage Temperature Range | −65°C to +150°C | | |

## electrical characteristics (Note 3)

| PARAMETER | CONDITIONS | LF198/LF298 | | | LF398 | | | UNITS |
|---|---|---|---|---|---|---|---|---|
| | | MIN | TYP | MAX | MIN | TYP | MAX | |
| Input Offset Voltage, (Note 6) | $T_j = 25°C$ | | 1 | 3 | | 2 | 7 | mV |
| | Full Temperature Range | | | 5 | | | 10 | mV |
| Input Bias Current, (Note 6) | $T_j = 25°C$ | | 5 | 25 | | 10 | 50 | nA |
| | Full Temperature Range | | | 75 | | | 100 | nA |
| Input Impedance | $T_j = 25°C$ | | $10^{10}$ | | | $10^{10}$ | | Ω |
| Gain Error | $T_j = 25°C, R_L = 10k$ | | 0.002 | 0.005 | | 0.004 | 0.01 | % |
| | Full Temperature Range | | | 0.02 | | | 0.02 | % |
| Feedthrough Attenuation Ratio at 1 kHz | $T_j = 25°C, C_h = 0.01\mu F$ | 86 | 96 | | 80 | 90 | | dB |
| Output Impedance | $T_j = 25°C$, "HOLD" mode | | 0.5 | 2 | | 0.5 | 4 | Ω |
| | Full Temperature Range | | | 4 | | | 6 | Ω |
| "HOLD" Step, (Note 4) | $T_j = 25°C, C_h = 0.01\mu F, V_{OUT} = 0$ | | 0.5 | 2.0 | | 1.0 | 2.5 | mV |
| Supply Current, (Note 6) | $T_j = 25°C$ | | 4.5 | 5.5 | | 4.5 | 6.5 | mA |
| Logic and Logic Reference Input Current | $T_j = 25°C$ | | 2 | 10 | | 2 | 10 | μA |
| Leakage Current into Hold Capacitor (Note 6) | $T_j = 25°C$, (Note 5) Hold Mode | | 30 | 100 | | 30 | 200 | pA |
| Acquisition Time to 0.1% | $\Delta V_{OUT} = 10V, C_h = 1000$ pF | | 4 | | | 4 | | μs |
| | $C_h = 0.01\mu F$ | | 20 | | | 20 | | μs |
| Hold Capacitor Charging Current | $V_{IN} - V_{OUT} = 2V$ | | 5 | | | 5 | | mA |
| Supply Voltage Rejection Ratio | $V_{OUT} = 0$ | 80 | 110 | | 80 | 110 | | dB |
| Differential Logic Threshold | $T_j = 25°C$ | 0.8 | 1.4 | 2.4 | 0.8 | 1.4 | 2.4 | V |

**Note 1:** The maximum junction temperature of the LF198 is 150°C, for the LF298, 115°C, and for the LF398, 100°C. When operating at elevated ambient temperature, the TO-5 package must be derated based on a thermal resistance $(\theta)_{jA}$ of 150°C/W.

**Note 2:** Although the differential voltage may not exceed the limits given, the common-mode voltage on the logic pins may be equal to the supply voltages without causing damage to the circuit. For proper logic operation, however, one of the logic pins must always be at least 2V below the positive supply and 3V above the negative supply.

**Note 3:** Unless otherwise specified, the following conditions apply. Unit is in "sample" mode, $V_S = ±15V$, $T_j = 25°C$, $-11.5V \leq V_{IN} \leq +11.5V$, $C_h = 0.01\mu F$, and $R_L = 10 k\Omega$. Logic reference voltage = 0V and logic voltage = 2.5V.

**Note 4:** Hold step is sensitive to stray capacitive coupling between input logic signals and the hold capacitor. 1 pF, for instance, will create an additional 0.5 mV step with a 5V logic swing and a 0.01μF hold capacitor. Magnitude of the hold step is inversely proportional to hold capacitor value.

**Note 5:** Leakage current is measured at a *junction* temperature of 25°C. The effects of junction temperature rise due to power dissipation or elevated ambient can be calculated by doubling the 25°C value for each 11°C increase in chip temperature. Leakage is guaranteed over full input signal range.

**Note 6:** These parameters guaranteed over a supply voltage range of ±5 to ±18V.

## typical performance characteristics

Aperture Time*

Capacitor Hysteresis

Dynamic Sampling Error

*See definition

2

---

Reprinted with permission from National Semiconductor Corp., Santa Clara, CA.

## typical performance characteristics (con't)

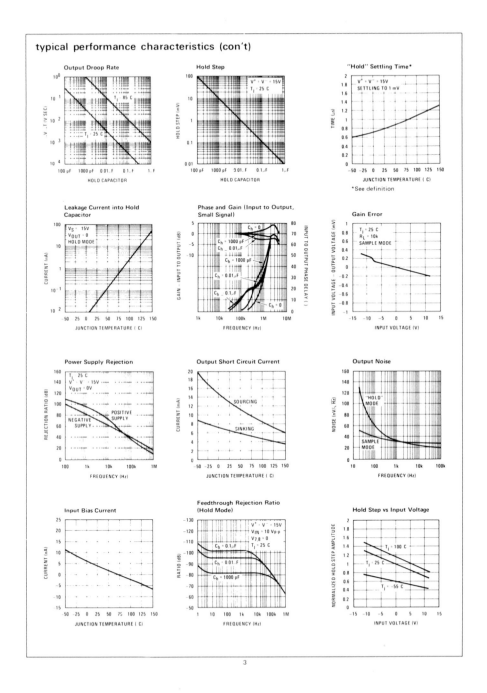

3

Reprinted with permission from National Semiconductor Corp., Santa Clara, CA.

## application hints

### Hold Capacitor

Hold step, acquisition time, and droop rate are the major trade-offs in the selection of a hold capacitor value. Size and cost may also become important for larger values. Use of the curves included with this data sheet should be helpful in selecting a reasonable value of capacitance. Keep in mind that for fast repetition rates or tracking fast signals, the capacitor drive currents may cause a significant temperature rise in the LF198.

A significant source of error in an accurate sample and hold circuit is dielectric absorbtion in the hold capacitor. A mylar cap, for instance, may "sag back" up to 0.2% after a quick change in voltage. A long "soak" time is required before the circuit can be put back into the hold mode with this type of capacitor. Dielectrics with very low hysteresis are polystyrene, polypropylene, and Teflon. Other types such as mica and polycarbonate are not nearly as good. Ceramic is unusable with $> 1\%$ hysteresis. The advantage of polypropylene over polystyrene is that it extends the maximum ambient temperature from $85°C$ to $100°C$. For more exact data, see the curve labeled dielectric absorption error vs sample time. The hysteresis numbers on the curve are final values, taken after full relaxation. The hysteresis error can be significantly reduced if the output of the LF198 is digitized quickly after the hold mode is initiated. The hysteresis relaxation time constant in polypropylene, for instance, is $10-50$ ms. If A-to-D conversion can be made within 1 ms, hysteresis error will be reduced by a factor of ten.

### DC and AC Zeroing

DC zeroing is accomplished by connecting the offset adjust pin to the wiper of a 1 k$\Omega$ potentiometer which has one end tied to $V^+$ and the other end tied through a resistor to ground. The resistor should be selected to give $\approx 0.6$ mA through the 1k potentiometer.

AC zeroing (hold step zeroing) can be obtained by adding an inverter with the adjustment pot tied input to output. A 10 pF capacitor from the wiper to the hold capacitor will give $\pm 4$ mV hold step adjustment with a 0.01$\mu$F hold capacitor and 5V logic supply. For larger logic swings, a smaller capacitor ($<$ 10 pF) may be used.

### Logic Rise Time

For proper operation, logic signals into the LF198 must have a minimum dV/dt of 0.2 V/$\mu$s. Slower signals will cause excessive hold step. If a R/C network is used in front of the logic input for signal delay, calculate the slope of the waveform at the threshold point to ensure that it is at least 0.2 V/$\mu$s.

### Sampling Dynamic Signals

Sample error due to moving input signals probably causes more confusion among sample-and-hold users than any other parameter. The primary reason for this is that many users make the assumption that the sample and hold amplifier is truly locked on to the input signal while in the sample mode. In actuality, there are finite phase delays through the circuit creating an input-output differential for fast moving signals. In addition, although the output may have settled, the hold capacitor has an additional lag due to the 300$\Omega$ series resistor on the chip. This means that at the moment the "hold" command arrives, the hold capacitor voltage may be somewhat different than the actual analog input. The effect of these delays is opposite to the effect created by delays in the logic which switches the circuit from sample to hold. For example, consider an analog input of 20 Vp-p at 10 kHz. Maximum dV/dt is 0.6 V/$\mu$s. With no analog phase delay and 100 ns logic delay, one could expect up to $(0.1\mu s)(0.6V/\mu s) = 60$ mV error if the "hold" signal arrived near maximum dV/dt of the input. A positive-going input would give a $\pm 60$ mV error. Now assume a 1 MHz (3 dB) bandwidth for the overall analog loop. This generates a phase delay of 160 ns. If the hold capacitor sees this exact delay, then error due to analog delay will be $(0.16\mu s)(0.6 V/\mu s) = -96$ mV. Total output error is $+60$ mV (digital) $-96$ mV (analog) for a total of $-36$ mV. To add to the confusion, analog delay is proportional to hold capacitor value while digital delay remains constant. A family of curves (dynamic sampling error) is included to help estimate errors.

A curve labeled <u>Aperture Time</u> has been included for sampling conditions where the input is steady during the sampling period, but may experience a sudden change nearly coincident with the "hold" command. This curve is based on a 1 mV error fed into the output.

A second curve, <u>Hold Settling Time</u> indicates the time required for the output to settle to 1 mV after the "hold" command.

### Digital Feedthrough

Fast rise time logic signals can cause hold errors by feeding externally into the analog input at the same time the amplifier is put into the hold mode. To minimize this problem, board layout should keep logic lines as far as possible from the analog input. Grounded guarding traces may also be used around the input line, especially if it is driven from a high impedance source. Reducing high amplitude logic signals to 2.5V will also help.

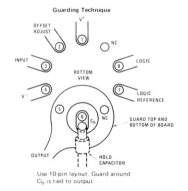

**Guarding Technique**

Use 10-pin layout. Guard around $C_h$ is tied to output.

4

Reprinted with permission from National Semiconductor Corp., Santa Clara, CA.

# IC and Semiconductor Lead Assignments

# Appendix D

Included in this appendix are the pin assignments of the most common series 74 TTL integrated circuits, lead assignments for diodes and transistors, and standard op amp pin assignments.

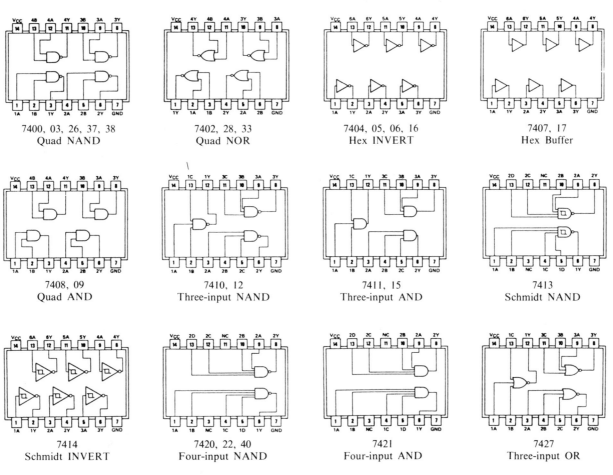

7400, 03, 26, 37, 38
Quad NAND

7402, 28, 33
Quad NOR

7404, 05, 06, 16
Hex INVERT

7407, 17
Hex Buffer

7408, 09
Quad AND

7410, 12
Three-input NAND

7411, 15
Three-input AND

7413
Schmidt NAND

7414
Schmidt INVERT

7420, 22, 40
Four-input NAND

7421
Four-input AND

7427
Three-input OR

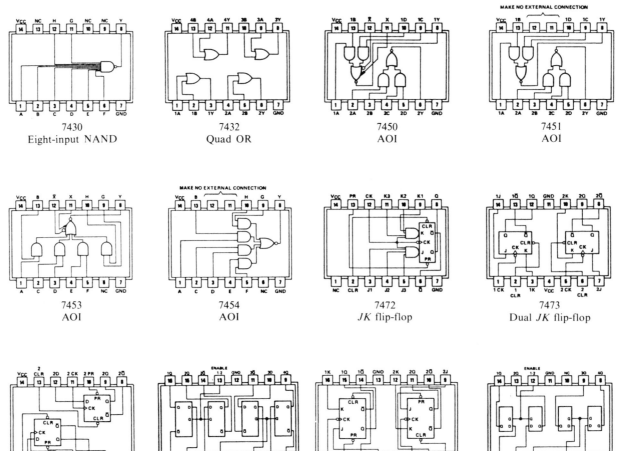

7430
Eight-input NAND

7432
Quad OR

7450
AOI

7451
AOI

7453
AOI

7454
AOI

7472
*JK* flip-flop

7473
Dual *JK* flip-flop

7474
Dual D flip-flop

7475
Quad latch

7476
Dual *JK* flip-flop

7477
Quad latch

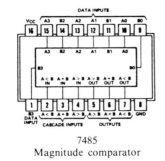

7485
Magnitude comparator

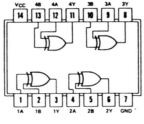

7486
Exclusive-OR

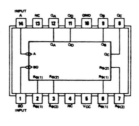

7490
Decade counter

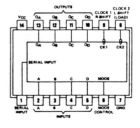

7495
4-bit Shift Register

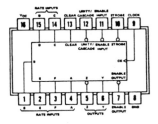

7497
Binary rate multiplier

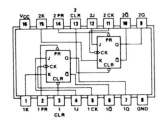

74110
*JK* flip-flop

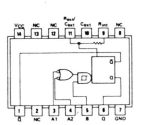

74111
Dual *JK* flip-flop

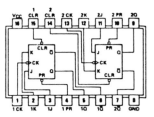

74112
Dual *JK* flip-flop

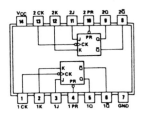

74113
Dual *JK* flip-flop

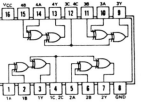

74114
Dual *JK* flip-flop

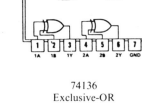

74121
Monostable multivibrator

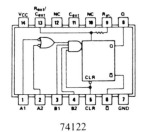

74122
Monostable

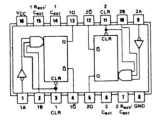

74123
Dual monostable

Exclusive-OR/NOR
74135

74136
Exclusive-OR

74160, 161, 162, 163
Decade counters (160, 162)
Binary counters (161, 163)

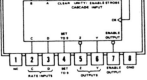

74167
Decade rate multiplier

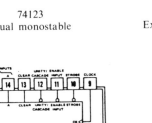

74168, 169
Up-down counter
decade (168), binary (169)

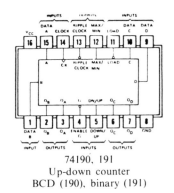

74190, 191
Up-down counter
BCD (190), binary (191)

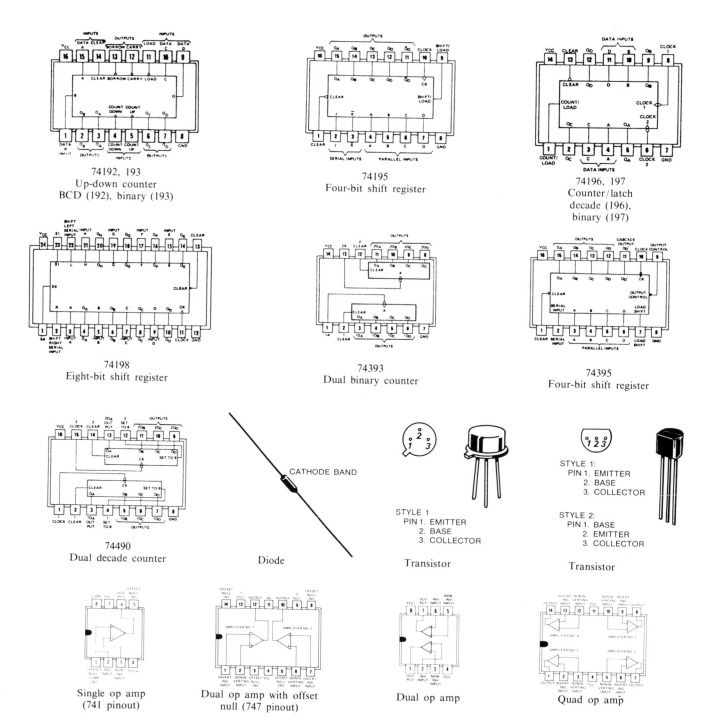

74192, 193
Up-down counter
BCD (192), binary (193)

74195
Four-bit shift register

74196, 197
Counter/latch
decade (196),
binary (197)

74198
Eight-bit shift register

74393
Dual binary counter

74395
Four-bit shift register

74490
Dual decade counter

Diode

Transistor

Transistor

Single op amp
(741 pinout)

Dual op amp with offset
null (747 pinout)

Dual op amp

Quad op amp

# Bibliography

## Chapter 1

1. Brophy, James J. *Basic Electronics for Scientists.* New York: McGraw-Hill Book Co., 1977.

2. Malmstadt, H.V., Enke, C.G., and Crouch, S.R. *Electronic Measurements for Scientists.* Menlo Park, CA: Benjamin/Cummings Publishing Co., Inc., 1974.

3. Oliver, B.M., and Cage, J.M. *Electronic Measurements and Instrumentation.* New York: McGraw-Hill Book Co., 1971.

4. Smith, Ralph J. *Circuits, Devices and Systems.* 3rd ed. New York: John Wiley & Sons, 1976.

5. Su, Kendall L. *Fundamentals of Circuits, Electronics and Signal Analysis.* Boston: Houghton Mifflin, 1978.

## Chapter 2

1. Bracewell, Ron. *The Fourier Transform and Its Applications.* New York: McGraw-Hill Book Co., 1965.

2. Brophy, James J. *Basic Electronics\for Scientists.* New York: McGraw-Hill Book Co., 1977.

3. Malmstadt, H.V., Enke, C.G., and Crouch, S.R. *Electronic Measurements for Scientists.* Menlo Park, CA: Benjamin/Cummings Publishing Co., Inc., 1974.

4. Oliver, B.M., and Cage, J.M. *Electronic Measurements and Instrumentation.* New York: McGraw-Hill Book Co., 1971.

5. Smith, Ralph J. *Circuits, Devices and Systems.* 3rd ed. New York: John Wiley & Sons, 1976.

6. Su, Kendall L. *Fundamentals of Circuits, Electronics and Signal Analysis.* Boston: Houghton Mifflin, 1978.

## Chapter 3

1. Angelo, E. James, Jr. *Electronics: BJT's, FET's, and Microcircuits.* New York: McGraw-Hill Book Co., 1969.

2. Brophy, James J. *Basic Electronics for Scientists.* New York: McGraw-Hill Book Co., 1977.

3. Hnatek, Eugene R. *A User's Handbook of Integrated Circuits.* New York: Wiley-Interscience, 1973.

4. Millman, Jacob. *Microelectronics: Digital and Analog Circuits and Systems.* New York: McGraw-Hill Book Co., 1979.

5. Su, Kendall L. *Fundamentals of Circuits, Electronics and Signal Analysis.* Boston: Houghton Mifflin, 1978.

## Chapter 4

1. Brophy, James J. *Basic Electronics for Scientists*. New York: McGraw-Hill Book Co., 1977.

2. Carr, Joseph J. *Digital Interfacing with an Analog World*. Blue Ridge Summit, PA: TAB Books, 1978.

3. Diefenderfer, A. James. *Principles of Electronic Instrumentation*, 2nd ed. Philadelphia: W.B. Saunders Co., 1979.

4. Geddes, L.A., and Baker, L.E. *Principles of Applied Biomedical Instrumentation*. New York: John Wiley & Sons, 1968.

5. Lion, Kurt S. *Elements of Electrical and Electronic Instrumentation*. New York: McGraw-Hill Book Co., 1975.

6. Oliver, Frank J. *Practical Instrumentation Transducers*. New York: Hayden Book Co., 1971.

7. *Optoelectronics Applications Manual*. New York: McGraw-Hill Book Co., 1977.

8. *Photomultiplier Manual*. RCA Corporation, 1970.

9. Webster, John G. *Medical Instrumentation*. Boston: Houghton Mifflin, 1978.

## Chapter 5

1. Berlin, Howard M. *The Design of Active Filters with Experiments*. Derby, CT: E & L Instruments, Inc., 1977.

2. Berlin, Howard M. *The Design of Operational Amplifier Circuits, with Experiments*. Derby, CT: E & L Instruments, Inc., 1977.

3. Brophy, James J. *Basic Electronics for Scientists*. New York: McGraw-Hill Book Co., 1977.

4. Carr, Joseph J. *Digital Interfacing with an Analog World*. Blue Ridge Summit, PA: TAB Books, 1978.

5. Glaser, Arthur B., and Subak-Sharpe, Gerald E. *Integrated Circuit Engineering*. Reading, MA: Addison-Wesley Publishing Co., 1977.

6. Graeme, Jerald G. *Applications of Operational Amplifiers, Third-Generation Techniques*. New York: McGraw Hill Book Co., 1973.

7. Graeme, Jerald G. *Designing with Operational Amplifiers, Applications Alternatives*. New York: McGraw-Hill Book Co., 1977.

8. Graeme, Jerald G., Tobey, Gene E., and Huelsman, Lawrence P. *Operational Amplifiers, Design and Applications*. New York: McGraw-Hill Book Co., 1971.

9. Hnatek, Eugene R. *A User's Handbook of Integrated Circuits*. New York: Wiley-Interscience, 1973.

10. Jung, Walter G. *IC Op-Amp Cookbook*. Indianapolis, IN: Howard W. Sams & Co., Inc., 1974.

11. Millman, Jacob. *Microelectronics: Digital and Analog Circuits and Systems*. New York: McGraw-Hill Book Co., 1979.

12. Oliver, B.M., and Cage, J.M. *Electronic Measurements and Instrumentation*. New York: McGraw-Hill Book Co., 1971.

13. Su, Kendall L. *Fundamentals of Circuits, Electronics and Signal Analysis*. Boston: Houghton Mifflin, 1978.

14. Taub, Herbert, and Schilling, Donald. *Digital Integrated Electronics*. New York: McGraw-Hill Book Co., 1977.

## Chapter 6

1. Berlin, Howard M. *The Design of Operational Amplifier Circuits, with Experiments*. Derby, CT: E & L Instruments, Inc., 1977.

2. Carr, Joseph J. *Digital Interfacing with an Analog World*. Blue Ridge Summit, PA: TAB Books, 1978.

3. Carr, William N., and Mize, Jack P. *MOS/LSI Design and Application*. New York: McGraw-Hill Book Co., 1972.

4. Graeme, Jerald G. *Applications of Operational Amplifiers, Third Generation Techniques*. New York: McGraw-Hill Book Co., 1973.

5.  Graeme, Jerald G., Tobey, Gene E., and Huelsman, Lawrence, P. *Operational Amplifiers, Design and Applications.* New York: McGraw-Hill Book Co., 1971.

6.  Jung, Walter G. *IC Converter Cookbook.* Indianapolis, IN: Howard W. Sams & Co., Inc., 1978.

7.  Taub, Herbert, and Schilling, Donald. *Digital Integrated Electronics.* New York: McGraw-Hill Book Co., 1977.

## Chapter 7

1.  Angelo, E. James, Jr. *Electronics: BJT's, FET's, and Microcircuits.* New York: McGraw-Hill Book Co., 1969.

2.  Brophy, James J. *Basic Electronics for Scientists.* New York: McGraw-Hill Book Co., 1977.

3.  Geddes, L.A., and Baker, L.E. *Principles of Applied Biomedical Instrumentation.* New York: John Wiley & Sons, 1968.

4.  Glaser, Arthur B., and Subak-Sharpe, Gerald E. *Integrated Circuit Engineering.* Reading, MA: Addison-Wesley Publishing Co., 1977.

5.  Graeme, Jerald G., Tobey, Gene E., and Huelsman, Lawrence P. *Operational Amplifiers, Design and Applications.* New York: McGraw-Hill Book Co., 1971.

6.  Millman, Jacob. *Microelectronics: Digital and Analog Circuits and Systems.* New York: McGraw-Hill Book Co., 1979.

7.  Smith, Ralph J. *Circuits, Devices and Systems*, 3rd ed. New York: John Wiley & Sons, 1976.

8.  Taub, Herbert, and Schilling, Donald. *Digital Integrated Electronics.* New York: McGraw-Hill Book Co., 1977.

## Chapter 8

1.  Berlin, Howard M. *The Design of Active Filters with Experiments.* Derby, CT: E & L Instruments, Inc., 1977.

2.  Berlin, Howard M. *The Design of Operational Amplifier Circuits, with Experiments.* Derby, CT: E & L Instruments, Inc., 1977.

3.  Bracewell, Ron. *The Fourier Transform and Its Applications.* New York: McGraw-Hill Book Co., 1965.

4.  Carr, Joseph J. *Digital Interfacing with an Analog World.* Blue Ridge Summit, PA: TAB Books, 1978.

5.  Glaser, Arthur B., and Subak-Sharpe, Gerald E. *Integrated Circuit Engineering.* Reading, MA: Addison-Wesley Publishing Co., 1977.

6.  Graeme, Jerald G. *Applications of Operational Amplifiers, Third-Generation Techniques.* New York: McGraw-Hill Book Co., 1973.

7.  Graeme, Jerald G. *Designing with Operational Amplifiers, Applications Alternatives.* New York: McGraw-Hill Book Co., 1977.

8.  Graeme, Jerald G., Tobey, Gene E., and Huelsman, Lawrence P. *Operational Amplifiers, Design and Applications.* New York: McGraw-Hill Book Co., 1971.

9.  Hnatek, Eugene R. *A User's Handbook of Integrated Circuits.* New York: Wiley-Interscience, 1973.

10.  Jung, Walter G. *Unique IC Op-Amp Applications.* Indianapolis, IN: Howard W. Sams & Co., Inc., 1975.

11.  Jung, Walter G. *IC Op-Amp Cookbook.* Indianapolis, IN: Howard W. Sams & Co., Inc., 1974.

12.  Meiksin, Z.H. and Thackray, P.C. *Electronic Design with Off-the-Shelf Integrated Circuits.* West Nyack, NY: Parker Publishing Co., 1980.

13.  Temes, Gabor C., and Mitra, Sanjit K. *Modern Filter Theory and Design.* New York: Wiley-Interscience, 1973.

14.  VanDer Ziel, Albert. *Noise in Measurements.* New York: Wiley-Interscience, 1976.

## Chapter 9

**1.** Berlin, Howard M. *The Design of Active Filters with Experiments.* Derby, CT: E & L Instruments, Inc., 1977.

**2.** Brophy, James J. *Basic Electronics for Scientists.* New York: McGraw-Hill Book Co., 1977.

**3.** Carr, Joseph J. *Digital Interfacing with an Analog World.* Blue Ridge Summit, PA: TAB Books, 1978.

**4.** Glaser, Arthur B., and Subak-Sharpe, Gerald E. *Integrated Circuit Engineering.* Reading, MA: Addison-Wesley Publishing Co., 1977.

**5.** Graeme, Jerald G. *Applications of Operational Amplifiers, Third-Generation Techniques.* New York: McGraw-Hill Book Co., 1973.

**6.** Graeme, Jerald G. *Designing with Operational Amplifiers, Applications Alternatives.* New York: McGraw-Hill Book Co., 1977.

**7.** Graeme, Jerald G., Tobey, Gene E., and Huelsman, Lawrence P. *Operational Amplifiers, Design and Applications.* New York: McGraw-Hill Book Co., 1971.

**8.** Hilburn, John L., and Johnson, David E. *Manual of Active Filter Design.* New York: McGraw-Hill Book Co., 1973.

**9.** Hnatek, Eugene R. *A User's Handbook of D/A and A/D Converters.* New York: Wiley-Interscience, 1976.

**10.** Jung, Walter G. *IC Converter Cookbook.* Indianapolis, IN: Howard W. Sams & Co., Inc., 1978.

**11.** Lancaster, Don. *Active Filter Cookbook.* Indianapolis, IN: Howard W. Sams & Co., Inc., 1975.

**12.** Oliver, B.M., and Cage, J.M. *Electronic Measurements and Instrumentation.* New York: McGraw-Hill Book Co., 1971.

**13.** Titus, Jonathan A., Titus, Christopher A., Rony, Peter R., and Larsen, David G. *The Bugbook VII: Microcomputer-Analog Converter Software and Hardware Interfacing.* Derby, CT: E & L Instruments, Inc., 1978.

## Chapter 10

**1.** Arnold, James T. *Simplified Digital Automation with Microprocessors.* New York: Academic Press, 1979.

**2.** Blakeslee, Thomas R. *Digital Design with Standard MSI and LSI,* 2nd ed. New York: John Wiley & Sons, 1979.

**3.** Brophy, James J. *Basic Electronics for Scientists.* New York: McGraw-Hill Book Co., 1977.

**4.** Millman, Jacob. *Microelectronics: Digital and Analog Circuits and Systems.* New York: McGraw-Hill Book Co., 1979.

**5.** Morris, Robert L., and Miller, John R. *Designing with TTL Integrated Circuits.* New York: McGraw-Hill Book Co., 1971.

**6.** Su, Kendall L. *Fundamentals of Circuits, Electronics and Signal Analysis.* Boston: Houghton Mifflin, 1978.

**7.** Taub, Herbert, and Schilling, Donald. *Digital Integrated Electronics.* New York: McGraw-Hill Book Co., 1977.

**8.** Wickes, William E. *Logic Design with Integrated Circuits.* New York: John Wiley & Sons, 1968.

## Chapter 11

**1.** Blakeslee, Thomas R. *Digital Design with Standard MSI and LSI,* 2nd ed. New York: John Wiley & Sons, 1979.

**2.** Carr, William N., and Mize, Jack P. *MOS/LSI Design and Application.* New York: McGraw-Hill Book Co., 1972.

**3.** Glaser, Arthur B., and Subak-Sharpe, Gerald E. *Integrated Circuit Engineering.* Reading, MA: Addison-Wesley Publishing Co., 1977.

**4.** Hnatek, Eugene R. *A User's Handbook of Integrated Circuits.* New York: Wiley-Interscience, 1973.

**5.** Lancaster, Don. *CMOS Cookbook.* Indianapolis, IN: Howard W. Sams & Co., Inc., 1977.

6.  Larsen, David G., and Rony, Peter R. *The Bugbook IIA: Interfacing and Scientific Data Communication Experiments Using the UART and 20 mA Current Loops.* Derby, CT: E & L Instruments, Inc., 1975.

7.  McGlynn, Daniel R. *Microprocessors, Technology, Architecture and Applications.* New York: Wiley-Interscience, 1976.

8.  Meiksin, Z.H., and Thackray, P.C. *Electronic Design with Off-the-Shelf Integrated Circuits.* West Nyack, NY: Parker Publishing Co., 1980.

9.  Morris, Robert L., and Miller, John R. *Designing with TTL Integrated Circuits.* New York: McGraw-Hill Book Co., 1971.

10. *Optoelectronics Applications Manual.* New York: McGraw-Hill Book Co., 1977.

11. Su, Kendall L. *Fundamentals of Circuits, Electronics and Signal Analysis.* Boston: Houghton Mifflin, 1978.

12. Taub, Herbert, and Schilling, Donald. *Digital Integrated Electronics.* New York: McGraw-Hill Book Co., 1977.

13. Millman, Jacob. *Microelectronics: Digital and Analog Circuits and Systems.* New York: McGraw-Hill Book Co., 1979.

# Chapter 12

1.  Arnold, James T. *Simplified Digital Automation with Microprocessors.* New York: Academic Press, 1979.

2.  Barna, Arpad, and Porat, Dan I. *Introduction to Microcomputers and Microprocessors.* New York: Wiley-Interscience, 1976.

3.  Bibbero, Robert J. *Microprocessors in Instruments and Control.* New York: Wiley-Interscience, 1977.

4.  Blakeslee, Thomas R. *Digital Design with Standard MSI and LSI,* 2nd ed. New York: John Wiley & Sons, 1979.

5.  Camp, R.C., Smay, T.A., and Triska, C.J. *Microprocessor Systems Engineering.* Portland, OR: Matrix Publishers, 1979.

6.  Carr, William N., and Mize, Jack, P. *MOS/LSI Design and Application.* New York: McGraw-Hill Book Co., 1972.

7.  Givone, D.D., and Roesser, R.P. *Microprocessors/Microcomputers: An Introduction.* New York: McGraw-Hill Book Co., 1980.

8.  Korn, Granino A. *Microprocessors & Small Digital Computer Systems for Engineers and Scientists.* New York: McGraw-Hill Book Co., 1977.

9.  Lesea, Austin, and Zaks, Rodney. *Microprocessor Interfacing Techniques,* 2d ed. Berkeley, CA: Sybex, Inc., 1978.

10. McGlynn, Daniel R. *Microprocessors, Technology, Architecture and Applications.* New York: Wiley-Interscience, 1976.

11. Ogdin, Carol Anne. *Microcomputer Design.* Englewood Cliffs, NJ: Prentice-Hall, 1978.

12. Ogdin, Carol Anne. *Software Design for Microcomputers.* Englewood Cliffs, NJ: Prentice-Hall, 1978.

13. Osborne, A. *An Introduction to Microcomputers.* Berkeley, CA: A. Osborne Associates, 1975.

14. Pooch, U., and Chattergy, R. *Designing Microcomputer Systems.* Rochelle Park, NJ: Hayden Book Co., 1979.

15. Rony, P., Larson, D., and Titus, J. *The 8080A Bugbook.* Indianapolis, IN: Howard W. Sams & Co., Inc., 1977.

16. Soucek, Branko. *Microprocessors and Microcomputers.* New York: Wiley-Interscience, 1976.

17. Taub, Herbert, and Schilling, Donald. *Digital Integrated Electronics.* New York: McGraw-Hill Book Co., 1977.

18. Zaks, Rodney. *Microprocessors from Chips to Systems.* Berkeley, CA: Sybex, Inc., 1977.

# Chapter 13

1. Arnold, James T. *Simplified Digital Automation with Microprocessors*. New York: Academic Press, 1979.

2. Artwick, Bruce A. *Microcomputer Interfacing*. Englewood Cliffs, NJ: Prentice-Hall, 1980.

3. Barna, Arpad, and Porat, Dan I. *Introduction to Microcomputers and Microprocessors*. New York: Wiley-Interscience, 1976.

4. Carr, Joseph J. *Digital Interfacing with an Analog World*. Blue Ridge Summit, PA: TAB Books, 1978.

5. Camp, R.C., Smay, T.A., and Triska, C.J. *Microprocessor Systems Engineering*. Portland, OR: Matrix Publishers, 1979.

6. Goldsbrough, P.F. *The Bugbook IV: Microcomputer Interfacing with the 8255 PPI Chip*. Indianapolis, IN: Howard W. Sams & Co., Inc., 1979.

7. Graeme, Jerald G., Tobey, Gene E., and Huelsman, Lawrence P. *Operational Amplifiers, Design and Applications*. New York: McGraw-Hill Book Co., 1971.

8. Hnatek, Eugene R. *A User's Handbook of D/A and A/D Converters*. New York: Wiley-Interscience, 1976.

9. Jung, Walter G. *IC Converter Cookbook*. Indianapolis, IN: Howard W. Sams & Co., Inc., 1978.

10. Larsen, David G., and Rony, Peter R. *The Bugbook IIA: Interfacing and Scientific Data Communication Experiments Using the UART and 20 mA Current Loops*. Derby, CT: E & L Instruments, Inc., 1975.

11. Lesea, Austin, and Zaks, Rodney. *Microprocessor Interfacing Techniques*, 2d ed. Berkeley, CA: Sybex Inc., 1978.

12. Meiksin, Z.H., and Thackray, P.C. *Electronic Design with Off-the-Shelf Integrated Circuits*. West Nyack, NY: Parker Publishing Co., 1980.

13. Millman, Jacob. *Microelectronics: Digital and Analog Circuits and Systems*. New York: McGraw-Hill Book Co., 1979.

14. Taub, Herbert, and Schilling, Donald. *Digital Integrated Electronics*. New York: McGraw-Hill Book Co., 1977.

15. Titus, Jonathan A., Titus, Christopher A., Rony, Peter R., and Larsen, David G. *The Bugbook VII: Microcomputer-Analog Converter Software and Hardware Interfacing*. Derby, CT: E & L Instruments, Inc., 1978.

16. Wicker, David. "Programmable Interval Timer," *Digital Design*. **10** (No. 5), 32 (1980).

17. Zaks, Rodney. *Microprocessors from Chips to Systems*. Berkeley, CA: Sybex, Inc., 1977.

18. Zuch, Eugene L., ed. *Data Acquisition and Conversion Handbook: A Technical Guide to A/D and D/A Converters and Their Applications*. Mansfield, MA: Datel-Intersil, 1979.

# Chapter 14

1. Arnold, James T. *Simplified Digital Automation with Microprocessors*. New York: Academic Press, 1979.

2. Betty, K.R., and Horlick, Gary. "Frequency Response Plots for Savitsky-Golay Filter Functions," *Analytical Chemistry*. **49**, 351 (1977).

3. Bibbero, Robert J. *Microprocessors in Instruments and Control*. New York: Wiley-Interscience, 1977.

4. Blakeslee, Thomas R. *Digital Design with Standard MSI and LSI*, 2nd ed. New York: John Wiley & Sons, 1979.

5. Bracewell, Ron. *The Fourier Transform and Its Applications*. New York: McGraw-Hill Book Co., 1965.

6. Carr, Joseph J. *Digital Interfacing with an Analog World*. Blue Ridge Summit, PA: TAB Books, 1978.

7. Carrick, A. *Computers and Instrumentation*. London: Hayden, 1979.

8. Cooper, George R., and McGillem, Clare D. *Methods of Signal and System Analysis*. New York: Holt, Rinehart & Winston, 1967.

9. Cunningham, John E. *Handbook of Remote Control and Automation Techniques*. Blue Ridge Summit, PA: TAB Books, 1978.

10. Darland, E.J., Leroi, G.E., and Enke, C.G. "Maximum Efficiency Pulse Counting in Computerized Instrumentation," *Analytical Chemistry*, **52**, 714 (1980).

11. Darland, E.J., Leroi, G.E., and Enke, C.G. "Pulse (Photon) Counting: Determination of Optimum Measurement System Parameters," *Analytical Chemistry*, **51**, 240 (1979).

12. Deming, S.N., and Morgan, S.L. "Simplex Optimization of Variables in Analytical Chemistry," *Analytical Chemistry*, **45**, (No. 3) 278A (1973).

13. Enke, C.G., and Nieman, T.A. "Signal-to-Noise Ratio Enhancement by Least-Squares Polynomial Smoothing," *Analytical Chemistry*, **48** (No. 8) 705A (1976).

14. Griffiths, Peter R., ed. *Transform Techniques in Chemistry*. New York: Plenum Press, 1978.

15. Hieftje, G.M. "Signal-to-Noise Enhancement Through Instrumental Techniques. Part I. Signals, Noise, and S/N Enhancement in the Frequency Domain," *Analytical Chemistry*, **44** (No. 6) 81A (1972); "Part II. Signal Averaging, Boxcar Integration and Correlation Techniques," *Analytical Chemistry*, **44** (No. 7) 69A (1972).

16. Horlick, Gary. "Digital Data Handling of Spectra Using Fourier Transformation," *Analytical Chemistry*, **44**, 943 (1972).

17. Horlick, Gary. "Detection of Spectral Information Utilizing Cross-Correlation Techniques," *Analytical Chemistry*, **45**, 319 (1973).

18. Ingle, J.D., and Crouch, S.R. "Pulse Overlap Effects on Linearity and Signal-to-Noise Ratio in Photon Counting Systems," *Analytical Chemistry*, **44**, 777 (1972).

19. Kelly, P.C., and Horlick, Gary. "Practical Considerations for Digitizing Analog Signals," *Analytical Chemistry*, **45**, 518 (1973).

20. Malmstadt, H.V., Enke, C.G., and Crouch, S.R. *Electronic Measurements for Scientists*. Menlo Park, CA: Benjamin/Cummings Publishing Co., Inc., 1974.

21. Savitsky, Abraham, and Golay, Marcel, J.E. "Smoothing and Differentiation of Data by Simplified Least Squares Procedures," *Analytical Chemistry*, **36**, 1627 (1964).

22. Stanley, William D. *Digital Signal Processing*. Reston, VA: Reston Publishing Co., 1975.

23. Temes, Gabor C., and Mitra, Sanjit K. *Modern Filter Theory and Design*. New York: Wiley-Interscience, 1973.

24. Van der Ziel, Albert. *Noise in Measurements*. New York: Wiley-Interscience, 1976.

25. Zaks, Rodney. *Microprocessors from Chips to Systems*. Berkeley, CA: Sybex, Inc., 1977.

## Appendix A

1. Ott, Henry W. *Noise Reduction Techniques in Electronic Systems*. New York: Wiley-Interscience, 1976.

2. Morrison, Ralph. *Grounding and Shielding Techniques in Instrumentation*. New York: Wiley-Interscience, 1967.

3. Oliver, Frank J. *Practical Instrumentation Transducers*. New York: Hayden Book Co., 1971.

4. Buus, R.G. "Electrical Interference," *Design Technology*, vol. 1. Englewood Cliffs, NJ: Prentice-Hall, 1970.

# Appendix B

1. Giacoletto, L.J. *Electronics Designers Handbook,* 2nd ed. New York: McGraw-Hill Book Co., 1977.

2. *Technical Manual and Catalog.* Catsworth, CA: Bishop Graphics, Inc.

3. Jones, Thomas H. *Electronic Components Handbook.* Reston, VA: Reston Publishing Co., 1979.

4. DeMaw, Douglas, ed. *ARRL Electronics Data Book.* Newington, CT: American Radio Relay League, 1976.

5. Jung, Walter, and March, Richard. "Picking Capacitors," *Audio,* Feb., 1980, 51–62 and March, 1980, 50–60.

# INDEX